Nematode Disease Complexes in Agricultural Crops

Nematode Disease Complexes in Agricultural Crops

Edited by

Mujeebur Rahman Khan

CABI

CABI is a trading name of CAB International

CABI
Nosworthy Way
Wallingford
Oxfordshire OX10 8DE
UK

Tel: +44 (0)1491 832111
E-mail: info@cabi.org
Website: www.cabi.org

CABI
200 Portland Street
Boston
MA 02114
USA

T: +1 (617)682-9015
E-mail: cabi-nao@cabi.org

A catalogue record for this book is available from the British Library, London, UK.

ISBN-13: 9781800625204 (hardback)
9781800625211 (ePDF)
9781800625228 (ePub)

DOI: 10.1079/9781800625228.0000

Commissioning Editor: Rebecca Stubbs
Editorial Assistant: Helen Elliott
Production Editor: Rosie Hayden

Typeset by Exeter Premedia Services Pvt Ltd, Chennai, India
Printed and bound in the UK by CPI Group (UK) Ltd, Croydon, CR0 4YY

Dedicated to

Professor George Francis Atkinson
(1854-1918)
Discovered the synergistic interaction between plant parasitic nematodes and fungi

Professor Mohammad Wajid Khan
(1942-2001)
Explored agricultural significance, and contributed a monumental book on nematode interactions.

Contents

About the Editor

Mujeebur Rahman Khan is a Professor and Chairperson at the Department of Plant Protection, Aligarh Muslim University (AMU), India, and has served as the Dean, Faculty of Agricultural Sciences, AMU. He obtained PhD in Botany (Plant Pathology and Nematology) from AMU, and worked as postdoc at the North Carolina State University, California Department of Food and Agriculture, USA, and the Commonwealth Institute of Parasitology, UK. He has to his credit more than 300 research publications and review articles, 20 books (CABI, Elsevier, Brill, CRC/Science, Springer, NOVA Science, Cambridge, Oxford and IBH, etc.), two patents on biopesticides (USA and India), and 10 major research projects. He has guided around 20 PhD and 55 MSc students. Dr Khan is a Fellow of the International Society of Environmental Botanists, National Academy of Agricultural Sciences (India), and Indian Phytopathological Society, and served as Editor of Indian Phytopathology. He has received several awards, importantly, Dr K.C. Mehta Award (NAAS), B.B. Mundukar Award (IPS), Outstanding Scientist Award (AMU), Professor H.L. Chakraborti Award (ISCA), Outstanding Scientist Award (SPPS), Professor H.M. Shah Memorial (NSI), and Best Research Project Award (AMU). His research interest includes management of mono- and multi-pathogenic plant diseases, impact of climate change and environmental contamination on plant disease development, and biosynthesis of nanoparticles and their effects on plants and microbes.

Contributors

Abdou Lahat Mbaye, College of Agriculture and Environmental Sciences, Mohammed VI Polytechnic University, Morocco (mbaye.abdoulahat@um6p.ma)

Abolfazl Hajihassani, Entomology and Nematology Department, Fort Lauderdale Research and Education Center, University of Florida, Davie, Florida, USA (ahajihassani@ufl.edu)

Aliou Moussa Diouf, Institute of Biological Sciences (ISSB), Mohammed VI Polytechnic University, Morocco (diouf.moussa@um6p.ma)

Awol Seid, Department of Plant Sciences, College of Agriculture, Wollo University, Ethiopia (awolseid07@gmail.com)

B.B. Westerdahl, Department of Entomology and Nematology, University of California, USA (bbwesterdahl@ucdavis.edu)

Bala Rathinasabapathi, Horticultural Sciences Department, 2247 Fifield Hall, Hull Road, University of Florida, Gainesville, USA (brath@ufl.edu)

Carla S. Pimentel, Forest Research Center (CEF), School of Agriculture, University of Lisbon, Portugal (carlapimentel@isa.ulisboa.pt)

Conard Bonsi, Department of Plant Science, Tuskegee University, Tuskegee, USA

Dayani Withanage, Olds College Centre for Innovation, Alberta, Canada (dwithanage@oldscollege.ca)

Deliang Peng, State Key Laboratory for Biology of Plant Diseases and Insect Pests, Institute of Plant Protection, Chinese Academy of Agricultural Sciences, Beijing, China (dlpeng@ippcaas.cn; pengdeliang@caas.cn)

Dominick Padilla, Horticultural Sciences Department, 2247 Fifield Hall, Hull Road, University of Florida, Gainesville, USA (dpadilla@ufl.edu)

Ebrahim Shokoohi, Department of Biochemistry, Microbiology, and Biotechnology, University of Limpopo, South Africa (Ebrahim.shokoohi@ul.ac.za)

Edgard Henrique Costa Silva, University of Western São Paulo, Unoeste, School of Agricultural Sciences, São Paulo, Brazil (edgarddragde@gmail.com)

Frank Kagoda, Crops and Natural Resources Program, NARO-Buginyanya ZARDI, Mbale, Uganda (fkagoda@gmail.com)

George Bird, Department of Entomology, College of Agriculture and Natural Resources, Michigan State University, USA (birdg@msu.edu)

Gerhard Du Preez, Unit for Environmental Sciences and Management, North-West University, Potchefstroom, South Africa (Gerhard.DuPreez@nwu.ac.za)

Giada d'Errico, Research Centre for Plant Protection and Certification, Council for Agricultural Research and Economics (CREA-DC), Florence, Italy (giada.derrico20@gmail.com)

Gregory C. Bernard, Department of Agricultural and Environmental Sciences, Tuskegee University, Tuskegee, USA (gbernard@tuskegee.edu)

Guiping Yan, Department of Plant Pathology, North Dakota State University, USA (guiping.yan@ndsu.edu)

Horacio D. Lopez Nicora, Department of Plant Pathology, The Ohio State University, Columbus, USA (lopez-nicora.1@osu.edu)

Irfan Ahmad, Department of Plant Protection, Faculty of Agricultural Sciences, Aligarh Muslim University, India (ahmadirfan8923@gmail.com)

Mahfouz M.M. Abd-Elgawad, Plant Pathology Department, National Research Centre, Giza, Egypt (mahfouzian2000@yahoo.com)

Marceline Egnin, Department of Plant Science, Tuskegee University, Tuskegee, USA

Milad Rashidifard, Department of Zoology and Entomology, University of the Free State, South Africa (RashidifardM@ufs.ac.za)

Mohd Shahid Anwar Ansari, Department of Plant Protection, Faculty of Agricultural Sciences, Aligarh Muslim University, India (ansari.shahid0786@gmail.com)

Mujeebur Rahman Khan, Department of Plant Protection, Faculty of Agricultural Sciences, Aligarh Muslim University, India (mrkhan777in@yahoo.co.in)

Mustapha Barakate, Laboratory of Microbial Biotechnology, Cadi Ayyad University, Morocco (mustapha.barakate@um6p.ma)

Natsumi Kanzaki, Kansai Research Centre, Forestry and Forest Products Research Institute, Kyoto, Japan (natsumikanzaki@hotmail.com; nkanzaki@affrc.go.jp)

Rachid Lahlali, Molecular Plant Pathology & Microbiology, Département de Protection des Plantes, Meknès, Morocco (rlahlali@enameknes.ac.ma)

Rizwan Ali Ansari, Department of Plant Protection, Faculty of Agricultural Sciences, Aligarh Muslim University, India (rizwans.ansari@gmail.com)

Samad Ashrafi, Institute for Crop and Soil Science, Julius Kühn Institute, Germany; Department of Zoology and Entomology, University of the Free State, South Africa (samad.ashrafi@julius-kuehn.de)

Shabeg S. Briar, Research Agronomist, Olds College Centre for Innovation, Alberta, Canada (sbriar@oldscollege.ca)

Sunil Singh, School of Pacific Arts, The University of the South Pacific, Fiji (sunil.singh@usp.ac.fj)

Tanveer Fatima Rizvi, Department of Plant Protection, Faculty of Agricultural Sciences, Aligarh Muslim University, India (tanveer.124000@gmail.com)

Yao A. Kolombia, International Institute of Tropical Agriculture, Ibadan, Nigeria (Y.Kolombia@cgiar.org)

Zafar Handoo, Mycology and Nematology Genetic Diversity and Biology Laboratory, USDA, USA (zafar.handoo@usda.gov)

Ziaul Haque, Department of Plant Protection, Faculty of Agricultural Sciences, Aligarh Muslim University, India (zia_haq07@yahoo.com)

Zineb Belabess, Plant Protection Laboratory, Regional Center of Agricultural Research of Meknes, Morocco (zineb.belabess@inra.ma)

Zineb Rchiad, Biotechnology Research Unit, Regional Center of Agricultural Research of Rabat, Morocco; Mohammed VI Polytechnic University, Ben Guerir, Morocco (zineb.rchiad@um6p.ma)

Foreword

The system of crop husbandry includes management of soil-borne plant diseases to mitigate damage and resulting yield reduction of economically important crops. Plant-parasitic nematodes have long been recognised as a major component of crop diseases and a threat to global food security, and there is a considerable body of international research on important aspects of nematode biology. Control and management of these pests have previously been predicated on the use of toxic chemicals, but now alternative methods are needed. Although, by themselves, plant-parasitic nematodes cause considerable damage that may even result in total crop loss, it is unrealistic for management purposes to focus on nematode-plant interactions as an isolated component of the soil ecosystem. Thus, plant-parasitic nematodes cannot be considered in isolation; they are one component of multitrophic disease complexes. These complexes involve two or more pathogens, such as nematodes, fungi, bacteria, viruses and insects. For example, some plant-parasitic nematodes are vectors of pathogenic fungi and viruses, some enhance the aggressiveness of certain pathogens (e.g. *Fusarium*), and some species of nematodes (e.g. *Bursaphelenchus xylophilus* and *B. cocophilus*) rely on vectors for their spread. Nematode interaction with other pathogens is of considerable importance, and some complexes have often been under-researched or even neglected. Therefore, there are important research thrusts: the need to look for environmentally acceptable management strategies of plant diseases that reduce or eliminate reliance on chemicals, and the need to examine nematodes as a component of disease complexes.

The book, *Nematode Disease Complexes in Agricultural Crops*, edited by Professor Mujeebur Rahman Khan of Aligarh Muslim University, India, is a timely contribution to the study of plant-parasitic nematodes by broadening the coverage to examine the interaction of nematodes with other components of disease complexes. Principles and mechanisms of disease complexes are covered in separate chapters, and the molecular basis, including the role of effectors and elicitors, is examined. Interaction of nematodes with fungi, bacteria, plant viruses and insects, and between different species of nematodes, are each reviewed in separate chapters. The overview on management of disease complexes brings the information in preceding chapters together. The final three chapters move away from disease complexes *per se* by reviewing the role of nematodes as bioindicators and the effects of air pollution on nematode infection of plants. The final chapter is an important review of the overall effects of climate change on nematode disease complexes.

Above all, the underlying message from this book is the requirement for integrated research to ensure biosecurity in agriculture. This will require collaboration between scientists of different disciplines to ensure the isolated 'research box' labelled 'plant-parasitic nematodes' is unlocked to allow integration with other research disciplines to enable effective research into disease complexes. This book underpins and justifies such collaborative research.

Professor Roland N. Perry
School of Health, Medicine and Life Sciences,
University of Hertfordshire,
United Kingdom

Preface

Nematodes are potential pests of agricultural crops, and inflict economic loss to them. Besides causing direct damage to plants, nematodes interact with other pathogens present in the soil or on the plant surface, and often help them in the infection process. One of the serious consequences of nematode interactions is the increase in the aggressiveness of fungi and bacteria or partial or full loss of host resistance. Plant nematodes also act as vector for fungi, bacteria and viruses, and interact with insects in different ways. The synergistic interaction between nematodes and pathogens leads to formation of disease complex.

The disease complexes are of immense economic significance as well as concern to plant protectionists because of their greater plant damage potential, and frequent occurrence in several important crops. The management of disease complexes becomes complicated and challenging due to its multi-pathogenic nature. The interaction aspects of nematodes have been widely recognized by plant pathologists since 1993 when Prof. M. Wajid Khan, my mentor and philosophical guide, contributed a monumental book on this aspect. However, despite serious consequences of disease complexes resulting due to nematode interactions with other pathogens, limited innovative research efforts on their management have have been undertaken, and hardly any books on this aspect have been published during the last 30 years. The idea to bring out a book specifically on this aspect was conceived to fill the gap of knowledge, and to offer complete and up-to-date information on disease complexes on different agricultural crops along with conventional and innovative management strategies. I appreciate that CABI, realizing the significance, approved the present book on disease complexes.

The book, *Nematode Disease Complexes in Agricultural Crops* is spread over 16 chapters, which are contributed by expert nematologists specializing in the respective areas. The book offers updated and detailed information on the agricultural significance, aetiology and epidemiology of important disease complexes and their management strategies based on integration of conventional and innovative approaches. The role of nematodes in the formation of disease complexes and the mechanisms involved, along with the impact of environmental contamination and climate change on nematodes and the disease complexes, are discussed in detail. The book provides applied and field-based information on the synergistic interaction of nematodes with other pathogens, which is largely scattered and not available in one place for students and researchers. The book also offers practical solutions to handle the disease complexes, and additionally highlights the limitations and challenges in the existing management technologies. This book serves as an up-to-date and elaborated reference source to agriculture researchers, teachers, scientists, undergraduate/postgraduate students, plant nematologists, plant pathologists, entomologists, plant protectionists, agronomists, horticulturalists, helminthologists, extension workers and NGOs.

In view of the highly specialized nature of the subject, it was a real challenge to find relevant expert nematologists for contributing the chapters. I am grateful to the global nematology fraternity for supporting overwhelmingly the present academic venture. I wish to express my warm thanks to all 39 learned nematologists from some 17 countries for contributing informative and exhaustive chapters in the book. I sincerely thank Professor Roland N. Perry for providing the Foreword for this book. I would like to thank CAB International, UK for offering the best help and cooperation in resolving the issues that arose during the publication of this volume with a special mention of Rebecca Stubbs, Helen Elliott, Ray Loughlin and Rosie Hayden. I earnestly appreciate the unwavering and selfless help offered by my students, Mr. Irfan Ahmad, Mr. M. Shahid A. Ansari and Dr. M. Haniph Shah throughout the publication process of this book. I am highly thankful to my wife Huma, daughter Ayesha and son Ahzam for the support, care and affections they showered upon me, and apologize for neglect and loss they suffered due to the extra time I devoted to this book.

Finally and above all, I am extremely indebted to the Almighty Allah SWT for guiding me through all channels to work in cohesion and coordination to make this publication possible.

Mujeebur Rahman Khan
26 January, 2025

1 Plant-parasitic Nematodes Impacting Global Food Security

Mujeebur Rahman Khan*
Department of Plant Protection, Faculty of Agricultural Sciences, Aligarh Muslim University, Aligarh, India

Abstract
Plant nematodes pose a significant challenge to global food security. Their persistent nature, prevalent occurrence, non-specific host preferences and the concealed nature of the plant damage make them an important group of crop pests. Nematodes attack both major and minor agricultural crops as well as the forest flora in all kinds of agroclimates and ecosystems. Plant nematodes may inflict 5–21% yield loss with an average of around 13%, which usually remains unrecognized because nematodes mostly incite non-specific symptoms on above-ground parts which resemble nutrition deficiency. The nematode-induced crop damage is usually realized only after the harvest, when the crop yields significantly less than the expected productivity. Nematodes assume additional significance for having the special capability of aiding a number of soil-borne bacteria and fungi and enhancing the severity of their infections. Nematodes also serve as vectors for plant viruses and other pathogens. For these reasons, the management of phytonematodes becomes essential. In this regard it is most important to implement regular outreach activities to offer knowledge and training to growers to recognize nematode damage in crops, and to apply suitable management strategies. This chapter offers an insight into the impact of plant nematodes and their disease complexes on global food production.

Introduction

Nematodes represent a diverse and foundational lineage within Ecdysozoa, a superphylum inclusive of all organisms that undergo molting, and are bilaterally symmetrical, pseudocoelomic, possessing a cylindrical body conformation, morphologically like a 'tube within another tube,' and are zoologically called roundworms. The fluid-filled body cavity (pseudocoelom), epithelialized by mesoderm on one side and endoderm on the other, makes them unique from other worms and serves as a hydrostatic skeleton nestled between the alimentary canal and a layer of longitudinal muscles beneath the external cuticle stratum. The nematode body is typically three-layered, i.e. the cuticle, hypodermis and muscle layer, with the musculature being broken at the cords. The digestive system comprises an anterior end containing a mouth, pharynx and intestine, whereas the posterior end comprises a rectum and anus or cloaca. Reproductive structures are species-variable and encompass gonads and specialized ducts. The excretory system is not well developed and consists of gland cells or excretory canal(s). Nematodes possess a primary nervous system consisting of a circum-esophageal commissure (nerve ring or 'nematode brain'), nerve cords and ganglia, orchestrating sensory perception and motor responses with a uniqueness that muscle

*Corresponding author: mrkhan777in@yahoo.co.in

 Nematode Disease Complexes in Agricultural Crops (ed. Mujeebur Rahman Khan)
DOI: 10.1079/9781800625228.0001

cells send extensions to the nerves. Nematode body postures may be straight, curved, spiral, etc. (Fig. 1.1).

Nematodes are a captivating group of animals, richly diverse and ubiquitous in nature, and are found in virtually every ecosystem on Earth, including terrestrial and aquatic habitats (Sasser, 1989; Khan, 2008). They are basically aquatic organisms but have adopted terrestrial habitats, constitute 80–90% of the multicellular animals and inhibit all kinds of ecosystems in soil (Khan and Jairajpuri, 2010a, b, 2012; Sikora *et al.*, 2018b). Fortunately, only a small proportion of nematodes present in the soil possess the capability of parasitizing plants, but their agricultural implications are tremendous, significantly affecting global food production (Khan, 2023a, b). In addition, plant-parasitic nematodes also contribute to the nutrient cycling, biology and ecology of the soil. Plant-parasitic nematodes have undergone remarkable evolutionary development, resulting in the formation of the most characteristic feature, a hollow protrusible stylet that enables them to penetrate the host tissue and to establish feeding.

The plant-parasitic nematodes constitute one of the most notorious groups of pests, and attack all types of crops in every agroecosystem (Fig. 1.2). Since plant-parasitic nematode populations grow slowly and gradually, crops growers are unable to realize their slow and hidden damage to the crops until the yield decline becomes frequent and obvious (Sasser, 1989; Khan *et al.*, 2019a, b). Among the major plant pathogens, the crop damage caused by nematodes stands fourth after fungi, bacteria and viruses (Fig. 1.2). Nematode infestations generally result in 5–21% yield loss to agricultural crops with an average of 13%. Several estimates for cumulative annual monetary losses for agricultural crops due to plant-parasitic nematodes have been made which have gradually increased with the progress of time, apparently due to inflation and realization of the nematode damage, e.g. US$80 billion (Agrios, 2005), US$100 billion (Khan, 2008; Sikora *et al.*, 2018b), US$157 billion (Mendoza-de Gives, 2022), US$175 billion (Tóthné Bogdányi *et al.*, 2021) and US$200 billion (Khan, 2023a). Major

Fig. 1.1. Suspension of nematodes isolated from a crop field soil.

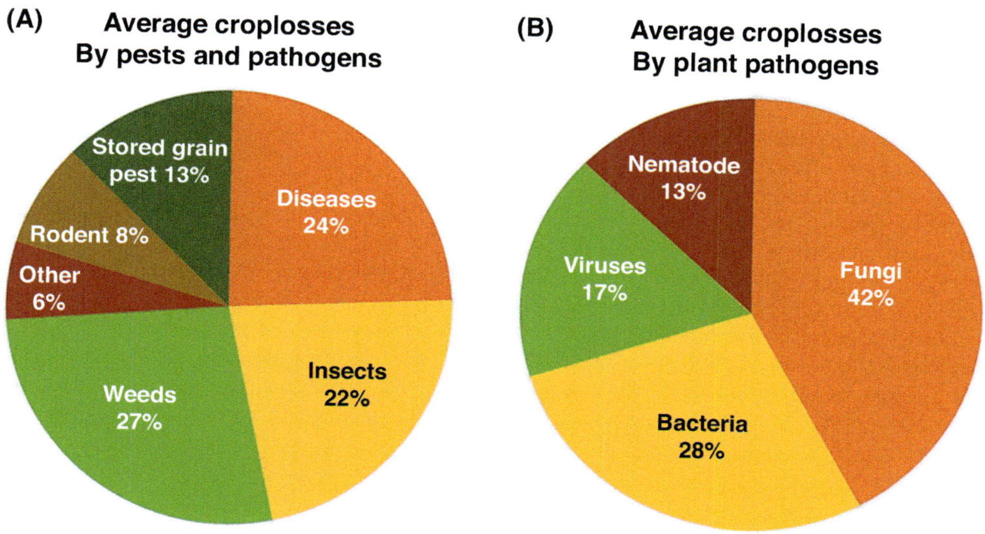

Fig. 1.2. Pie charts showing average crop losses caused by (A) pests and pathogens and (B) plant pathogens (Khan and Jairajpuri, 2010c).

loss to crop production is inflicted by some 15 genera, including *Meloidogyne*, *Pratylenchus*, *Heterodera*, *Ditylenchus*, *Globodera*, *Tylenchulus*, *Xiphinema*, *Radopholus*, *Helicotylenchus*, *Rotylenchulus*, *Tylenchorhynchus*, *Hoplolaimus*, *Aphelenchoides*, *Bursaphelenchus* and *Belonolaimus* (Khan, 2023a). In addition to directly damaging the host, plant nematodes also help other pathogens and enhance their infection; such diseases are called disease complexes (Khan *et al.*, 2023e). Often, nematode disease complexes involve the simultaneous presence of different nematode species along with phytopathogenic fungi or bacteria, each contributing to the overall disease severity (Powell, 1971, 1979; Khan, 1993).

Plant nematodes invade various parts of the plant, including the roots, stem, leaves, crown, inflorescence, flower and developing seeds (Perry and Moens, 2013); however, underground parts are the most preferred tissue for their parasitism because of their soil-borne nature (Khan *et al.*, 2021). The severity of crop damage depends on several factors such as the vulnerability of the cultivar, the soil population of the nematode species and the prevailing environmental conditions (Khan, 2016). The nematode parasitism often leads to a gradual decline in plant health without any visible specific symptoms. The most common non-specific symptoms of nematode damage include stunting of plant growth and leaf yellowing as well as quantitative and qualitative yield reductions (Khan, 2023c; Khan and Quintanilla, 2023). However, some nematode genera cause specific diseases, such as molya in wheat (*Heterodera avenae*), ufra in rice (*Ditylenchus angustus*), root rot in maize (*Pratylenchus zeae*), and root knot in vegetables and pulses (*Meloidogyne* spp.). These diseases are responsible for heavy losses to agricultural production (Luc *et al.*, 2005). The plant stunting and leaf yellowing induced by nematodes resemble those caused by nutritional deficiency in the field. This confuses growers and complicates the nematode management strategy (Khan, 2005, 2007). With such symptoms, farmers apply fertilizers, which promote biomass production providing greater root mass to nematodes for invasion and feeding, and as a result nematode multiplication and the resulting yield losses are further aggravated. The average yield losses caused by plant nematodes range from 5 to 21% across various crops, although this figure can vary significantly with the nematode species and density, host species, soil type and climatic conditions (Khan *et al.*, 2009). In some specific geographic areas, the crop losses can be more severe, with a yield decline of 80–90% in individual fields. In extreme cases, the crop harvest may have no economic value. Sometimes farmers plough the infested crop because of complete cessation of plant growth, no hope for their recovery and expected zero yield (Khan *et al.*, 2023c).

In addition, plant nematodes have a unique capability of interacting synergistically with phytofungi and bacteria (Khan, 1993), causing the disease complexes (Powell, 1971, 1979). They exacerbate or facilitate fungal and bacterial infections, or may break the cultivar resistance. The hairy root caused by *Agrobacterium rhizogenes* in rose plants may be of limited concern, but its severity escalates greatly in co-infection with *Pratylenchus vulnus* (Sitaramaiah and Pathak, 1993). Similarly, the cultivars of vegetables, pulses, cotton and tobacco resistant to fusarium wilt can lose the resistance when plant-parasitic nematodes invade the roots. Wilt disease complex (Francl and Wheeler, 1993; Khan and Sharma, 2020) and root-rot disease complex (Evans and Haydock, 1993) are most devasting, highly damaging and frequently occurring in vegetable and pulse crops, in particular. Plant-parasitic nematodes also serve as vectors for a range of bacteria, fungi and viruses. For instance, the infective juveniles (J_2) of *Anguina tritici* carry *Rathayibacter* (= *Clavibacter*) *tritici* and the fungus *Dilophospora alopecuri* to the shoot of wheat plants (Khan and Dasgupta, 1993). Nematode species from two key families of the order Dorylaimida, namely Longidoridae and Trichodoridae, transmit plant viruses (Hao *et al.*, 2012). *Xiphinema*, *Longidorus* and *Paralongidorus* (Longidoridae) transmit nepoviruses, i.e. Arabis mosaic viruses, grapevine fanleaf virus, tomato black ring virus, etc., whereas members of the family Trichodoridae (*Trichodorus* and *Paratrichodorus*) transmit tobraviruses, i.e. pea early browning virus, tobacco rattle viruses and pepper ringspot virus (Taylor and Brown, 1997; Visser *et al.*, 1999; Hao *et al.*, 2012). Plant nematodes are generally non-host-specific parasites of crops and have a wide host range. For example, *Meloidogyne incognita*, *Pratylenchus penetrans*, *Rotylenchulus reniformis*, *Radopholus similis*, etc., are polyphagous, attacking hundreds of plant species (Khan, 2023a). The major nematodes problems in different plant groups are summarized below.

Cereals

Cereals constitute the most important part of human food, usually amounting to 60–80% of the total food intake. The major cereals include wheat, rice, maze, barley, oat, etc. Plant nematodes attack all kinds of cereals and inflict 7–77% yield decline, varying with the host variety, nematode species, its population and environmental conditions (Owen *et al.*, 2023). The cyst nematode *Heterodera* (Rocha and Duggal, 2023), lesion nematode *Pratylenchus* (Mokrini *et al.*, 2023), root-knot nematode *Meloidogyne* (Khan *et al.*, 2023a, 2023b), stem and bulb nematode *Ditylenchus* (Dash *et al.*, 2023), and leaf and bud nematode *Aphelenchoides* (Westerdahl and Bello, 2023) are the major nematode genera attacking cereal crops (Dababat and Fourie, 2018). The important cyst-forming nematode pests of cereal crops are cereal cyst nematodes (*Heterodera avenae* and *H. filipjevi*) (Fig. 1.3A and B), mediterranean cereal cyst nematode (*H. latipons*), sorghum cyst nematode (*H. sorghi*), corn cyst nematode (*H. zeae*), rice cyst nematode (*H. oryzicola*), Australian cereal cyst nematode (*H. australis*), Chinese cereal cyst nematode/Sturhan's cyst nematode (*H. sturhani*) and barley cyst nematode (*H. hordecalis*) (Turner and Subbotin, 2006; Rocha and Duggal, 2023). *Heterodera* spp. have been reported to attack almost all of the cereal and millet crops, including wheat, maize, rice, barley, oat, sorghum, pear millets, etc., and may inflict up to 80% yield decline under extreme situations (Wu *et al.*, 2014). Among *Pratylenchus* spp., *P. penetrans*, *P. thornei*, *P. zeae*, *P. goodeyi*, *P. delattrei*, *P. coffee*, *P. neglectus* and *P. scribneri* are important lesion nematode species in cereals (Mokrini *et al.*, 2019). The lesion nematodes are globally distributed in cereal cultivations, especially in wheat, maize, oat, barley, etc. (Owen *et al.*, 2023), and suppress their yields by 10–19% (Nicol and Rivoal, 2008) or up to 85% under severe infestation (Mokrini *et al.*, 2019).

Meloidogyne spp. causing root knot is an important pest of all kinds of crops including cereals, with *M. graminicola*, *M. incognita*, *M. nasi*, *M. javanica*, *M. oryzicola*, *M. artiellia*, *M. kikuyensis*, *M. graminis*, etc., being widely distributed species

Fig. 1.3. Nematode diseases in cereals: (A–C) root knot in rice and (D) ear rot in wheat. [Figure used with permission from NASI, India.]

in cereal fields causing significant yield declines throughout the world (Sikora, 1988; Swarup and Sossa-Moss, 1990) (Fig. 1.3B). *Ditylenchus* spp., especially *D. dipsaci*, are prevalent stem and bulb nematodes in cereals such as rye and oat (McDonald and Nicol, 2005). The nematode is widely distributed in Europe, South and North America and Africa (Plowright *et al.*, 2002). The other important species is *D. angustus*, which causes ufra in rice that is prevalent in deep-water rice in Asia (Bora and Rahman, 2010) and inflicts considerable yield decline in the region of 5–50%, with up to 100% in hot areas (Prasad *et al.*, 2010). The leaf and bud nematode *Aphelenchoides bessyi* is another common pest in rice and usually suppresses the yield by 10–17% (Khan *et al.*, 2023c). In addition, *Hirschmanniella*, *Hoplolaimus*, *Xiphinema*, *Trichodorus*, *Belonolaimus*, etc., are potentially damaging nematodes, and may inflict significant damage to cereal crops when their infestation levels are high (McDonald and Nicol, 2005). *Tylenchorhynchus vulgaris* is one of the commonly occurring ectoparasites in sorghum, oat, barley, wheat, pearl millet and other cereals, causing considerable damage to these crops (Dababat and Fourie, 2018; Owen *et al.*, 2023).

Disease complex

Plant nematodes also contribute significantly in forming disease complexes by synergistically interacting with plant-pathogenic fungi and bacteria. The second-stage juveniles of *A. tritici* carry the spores of *D. alopecuri* (fungus) and *R. tritici* (bacteria) to the shoot meristem causing ear twist in wheat (Barbetti and Riley, 2006) (Fig. 1.3D), and yellow slime ('tundu' disease) in wheat and barley (Ahuja and Somvanshi, 2021), respectively. The nematode also vectors another bacterium, *Rathayibacter toxicus*, in ryegrass that leads to annual ryegrass toxicity, which causes a fatal disease in cattle grazing on the diseased ryegrass (Murray *et al.*, 2017).

Pulse Crops

Pulse crops are highly susceptible to phytonematodes and exhibit 20–35% yield reductions (Khan, 2005). Numerous nematode species, including *M. incognita*, *M. javanica*, *M. arenaria*, *R. reniformis*, *P. thornei*, *Heterodera* spp., *D. dipsaci*, *Paratrichodorus*, *Xiphinema*, *Tylenchorhynchus*, *Helicotylenchus* spp., are frequently encountered in pulse cultivation (Ali *et al.*, 2010; Briar *et al.*, 2023). The nematode infestations lead to different types of symptoms such as galling, root decay, root proliferation and suppression in nodulation (Khan *et al.*, 2016a, b) (Fig. 1.4) and small and partially filled pods (Ali *et al.*, 2010; Gerrit *et al.*, 2013). The infection of *Heterodera ciceri* (chickpea), *H. cajani* (pigeon pea) and *H. glycines* (soybean) causes stunting and yellowing of plants, and cysts can be seen attached on the root surface (Askary, 2017).

Disease complex

In pulse crops, wilt disease complex (Khan *et al.*, 2011) and root-rot disease complex (Evans and Haydock, 1993) are commonly encountered causing severe yield losses, and sometimes complete crop failure may occur (Kumari and Katoch, 2020). The wilt disease complexes in pulses such as chickpea, lentil, cowpea, pigeon pea, etc., involving *Meloidogyne* spp. and *Fusarium* spp. have been a factor in the continuous decline in pulse production in many countries (Singh and Goswami, 2001; Khan, 2005; Ali *et al.*, 2010; Kumar *et al.*, 2017). The important disease complexes in pulses occur due to the interaction of *M. artiellia* and *F. oxysporum* f. sp. *ciceris* race 5 (Castillo *et al.*, 2003), *M. incognita* and *F. oxysporum* f. sp. *ciceri*/*F. udum* on chickpea/pigeon pea (Khan *et al.*, 2011), *M. incognita* and *Macrophomina phaseolina* in lentil (Singh and Azam, 2010), green gram (Ahmad and Khan, 2021) (Fig. 1.5), and *H. cajani* and *F. udum* in pigeon pea (Maurya *et al.*, 2020). The root-rot complex incited by *P. thornei* and *Rhizoctonia solani* is highly damaging to chickpea (Khan and Sharma, 2020).

Industrial Crops

Industrial crops, including sugar-yielding crops (sugarcane and sugarbeet), beverage crops (tea and coffee), spices, crops, forage crops, medicinal plants, fiber and wood-yielding plants, etc., are attacked by several nematodes, and often exhibit a yield decline of economic significance (Khan and Jairajpuri, 2010b).

Sugar-yielding crops

Infestation with *Pratylenchus*, *Hoplolaimus*, *Tylenchorhynchus*, *Meloidogyne*, *Helicotylenchus*, etc., commonly occurs in sugarcane fields (Westerdahl *et al.*, 2023). These nematodes may suppress the weight by up to 15% or more (Spaull and Cadet, 1990). Necrosis and galling on the roots of sugarcane is a common symptom incited

Fig. 1.4. Root knot incited by *Meloidogyne* spp. in (A) chickpea, (B) pigeon pea and (C) mung bean. [(A, B) M.R. Khan. (C) M.R. Khan and I. Ahmad].

by *P. zeae* and *M. incognita*, respectively (Fig. 1.6A and B). The decay and heavy loss of root hair in sugarcane are caused by *Tylenchorhynchus* species (Cadet and Spaull, 2005).

The major nematodes which attack sugarbeet crops are *Heterodera schachtii* (Fig. 1.6C), *Meloidogyne* (Fig. 1.6D), *Ditylenchus destructor*, *Trichodorus* and *Paratrichodorus* spp., etc. (Westerdahl *et al.*, 2023). *H. schachtii* is a serious and prevalent nematode problem in beet cultivation in Europe and America (Naderi *et al.*, 2020) and may reduce crop productivity by 11–19% (Hauer *et al.*, 2016). The presence of creamy cysts on the roots can be easily recognized under field conditions (Bohlmann, 2015). *Meloidogyne* spp. incite discernible galls on the beet roots. Severe necrosis of beet root and significant yield loss can be seen due to infection with *D. destructor* (Sturhan and Brzeski, 2020).

Disease complex

Some disease complexes involving nematodes and fungi have been reported in sugarbeet. Whitney (1974) observed severe root decay of sugarbeet roots due to *Pythium ultimum* in the presence of *H. schachtii*. Similarly, root rot due to infection of

Fig. 1.5. Photographs showing the synergistic interaction between root-knot nematode *Meloidogyne incognita* and root-rot fungus *Macrophomina phaseolina* on green gram cv. SML-668. (A) Healthy root system with extensive nodulation (control). (B) Extensive galling caused by the nematode and the root system devoid of nodules. (C) Rotting on the root and girdling of the basal part of the stem caused by the root-rot fungus. (D) Extensive rotting of basal stem, roots and galls due to concomitant inoculation with the fungus and nematode. [(A-D) M.R. Khan and I. Ahmad].

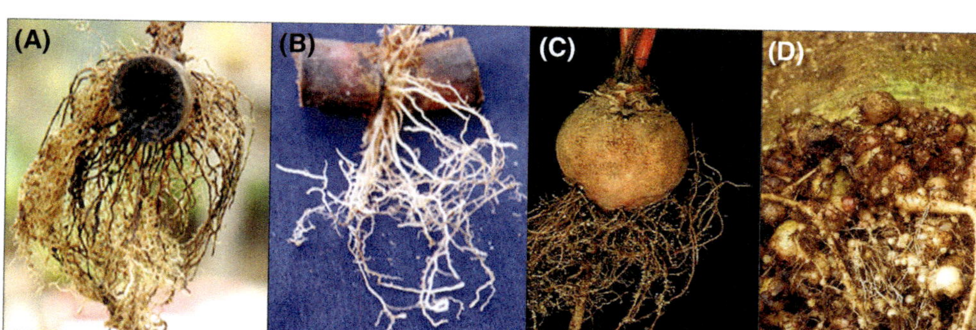

Fig. 1.6. Sugarcane roots show necrosis caused by *Pratylenchus zeae* (A) and galling by *Meloidogyne* sp.(B). Sugarbeet roots show cysts of *Heterodera* (C) and galls by *Meloidogyne* sp. (D). [(A, B) Photographs courtesy of Sankar Narayanan, NASI, India. (C, D) Figure used with permission from IPM images (Bugwood).]

Curvularia lunata was increased in co-infection with *M. javanica* (Khurana and Singh, 1971).

Beverage crops

The common nematode infestants in tea include *M. incognita*, *M. exigua*, *Pratylenchus coffeae* and *P. brachyurus* worldwide (Gnanapragasam and Mohotti, 2005). Nematode infestation in the tea may inflict 4–100% yield decline (Gnanapragasam and Mohotti, 2018). The infection with *Meloidogyne* spp. causes root galls in tea (Fig. 1.7A) and may suppress the yield up to 85% (Liyanage *et al.*, 2002). *P. coffeae*, *P. brachyurus*, *P. zeae*, etc., may cause decay of feeder roots of tea plants (Dasgupta *et al.*, 2010). Arabica and robusta coffee plantations are invaded by *P. coffeae*, *R. similis*, *R. reniformis* and *Hemicriconemoides* spp. (Souza, 2008; Villain *et al.*, 2018). The 'die back' or 'spreading decline' of coffee plantations caused by *P. coffeae* is a serious disease (Subramanian *et al.*, 2010). *M. exigua*, *M. brevicaudata* and *M. incognita* incite galls and decay in coffee roots (Fig. 1.7B). *M. exigua* may suppress the bean yield by as much as 45% (Barbosa *et al.*, 2004). *R. reniformis* (Lordello, 1980; Bridge, 1988) and *Hemicriconemoides* spp. (Kumar, 1984) may also assume economic significance in coffee plantations when they occur at high populations.

Oil-yielding crops

Oil-yielding crops including soybean, peanut and coconut are quite susceptible to attack by plant nematodes. The burrowing nematode *R. similis*, red ring nematode *Bursaphelenchus cocophilus*, root-knot nematode *M. incognita*, root-lesion nematode *Pratylenchus* (Castillo and Vovlas, 2007) and spiral nematode, *Helicotylenchus* spp. (Singh and Campos, 2023) are major nematode species infesting oil palms and cause significant declines in oil yield (Kanzaki *et al.*, 2023). Generally, rapeseed/mustard escape nematode attack except for a few nematodes such as *Tylenchorhynchus brassicae* which may cause stunting of roots. Peanut and soybean are highly susceptible to several nematodes. Infestations of *M. arenaria*, *P. brachyurus*, *Aphelenchoides arachidis*, *T. brevilineatus*, *Belonolaimus longicaudatus*, *Criconemoides omatus* and *Ditylenchus africanus* are commonly encountered in peanut fields (Singh and Campos, 2023). These nematodes cause deshaping and discoloration of pods with up to 20–60% yield declines (Dickson and De Waele, 2005). Similarly, soybean crop is attacked by *H. glycines*, *H. goettingiana*, *M.*

Fig. 1.7. Root knots incited by (A) *Meloidogyne incognita* and (B) *M. exigua* in tea and coffee roots, respectively. [(A) Figure used with permission from IPM images (Bugwood). (B) Figure used with permission from NASI, India.]

javanica, M. incognita, R. reniformis, P. penetrans, P. coffeae, Hoplolaimus columbus, etc. (Fourie *et al.*, 2015). The soyabean cyst nematode *H. glycines* is a major and prevalent nematode pest in soybean cultivations. The nematode infestations cause notable suppression of shoot and root growth and pod formation in soybean (Lima *et al.*, 2017) (Fig. 1.8C). The burrowing nematode *R. similis* is a common nematode pest in palm throughout the world (Nicol *et al.*, 2011; Baby *et al.*, 2017). The nematode causes growth suppression of palms. The infected palms exhibit undersized leaves with yellowing, flowering delayed and smaller fruits. The lateral roots become necrotic (Sikora *et al.*, 2018b) (Fig. 1.8E). *B. cocophilus* (red ring nematode) causes a serious disease in palms, but its occurrence is restricted to South and North America (Giblin-Davis *et al.*, 2013). This nematode incites a characteristic internal symptom of a red ring visible under the cross-section of the palm trunk (Fig. 1.8D). The nematode is transmitted by palm weevils, *Rhynchophorus palmarum* (Gerber and Giblin-Davis, 1990).

Disease complex

The disease complexes can be commonly seen in soybean and peanut crops (Singh and Campos, 2023). Root-rot complexes involving *H. glycines* or *Meloidogyne* spp. and *Fusarium* spp. (Ross, 1964; Arias, 2012), *H. glycines* and *Fusarium solani* or *F. tricinctum* (Yan and Nelson Jr, 2021), *H. glycines* and *M. phaseolina* (Pazan *et al.*, 2020) or *Calonectria crotalariae* (Overstreet *et al.*, 1990) have been reported in soybean. The sudden death syndrome caused by *H. glycines* and *F. virguliforme* is a serious disease complex in soybean cultivation (Westphal *et al.*, 2014). The disease complex caused by *Phytophthora sojae* and *H. glycines* is also important (Chowdhury *et al.*, 2022). The root-rot disease complex in peanut is formed due to the interaction of *M. arenaria* and *Aspergillus flavus* (Patel *et al.*, 1986), and *M. incognita* and *R. solani* (Abdel-Momen *et al.*, 1998). The root-rot complex in peanut is caused by *P. brachyurus* and *Sclerotium rolfsii* (Good *et al.*, 1958), and *M. arenaria* and *P. myriotyrum* and

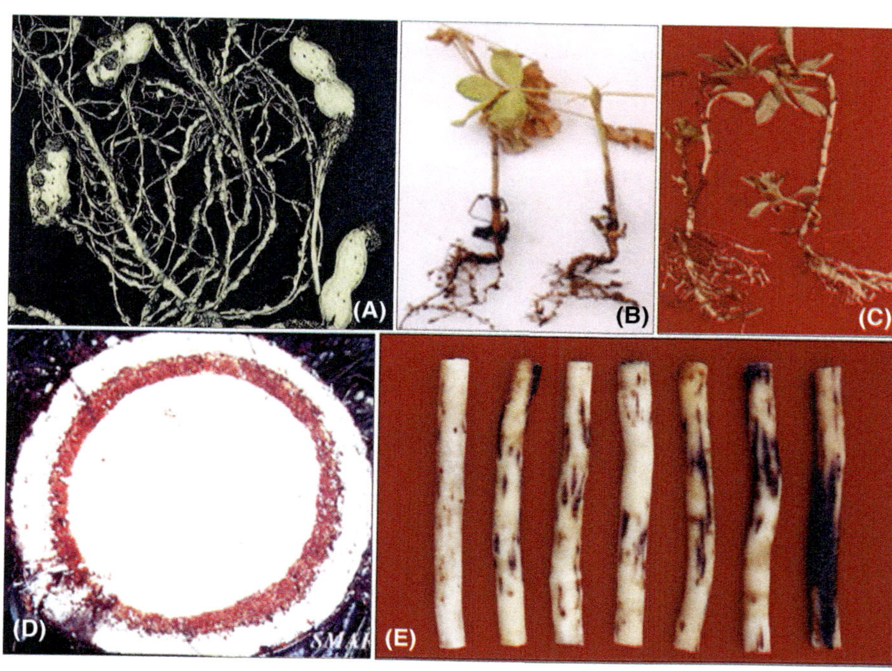

Fig. 1.8. Root knots in (A, B) peanut (*Meloidogyne arenaria*) and (C) soybean (*M. incognita*). (D) Cross-section of coconut trunk showing red ring of coconut incited by *Bursaphelenchus cocophilus*. (E) Necrosis and rotting of coconut roots incited by *Radopholus similis*. [(A, B) Sasser, J.N. (1971). (C) A.A. Khan. (D) V.K. Sosamma and P.K. Koshi. (E) H. Ferris, source Nemaplex.UCDavis.edu.]

M. arenaria and *R. solani* (Garcia and Mitchell, 1975). *Cylindrocarpon* spp. such as *C. effusum*, *C. clavatum* and *C. lucidum* are of minor significance in coconut; however, cylindrocarpon wilt (root rot) becomes severe in plants co-infected by *R. similis* (Koshy and Sosamma, 1987). Red ring disease is also a complex disease, where the palm weevil (*R. palmarum*) transmits the nematode (Giblin-Davis *et al.*, 2003). The palm is killed by the joint action of the nematode causing wilting and growth suppression (Khan, 2020) and the insect may cause toppling of the coconut crown.

Fiber-yielding crops

Important fiber crops include cotton, jute, ramie and flax, which are all susceptible to nematodes (Banu, 2010; Bora *et al.*, 2010). In cotton fields, infestations of *M. incognita*, *R. reniformis*, *P. brachyurus* and *H. columbus* are commonly encountered. These nematodes are reported to suppress plant growth, lint yield and boll size in cotton (Davis *et al.*, 2018; Lahlali *et al.*, 2023). The root-knot, root-lesion and reniform nematodes may reduce cotton yield by 14–45%, 9–17% and 10–25%, respectively (Banu, 2010). Jute fields are commonly infested with *M. incognita*, *M. javanica*, *R. reniformis*, *Helicotylenchus dihystera*, *Hoplolaimus indicus* and *Tylenchorhynchus annulatus* (Neog, 2021). These nematodes may suppress jute fiber yield up to 59%. Nematode symptoms in cotton and jute include the stunting of plants, leaf yellowing and incipient wilting especially during hotter periods of the day. Ramie and flax are also susceptible to root-knot, root-lesion, lance and spiral nematodes, exhibiting significant suppression in fiber yield, both qualitatively and quantitively (Bora *et al.*, 2010) (Fig. 1.9).

Disease complex

The concomitant infection with *Meloidogyne* spp. and *F. oxysporum* f. sp. *vasinfectum* results in a most devastating wilt disease complex in cotton (Rothrock *et al.*, 2015; da Silva *et al.*, 2019; Khan and Sharma, 2020). Other nematode species such as *B. longicaudatus* and *R. reniformis* are also found interacting with *F. oxysporum* and forming the wilt disease complex in cotton (Seinhorst and Kuniyasu, 1971). The reniform nematode *R.*

Fig. 1.9. Root knots incited by *Meloidogyne* spp. in (A) cotton and (B) jute seedlings. [(A) Figure used with permission from IPM images (Bugwood). (B) Figure used with permission from NASI, India.]

reniformis and the sting nematode *Belonolaimus gracilis* may greatly increase wilt incidence in fungus-susceptible cotton varieties (Robinson, 2007). *R. solani*, *Thielaviopsis basicola* and *P. ultimum* incited pre- and post-emergence damping-off of cotton causing necrotic lesions on the hypocotyl and reducing root vigor of the surviving seedlings. *Meloidogyne* spp. are reported to further increase the severity of the aforementioned diseases in cotton (Wagner *et al.*, 2022). Black root rot caused by *T. basicola* was found to be more severe in cotton, and the plants grew slowly due to co-infestation with *Meloidogyne* spp. in field. The root-knot, reniform and lance nematodes are also reported to synergize with *Alternaria* spp., *Glomerella gossypii* and *F. oxysporum* to cause severe seedling diseases in cotton (McLean and Gazaway, 2002). *Ralstonia solanacearum*, *Rhizoctonia bataticola* and *M. incognita* are the primary pathogens associated with jute and their co-occurrence significantly increases disease severity (Mandal and Mishra, 2001; Sarkar and Gawande, 2016).

Spices

Black pepper, cardamon, ginger, turmeric, nutmeg, vanilla, clove, etc., are important spice crops which harbor a number of nematodes, among them *R. similis*, *Meloidogyne* and *Pratylenchus* prevalently occur in spice plantations (Pervez *et al.*, 2023). In black pepper, cardamon and ginger, *R. similis* causes root rot and leaf yellowing (Reddy, 2014; Pervez, 2018). Root knot caused by *M. incognita* (Fig. 1.10) and root necrosis by *P. coffee* are other important diseases in spice crops, and are responsible for up to 46% yield reductions (Eapen and Pandey, 2018).

Disease complex

The slow decline disease complex in black pepper is formed by the concomitant infection of *M. incognita* and *Phytophthora capsici* (Eapen, 2010). The rhizome rot of ginger caused by the *Pythium aphanidermatum* became severe in the presence of *M. incognita* or *P. coffeae* (Dohroo, 2016). The root-rot disease complex caused by *R. similis* and *R. solani* is an important constraint in black pepper cultivation (Pervez *et al.*, 2023).

Horticultural Crops

The soft and succulent body and their cultivation under optimum soil moisture render horticultural crops highly susceptible to nematode infestation (Khan and Jairajpuri, 2012). The roots, bulb, corm, stem, leaves, flowers and fruits of vegetable crops (Gowda *et al.*, 2019; Phani *et al.*, 2024),

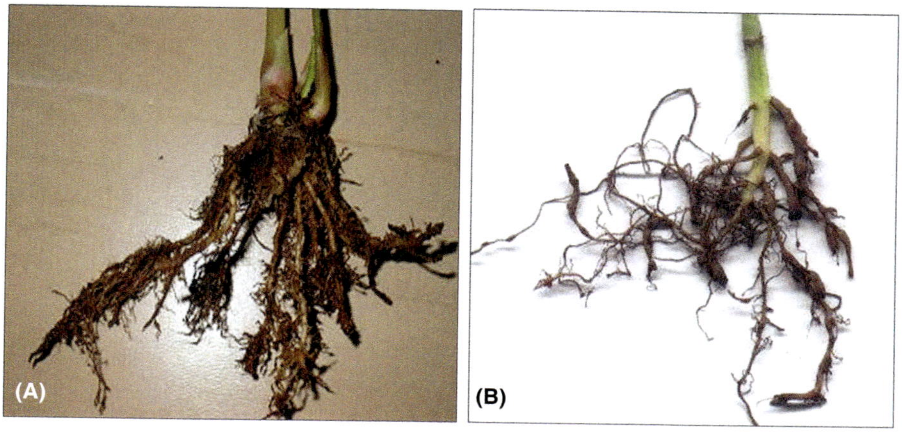

Fig. 1.10. Root knots incited by *Meloidogyne* spp. on (A) cardamom and (B) black pepper. [Figure courtesy of S.J. Eapen, NASI, India.]

ornamental plants (Howland and Quintanilla, 2023), fruit crops (Bahadur, 2021) and medicinal plants (Pandey *et al.*, 2024) are commonly attacked by phytonematodes, leading to 9–21% yield reductions (Khan and Quintanilla, 2023). Crop damages to horticultural crops due to nematodes are summarized below.

Vegetable crops

Vegetables constitute a highly diverse group of crops which may be categorized on the basis of family, such as solanaceous vegetables, cucurbitaceous vegetables, etc. The solanaceous vegetables, such as potato, tomato, eggplant, pepper, etc., are highly susceptible to plant nematodes, especially *Meloidogyne*, *Pratylenchus*, *Globodera*, *Ditylenchus*, *R. reniformis*, *Trichodorus*, *Helicotylenchus*, *Tylenchorhynchus*, etc. (Khan *et al.*, 2012b; Desaeger *et al.*, 2023). *M. incognita*, *M. javanica* and *M. hapla* incite typical galls on the underground parts of all kinds of vegetable plants (Khan and Khan, 1994a, b) (Fig. 1.11) and cause 5–43% yield decline in vegetables (Sasser, 1989). *Globodera* is a serious nematode, infesting potato, tomato and eggplant, and is generally prevalent under cool climates (Niere and Karuri, 2018). The root-lesion nematode and reniform nematode cause necrosis and decay of lateral roots, and suppress the plant growth and yield of hundreds of plant species (Desaeger *et al.*, 2023). Ectoparasitic nematodes such as *Tylenchorhynchus*, *Helicotylenchus*, *Belonolaimus*, *Xiphinema*, *Trichodorus*, etc., are also common infestants in vegetables crops (Chitambar *et al.*, 2018).

Cucurbitaceous vegetables constitute 5% of the vegetable production. *Meloidogyne*, *Pratylenchus*, *R. reniformis* and ectoparasitic nematodes are important pests of cucurbits and cause 10–25% yield declines (Khan *et al.*, 2012a, 2023d). The four major *Meloidogyne* spp. are the most widely distributed and destructive nematodes to cucurbitaceous vegetables, and cause severe yield reductions, but farmers usually fail to recognize the nematode attack, although it incites characteristic symptoms (Khan *et al.*, 2012a; Mohammed and Khan, 2021). The above-ground symptoms resemble nutrition deficiency, whereas on the underground parts, the nematode causes tumor-like large fleshy galls on lateral roots and also stem galls at nodes touching the ground (Khan and

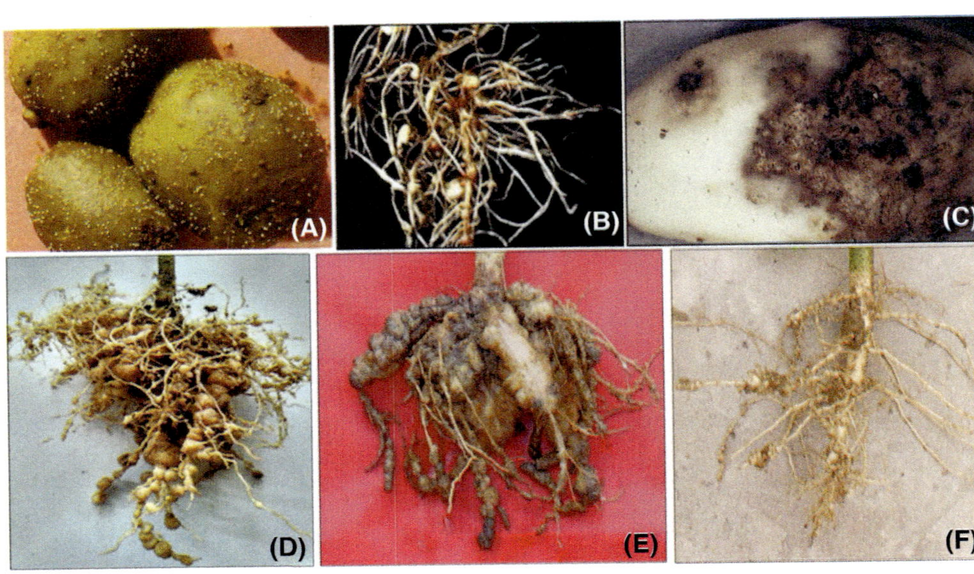

Fig. 1.11. (A) *Globodera* cysts attached on potato tuber. (B) Root knots incited by *Meloidogyne* spp. and (C) tuber rot by *Ditylenchus destructor* on potato. Root knots on (D) tomato, (E) eggplant and (F) okra. [Figure used with permission from IPM images (Bugwood).]

Khan, 1998) (Fig. 1.12A and E). Sometimes the root galls resemble string beads, as seen in bottle gourd, pumpkin, etc.

Peas and beans form an important group of vegetables. They are attacked by *Meloidogyne*, *Heterodera*, *Pratylenchus*, *Radopholus*, *Rotylenchulus*, *Helicotylenchus*, *Tylenchorhynchus*, etc. (Ganguly *et al.*, 2012; Correia *et al.*, 2023) and cause significant yield decline (Khan *et al.*, 2012c; Sikora *et al.*, 2018a, b). The nematodes also suppress root nodulation which leads to a significant decrease in the protein content of the grains (Khan *et al.*, 2016a, b). *M. incognita*, *M. javanica*, *M. hapla* and *M. arenaria* are the most prevalent and notorious nematode pests of pea, cowpea, French bean, broad bean and lima bean, causing 5–30% yield declines (Briar *et al.*, 2023) (Fig. 1.12B–D). The reniform nematode is another common infestant in legume fields; however, its infection is overlooked due to concomitant occurrence with *Meloidogyne* spp. (Khan, 2023b). The nematode, besides reducing plant growth and yield, is also reported to suppress seed germination and seedling development (Ali *et al.*, 2010). *Heterodera cajani*, *H. goettingiana*, etc., attack pea, broad bean, lima bean, etc., and can reduce the yield by 20–50% (Sikora and Fernandez, 2005; Ganguly *et al.*, 2012). The root system of *Heterodera*-infected plants shows white females and cysts, whereas the foliage turns reddish yellow with stunted growth bearing less flowering, and producing empty and/or fewer pods (Sikora *et al.*, 2018a, b).

Leafy vegetables such as spinach, lettuce, chenopodium, fenugreek, dill soa, etc., contribute significantly in enhancing the nutritional value of the human diet by providing vital minerals, vitamins and roughage. Nematode infestation in leafy vegetable crops is often neglected, although it is quite common. Spinach (*Spinacea oleracea*) has been found harboring several nematodes, importantly cyst nematode, root-lesion nematode, root-knot nematode, reniform nematode, etc. (Correll *et al.*, 1994). *Meloidogyne* spp. cause discernible galls on the roots of leafy vegetables (Khan and Rizvi, 2013) (Fig. 1.13). Over 50% incidence of root-knot disease has been recorded in leafy vegetable fields (Anamika *et al.*, 2010). Another commonly occurring nematode in leafy vegetables is *P. penetrans*, and its infection causes leaf chlorosis, undersized leaves and wilting of foliage, and greatly reduces the market value of these vegetables (Srivastava *et al.*, 2012). The cyst nematode *H. schachtii* attacks leafy vegetable plants of the Chenopodiaceae and Brassicaceae, and induces significant suppression in foliage production varying with the population level and host species (Srivastava *et al.*, 2012). The yield loss in broccoli, cabbage and cauliflower (Brassicaceae) and sugarbeet (Chenopodiaceae) may reach 22–60% due to *H. schachtii*. Some ectoparasitic nematodes such as *T. brassicae*, *Belonolaimus* spp. and *Dolichodorus* have also been found to invade the brassica vegetables (Pankaj and Mhatre, 2012).

Disease complex

Heterodera, *Meloidogyne*, *Rotylenchulus*, *Ditylenchus*, *Pratylenchus*, *Hoplolaimus*, etc., are often seen interacting with phytopathogenic fungi and bacteria including *Fusarium*, *Sclerotium*, *Alternaria*, *Rhizoctonia*, *R.*

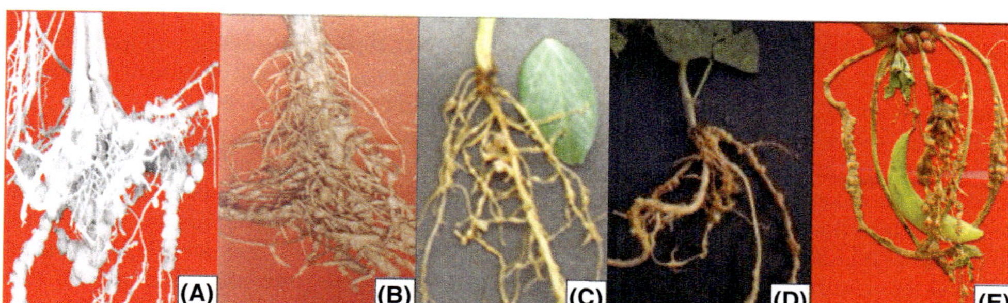

Fig. 1.12. *Meloidogyne* galls on the roots of (A) bottle gourd, (B) broad bean, (C) pea, (D) cowpea and (E) sponge gourd (galls on the stem and roots).

Fig. 1.13. Leafy vegetables showing severe root galls: (A) spinach, (B) lettuce and (C) Chinese cabbage. [(A, B) Figure used with permission from NASI, India.]

solanacearum, etc., on vegetable crops, causing severe disease symptoms and significantly greater reduction in plant growth and yield (Archana *et al.*, 2023; Habteweld *et al.*, 2024; Kashyap *et al.*, 2024). *Meloidogyne*, *Heterodera*, *Pratylenchus* and *Globodera* spp. cause root-rot and wilt disease complexes in association with *R. solani*, *Verticillium dahliae* and *R. solanacearum* (Evans and Haydock, 1993; Francl and Wheeler, 1993). The commonly occurring disease complexes in vegetable crops are formed due to synergistic interaction between *M. incognita* and *F. oxysporum* f. sp. *lycopersici* (Khan and Akram, 2000; Kumar *et al.*, 2017; Meena *et al.*, 2021), *P. penetrans* and *Verticillium rolfsii* on tomato (Scholte and s'Jacob, 1989), *P. penetrans* and *V. sclerotionis* on potato (Rowe *et al.*, 1985), *M. incognita* and *F. oxysporum* (Smits and Noguera, 1982) and *M. incognita* and *R. solanacearum* or *Phomopsis vexans* on eggplant (Khan and Siddiqui, 2017), *M. incognita* and *F. oxysporum* f. sp. *melongenae* on *Solanum aethiopicum* (Ndifon, 2023), *M. incognita* and *P. aphanidermatum* (Hasan, 1988), *M. incognita* and *R. solani* on okra (Golden and Van Gundy, 1975), and *P. penetrans* and *V. dahliae* (Parrado, 2024) and *G. rostochiensis* and *R. solani* on potato (Dunn and Hughes, 1964; Back *et al.*, 2006, 2010; Bhattarai *et al.*, 2010). Potato early dying complex is a very serious disease that develops due to the interaction of *P. penetrans*, *P. scribneri* and *V. dahliae* (Wheeler *et al.*, 1992).

Mushrooms are a popular vegetable and are an excellent source of nutrition. Some mycrobivore nematodes are important constraints in mushroom cultivation (Beyer and Rinker, 2023). *Aphelenchoides* and *Ditylenchus* species, especially *A. composticola* and *D. myceliophagus*, are serious pests of mushrooms such as *Agaricus* spp. in compost. These nematodes feed on the mycelium of white button mushroom in particular (Bajaj and Kanwar, 2012) and significantly reduce button production (Hesling, 1972).

Fruit crops

Fruit crops constitute a major share in global horticulture production. A number of nematodes attack roots fruit crops and cause 9–28% or greater yield decline (Mokrini *et al.*, 2023; Shokoohi *et al.*, 2023). *Radopholus*, *Pratylenchus*, *Meloidogyne*, *Tylenchorhynchus*, *Rotylenchulus*, *Helicotylenchus*, *Xiphinema*, *Trichodorus*, *Paratrichodorus*, etc., are important nematode pests in banana, grapes, citrus, apples, peaches, plumes, berries, pineapple, etc. (El-Borai and Duncan, 2005). *R. similis*, *P. coffeae* and *H. multicinctus* cause rhizome rot or root rot in banana (Fig. 1.14), sometimes leading to toppling of the tree and reducing the yield by up to 40% or more (Sheela and Sundararaju, 2012).

Citrus fruits such as mandarins, oranges, lemons, limes, etc., are popularly known for their high vitamin C content. The fruit trees are commonly attacked by *Tylenchulus semipenetrans* (citrus nematode), *Radopholus citrophilus* (citrus burrowing nematode) and *Helicotylenchus multicinctus* (spiral nematode) (Shokoohi and Duncan, 2018), and may cause a decline of 8–100%, 11–68% and 6–21% in

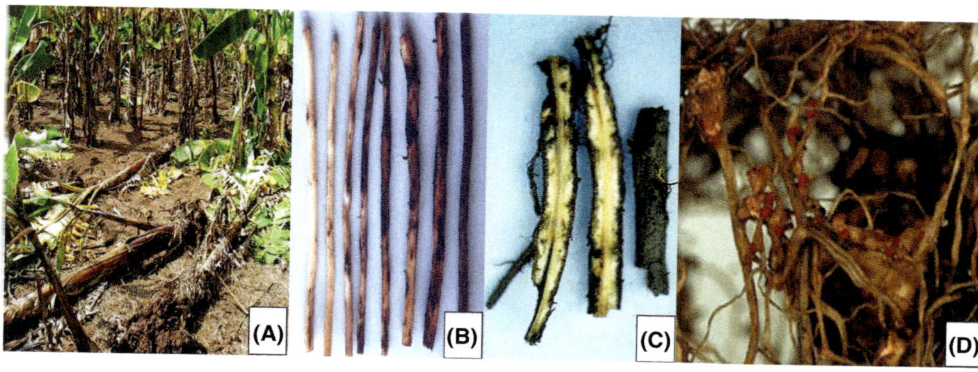

Fig. 1.14. Banana trees infected with *Radopholus similis* showing (A) toppling and (B) necrosis and decay of feeder roots of banana incited by the nematodes. (C) Root necrosis caused by *Helicotylenchus multicinctus* on banana roots. (D) Root knot caused by *Meloidogyne* spp. on grapes. [(A, B) Figure used with permission from NASI, India. (C) Figure courtesy of CABI. (D) Figure courtesy of Daniel Ng, University of California Davis.]

yield, respectively (Duncan and Cohn, 1990). Similarly, pineapple is susceptible to *R. reniformis*, *Meloidogyne*, *Pratylenchus*, *Helicotylenchus*, etc. (Sipes and Chinnasri, 2018) and may exhibit a reduction of 21–100% in yield, which varies with the nematode species, its population density and crop age (Sarah, 1986; Mohilal and Khan, 2012). The infestation of pineapple with *R. reniformis* can reduce crop yield up to 40% at the first fruiting or up to 80–100% in subsequent ratoon crops (complete crop failure), whereas *Pratylenchus* and *Meloidogyne* may cause yield declines of 30–80% (Sarah, 1986) and 10–35% (Mohilal and Khan, 2012), respectively. Other fruits crop, i.e. papaya, grapes, plums, etc., are also highly susceptible to *Meloidogyne*, *Pratylenchus*, *Rotylenchulus*, *Helicotylenchus*, *Trichodorus*, *Tylenchorhynchus*, etc., and exhibit significant yield losses (Mokrini *et al.*, 2023).

Disease complex

Some disease complexes have been reported in fruits crops. Panama wilt in banana becomes severe when *R. similis* and *Fusarium oxysporum* f. sp. *cubense* co-occur infecting a common host plant (Sheela and Sundararaju, 2012). The root-rot disease complex in papaya assumes high severity due to the interaction of *R. reniformis* and *Phytophthora parasitica* (Rao *et al.*, 2016). Similarly, *P. penetrans* and *Vertcillium* spp. in strawberry form root-rot disease complex (McKinley and Talboys, 1979; Bélair *et al.*, 2018). However, interactions between *M. incognita* and *Fusarium nirenbergiae*, *M. incognita* and *Neocosmospora* sp., *M. javanica* and *F. nirenbergiae*, and *M. javanica* and *Neocosmospora* sp. were not synergistic on passion fruit *Passiflora edulis* (Mangeiro *et al.*, 2022). However, synergistic interaction was recorded between *M. incognita* and *F. oxysporum* f. sp. *passiflorae* in the seedlings of grafted passion fruit (Silva, 2023).

Ornamental crops

Floriculture is an important commercial branch of horticulture, but nematode attack is generally ignored in this group of horticultural crops (Crow *et al.*, 2023). Moreover, a number of ornamental plants are grown under protected cultivation, where the risk of nematode attack becomes higher (Khan *et al.*, 2023b). Commercial flower plants such as roses, tulips, gladioli, chrysanthemums, tuberose, jasmine, asters, lilies, etc., are mostly susceptible to species from *Meloidogyne*, *Ditylenchus*, *Aphelenchoides*, *Pratylenchus*, *Belonolaimus*, *Trichodorus*, *Xiphinema*, etc. (Devappa *et al.*, 2012; Khan *et al.*, 2017), and exhibit significant growth reduction and a decrease in the number and size of flowers (Crow *et al.*, 2023). Seasonal plants such as balsam, hollyhock, petunia, poppy, kochia, verbena, etc., show a high degree of susceptibility to root-knot nematodes (Fig. 1.15). Besides flower plants, nematodes also potentially attack turfgrasses (Jonathan *et al.*, 2012). Infestations of ectoparasites such as *Belonolaimus*, *Helicotylenchus*, *Criconemoides*, *Trichodorus*, *Tylenchorhynchus*, etc., are commonly encountered in turfgrasses and lead to significant damage to the grass. Infestation with *B. longicaudatus* poses a serious nematode problem in turf, causing chlorosis and drying of the grass in patches (Crow *et al.*, 2023). The root-knot nematode infects the roots of gladiolus cultivars (Khan and Mustafa, 2005; Devappa *et al.*, 2012) and also causes the wilt disease complex in association with *F. oxysporum* f. sp. *gladioli* (Khan *et al.*, 2017).

Forage crops

Forage crops form an important source of fodder for cattle, and include sorghum, cowpea, berseem, alfalfa, etc. Hasan (2010) summarized that forage crops are susceptible to *Meloidogyne*, *Pratylenchus*, *Ditylenchus*, *Heterodera*, *Tylenchorhynchus*, *Helicotylenchus*, *Longidorus*, *Xiphinema*, *Trichodorus*, etc. (Fig. 1.16). In addition to using chopped dry plants of maize as a fodder, green plants are also used to feed cattle. Maize plants are also susceptible to the nematodes mentioned above (Parihar and Siddiqui, 2010). Nematode infestations may suppress the plant growth of forage crops by 8–59%, which may vary with the nematode population, host species and environmental conditions.

Disease complex

In forage crops, a number of disease complexes involving *M. hapla* and *Phytophthora megasperma*/*F. oxysporum* (Griffin *et al.*, 1988; Griffin and Thyr, 1988), *D. dipsaci* and *F. oxysporum* (Griffin, 1990), *Meloidogyne*/*Pratylenchus*/*X hiphinema* and *P. ultimum* in alfalfa (Townshend, 1984), *M. javanica* and *F. oxysporum* (Regmi *et al.*, 2022), *M. incognita*/*R. reniformis* and *R. solani*

Fig. 1.15. Root knot incited by *Meloidogyne incognita* in (A) balsam, (B) kochea and (C) hollyhock. (D) Interveinal necrosis of chrysanthemum leaves caused by *Aphelenchoides ritzemabosi*. [(A–C) Figure used with permission from NASI, India. (D) Figure courtesy of CABI.]

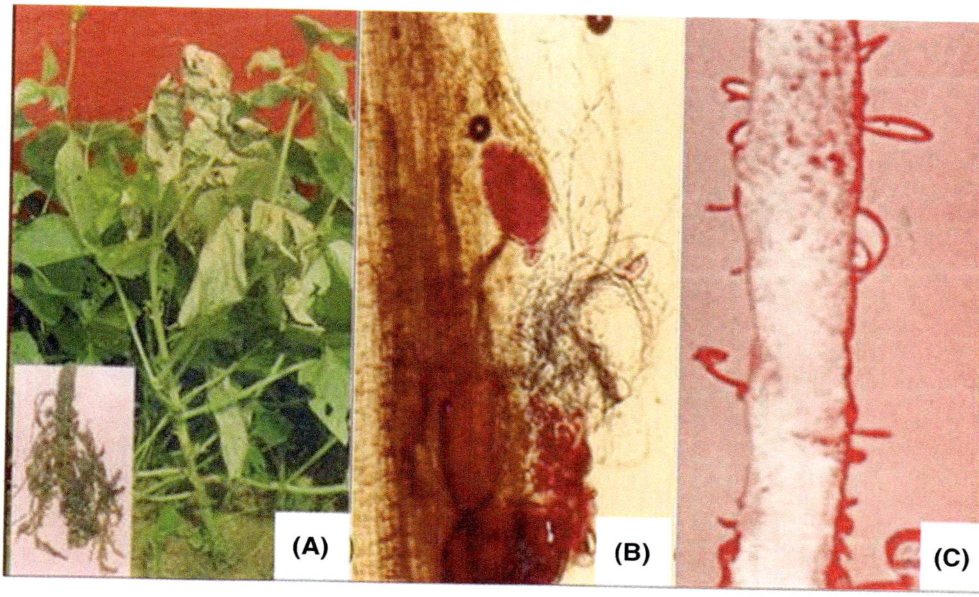

Fig. 1.16. Roots of cowpea showing (A) galls incited by *Meloidogyne* spp., (B) female/cyst of *Heterodera cajani* and (C) invasion of *Helicotylenchus dihystera*. [Figure used with permission from NASI, India.]

in cowpea (Varshney *et al.*, 1987; Khan and Husain, 1989), *M. incognita/Pratylenchus/Hetero dera* and *R. solani* in sorghum (Cuarezma-Teran, 1985), *H. daverti* and *F. avenaceum* or *F. oxysporum* in clover (Nordmeyer and Sikora, 1983; Oro *et al.*, 2024), *M. trifoliophila/Helicotylenchus dihystera/Heterodera trifolii* and *Pythium irregulare/P. nebulosa/Colletotrichum coccodes/M. phaseolina/P. medicaginis/Phoma* sp. in white clover (Zahid *et al.*, 2002). *Hoplolaimus aegypti* and *Cephalosporium maydis/F. moniliforme* in maize (Koura and Satour, 1987), and *M. incognita* and *F. oxysporum* in cowpea (Krishna and Nisha, 2023) have been reported. These disease complexes have been found to severely reduce forage productivity (Hasan, 2010).

Medicinal plants

Due to the side effects of allopathic medicines, a significant shift towards a natural system of therapy using direct formulations from medicinal plants has gained popularity during the last few decades. Medicinal plants have been found to be susceptible to several nematodes including *M. incognita, M. javanica, P. thornei, T. vulgaris, R. reniformis, Trichodorus, Xiphinema*, etc. Root-knot nematode infestation may incite conspicuous root galls, suppress the plant growth and reduce the yield of medicinal plants such as henbanes, ashwagandha, brahmi, musli, mint, etc., by 25–30% (Eapen and Pandey, 2018) (Fig. 1.17). The invasion of *P. thornei* causes necrosis and decay of the roots (Pandey, 2010), whereas *T. vulgaris* causes severe stunting of root and shoot growth of medicinal plants (Pandey *et al.*, 2024).

Disease complex

Some multipathogenic diseases caused by nematodes and fungi have been observed in medicinal plants. Wilt in mint species caused by *V. dahliae* is a major disease (Johnson *et al.*, 2006), which becomes highly severe in the presence of the lesion nematode *P. penetrans* (Ingham and Merrifield, 1996; Johnson and Santo, 2001). The synergistic interaction between *Pratylenchus* spp. and *V. dahliae* results in a serious disease complex in mint (Wheeler *et al.*, 2019), and may cause complete crop failure leading to up to 80%

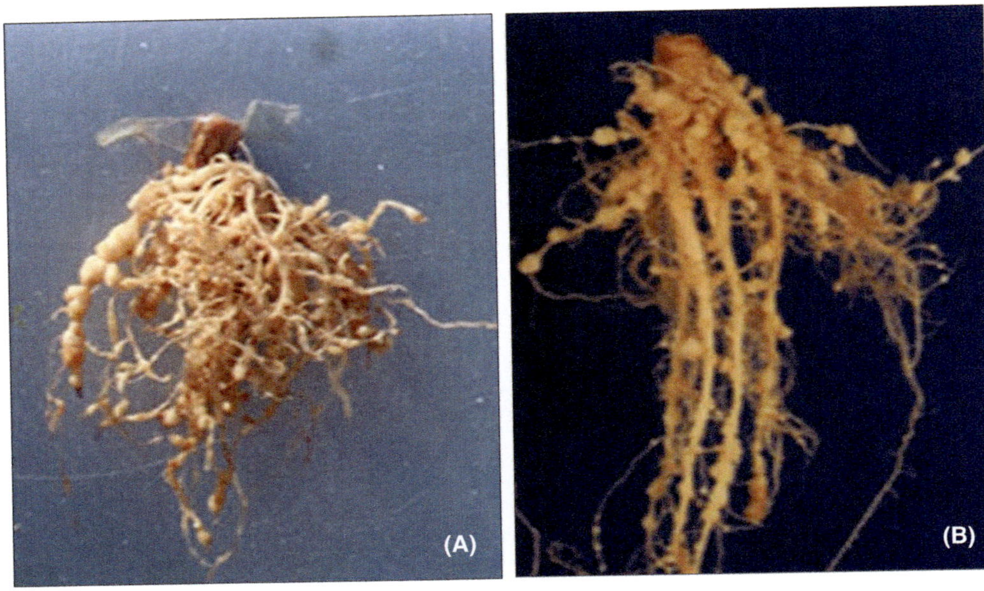

Fig. 1.17. Root galls on (A) Egyptian henbane (*Hyoscyamus muticus*) and (B) black henbane (*H. niger*). [Figure courtesy of R. Pandey, CIMAP, India.]

death of plants (Ingham and Merrifield, 1996; Parrado and Quintanilla, 2024).

Forest trees

Forest flora is an extremely neglected area with regard to nematode infestation (Khan, 2010; Khan and Khan, 2010a, b). Forest soils or plant parts are commonly found infested with *Meloidogyne*, *Pratylenchus*, *Radopholus*, *Bursaphelenchus*, *Helicotylenchus*, *Tylenchorhynchus*, *Hemicriconemoides*, *Xiphinema*, *Trichodorus* and *Paratrichodorus* (Khan, 2012; Pimentel *et al.*, 2023). Nematode infestations (except *Bursaphelenchus* spp.) in forest trees generally cause a slow decline in growth leading to drying and death of the tree in 5–10 years (Fig. 1.18). However, some diseases such as pine wilt (*Bursaphelenchus xylophilus*) may cause tree mortality within 1 year (Zheng and Khan, 2023). A few other nematodes such as *Fergusobia* and *Schistonchus* species causing leaf gall and flower gall disease in eucalypts and figs, respectively, represent serious diseases in forest trees (Khan, 2020). The above three nematodes are transmitted by insect vectors (see details in Chapter 12 on nematode–insect interaction).

Management of Plant Nematodes

Recent decades have witnessed massive growth in global food production in order to feed the continuously burgeoning human population, but the existing production rate has been found to be inadequate and millions of people have been facing huger-related issues. There is sufficient evidence which shows that around 15–25% crop yields are eaten away by various pests and pathogens. Hence, proper management of these biotic stresses may give an immediate 10–15% increase in crop productivity without increasing the crop area or cultivation. The insect damage or the symptoms of the diseases caused by fungi, bacteria and viruses become quite obvious and are recognized by crop growers, but nematodes are hidden parasites of plants, and generally do not incite recognizable symptoms. Plant nematodes, in addition to causing direct damage, also promote the formation of disease complexes with other pathogens (Khan, 1993;

Fig. 1.18. Pine trees showing reddening and drying of pine needles caused by *Bursaphelenchus xylophilus* which is transmitted by *Monochamus* sp. [Figure courtesy of Y. Zheng.]

Back *et al.*, 2002). The nematode disease complexes are extremely serious, as well as highly damaging to crops and challenging to manage (Khan *et al.*, 2023e). In view of the economic significance in crop production, it is essential that plant nematodes are managed properly. In this regard, a nematology extension program may prove extremely effective in minimizing nematode losses in crops. Hence, well-planned extension programs should be organized in villages and crop farms, especially in nematode hotspots, to familiarize crop growers with the damage potential of nematodes, and to offer them adequate knowledge of their recognition and management. Several methods of nematode management are available, but their usefulness and effectiveness vary with different factors. For example, physical methods such as soil solarization, hot water treatment, steaming, etc., and cultivation practices such as host resistance, trap crop, intercropping, crops rotation, etc., may effectively suppress soil populations of nematodes, if initiated before planting a crop. Diversified cropping should be particularly emphasized during the extension and training programs.

Biocontrol using naturally occurring antagonists may also reduce nematode density in the soil (Viaene *et al.*, 2013; Khan, 2016). In recent years, the biocontrol method has gained popularity because of its effective and eco-friendly nature (Khan and Mohiddin, 2023). There are two major groups of biocontrol agents: true antagonists and plant growth-promoting microorganisms (Khan *et al.*, 2009; Khan *et al.*, 2019a). Both these groups can be used to control nematode disease complexes in crop plants. Commercial formulations of *Aspergillus*, *Purpureocillium*, *Trichoderma*, *Penicillium*, *Pochonia*, *Pseudomonas* and *Bacillus* may also be applied to control nematode infestations and the disease complexes. For example, soil application of *Trichoderma*, *Pseudomonas* or *Bacillus* biopesticides in fields and orchards before or at the time of planting may give effective control of plant nematodes as well as the disease complexes in a sustainable way (Khan and Mohiddin, 2023; Quintanilla and Fazlabadi, 2023). *Trichoderma* biopesticides in particular are commercially available and can give satisfactory control of nematode and fungal diseases (Khan *et al.*, 2018a).

However, the ideal management module against disease complexes may be evolved by integrating effective, mutually compatible, sustainable methods or measures. The integrated nematode management (INM) module may include removing the root stubs and crop debris immediately after the harvest, deep ploughing followed by soil solarization, rotation with non-host crops or resistant cultivars, use of healthy planting material and application of biopesticides on the planting material may prevent the nematode and fungus/bacteria inocula carried form the previous crop. Green manuring with sun hemp, alfalfa, crotalaria or interculture with mustard may be done considering the economics and viability of the intercropping module. Further, growing an antagonistic plant such as marigold as a sole or interculture crop along with soil application of *Trichoderma*, *Pseudomonas* or *Bacillus* biopesticides may give good results in reducing the inocula of soil-borne nematodes, fungi and bacteria, offering an effective tactic to manage disease complexes in agricultural crops.

Conclusion and Future Prospects

Plant nematodes invade all kinds of agricultural crops, but their infestation is often unrecognized by growers as well as by scientists and planners. Hence, concerted efforts should be undertaken to strengthen nematode extension and advisory services especially to marginal farmers to acquaint them with the phytonematodes, their crop damage potential and the benefit in their management. Active functioning of a nematode extension service may also help in informing farmers about the potential role of nematodes in enhancing the severity of other diseases (disease complexes), and the serious and devastating nature of disease complexes, for which farmers have been abandoning the cultivation of pulses and vegetables in many areas. The farmers should be familiarized with the advantage of diversified cropping and be advised to follow regular crop rotations. If rotation crop or resistant/tolerant cultivars are not economically viable or socially feasible, adequate nematode management measures must be implemented systematically from the beginning to suppress the pathogen inocula in the soil at the pre-plant stage. Further, disregard of pesticides and use of INM should be advocated because of its economic viability and ecological sustainability. The biopesticides and the biofertilizers should be integrated as a general cultural practice. Another important aspect that may prove highly effective in dealing with plant nematodes includes diversified agriculture. Farmers must be convinced to avoid continuous monoculture unless compelled by agroclimatic limitations. State initiatives such as offering cash and subsidies on seeds, fertilizers, attractive minimum sale price, etc., may motivate farmers to adopt diversified cropping. This practice may prove a milestone in the management of nematodes as well as their disease complexes in agricultural crops.

References

Abdel-Momen, S.M., Simpson, C.E. and Starr, J.L. (1998) Resistance of interspecific *Arachis* breeding lines to *Meloidogyne javanica* and an undescribed *Meloidogyne* species. *Journal of Nematology* 30(3), 341.

Agrios, G.N. (2005) *Plant Pathology*, 5th edn. Elsevier Academic Press, Burlington, MA.

Ahmad, I. and Khan, M.R. (2021) Biocontrol of fungus–nematode disease complex in mungbean. In: *Souvenir & Abstracts, IPS-National e-Conference on Plant Health and Food Security: Challenges and Opportunities*, New Delhi, pp. 61–62.

Ahuja, A. and Somvanshi, V.S. (2021) Diagnosis of plant-parasitic nematodes using loop-mediated isothermal amplification (LAMP): A review. *Crop Protection* 147, 105459. DOI: 10.1016/j.cropro.2020.105459.

Ali, S.S., Naimuddin, A.M. and Ali, M. (2010) Nematode infestation in pulses. In: Khan, M.R. and Jairajpuri, M.S. (eds) *Nematode Infestations Part I: Food Crops*. National Academy of Sciences, Allahabad, India, pp. 288–325.

Anamika, A., Simon, S. and Devi, J. (2010) Survey report on incidence and intensity of root-knot nematode (*Meloidogyne incognita*) on spinach (*Spinacea oleracea*) in U.P and Bihar. *European Journal of Applied Sciences* 2, 51–54.

Archana, T.S., Kumar, D., Kumar, V. and Shukuru, B.N. (2023) Root-knot disease complex: An interactive perspective with microorganisms. In: Ahmad, F. and Blázquez, G.N. (eds) *Root-Galling Disease of Vegetable Plants*. Springer, Singapore, pp. 237–251.

Arias, M.M.D. (2012) *Fusarium* species infecting soybean roots: Frequency, aggressiveness, yield impact and interaction with the soybean cyst nematode. PhD dissertation, Iowa State University, Ames, IA.

Askary, T.H. (2017) Diversity of plant parasitic nematodes in pulses. In: Ansari, A.A., Gill, S.S., Abbas, Z.K. and Naeem, M. (eds) *Plant Biodiversity: Monitoring, Assessment and Conservation*. CAB International, Wallingford, UK, pp. 239–274.

Baby, B.M., Eapen, S.J., Martin, A. and Rosana, O.B. (2017) Radobase – a database on burrowing nematodes infesting tropical and subtropical crops. *Indian Journal of Nematology* 47(2), 197–200.

Back, M., Haydock, P. and Jenkinson, P. (2006) Interactions between the potato cyst nematode *Globodera rostochiensis* and diseases caused by *Rhizoctonia solani* AG3 in potatoes under field conditions. *European Journal of Plant Pathology* 114(2), 215–223. DOI: 10.1007/s10658-005-5281-y.

Back, M., Jenkinson, P., Deliopoulos, T. and Haydock, P. (2010) Modifications in the potato rhizosphere during infestations of *Globodera rostochiensis* and subsequent effects on the growth of *Rhizoctonia solani*. *European Journal of Plant Pathology* 128, 459–471.

Back, M.A., Haydock, P.P.J. and Jenkinson, P. (2002) Disease complexes involving plant parasitic nematodes and soilborne pathogens. *Plant Pathology* 51(6), 683–697. DOI: 10.1046/j.1365-3059.2002.00785.x.

Bahadur, A. (2021) Nematodes diseases of fruits and vegetables crops in India. In: Belle, C. and Kaspary, T.T. (eds) *Nematodes – Recent Advances, Management and New Perspectives*. InTechOpen, London. DOI: 10.5772/intechopen.98850.

Bajaj, H.K. and Kanwar, R.S. (2012) Nematode infestation in mushrooms. In: Khan, M.R. and Jairajpuri, M.S. (eds) *Nematode Infestations Part III: Horticultural Crops*. National Academy of Sciences, Allahabad, India, pp. 356–381.

Banu, G. (2010) Nematode infestation in cotton. In: Khan, M.R. and Jairajpuri, M.S. (eds) *Nematode Infestations Part II: Industrial Crops*. National Academy of Sciences, Allahabad, India, pp. 256–288.

Barbetti, M.J. and Riley, I.T. (2006) Field application of *Dilophospora alopecuri* to manage annual *Ryegrass toxicity* caused by rathayibacter toxicus. *Plant Disease* 90(2), 229–232. DOI: 10.1094/PD-90-0229.

Barbosa, D.H.S.G., Vieira, H.D., Souza, R.M., Viana, A.P. and Silva, C.P. (2004) Field estimates of coffee yield losses and damage threshold by *Meloidogyne exigua*. *Nematologia Brasileira* 28(1), 49–54.

Bélair, G., Coulombe, J. and Dauphinais, N. (2018) Management of *Pratylenchus penetrans* and *Verticilllium* symptoms in strawberry. *Phytoprotection* 98(1), 1–3. DOI: 10.7202/1046783ar.

Beyer, D.M. and Rinker, D. (2023) Nematode problems in mushroom cultivation and their sustainable management. In: Khan, M.R. and Quintanilla, M. (eds) *Nematode Diseases of Crops and Their Sustainable Management*. Academic Press, Cambridge, MA, pp. 337–347.

Bhattarai, S., Haydock, P.P.J., Back, M.A., Hare, M.C. and Lankford, W.T. (2010) Interactions between field populations of the potato cyst nematode *Globodera pallida* and *Rhizoctonia solani* diseases of potatoes under controlled environment and glasshouse conditions. *Nematology* 12(5), 783–790. DOI: 10.1163/138855410X12631974516235.

Bohlmann, H. (2015) Introductory chapter on the basic biology of cyst nematodes. *Advances in Botanical Research* 73, 33–59.

Bora, B.C. and Rahman, M.F. (2010) Stem nematode infestation in rice. In: Khan, M.R. and Jairajpuri, M.S. (eds) *Nematode Infestations Part II: Industrial Crops*. National Academy of Sciences, Allahabad, India, pp. 123–139.

Bora, B.C., Gogoi, B.B., Khan, M.R., Khan, U., Khan, M.M.M. *et al.* (2010) Nematode infestation in jute and other bast fibre crops. In: Khan, M.R. and Jairajpuri, M.S. (eds) *Nematode Infestations Part II: Industrial Crops*. National Academy of Sciences, Allahabad, India, pp. 289–304.

Briar, S.S., Khan, M.R. and Zwart, R. (2023) Nematode problems in pulse crops and their sustainable management. In: Khan, M.R. and Quintanilla, M. (eds) *Nematode Diseases of Crops and Their Sustainable Management*. Academic Press, Cambridge, MA, pp. 183–204.

Bridge, J. (1988) Plant-parasitic nematode problems in the Pacific Islands. *Journal of Nematology* 20(2), 173–183.

Cadet, P. and Spaull, V.W. (2005) Nematode parasites of sugarcane. In: Luc, M., Sikora, R.A. and Bridge, L. (eds) *Plant Parasitic Nematodes in Subtropical and Tropical Agriculture*, 2nd edn. CAB International, Wallingford, UK, pp. 645–674. DOI: 10.1079/9780851997278.0645.

Castillo, P. and Vovlas, N. (2007) Pratylenchus *(Nematoda: Pratylenchidae): Diagnosis, Biology, Pathogenicity and Management*. E.J. Brill, Leiden, The Netherlands. DOI: 10.1163/ej.9789004155640.i-523.

Castillo, P., Navas-Cortés, J.A., Gomar-Tinoco, D., Vito, M. and Jiménez-Díaz, R.M. (2003) Interactions between *Meloidogyne artiellia*, the cereal and legume root-knot nematode, and *Fusarium oxysporum* f. sp. *ciceris* race 5 in chickpea. *Phytopathology* 93(12), 1513–1523. DOI: 10.1094/PHYTO.2003.93.12.1513.

Chitambar, J.J., Westerdahl, B.B. and Subbotin, S.A. (2018) Plant parasitic nematodes in California agriculture. In: Subbotin, S.A. and Chitambar, J.J. (eds) *Plant Parasitic Nematodes in Sustainable Agriculture of North America. Volume 1 –Canada, Mexico and Western USA*. Springer, Dordrecht, The Netherlands, pp. 131–192.

Chowdhury, R.N., Okello, P.N. and Byamukama, E. (2022) Examining the interaction between *Phytophthora sojae* and soybean cyst nematode on soybean (*Glycine max*). *Plants* 11(4), 560. DOI: 10.3390/plants11040560.

Correia, V.R., Khan, M.R. and Ahamad, F. (2023) Nematode problems in vegetable legumes and their sustainable management. In: Khan, M.R. and Quintanilla, M. (eds) *Nematode Diseases of Crops and Their Sustainable Management*. Academic Press, Cambridge, MA, pp. 311–336.

Correll, J.C., Morelock, T.E., Black, M.C., Koike, S.T., Brandenberger, L.P. *et al.* (1994) Economically important diseases of spinach. *Plant Disease* 78(7), 653. DOI: 10.1094/PD-78-0653.

Crow, W.T., Mitkowski, N.A. and LaMondia, J.A. (2023) Nematode problems in ornamentals and turf and their sustainable management. In: Khan, M.R. and Quintanilla, M. (eds) *Nematode Diseases of Crops and Their Sustainable Management*. Academic Press, Cambridge, MA, pp. 655–683.

Cuarezma-Teran, J.A. (1985) Nematodes and fungi associated with a sorghum root disease complex. *Dissertation Abstracts International, B. Sciences and Engineering* 45(9), 2759.

da Silva, M.B., Davis, R.F., Doan, H.K., Nichols, R.L., Kemerait, R.C. *et al.* (2019) Fusarium wilt of cotton may commonly result from the interaction of *Fusarium oxysporum* f. sp. *vasinfectum* with *Belonolaimus longicaudatus*. *Journal of Nematology* 51, 1–10. DOI: 10.21307/jofnem-2019-015.

Dababat, A.A. and Fourie, H. (2018) Nematode parasites of cereals. In: Sikora, R.A., Coyne, D., Hallmann, J. and Timper, P. (eds) *Plant Parasitic Nematodes in Subtropical and Tropical Agriculture*, 3rd edn. CAB International, Wallingford, UK, pp. 163–221.

Dasgupta, M.K., Mukherjee, B., Bhattacharya, C., Khan, M.R. and Ghosh, S. (2010) Nematode infestation in tea. In: Khan, M.R. and Jairajpuri, M.S. (eds) *Nematode Infestations Part II: Industrial Crops*. National Academy of Sciences, Allahabad, India, pp. 213–255.

Dash, M., Somvanshi, V.S. and Walia, R.K. (2023) Stem and bulb nematodes in agricultural crops and their management by biological and biotechnological methods. In: Khan, M.R. (ed.) *Novel Biological and Biotechnological Applications in Plant Nematode Management*. Springer, Singapore, pp. 341–358. DOI: 10.1007/978-981-99-2893-4_15.

Davis, A.P., Wilkinson, T., Williams, J., Baena, S. and Moat, J. (2018) *Coffee Atlas of Ethiopia*. Kew Publishing, London.

Desaeger, J., Khan, M.R. and Silva, E.H.C. (2023) Nematode problems in tomato, okra, and other common vegetables, and their sustainable management. In: Khan, M.R. and Quintanilla, M. (eds) *Nematode Diseases of Crops and Their Sustainable Management*. Academic Press, Cambridge, MA, pp. 223–250. DOI: 10.1016/B978-0-323-91226-6.00013-4.

Devappa, V., Khan, M.R., Jahagirdar, S. and Shrikant, K. (2012) Nematode infestation in carnation, crossandra, chrysanthemum, gladiolus, jasmine, rose and tuberose. In: Khan, M.R. and Jairajpuri, M.S. (eds) *Nematode Infestations Part III: Horticultural Crops*. National Academy of Sciences, Allahabad, India, pp. 445–463.

Dickson, D.W. and De Waele, D. (2005) Nematode parasites of peanut. In: Luc, M., Sikora, R.A. and Bridge, L. (eds) *Plant Parasitic Nematodes in Subtropical and Tropical Agriculture*, 2nd edn. CAB International, Wallingford, UK, pp. 393–436. DOI: 10.1079/9780851997278.0393.

Dohroo, N.P. (2016) Diseases of ginger. In: Ravindran, P.N. and Nirmal Babu, K. (eds) *Ginger*. CRC Press, Boca Raton, FL, pp. 325–360.

Duncan, L.W. and Cohn, E. (1990) Nematode parasites of citrus. In: Luc, M., Sikora, R.A. and Bridge, J. (eds) *Plant Parasitic Nematodes in Subtropical and Tropical Agriculture*. CAB International, Wallingford, UK, pp. 321–346.

Dunn, E. and Hughes, W.A. (1964) Interrelationship of the potato root eelworm, *Heterodera rostochiensis* Woll., *Rhizoctonia solani* Kuhn and *Colletotrichum atramentarium* (B. and Br.) Toub., on the growth of the tomato plant. *Nature* 201(4917), 413–414. DOI: 10.1038/201413b0.

Eapen, S.J. (2010) Nematode infestation in spice cultivation. In: Khan, M.R. and Jairajpuri, M.S. (eds) *Nematode Infestations Part II: Industrial Crops*. National Academy of Sciences, Allahabad, India, pp. 82–119.

Eapen, S.J. and Pandey, R. (2018) Nematode parasites of spices and medicinal plants. In: Luc, M., Sikora, R.A., Coyne, D., Hallmann, J. and Timper, P. (eds) *Plant Parasitic Nematodes in Subtropical and Tropical Agriculture*, 3rd edn. CAB International, Wallingford, UK, pp. 755–794. DOI: 10.1079/9781786391247.0755.

El-Borai, F.E. and Duncan, L.W. (2005) Nematode parasites of subtropical and tropical fruit tree crops. In: Luc, M., Sikora, R.A. and Bridge, L. (eds) *Plant Parasitic Nematodes in Subtropical and Tropical Agriculture*, 2nd edn. CAB International, Wallingford, UK, pp. 467–492. DOI: 10.1079/9780851997278.0467.

Evans, K. and Haydock, P.P.J. (1993) Interaction of nematodes with root-rot fungi. In: Khan, M.W. (ed.) *Nematode Interactions*. Chapman & Hall, London, pp. 104–133. DOI: 10.1007/978-94-011-1488-2_6.

Fourie, H., De Waele, D., Mc Donald, A.H., Mienie, C., Marais, M. *et al.* (2015) Nematode pests threatening soybean production in South Africa, with reference to *Meloidogyne*. *South African Journal of Science* 111(9/10), 9. DOI: 10.17159/sajs.2015/20140212.

Francl, L.J. and Wheeler, T.A. (1993) Interaction of plant-parasitic nematodes with wilt inducing fungi. In: Khan, M.W. (ed.) *Nematode Interactions*. Chapman & Hall, London, pp. 79–103. DOI: 10.1007/978-94-011-1488-2_5.

Ganguly, A.K., Sharma, H.K. and Singh, S. (2012) Nematode infestation in vegetable legumes. In: Khan, M.R. and Jairajpuri, M.S. (eds) *Nematode Infestations Part III: Horticultural Crops*. National Academy of Sciences, Allahabad, India, pp. 108–139.

Garcia, R. and Mitchell, D.J. (1975) Interactions of *Pythium myriotylum* with *Fusarium solani*, *Rhizoctonia solani* and *Meloidogyne arenaria* in pre-emergence damping-off of peanut. *Plant Disease Reporter* 59, 665–669. DOI: 10.1094/Phyto-65-832.

Gerber, K. and Giblin-Davis, R.M. (1990) Association of the red ring nematode and other nematode species with the palm weevil, *Rhynchophorus palmarum*. *Journal of Nematology* 22(2), 143.

Gerrit, K., Wesemael, W. and Moens, M. (2013) Root-knot nematodes. In: *Plant Nematology*, 2nd edn. CAB International, Wallingford, UK. DOI: 10.1079/9781800622456.0003.

Giblin-Davis, R.M., Davies, K.A., Morris, K. and Thomas, W.K. (2003) Evolution of parasitism in insect-transmitted plant nematodes. *Journal of Nematology* 35(2), 133–141.

Giblin-Davis, R.M., Faleiro, J.R., Jacas, J.A., Peña, J.E. and Vidyasagar, P.S.P.V. (2013) Biology and management of the red palm weevil, *Rhynchophorus ferrugineus*. In: Peña, J.E. (ed.) *Potential Invasive Pests of Agricultural Crops*. CAB International, Wallingford, UK, pp. 1–34.

Gnanapragasam, N.C. and Mohotti, K.M. (2005) Nematode parasites of tea. In: Luc, M., Sikora, R.A. and Bridge, L. (eds) *Plant Parasitic Nematodes in Subtropical and Tropical Agriculture*, 2nd edn. CAB International, Wallingford, UK, pp. 581–609.

Gnanapragasam, N.C. and Mohotti, K.M. (2018) Nematode parasites of tea. In: Sikora, R.A., Coyne, D., Hallmann, J. and Timper, P. (eds) *Plant Parasitic Nematodes in Subtropical and Tropical Agriculture*, 3rd edn. CAB International, Wallingford, UK, pp. 584–616.

Golden, J.K. and Van Gundy, S.D. (1975) A disease complex of okra and tomato involving the nematode, *Meloidogyne incognita*, and the soil-inhabiting fungus *Rhizoctonia Solani*. *Phytopathology* 65(3), 265. DOI: 10.1094/Phyto-65-265.

Good, J.M., Boyle, L.W. and Hammons, R.D. (1958) Studies of *Pratylenchus brachyurus* on peanuts. *Phytopathology* 48, 530–535.

Gowda, M.T., Rai, A.B. and Singh, B. (2019) Root knot nematodes menace in vegetable crops and their management in India: A review. *Vegetable Science* 46(1 and 2), 1–16. DOI: 10.61180/aszcm525.

Griffin, G.D. (1990) Pathological relationship of *Ditylenchus dipsaci* and *Fusarium oxysporum* f. sp. *medicaginis* on alfalfa. *Journal of Nematology* 22, 333–336.

Griffin, G.D. and Thyr, B.D. (1988) Interaction of *Meloidogyne hapla* and *Fusarium oxysporum* f. sp. *medicaginis* on alfalfa. *Phytopathology* 78, 421.

Griffin, G.D., Gray, F.A. and Johnson, D.A. (1988) Effect of *Meloidogyne hapla* on resistance and susceptibility of alfalfa to *Phytophthora megasperma* f. sp. *medicaginis*. In: Montray, J.B. and Elgin, J.H. (eds) *Report of the 31st North American Alfalfa Improvement Conference*, Beltsville, MD, June 19–23, p. 23.

Habteweld, A., Kantor, M., Kantor, C. and Handoo, Z. (2024) Understanding the dynamic interactions of root-knot nematodes and their host: Role of plant growth promoting bacteria and abiotic factors. *Frontiers in Plant Science* 15, 1377453. DOI: 10.3389/fpls.2024.1377453.

Hao, Z., Fayolle, L., Tuinen, D., Chatagnier, O., Li, X. *et al*. (2012) Local and systemic mycorrhiza-induced protection against the ectoparasitic nematode *Xiphinema index* involves priming of defence gene responses in grapevine. *Journal of Experimental Botany* 63(10), 3657–3672. DOI: 10.1093/jxb/ers046.

Hasan, A. (1988) Interaction between *Pratylenchus coffeae* and *Pythium aphanidermatum* and/or *Rhizoctonia solani* on chrysanthemum. *Journal of Phytopathology* 123, 227–232. DOI: 10.1111/j.1439-0434.1988.tb04472.x.

Hasan, N. (2010) Nematode infestation in forage crops. In: Khan, M.R. and Jairajpuri, M.S. (eds) *Nematode Infestations Part II: Industrial Crops*. National Academy of Sciences, Allahabad, India, pp. 335–394.

Hauer, M., Koch, H.J., Krüssel, S., Mittler, S. and Märländer, B. (2016) Integrated control of *Heterodera schachtii* Schmidt in central Europe by trap crop cultivation, sugar beet variety choice and nematicide application. *Applied Soil Ecology* 99, 62–69. DOI: 10.1016/j.apsoil.2015.11.017.

Hesling, J.J. (1972) Nematode pests of mushrooms. In: Webster, J.M. (ed.) *Economic Nematology*. Academic Press, London, pp. 435–468.

Howland, A.D. and Quintanilla, M. (2023) Plant-parasitic nematodes and their effects on ornamental plants: A review. *Journal of Nematology* 55(1), 20230007. DOI: 10.2478/jofnem-2023-0007.

Ingham, R.E. and Merrifield, K. (1996) Biology and management of nematodes in mint. IPPC Publication no.996. Integrated Plant Protection Center, Oregon StateUniversity, Corvallis, OR.

Johnson, D.A. and Santo, G.S. (2001) Development of wilt in mint in response to infection by two pathotypes of *Verticillium dahliae* and co-infection by *Pratylenchus penetrans*. *Plant Disease* 85(11), 1189–1192. DOI: 10.1094/PDIS.2001.85.11.1189.

Johnson, D.A., Zhang, H. and Alldredge, J.R. (2006) Spatial pattern of *Verticillium* wilt in commercial mint fields. *Plant Disease* 90(6), 789–797. DOI: 10.1094/PD-90-0789.

Jonathan, E.I., Cannayane, I. and Kavitha, P.G. (2012) Nematode infestation in turf grasses. In: Khan, M.R. and Jairajpuri, M.S. (eds) *Nematode Infestations Part III: Horticultural Crops*. National Academy of Sciences, Allahabad, India, pp. 464–483.

Kanzaki, N., Banu, G. and Anes, K.M. (2023) Nematode problems in palms and their sustainable management. In: Khan, M.R. and Quintanilla, M. (eds) *Nematode Diseases of Crops and Their Sustainable Management*. Academic Press, Cambridge, MA, pp. 427–455. DOI: 10.1016/B978-0-323-91226-6.00017-1.

Kashyap, D., Parveen, A., Khan, M. and Ahamad, L. (2024) Opportunistic fungi, plant, and nematode interactions in vegetable crops. In: Aktar, M.S. (ed.) *Opportunistic Fungi, Nematode and Plant Interactions: Interplay and Mechanisms*. Springer, Singapore, pp. 33–51.

Khan, M.W. (1993) Mechanism of interactions between nematodes and other plant pathogens. In: Khan, M.W. (ed.) *Nematode Interactions*. Chapman & Hall, London, pp. 55–78.

Khan, M.R. (2005) *Biological Control of Fusarial Wilt and Root-Knot of Legumes*. Department of Biotechnology, Ministry of Science and Technology, Government of India Publication, New Delhi.

Khan, M.R. (2007) Prospects of microbial control of root-knot nematodes infecting vegetable crops. In: Sharma, M. and Singh, H.B. (eds) *Biotechnology Plant Health Management*. International Book Distributing, Co, Lucknow, India, pp. 643–665.

Khan, M.R. (2008) *Plant Nematodes – Methodology, Morphology, Systematics, Biology and Ecology*. Science Publishers, Enfield, NH. DOI: 10.1201/b10766.

Khan, M.R. (2010) Emerging nematode problems of forest trees and their management. In: *National Conference on Innovations in Nematological Researchfor Agricultural Sustainability – Challenges and a Roadmap Ahead*, TNAU, Coimbatore, February 23–25, p. 50.

Khan, M.R. (2012) Nematodes, an emerging threat to global forests: Assessment and management. *Plant Pathology Journal* 11(4), 99–113. DOI: 10.3923/ppj.2012.99.113.

Khan, M.R. (2016) Nematode biocontrol agents: Diversity and effectiveness against phytonematodes in sustainable crop protection. Indian Phytopathology 69(4s), 453–463.

Khan, M.R. (2020) Nematode infestation, a potential threat to Indian forests. Indian Phytopathology 73(3), 397–414. DOI: 10.1007/s42360-020-00228-9.

Khan, M.R. (2023a) Nematode pests of agricultural crops, a global overview. In: Khan, M.R. (ed.) *Novel Biological and Biotechnological Applications in Plant Nematode Management*. Springer, Singapore, pp. 3–45.

Khan, M.R. (2023b) Plant nematodes, an underestimated constraint in the global food production. In: Khan, M.R. and Quintanilla, M. (eds) *Nematode Diseases of Crops and Their Sustainable Management*. Academic Press, Cambridge, MA, pp. 3–26. DOI: 10.1016/B978-0-323-91226-6.00009-2.

Khan, M.R. (ed.) (2023c) *Novel Biological and Biotechnological Applications in Plant Nematode Management*. Springer, Singapore. DOI: 10.1007/978-981-99-2893-4.

Khan, M.R. and Akram, M. (2000) Effects of certain antagonistic fungi and rhizobacteria on wilt disease complex of tomato caused by *Meloidogyne incognita* and *Fusarium oxysporum* f. sp. *lycopersici*. *Nematologia Mediterranea* 28(2), 139–144.

Khan, M.W. and Dasgupta, M.K. (1993) The concept of interaction. In: Khan, M.W. (ed.) *Nematode Interactions*. Chapman & Hall, London, pp. 42–54.

Khan, T.A. and Husain, S.L. (1989) Relative resistance of six cowpea cultivars as affected by the concomitance of two nematodes and a fungus. *Nematologia Mediterranea* 17, 39–41.

Khan, M.R. and Jairajpuri, M.S. (eds) (2010a) *Nematode Infestations Part I: Food Crops*. National Academy of Sciences, Allahabad, India.

Khan, M.R. and Jairajpuri, M.S. (eds) (2010b) *Nematode Infestations Part II: Industrial Crops*. National Academy of Sciences, Allahabad, India.

Khan, M.R. and Jairajpuri, M.S. (2010c) Nematode infestation in food crops – national scenario. In: Khan, M.R. and Jairajpuri, M.S. (eds) *Nematode Infestations Part I: Food Crops*. National Academy of Sciences, Allahabad, India, pp. 1–16.

Khan, M.R. and Jairajpuri, M.S. (2012) Nematode infestation in horticultural crops – national scenario. In: Khan, M.R. and Jairajpuri, M.S. (eds) *Nematode Infestations Part III: Horticultural Crops*. National Academy of Sciences, Allahabad, India, pp. 1–30.

Khan, M.R. and Khan, M.W. (1994a) Effects of simulated acid rain and root-knot nematode on tomato. *Plant Pathology* 43, 41–49.

Khan, M.R. and Khan, M.W. (1994b) Single and interactive effects of root-knot nematode and coal-smoke on okra. *New Phytologist* 126, 337–342.

Khan, M.R. and Khan, M.W. (1998) Biomanagement of fungus–nematode wilt disease complexes of pulses. *Botanica* 48, 150–156.

Khan, M.R. and Khan, U. (2010a) Nematode infestation in forest trees. In: Khan, M.R. and Jairajpuri, M.S. (eds) *Nematode Infestations Part II: Industrial Crops*. National Academy of Sciences, Allahabad, India, pp. 395–426.

Khan, M.R. and Khan, U. (2010b) Nematode infestation in Indian forests, a potential threat. In: *Abstract in National Forestry Conference*, Forest Research Institute, Dehradun, India, pp. 164–165.

Khan, M.R. and Mohiddin, F.A. (2023) Biocontrol strategies for nematode management, an overview. In: Khan, M.R. (ed.) *Novel Biological and Biotechnological Applications in Plant Nematode Management*. Springer, Singapore, pp. 113–131.

Khan, M.R. and Mustafa, U. (2005) Corm rot and yellows of gladioli and its biomanagement. *Phytopathologia Mediterranea* 44, 30–37.

Khan, M.R. and Quintanilla, M. (eds) (2023) *Nematode Diseases of Crops and Their Sustainable Management*. Academic Press, Cambridge, MA.

Khan, M.R. and Rizvi, T.F. (2013) Pathogenicity of *Meloidogyne incognita* on spinach, dill soa and fenugreek, and effect of *Trichoderma* species. *Annals of Plant Protection Sciences* 21(1), 143–147.

Khan, M. and Siddiqui, Z.A. (2017) Interactions of *Meloidogyne incognita*, *Ralstonia solanacearum* and *Phomopsis vexans* on eggplant in sand mix and fly ash mix soils. *Scientia Horticulturae* 225, 177–184. DOI: 10.1016/j.scienta.2017.06.016.

Khan, M.R. and Sharma, R.K. (2020) Fusarium–nematode wilt disease complexes, etiology and mechanism of development. *Indian Phytopathology* 73(4), 615–628. DOI: 10.1007/s42360-020-00240-z.

Khan, M.R., Altaf, S., Mohiddin, F.A., Khan, U. and Anwer, A. (2009) Biological control of plant nematodes with phosphate solubilizing microorganisms. In: Khan, M.S. and Zaidi, A. (eds) *Phosphate Solubilizing Microbes for Crop Improvement*. Nova Science Publishers, Hauppauge, NY, pp. 395–426.

Khan, M.R., Majid, S., Mohidin, F.A. and Khan, N. (2011) A new bioprocess to produce low cost powder formulations of biocontrol bacteria and fungi to control fusarial wilt and root-knot nematode of pulses. *Biological Control* 59(2), 130–140. DOI: 10.1016/j.biocontrol.2011.04.007.

Khan, M.R., Khan, A.A. and Khan, U. (2012a) Nematode infestation in cucurbitaceous vegetables. In: Khan, M.R. and Jairajpuri, M.S. (eds) *Nematode Infestations Part III: Horticultural Crops*. National Academy of Sciences, Allahabad, India, pp. 158–181.

Khan, M.R., Reshu, S.S. and Shahid, S. (2012b) Nematode infestation in solanaceous vegetables. In: Khan, M.R. and Jairajpuri, M.S. (eds) *Nematode Infestations Part III: Horticultural Crops*. National Academy of Sciences, Allahabad, India, pp. 31–77.

Khan, M.R., Umamaheswari, R. and Khan, M.M. (2012c) Nematode infestation in beet and beans. In: Khan, M.R. and Jairajpuri, M.S. (eds) *Nematode Infestations Part III: Horticultural Crops*. National Academy of Sciences, Allahabad, India, pp. 210–235.

Khan, M.R., Mohiddin, F.A. and Ahamad, F. (2016a) Inoculant rhizobia suppressed root-knot disease, and enhanced plant productivity and nutrient uptake of some field-grown food legumes. *Acta Agriculturae Scandinavica, Section B – Soil & Plant Science* 68(2), 166–174. DOI: 10.1080/09064710.2017.1374448.

Khan, M.R., Mohidin, F.A., Khan, U. and Ahamad, F. (2016b) Native *Pseudomonas* spp. suppressed the root-knot nematode in *in-vitro* and *in-vivo*, and promoted the nodulation and grain yield in the field grown mungbean. *Biological Control* 101, 159–168. DOI: 10.1016/j.biocontrol.2016.06.012.

Khan, M.R., Shahid, S., Mohidin, F.A. and Mustafa, U. (2017) Interaction of *Fusarium oxysporum* f. sp. *gladioli* and *Meloidogyne incognita* on gladiolus cultivars and its management through corm treatment with biopesticides and pesticides. *Biological Control* 115, 95–104. DOI: 10.1016/j.biocontrol.2017.09.010.

Khan, M.R., Ahmad, I. and Ahamad, F. (2018a) Effect of pure culture and culture filtrates of *Trichoderma* species on root-knot nematode, *Meloidogyne incognita* infesting tomato. *Indian Phytopathology* 71(2), 265–274. DOI: 10.1007/s42360-018-0031-1.

Khan, M.R., Mohiddin, F.A. and Ahamad, F. (2018b) Inoculant rhizobia suppressed root-knot disease, and enhanced plant productivity and nutrient uptake of some field-grown food legumes. *Acta Agriculturae Scandinavica, Section B – Soil & Plant Science* 68(2), 166–174. DOI: 10.1080/09064710.2017.1374448.

Khan, M.R., Ansari, R.A., Rizvi, T.F. and Rafeeq, K.A.M. (2019a) Nematode problem in vegetable crops and its biomanagement. In: Khan, M.R., Mukhopadhyay, A.N., Pandey, R.N., Thakur, M.P., Singh, D. *et al.* (eds) *Bio-Intensive Approaches: Application and Effectiveness in the Management of Plant Nematodes, Insects and Weeds*. Today & Tomorrow's Printers and Publishers, New Delhi, pp. 67–99.

Khan, M.R., Goswami, B.K., Singh, N., Akram, M. and Ahmad, I. (2019b) Microbial management of plant parasitic nematodes. In: Khan, M.R., Mukhopadhyay, A.N., Pandey, R.N., Thakur, M.P., Singh, D. *et al.* (eds) *Bio-Intensive Approaches: Application and Effectiveness in the Management of Plant Nematodes, Insects and Weeds*. Today & Tomorrow's Printers and Publishers, New Delhi, pp. 1–47.

Khan, M.R., Ahamad, I. and Shah, H. (2021) Emerging important nematode problems in field crops and their management. In: Singh, K.P., Jahagirdar, S. and Sarma, B.K. (eds) *Emerging Trends in Plant Pathology*. Springer, Dordrecht, The Netherlands, pp. 33–62. DOI: 10.1007/978-981-15-6275-4_3.

Khan, M.R., Ahmad, I., Ansari, M.S.A. and Shah, M.H. (2023a) Root-knot nematodes in cereal and pulse crops, and their management by novel biological and biotechnological approaches. In: Khan, M.R. (ed.) *Novel Biological and Biotechnological Applications in Plant Nematode Management*. Springer, Singapore, pp. 289–311.

Khan, M.R., Ahamad, F., Rizvi, T.F. and Akram, M. (2023b) Root-knot nematodes in vegetables and ornamentals and their management by novel biological and biotechnological tools. In: Khan, M.R. (ed.) *Novel Biological and Biotechnological Applications in Plant Nematode Management*. Springer, Singapore, pp. 261–288. DOI: 10.1007/978-981-99-2893-4_12.

Khan, M.R., Haque, Z., Ahamad, F. and Shah, M.H. (2023c) Nematode problems in rice and their sustainable management. In: Khan, M.R. and Quintanilla, M. (eds) *Nematode Diseases of Crops and Their Sustainable Management*. Academic Press, Cambridge, MA, pp. 133–166. DOI: 10.1016/B978-0-323-91226-6.00001-8.

Khan, M.R., Luca, R. and Mohammed, R.K. (2023d) Nematode problems in cucurbits and their sustainable management. In: Khan, M.R. (ed.) *Novel Biological and Biotechnological Applications in Plant Nematode Management*. Springer, Singapore, pp. 261–288. DOI: 10.1007/978-981-99-2893-4_7.

Khan, M.R., Mohiddin, F.A., Haque, Z. and Sharma, R.K. (2023e) Disease complexes and their sustainable management. In: Khan, M.R. and Quintanilla, M. (eds) *Nematode Diseases of Crops and*

Their Sustainable Management. Academic Press, Cambridge, MA, pp. .65–93. DOI: 10.1016/B978-0-323-91226-6.00016-X.

Khurana, S.M.P. and Singh, S. (1971) Interrelationship of a fungus, *Curvularia lunata* with root-knot nematode, *Meloidogyne javanica* in sugarcane seedling blight. *Japanese Journal of Phytopathology* 37(4), 313–315. DOI: 10.3186/jjphytopath.37.313.

Koshy, P.K. and Sosamma, V.K. (1987) Pathogenicity of *Radopholus similis* on coconut (*Cocos nucifera* L.) seedlings under greenhouse and field conditions. *Indian Journal of Nematology* 17, 108–118.

Koura, F.H. and Satour, M.M. (1987) Interaction of *Hoplolaimus aegypti* with soil pathogenic fungi *Cephalosporium maydis* and *Fusarium moniliforme* in the root rot complex of maize. *Annals of Agricultural Science, Ain-Shams University* 32, 1849–1855.

Krishna, K. and Nisha, M. (2023) Interaction of *Meloidogyne incognita* and *Fusarium oxysporum* on vegetable cowpea (*Vigna unguiculata* (L.) Walp. *Pest Management in Horticultural Ecosystems* 29(1), 144–150.

Kumar, A.C. (1984) Investigations on “Cannoncadoo dieback” in coffee. *Journal of Caffeine Research* 14, 85–103.

Kumar, N., Bhatt, J. and Sharma, R.L. (2017) Interaction between *Meloidogyne incognita* with *Fusarium oxysporum* f. sp. *lycpersici* on tomato. *International Journal of Current Microbiology and Applied Sciences* 6(8), 1770–1776. DOI: 10.20546/ijcmas.2017.608.209.

Kumari, N. and Katoch, S. (2020) Wilt and root rot complex of important pulse crops: Their detection and integrated management. In: Singh, B.P., Singh, G., Kumar, K., Nayak, S.C. and Srinivasa, N. (eds) *Management of Fungal Pathogens in Pulses: Current Status and Future Challenges*. Springer, Dordrecht, The Netherlands, pp. 93–119. DOI: 10.1007/978-3-030-35947-8_6.

Lahlali, R., Banu, G., Mokrini, F., Ibrahim, D.S. and Belabess, Z. (2023) Nematode problems in cotton and their sustainable management. In: Khan, M.R. and Quintanilla, M. (eds) *Nematode Diseases of Crops and Their Sustainable Management*. Academic Press, Cambridge, MA, pp. 531–564. DOI: 10.1016/B978-0-323-91226-6.00004-3.

Lima, F.S., Correa, V.R., Nogueira, S.R. and Santos, P.R. (2017) Nematodes affecting soybean and sustainable practices for their management. In: Kasia, M. (ed.) *Soybean – The Basis of Yield, Biomass and Productivity*. InTechOpen, London, pp. 95–110. DOI: 10.5772/67030.

Liyanage, D.D., Jayaratne, U.P. and Vitarana, S.I. (2002) *Nematode Pests of Tea*. Interface Science, Sri Lanka.

Lordello, L.G.E. (1980) Estado atual do nematoide reniforme como parasita do cafeeiro. *Revisita de Agricultura (Piracicaba, Brazil)* 55, 62.

Luc, M., Bridge, J. and Sikora, R.A. (2005) Reflections on nematology in tropical and subtropical agriculture. In: Luc, M., Sikora, R.A. and Bridge, J. (eds) *Plant Parasitic Nematodes in Subtropical and Tropical Agriculture*, 2nd edn. CAB International, Wallingford, UK, pp. 1–10. DOI: 10.1079/9780851997278.0001.

Mandal, R.K. and Mishra, C.D. (2001) Role of different organisms in inducing Hooghly wilt symptom in jute. *Environmental Ecology* 19(4), 969–972. DOI: 10.1515/9783110849035.bm.

Mangeiro, M.Z., Nunes, R.A., Vieira, O.L., Mussi-Dias, V., Viana, A.P. *et al.* (2022) Characterizing the interaction between root-knot nematodes and soil-borne fungi which are pathogenic to passion fruit (*Passiflora edulis*). *Journal of Nematology* 54(1). DOI: 10.2478/jofnem-2022-0023.

Maurya, A.K., John, V., Murmu, R., Simon, S. and Pant, H. (2020) An overview of *Fusarium udum* and *Heterodera cajani* interactions in pigeon pea (*Cajanus cajan*). *Current Research and Innovations in Plant Pathology* 9(6), 98–112.

McDonald, A.H. and Nicol, J.M. (2005) Nematode parasites of cereals. In: Luc, M., Sikora, R.A. and Bridge, L. (eds) *Plant Parasitic Nematodes in Subtropical and Tropical Agriculture*, 2nd edn. CAB International, Wallingford, UK, pp. 131–191.

McKinley, R.T. and Talboys, P.W. (1979) Effects of *Pratylenchus penetrans* on development of strawberry wilt caused by *Verticillium dahliae*. *Annals of Applied Biology* 92, 347–357.

McLean, K.S. and Gazaway, W.S. (2002) Nematode and fungal disease interactions on cotton. In: *2002 Beltwide Cotton Conferences*, Atlanta, GA, January 8–12, .

Meena, P., Chandrawat, B.S., Ahir, R.R., Maurya, S., Lakhran, L. *et al.* (2021) Effect of *Fusarium oxysporum* f. sp. *lycopersici* and *Meloidogyne incognita* interaction on wilt disease incidence in tomato. *International Journal of Chemical Studies* 9(1), 2621–2624. DOI: 10.22271/chemi.2021.v9.i1ak.11621.

Mendoza-de Gives, P. (2022) Soil-borne nematodes: Impact in agriculture and livestock and sustainable strategies of prevention and control with special reference to the use of nematode natural enemies. *Pathogens* 11(6), 640.

Mohammed, R.K.A. and Khan, M.R. (2021) Management of root-knot nematode in cucumber through seed treatment with multifarious beneficial microbes in polyhouse under protected cultivation. *Indian Phytopathology* 74(4), 1035–1043. DOI: 10.1007/s42360-021-00422-3.

Mohilal, N.M. and Khan, M.R. (2012) Nematode infestation in pineapple. In: Khan, M.R. and Jairajpuri, M.S. (eds) *Nematode Infestations Part III: Horticultural Crops*. National Academy of Sciences, Allahabad, India, pp. 288–321.

Mokrini, F., Viaene, N., Waeyenberge, L., Dababat, A.A. and Moens, M. (2019) Root-lesion nematodes in cereal fields: Importance, distribution, identification, and management strategies. *Journal of Plant Diseases and Protection* 126(1), 1–11. DOI: 10.1007/s41348-018-0195-z.

Mokrini, F., Laasli, S.E., Iraqi, D. and Lahlali, R. (2023) Nematode problems in tropical fruits and their sustainable management. In: Khan, M.R. and Quintanilla, M. (eds) *Nematode Diseases of Crops and Their Sustainable Management*. Academic Press, Cambridge, MA, pp. 351–374. DOI: 10.1016/B978-0-323-91226-6.00005-5.

Murray, T.D., Schroeder, B.K., Schneider, W.L., Luster, D.G., Sechler, A. *et al*. (2017) *Rathayibacter toxicus*, other *Rathayibacter* species inducing bacterial head blight of grasses, and the potential for livestock poisonings. *Phytopathology* 107(7), 804–815. DOI: 10.1094/PHYTO-02-17-0047-RVW.

Naderi, R., Esfahani, M.N. and Khaniki, G.B. (2020) Analysis of molecular characterizations of beets, *Beta vulgaris* in response to cyst nematodes, *Heterodera schachtii*. *Physiological and Molecular Plant Pathology* 112, 101518. DOI: 10.1016/j.pmpp.2020.101518.

Ndifon, E. (2023) Host–parasite interactions between *Solanum aethiopicum*, *Meloidogyne incognita*, and *Fusarium oxysporum* f. sp. *melongenae* as portrayed by disease traits and crop yield. *Journal of Agricultural Sciences* 33(3), 461–477.

Neog, P.P. (2021) Effect of crop sequences in combination with other ecofriendly practices for management of plant parasitic nematodes infecting jute (*Corchorus olitorius* L.). *Crop Research* 56(6), 363–368.

Nicol, J.M. and Rivoal, R. (2008) Global knowledge and its application for the integrated control and management of nematodes on wheat. In: Cianco, A. and Mukerji, K.G. (eds) *Integrated Management and Biocontrol of Vegetable and Grain Crops Nematodes*. Springer, Dordrecht, The Netherlands, pp. 251–294.

Nicol, J.M., Turner, S.J., Coyne, D.L., den Nijs, L., Hockland, S. *et al*. (2011) Current nematode threats to world agriculture. In: Jones, J.T., Gheysen, L. and Fenoll, C. (eds) *Genomics and Molecular Genetics of Plant–Nematode Interactions*. Springer, Dordrecht, The Netherlands, pp. 21–43. DOI: 10.1007/978-94-007-0434-3_2.

Niere, B. and Karuri, H. (2018) Nematode parasites of potato and sweet potato. In: Sikora, R.A., Coyne, D., Hallmann, J. and Timper, P. (eds) *Plant Parasitic Nematodes in Subtropical and Tropical Agriculture*, 3rd edn. CAB International, Wallingford, UK, pp. 222–251. DOI: 10.1079/9781786391247.0222.

Nordmeyer, O. and Sikora, R.A. (1983) Studies on the interaction between *Heterodera daverti, Fusarium avenaceum* and *F. Oxysporum* on *Trifolium subterraneum*. *Revue de Nematologie* 6, 193–198.

Oro, V., Stanisavljević, R. and Tabaković, M. (2024) First record of the clover cyst nematode *Heterodera daverti* in Serbia. In: *35th International Symposium of the European Societyof Nematologists*, European Society of Nematologists, April 15–19.

Overstreet, C., McGawley, E.C. and Russin, J.S. (1990) Interactions between *Calonectria crotalariae* and *Heterodera glycines* on soybean. *Journal of Nematology* 22(4), 496–505.

Owen, K., Walia, R.K., Yan, G. and Khan, M.R. (2023) Nematode problems in wheat and barley and their sustainable management. In: Khan, M.R. and Quintanilla, M. (eds) *Nematode Diseases of Crops and Their Sustainable Management*. Academic Press, Cambridge, MA, pp. 97–131. DOI: 10.1016/B978-0-323-91226-6.00026-2.

Pandey, R. (2010) Nematode infestation in medicinal plants. In: Khan, M.R. and Jairajpuri, M.S. (eds) *Nematode Infestations Part II: Industrial Crops*. National Academy of Sciences, Allahabad, India, pp. 305–334.

Pandey, R., Khan, M.R., Haque, Z. and Ahmad, I. (2024) Nematode problems in medicinal plants. In: Khan, M.R. and Haque, Z. (eds) *Diseases of Ornamental, Aromatic and Medicinal Plants*. Bentham Science, Sharjah, UAE, pp. 308–334.

Pankaj, J. and Mhatre, P. (2012) Nematode infestation in cruciferous vegetables. In: Khan, M.R. and Jairajpuri, M.S. (eds) *Nematode Infestations Part III: Horticultural Crops*. National Academy of Sciences, Allahabad, India, pp. 140–157.

Parihar, A. and Siddiqui, A.U. (2010) Nematode infestation in maize. In: Khan, M.R. and Jairajpuri, M.S. (eds) *Nematode Infestations Part II: Industrial Crops*. National Academy of Sciences, Allahabad, India, pp. 143–157.

Parrado, L.M. (2024) Investigating sustainable strategies to manage the root-lesion nematode *Pratylenchus penetrans* and the wilt-inducing fungus *Verticillium dahliae* in potato production. Doctoral dissertation, Michigan State University, East Lansing, MI.

Parrado, L.M. and Quintanilla, M. (2024) Plant-parasitic nematode disease complexes as overlooked challenges to crop production. *Frontiers in Plant Science* 15, 1439951. DOI: 10.3389/fpls.2024.1439951.

Patel, H.R., Vaishnav, M.U. and Dhruj, L.U. (1986) Effect of interaction between *Meloidogyne arenaria* and *Aspergillus flavus* on groundnut. *Indian Journal of Mycology and Plant Pathology* 16, 103–107.

Pazan, A., Jamali, S., Aghajani, M.A. and Maleki Ziarati, H. (2020) Interactions between cyst nematode (*Heterodera glycines*) and charcoal rot fungus (*Macrophomina phaselolina*) in susceptible and resistant soybean cultivars to the nematode. *Iranian Journal of Plant Protection Science* 51(2), 287–295.

Perry, R.N. and Moens, M. (2013) Introduction to plant-parasitic nematodes; modes of parasitism. In: Jones, J.T., Gheysen, L. and Fenoll, C. (eds) *Genomics and Molecular Genetics of Plant–Nematode Interactions*. Springer, Dordrecht, The Netherlands, pp. 3–20.

Pervez, R. (2018) Plant health management in major spice crops. In: Sharangi, A.B. (ed.) *Indian Spices*. Springer, Dordrecht, The Netherlands, pp. 205–247. DOI: 10.1007/978-3-319-75016-3_8.

Pervez, R., Rajkumarand Olabiyi, T.I. (2023) Nematode problems in major spice crops and their sustainable management. In: Khan, M.R. and Quintanilla, M. (eds) *Nematode Diseases of Crops and Their Sustainable Management*. Academic Press, Cambridge, MA, pp. 497–514. DOI: 10.1016/B978-0-323-91226-6.00025-0.

Phani, V., Gowda, M.T. and Dutta, T.K. (2024) Grafting vegetable crops to manage plant-parasitic nematodes: A review. *Journal of Pest Science* 97(2), 539–560. DOI: 10.1007/s10340-023-01658-w.

Pimentel, C., Khan, M.R., Zheng, Y. and Quintanilla, M. (2023) Nematode problems in forests and their sustainable management. In: Khan, M.R. and Quintanilla, M. (eds) *Nematode Diseases of Crops and Their Sustainable Management*. Academic Press, Cambridge, MA, pp. 457–493. DOI: 10.1016/B978-0-323-91226-6.00003-1.

Plowright, R.A., Caubel, G. and Mizen, K.A. (2002) *Ditylenchus* species. In: Starr, J.L., Cook, R. and Bridge, J. (eds) *Plant Resistance to Parasitic Nematodes*. CAB International, Wallingford, UK, pp. 107–139. DOI: 10.1079/9780851994666.0107.

Powell, N.T. (1971) Interactions between nematodes and fungi in disease complexes. *Annual Review of Phytopathology* 9(1), 253–274. DOI: 10.1146/annurev.py.09.090171.001345.

Powell, N.T. (1979) Internal synergisms among organisms inducing disease. In: Horsfall, J.G. and Cowling, E.B. (eds) *Plant Disease: An Advanced Treatise. Volume IV. How Pathogens Induce Disease*. Academic Press, New York, pp. 113–133.

Prasad, J.S., Somasekhar, N. and Varaprasad, K.S. (2010) Nematode infestation in paddy. In: Khan, M.R. and Jairajpuri, M.S. (eds) *Nematode Infestations Part II: Industrial Crops*. National Academy of Sciences, Allahabad, India, pp. 20–53.

Quintanilla, M. and Fazlabadi, R.Y. (2023) Methods of sustainable management of plant nematodes, limitations, and challenges for crop growers. In: Khan, M.R. and Quintanilla, M. (eds) *Nematode Diseases of Crops and Their Sustainable Management*. Academic Press, Cambridge, MA, pp. 55–63. DOI: 10.1016/B978-0-323-91226-6.00028-6.

Rao, M.S., Umamaheshwari, R., Prithi, K., Rajinikanth, R., Grace, G.N. *et al.* (2016) Role of biopesticides in the management of nematodes and associated diseases in horticultural crops. In: Hakeem, K.R., Aktar, M.S. and Abdulllah, S.N.A. (eds) *Plant, Soil and Microbes: Volume 1: Implications in Crop Science*. Springer, Cham, Switzerland, pp. 117–148. DOI: 10.1007/978-3-319-27455-3_7.

Reddy, P.P. (2014) Spice crops. In: *Biointensive Integrated Pest Management inHorticultural Ecosystems*. Springer, Dordrecht, The Netherlands, pp. 239–258. DOI: 10.1007/978-81-322-1844-9_19.

Regmi, H., Vallad, G.E., Hutton, S.F. and Desaeger, J. (2022) Interaction studies between *Meloidogyne javanica* and *Fusarium oxysporum* f. sp. *lycopersici* (Fol) Race 3 on differentisolines of tomato cv. Tasti Lee. *Journal of Nematology* 54(1), 20220018.

Robinson, A.F. (2007) Reniform in U.S. cotton: When, where, why, and some remedies. *Annual Review of Phytopathology* 45(1), 263–288. DOI: 10.1146/annurev.phyto.45.011107.143949.

Rocha, L.F. and Duggal, P. (2023) Management of cyst-forming nematodes in agricultural crops through novel biological and genetic engineering technologies. In: Khan, M.R. (ed.) *Novel Biological and Biotechnological Applications in Plant Nematode Management*. Springer, Singapore, pp. 313–339. DOI: 10.1007/978-981-99-2893-4_14.

Ross, J.P. (1964) Interaction of *Heterodera glycines* and *Meloidogyne incognita* on soybeans. *Phytopathology* 54, 304–307.

Rothrock, C.S., Woodward, J.E. and Kemerait Jr, R.C. (2015) Cotton diseases: Seed and seedling, bacterial blight, and vascular wilt. *Crops & Soils* 48(5), 32–41. DOI: 10.2134/cs2015-48-5-9.

Rowe, R.C., Riedel, R.M. and Martin, M.J. (1985) Synergistic interactions between *Verticillium dahliae* and *Pratylenchus penetrans* in potato early dying disease. *Phytopathology* 75(4), 412. DOI: 10.1094/Phyto-75-412.

Sarah, J.L. (1986) Influence of *Pratylenchus brachyurus* on the growth of development of pineapple in the Ivory Coast. *Revue de Nematologie* XVIII, 308–309.

Sarkar, S.K. and Gawande, S.P. (2016) Diseases of jute and allied fibre crops and their management. *Journal of Mycopathological Research* 54(3), 321–337.

Sasser, J.N. (1989) *Plant-Parasitic Nematodes: The Farmer's Hidden Enemy*. North Carolina State University Graphics, Raleigh, NC.

Sasser, J.N. (1971) An introduction to the plant nematode problem affecting world crops and a survey of current control methods. *Pflanzenschutz-Nachrichten* 24, 3–47.

Scholte, K. and s'Jacob, J.J. (1989) Synergistic interactions between *Rhizoctonia solani* Kühn, *Verticillium dahliae* Kleb., *Meloidogyne* spp. and *Pratylenchus neglectus* (Rensch) Chitwood & Oteifa, in potato. *Potato Research* 32(3), 387–395. DOI: 10.1007/BF02357852.

Seinhorst, J.W. and Kuniyasu, K. (1971) Interaction of *Pratylenchus penetrans* and *Fusarium oxysporum* forma *pisi* race 2 and of *Rotylenchus uniformis* and *F. oxysporum F. pisi* race 1 on peas. *Nematologica* 17, 444–452.

Sheela, M.S. and Sundararaju, P. (2012) Nematode infestation in banana. In: Khan, M.R. and Jairajpuri, M.S. (eds) *Nematode Infestations Part III: Horticultural Crops*. National Academy of Sciences, Allahabad, India, pp. 236–261.

Shokoohi, E. and Duncan, L.W. (2018) Nematode parasites of citrus. In: Sikora, R.A., Coyne, D., Hallmann, J. and Timper, P. (eds) *Plant Parasitic Nematodes in Subtropical and Tropical Agriculture*, 3rd edn. CAB International, Wallingford, UK, pp. 446–476. DOI: 10.1079/9781786391247.0446.

Shokoohi, E., Handoo, Z., Elshishka, M. and Salama, A.M. (2023) Nematode problems in temperate fruits and their sustainable management. In: Khan, M.R. and Quintanilla, M. (eds) *Nematode Diseases of Crops and Their Sustainable Management*. Academic Press, Cambridge, MA, pp. 375–426. DOI: 10.1016/B978-0-323-91226-6.00021-3.

Sikora, R.A. (1988) Plant-parasitic nematodes of wheat and barley in temperate and temperate semi-arid regions – a comparative analysis. In: Saxena, M.C., Sikora, R.A. and Srivastava, J.P. (eds) *Nematodes Parasitic to Cereals and Legumes in Temperate Semi-Arid Regions*. ICARDA, Aleppo, Syria, pp. 46–48.

Sikora, R.A. and Fernandez, E. (2005) Nematode parasites of vegetables. In: Luc, M., Sikora, R.A. and Bridge, L. (eds) *Plant Parasitic Nematodes in Subtropical and Tropical Agriculture*, 2nd edn. CAB International, Wallingford, UK, pp. 319–392. DOI: 10.1079/9780851997278.0319.

Sikora, R.A., Claudius-Cole, B. and Sikora, E.J. (2018a) Nematode parasites of food legumes. In: Sikora, R.A., Coyne, D., Hallmann, J. and Timper, P. (eds) *Plant Parasitic Nematodes in Subtropical and Tropical Agriculture*, 3rd edn. CAB International, Wallingford, UK, pp. 290–345. DOI: 10.1079/9781786391247.0290.

Sikora, R.A., Coyne, D., Hallmann, J. and Timper, P. (eds) (2018b) *Plant Parasitic Nematodes in Subtropical and Tropical Agriculture*, 3rd edn. CAB International, Wallingford, UK. DOI: 10.1079/9781786391247.0000.

Silva, A.M.G.B. (2023) Understanding the interaction between *Meloidogyne incognita* and *Fusarium oxysporum* f. sp. *passiflorae* in grafted passion fruit seedlings. Doctoral dissertation, Universidade de São Paulo, São Paulo, Brasil.

Singh, S. and Goswami, B.K. (2001) Studies on the management of disease-complex caused by root-knot nematode, *Meloidogyne incognita* and wilt fungus, *Fusarium oxysporum* on cowpea by neem cake and carbofuran. *Indian Journal of Nematology* 31(2), 122–125.

Singh, S. and Azam, T. (2010) Interactive effects of *Meloidogyne incognita* and *Macrophomina phaseolina* on lentil. *Indian Journal of Nematology* 40(1), 31–34.

Singh, S.K. and Campos, M.A. (2023) Nematode problems on oil seed crops and their sustainable management. In: Khan, M.R. and Quintanilla, M. (eds) *Nematode Diseases of Crops and Their Sustainable Management*. Academic Press, Cambridge, MA, pp. 205–219. DOI: 10.1016/B978-0-323-91226-6.00015-8.

Sipes, B.S. and Chinnasri, B. (2018) Nematode parasites of pineapple. In: Sikora, R.A., Coyne, D., Hallmann, J. and Timper, P. (eds) *Plant Parasitic Nematodes in Subtropical and Tropical Agriculture*, 3rd edn. CAB International, Wallingford, UK, pp. 717–737. DOI: 10.1079/9781786391247.0717.

Sitaramaiah, K. and Pathak, K.N. (1993) Nematode bacterial disease interaction. In: Khan, M.W. (ed.) *Nematode Interactions*. Chapman & Hall, London, pp. 232–250. DOI: 10.1007/978-94-011-1488-2_11.

Smits, B.G. and Noguera, R. (1982) Effect of *Meloidogyne incognita* on the pathogenicity of different isolates of *Fusarium oxysporum* on brinjal (*Solanum melongena* L.). *Agronomia Tropical* 32, 284–290.

Souza, R.M. (2008) Other coffee-associated nematodes. In: Souza, R.M. (ed.) *Plant-Parasitic Nematodes of Coffee*. Springer, Dordrecht, The Netherlands, pp. 209–222. DOI: 10.1007/978-1-4020-8720-2_11.

Spaull, V.W. and Cadet, P. (1990) Nematode parasites of sugarcane. In: Luc, M., Sikora, R.A. and Bridge, L. (eds) *Plant Parasitic Nematodes in Subtropical and Tropical Agriculture*. CAB International, Wallingford, UK, pp. 461–491.

Srivastava, A.N., Jaiswal, R.K. and Chawla, G. (2012) Nematode infestation in leafy vegetables. In: Khan, M.R. and Jairajpuri, M.S. (eds) *Nematode Infestations Part III: Horticultural Crops*. National Academy of Sciences, Allahabad, India, pp. 78–107.

Sturhan, D. and Brzeski, M.W. (2020) Stem and bulb nematodes, *Ditylenchus* spp. In: Nickle, W.R. (ed.) *Manual of Agricultural Nematology*. CRC Press, Boca Raton, FL, pp. 423–464. DOI: 10.1201/9781003066576-9.

Subramanian, S., Banu, G. and Umamaheshwari, R. (2010) Nematode infestation in coffee. In: Khan, M.R. and Jairajpuri, M.S. (eds) *Nematode Infestations Part II: Industrial Crops*. National Academy of Sciences, Allahabad, India, pp. 186–212.

Swarup, G. and Sossa-Moss, C. (1990) Nematode parasites of cereals. In: Luc, M., Sikora, R.A. and Bridge, J. (eds) *Plant Parasitic Nematodes in Subtropical and Tropical Agriculture*. CAB International, Wallingford, UK, pp. 109–136.

Taylor, C.E. and Brown, D.J.F. (1997) *Nematode Vectors of Plant Viruses*. CAB International, Wallingford, UK.

Tóthné Bogdányi, F., Boziné Pullai, K., Doshi, P., Erdős, E., Gilián, L.D. *et al.* (2021) Composted municipal green waste infused with biocontrol agents to control plant parasitic nematodes – a review. *Microorganisms* 9(10), 2130. DOI: 10.3390/microorganisms9102130.

Townshend, J.L. (1984) Inoculum densities of five plant parasitic nematodes in relation to alfalfa seedling growth. *Canadian Journal of Plant Pathology* 6(4), 309–312. DOI: 10.1080/07060668409501534.

Turner, S.J. and Subbotin, S.A. (2006) Cyst nematodes. In: Perry, R.N. and Moens, M. (eds) *Plant Nematology*. CAB International, Wallingford, UK, pp. 91–122.

Varshney, V.P., Khan, A.M. and Saxena, S.K. (1987) Interaction between *Meloidogyne incognita*, *Rhizoctonia solani* and *Rhizobium* species on cowpea. *International Nematology Network Newsletter* 4(3), 11–14.

Viaene, N., Coyne, D. and Davies, K.G. (2013) Biological and cultural management. In: Perry, R.N. and Moens, M. (eds) *Plant Nematology*, 2nd edn. CAB International, Wallingford, UK, pp. 383–410.

Villain, L., Salgado, S.L. and Phap, Q.T. (2018) Nematode parasites of coffee and cocoa. In: Sikora, R.A., Coyne, D., Hallmann, J. and Timper, P. (eds) *Plant Parasitic Nematodes in Subtropical and Tropical Agriculture*, 3rd edn. CAB International, Wallingford, UK, pp. 536–583.

Visser, P.B., Brown, D.J.F., Brederode, F.T and Bol, J.F. (1999) Nematode transmission of tobacco rattle virus serves as a bottleneck to clear the virus population from defective interfering RNAs. *Virology* 263(1), 155–165. DOI: 10.1006/viro.1999.9901.

Wagner, T.A., Duke, S.E., Davie, S.M., Magill, C. and Liu, J. (2022) Interaction of Fusarium wilt race 4 with root-knot nematode increases disease severity in cotton. *Plant Disease* 106(10), 2558–2562. DOI: 10.1094/PDIS-12-21-2725-SC.

Westerdahl, B.B. and Bello, O.M. (2023) Leaf and bud nematodes in agricultural crops and their management by biotechnological approaches. In: Khan, M.R. (ed.) *Novel Biological and Biotechnological Applications in Plant Nematode Management*. Springer, Singapore, pp. 359–376. DOI: 10.1007/978-981-99-2893-4_16.

Westerdahl, B.B., Khan, M.R., Nyström, Å.O. and Ahmad, I. (2023) Nematode problems in sugarcane and sugar beet and their sustainable management. In: Khan, M.R. and Quintanilla, M. (eds) *Nematode Diseases of Crops and Their Sustainable Management*. Academic Press, Cambridge, MA, pp. 565–594. DOI: 10.1016/B978-0-323-91226-6.00019-5.

Westphal, A., Li, C., Xing, L., McKay, A. and Malvick, D. (2014) Contributions of *Fusarium virguliforme* and *Heterodera glycines* to the disease complex of sudden death syndrome of soybean. *PLoS ONE* 9(6), e99529. DOI: 10.1371/journal.pone.0099529.

Wheeler, D.L., Scott, J., Dung, J.K.S. and Johnson, D.A. (2019) Evidence of a trans-kingdom plant disease complex between a fungus and plant-parasitic nematodes. *PLoS ONE* 14(2), e0211508. DOI: 10.1371/journal.pone.0211508.

Wheeler, T.A., Madden, L.V., Rowe, R.C. and Riedel, R.M. (1992) Modeling of yield loss in potato early dying caused by *Pratylenchus penetrans* and *Verticillium dahliae*. *Journal of Nematology* 24(1), 99–102.

Whitney, E.D. (ed.) (1974) Synergistic effect of *Pythium ultimum* and the additive effect of *P. aphanidermatum* with *Heterodera schachtii* on sugarbeet. *Phytopathology* 64(3), 380. DOI: 10.1094/Phyto-64-380.

Wu, W., Li, C., Ma, B., Shah, F., Liu, Y. *et al.* (2014) Genetic progress in wheat yield and associated traits in China since 1945 and future prospects. *Euphytica* 196(2), 155–168. DOI: 10.1007/s10681-013-1033-9.

Yan, H. and Nelson Jr, B. (2021) Effects of spore density and interaction with *Heterodera glycines* on soybean root rot caused by *Fusarium solani* and *F. tricinctum*. *Plant Disease* 105(9), 2426–2434. DOI: 10.1094/PDIS-09-20-1944-RE.

Zahid, M.I., Gurr, G.M., Nikandrow, A., Hodda, M., Fulkerson, W.J. *et al.* (2002) Effects of root- and stolon-infecting fungi on root-colonizing nematodes of white clover. *Plant Pathology* 51(2), 242–250. DOI: 10.1046/j.1365-3059.2002.00686.x.

Zheng, Y. and Khan, M.R. (2023) Pine wood nematode in coniferous forests and their management by novel biological and biotechnological interventions. In: Khan, M.R. (ed.) *Novel Biological and Biotechnological Applications in Plant Nematode Management*. Springer, Singapore, pp. 489–514. DOI: 10.1007/978-981-99-2893-4_22.

2 Importance and Principles of Nematode Disease Complexes

Horacio D. Lopez-Nicora[1]*, Edgard Henrique Costa Silva[2] and Giada d'Errico[3]
[1]Department of Plant Pathology, The Ohio State University, Columbus, OH, USA; [2]University of Western São Paulo, Unoeste, School of Agricultural Sciences, São Paulo, Brazil; [3]Research Centre for Plant Protection and Certification, Council for Agricultural Research and Economics (CREA-DC), Florence, Italy

Abstract

Nematodes and other pathogens often interact in disease complexes, leading to significant damage and economic loss in agricultural systems. In disease complexes, nematodes can interact synergistically or antagonistically with other pathogens such as fungi, bacteria and viruses, thereby influencing the severity and progression of the disease. For instance, when nematodes and fungi coexist, their combined impact on plant health can be greater than when either pathogen acts alone. This synergism can occur through various mechanisms, including increased root colonization, enhanced nutrient availability and altered plant defense responses. On the other hand, certain nematodes have been reported to suppress the growth and activity of other soil-borne pathogens such as bacteria or fungi. Such antagonistic interactions may involve the production of antimicrobial compounds or the activation of plant defense mechanisms. This chapter focusses on our understanding of the intricate interactions within disease complexes, which is crucial for developing effective management strategies and integrated pest management programs in agriculture.

Introduction

Plant nematodes, as microscopic organisms, are widely distributed in agricultural fields infecting all kinds of crops. The nematode infections can result in substantial yield losses, impacting agricultural productivity on a global scale. Plant nematodes, besides causing damage to crops on their own, also possess a significant capability of helping other pathogens in invading the hosts (Khan, 1993). The diseases which are formed due to the interactions of nematodes and other pathogens are commonly called disease complexes (Khan *et al.*, 2023c). An understanding of nematode disease complexes is vital for developing effective management strategies to protect agricultural crops and environmental health.

Plant-parasitic nematodes inflict extensive damage to crops worldwide, resulting in significant economic losses and food insecurity (Khan and Jairajpuri, 2010; Khan *et al.*, 2021). The Food and Agriculture Organization (FAO) estimated that nematodes cause annual losses of over US$100 billion globally (FAO, 2005). However, considering all major and minor crops, it is estimated that nematode-induced yield losses are worth US$200 billion (Khan, 2023a). Furthermore, managing nematode-related diseases requires substantial financial

*Corresponding author: lopez-nicora.1@osu.edu

 Nematode Disease Complexes in Agricultural Crops (ed. Mujeebur Rahman Khan)
DOI: 10.1079/9781800625228.0002

investments in procuring and administrating chemical treatments and carrying out breeding programs (Nicol *et al.*, 2011). Understanding nematode disease complexes is also crucial for maintaining environmental health. Chemical control measures, often employed against nematodes and other pathogens, can have adverse effects on non-target organisms and ecosystems. By comprehending the interactions within these complexes, researchers can develop more targeted and environmentally friendly approaches for managing nematode-related diseases, which are essential for promoting sustainable agriculture while minimizing detrimental impacts on the environment.

Moreover, nematode disease complexes have broader ecological implications. Certain nematode species can influence soil microbial communities, thereby altering nutrient cycling and overall soil health (Khan, 2023b). Additionally, nematode-induced plant diseases can disrupt plant community dynamics and biodiversity by influencing competitive relationships between plant species. These cascading effects emphasize the need for a holistic understanding of nematode disease complexes to preserve ecosystem stability. By unravelling the principles and significance of nematode disease complexes, researchers can develop targeted and sustainable management strategies, thereby contributing to the well-being of agriculture, the environment and global food security.

Significance of Nematode Disease Complexes

Plant-parasitic nematodes attack and injure a variety of crops, including cereals, vegetables, oilseed, fruit, fiber and beverage crops (Khan and Quintanilla, 2023). The distribution and prevalence of nematode diseases vary widely across different regions and cropping systems worldwide (Perry *et al.*, 2018). Plant nematodes are the most commonly occurring organisms in agricultural soils, representing 80–90% of the multicellular organisms present in soil (Khan, 2008). However, their distribution is influenced by factors such as climate, soil type, cropping history and agricultural practices. Plant nematodes are particularly prevalent in tropical and subtropical regions with warm and humid climates, as the conditions are more suitable for nematode reproduction and activity year-round, leading to higher nematode populations and increased disease pressure on crops (Haque and Khan, 2021). Nematode diseases are also common in temperate regions with milder climates, although their prevalence may vary seasonally (Evans and Haydock, 1993). In these regions, nematode populations tend to peak during the warmer months and decline during the colder seasons. Nematode diseases can occur in greenhouse and protected environments where crops are grown under controlled conditions. Infestations of root-knot nematodes, lesion nematodes and other nematodes in greenhouse crops, such as vegetables and ornamentals, have become increasingly common and severe, and lead to serious declines in crop growth and yield. Intensive cropping and limited soil management practices in greenhouses can further exacerbate nematode problems.

To invade the host, plant nematodes usually rely on the stylet, which is mainly used for making a perforation on the host surface or cells of the plant host (Vieira and Gleason, 2019). The nematode parasitism on the host plant may also break obstacles for other pathogens in invading the host (Khan and Dasgupta, 1993). A diverse range of organisms inhabit soil ecosystems worldwide, playing both beneficial and detrimental roles in crop production. Some of these organisms contribute to nutrient cycling and soil health, others pose significant threats to crops as plant pathogens or parasites. As nematodes and other plant pathogens occupy the same habitat, they are likely to interact with each other during parasitism, predation, mutualism, etc., either directly or indirectly (Topalović and Heuer, 2019; Topalović *et al.*, 2020). When different pathogens interact synergistically and share a common host plant, a disease complex is formed. Soil-borne pathogens are more likely to interact with nematodes as the soil is a rich ecological niche of microorganisms (Back *et al.*, 2002). A number of disease complexes involving plant nematodes, fungi or bacteria commonly occur in different crops worldwide. In most cases, the disease complexes involve root-knot nematodes and soil-borne fungi. In general, endoparasites are the most frequently associated nematodes in synergistic interaction with fungi and bacteria. However, other endoparasitic nematodes (*Globodera* spp., *Heterodera* spp., *Radopholus* and *Pratylenchus* spp.) or ectoparasitic nematodes (*Helicotylenchus*, *Tylenchorhynchus* and *Xiphinema*) also form the disease complexes (Siddiqui and Aziz, 2024).

Nematode disease complexes refer to situations where nematodes interact with other pathogens, including bacteria, fungi, viruses or other nematodes, leading to the development of a plant disease or an increase in severity of the disease (Jones *et al.*, 2013). Within these complexes, nematodes can act as vectors or facilitators, increasing the plant's susceptibility to other pathogens or directly causing damage through their feeding behavior (Fig. 2.1). The interactions within nematode disease complexes can occur through various mechanisms. For instance, nematodes may inflict wounds or perforations on the root surface which may serve as avenues for initial entry of secondary pathogens, as a result the host is more vulnerable to infections by fungal spores, bacterial cells, etc.

Fig. 2.1. Nematodes influence plant health through their relationship with other microbes in a variety of direct and indirect ways. (A) Nematodes, acting as biocontrol agents, exhibit distinct dietary preferences. Various nematode groups consume different types of organisms, such as fungi, bacteria, insects and even other nematodes. This predation can benefit plants by reducing the presence of potential plant pathogens or parasites in the soil environment. (B) Increased nutrient turnover. Nematode grazing of other microbes can stimulate nutrient turnover by releasing the nutrients contained within them to the environment. On the other hand, plant-parasitic nematodes break the cells of the roots, releasing their nutrient-rich contents. This increases the number of microbes in the local soil environment, which can then serve as a food source to bacterivore and fungivore nematodes. This creates a cycle that can increase the nutrient turnover in the rhizosphere. (C) Effects of wounding. Plant-parasitic nematodes enter the host plant leaving behind damaged cells. These can serve as points of entry for other plant pathogens which can form multipathogen complexes with the nematodes or establish an opportunistic infection. (D) Nematodes as pathogen vectors. Plant-pathogenic microbes can attach to the surface of nematodes and be transported closer to viable plant hosts. Some viruses are also vectored by nematodes. [Figure used with permission from Melanie Medina López.]

Additionally, nematodes introduce secretions into the host plant via their stylets, which may contain effector proteins that manipulate host defense mechanisms, ultimately facilitating the colonization of other pathogens (Siddique *et al.*, 2019). Moreover, certain nematodes act as vectors for particular plant viruses, resulting in a combination of nematode-induced damage and viral infection (Trudgill *et al.*, 2003).

Understanding the significance and principles of nematode disease complexes is crucial, as these interactions often lead to more severe and complex diseases compared to those caused by nematodes or other pathogens individually. The study of disease complexes helps us understand the intricate relationships between pathogens, their host plants and the broader ecosystem. This knowledge enables the development of integrated disease management strategies, vital for sustainable and effective control of plant diseases.

The first disease complex was reported in cotton, involving *Fusarium oxysporum* f. sp. *vasinfectum* and root-knot nematodes, *Meloidogyne* (Atkinson, 1892). It was observed that fusarium wilt in cotton became aggravated in the presence of *Meloidogyne* sp. These findings were later corroborated by Newsom and Martin (1953). *Fusarium–Meloidogyne* is one of the most documented disease complexes (Khan and Sharma, 2020), but several other fungi have also been found interacting with nematodes causing disease complexes (Evans and Haydock, 1993). Disease complexes resulting from the interaction between *Meloidogyne* and wilt-causing (mostly *Fusarium* and *Verticillium*) or root-rot-causing fungi (mostly *Rhizoctonia*, *Pythium*, *Phytophthora*, *Aspergillus*, *Macrophomina* and *Thielaviopsis*) are known to occur on a number of crops (Khan *et al.*, 2023c). Moreover, disease complexes between nematodes and plant-pathogenic bacteria have also been documented (mostly *Pseudomonas*, *Agrobacterium*, *Ralstonia* and *Clavibacter*) (Murfin *et al.*, 2012; Sitaramaiah and Pathak, 1993). Nematodes can also act as vectors for plant viruses (Weischer, 1993).

In a disease complex, nematodes usually undermine the host plant's ability to fight against the infection by fungi or bacteria. Thus, secondary fungi or bacteria may overcome the host defense strategies with the help of nematodes (Bergeson, 1972). Furthermore, nematodes can weaken the host plant and make it more susceptible to invasion by other pathogens or break the tolerance of a cultivar by modifying the host response.

Concept of Formation of Disease Complexes

Plant diseases in terms of etiology may be of two kinds: specific and complex. A disease with specific etiology is caused by a specific (single) pathogen, and it represents a typical disease triangle with three components, i.e. host, pathogen and the environment prevailing around both (Khan and Dasgupta, 1993). However, a disease of complex etiology has two or more pathogens in operation where the activity of one pathogen may influence the activity of the other pathogen. The alteration in the host plant induced by one pathogen is of great importance in the disease complex and it may influence the pathogenesis of the other pathogen. The pathogens involved in the disease complex infect the host sequentially or concomitantly. The resulting diseases are referred to as diseases of sequential etiology or diseases of concomitant etiology.

In a disease of sequential etiology, one pathogen infects and alters the host first, prior to invasion by the second pathogen. Hence, on the basis of the time of invasion, the two pathogens may be recognized as the primary pathogen and secondary pathogen (Powell, 1979). The primary pathogen invades the host prior to the invasion by the secondary pathogen, and plays a key role in the development of the disease complex. The primary pathogen modifies the host plant in such a way that the plant becomes more suitable for invasion by the secondary pathogen. The time interval in the sequential engagements may vary from a day to several weeks. A primary pathogen possesses specific, inherent and independent capabilities for causing diseases, whereas a secondary pathogen may or may not have this ability. However, under natural conditions, it is difficult to recognize a secondary pathogen. In concomitant etiology, the primary and secondary pathogens invade the host simultaneously (Khan *et al.*, 2023c).

In a disease complex when disease severity increases, it is due to biopredisposition, which may be defined as the predisposition caused by biotic factors. Predisposition is a modification caused by various factors in the expression of the host response to a pathogen. It is applicable in sequential etiology. In a disease of complex etiology, the nematode infection modifies the host in some way that benefits the plant-pathogenic fungi and bacteria (Powell, 1979). The nature and extent of biopredisposition in a disease of sequential etiology are determined by a primary pathogen, i.e. the pathogen that invades first. The mechanism of biopredisposition may be mechanical or physiological, or both. In biopredisposition, the host plant is altered and becomes more vulnerable to attack by another pathogen (Back *et al.*, 2002). Most bacterial pathogens require openings, either natural or artificial, for entry. The wounding of roots by nematodes generally helps bacterial plant pathogens. Physiological modifications in host plants are essential and mainly involved in biopredisposition.

Economic Significance of Disease Complexes

Nematode disease complexes are increasingly recognized as a major concern for global food security and agricultural sustainability (Mendoza-de Gives, 2022). The interaction between nematodes and other pathogens in crops not only affects plant health, but also has important socio-economic implications. A disease complex involving nematodes and other pathogens may result in significant yield losses and profitability deficit, which are the most significant and immediate impacts on a grower. Plant nematodes can negatively impact crop development, and reduce the quantity and quality of harvested products. However, under the disease complex situation the crop damage can become aggravated, leading to additional losses.

An accurate assessment of crop loss due to nematodes is challenging to estimate, as many factors can influence the degree of impact caused by the nematodes (Kumar *et al.*, 2020). There are several global estimates ranging from US$80 to US$200 billion (Abad *et al.*, 2008; Nicol *et al.*, 2011; Khan, 2023a). Furthermore, in many regions, nematodes occurrence is commonly misread as other problems, such as soil fertilization and water supply, as the above-ground symptoms may resemble the effects caused by these factors. Moreover, limited information is available on the economic significance of disease complexes involving nematodes and fungi or bacteria. However, Perveen *et al.* (2020) estimated yield loss of 30–85% in vegetable crops, sunflower and cotton when plants were infected with root-knot nematodes and root-rot pathogens. These authors argued that the degree of impact depends on the pathogens involved in the disease complex and on the soil type.

The yield and quality losses inflicted by disease complexes have serious implications for the food and nutritional security of rural and urban communities, as they can result in a decrease in the availability of quality and nutritious food, which can negatively affect the health of local populations, especially in developing countries in Africa, Latin America and Asia (Hassan *et al.*, 2013; Coyne *et al.*, 2018). Furthermore, income losses can lead farmers to resort to less sustainable livelihood strategies, exacerbating problems of environmental degradation and resource scarcity.

A significant impact on the environment may also result from disease complexes, mostly by the management tactics adopted. This usually means using more chemical pesticides, which can lead to the contamination of the environment, pesticide resistance and loss of biodiversity. Disease complexes can also significantly impact the livelihood and local economy of agricultural communities, as they can reduce the income of farmers and undermine their ability to invest in agricultural inputs, equipment and technologies that could improve their productivity and resilience. Furthermore, the impact on productivity and quality generated by disease complexes involving nematodes and other pathogens can affect the entire production chain, including processors, distributors and retailers, resulting in negative impacts on the local and regional economy.

Disease complexes involving nematodes may be affected by climate change, as it may have a direct influence on the distribution and population of pathogens. There are reports that climate change may contribute to the geographical distribution of nematodes as well as boost their reproduction, although they can also be negatively impacted (Dutta and Phani, 2023). Effective management of disease complexes involving nematodes and other pathogens requires approaches that consider not only the biological aspects of plant diseases, but also socio-economic and environmental aspects. These includes developing and promoting sustainable agricultural practices that strengthen soil health, crop genetic diversity and the resilience of agricultural systems to disease. Also, investments in research, extension and training are needed to enable farmers to adopt integrated disease management practices and reduce their vulnerability to the negative impacts of the interaction between nematodes and other pathogens.

Symptoms

Early diagnosis of the symptoms of the disease complex is of paramount significance for taking proactive measures to prevent further spread and damage to the crops. Early intervention helps preserve crop health and vigor, allowing plants to better withstand disease pressure and environmental stresses. Healthy plants are more resilient and better able to maintain yield and quality, even in the presence of disease complexes. Moreover, early detection helps prevent the spread of diseases complexes to other areas or neighboring fields. Plants with a disease complex exhibit the symptoms incited by the nematodes and the secondary pathogens described below.

Symptoms of nematode invasion

One of the most common and visible symptoms of nematode disease complexes in agricultural crops is damage to the root system. The sedentary endoparasitic nematode genera, such as *Meloidogyne*, *Heterodera* and *Globodera*, are often associated with severe disease complexes. Root-knot nematodes (*Meloidogyne* spp.) incite root galls, which can impede water and nutrient uptake and compromise plant health (Khan and Khan, 1994b). These galls are typically small and spherical, ranging in size from a few millimeters to several centimeters depending on the severity of infestation and the host plant species (Khan *et al.*, 2023a). Cyst nematodes (*Globodera* spp. and *Heterodera* spp.), on the other hand, cause the formation of cysts on the roots, which serve as protective structures for the nematode eggs (Bohlmann, 2015). These cysts are lemon or apple shaped and with a creamy to yellow coloration, attach to the root surface and can be easily recognized with the naked eye. As nematode populations increase, root symptoms may progress from localized lesions to widespread damage, resulting in reduced root biomass, impaired nutrient absorption and increased susceptibility to secondary infections.

While nematodes primarily affect plant roots, they can also induce mild chlorosis and wilting on above-ground parts of the plant (Khan and Khan, 1994a). In cases of severe infestation or interactions with secondary pathogens, the infested plants may show discernible stunting, chlorosis, wilting and leaf necrosis. These symptoms are indicative of physiological stress caused by root damage and impaired nutrient uptake, as well as disruptions to water and hormone transport within the plant. Nematode disease complexes can also manifest as agronomic symptoms that impact crop growth, development and yield. Infested plants may exhibit uneven growth and maturity, resulting in reduced stand uniformity and yield variability within fields. In addition to direct yield losses, nematode-infested crops may suffer from quality defects such as misshapen fruits, reduced sugar content and increased susceptibility to post-harvest diseases. These quality issues can have significant economic implications for growers, particularly in high-value crops such as fruits and vegetables where market standards are stringent. Furthermore, nematode-infested fields may require additional inputs such as fertilizers and irrigation to compensate for reduced nutrient uptake and water absorption by damaged roots, further increasing production costs and reducing profitability. Wilt and root rot are the major disease complexes in crops of agricultural significance.

Symptoms of wilt disease complex

Wilt disease is commonly caused by *Fusarium* or *Verticillium* species, and is also known as a vascular disease due obstruction in the xylem, impairing adequate uptake of water and nutrients (Francl and Wheeler, 1993). Both ectoparasites and endoparasites can be associated with wilt-inducing fungi, but the severity of wilting may vary with the nematode as their feeding behaviors are quite distinct. Ectoparasites interact less frequently with wilt-causing fungi, but there are some reports of *F. oxysporum* f. sp. *dianthi* and *F. oxysporum* f. sp. *vasinfectum* symptoms worsening in the presence of *Criconemella curvata* and *Belonolaimus longicaudatus* in carnation and cotton, respectively (Minton and Minton, 1966; Ferraz and Lear, 1976). Moreover, *Tylenchorhynchus* species have also been documented to be involved in disease complexes with *Fusarium*, *Pythium* and *Rhizoctonia* (Holdeman, 1956; Hasan and Bhaskar, 2004). In most cases of interaction between nematodes and wilt-inducing fungi the symptoms of wilting are aggravated, but the reproduction of the nematode is usually negatively influenced. Moreover, other outcomes such as early wilting symptoms, increased disease occurrence and/or severity and lower yield are frequently observed.

Symptoms of root-rot disease complex

Root-rot diseases can be caused by a variety of pathogens, especially from deuteromycetes, ascomycetes and oomycetes, but also some bacteria, and viruses in rare cases (Williamson-Benavides and Dhingra, 2021). The symptoms are more severe and noticeable if the root-rot-causing pathogen attacks the seed germination or during the early phase of plant development. Above-ground symptoms usually occur due to a high degree of damage to the root system, and usually infected plants do not recover and die. The common root symptoms include root lesions, browning and softening of root tips to the entire root system, yellowing and wilting of leaves, and stunting of plant growth, which can result in lower yield or even loss of the entire crop (Williamson-Benavides and Dhingra, 2021).

Etiological and Biological Aspects of Disease Complexes

Understanding the complex relationships that exist within nematode disease complexes is vital for effective management strategies (Figs 2.1 and 2.2). This section aims to explore the etiological and biological relationships between nematodes, plants and other pathogens to comprehend their impacts on plant health and to develop efficient disease control measures (Fig. 2.2).

Nematode–plant interactions

Plant nematodes interact directly and indirectly with the host plant and affect its health. Direct interaction occurs through feeding activities, which causes physical damage to the plant roots, leading to reduced nutrient uptake and impaired growth (Hussey and Grundler, 1998). The nematodes also exploit plant responses to infestation, inducing physiological changes and altering the host's defense mechanisms.

Nematode–microbe interactions

Nematodes often act as vectors for various plant pathogens, including viruses, bacteria and fungi, contributing to the spread of plant diseases. These pathogens reside on the nematode cuticle, within or outside the body, and may be introduced into plants through the secretions produced during the nematode feeding (Figs 2.1 and 2.2A). Studies have shown that dorylaim nematodes such as *Xiphinema* spp., *Longidorus* and *Trichodorus* spp. transmit certain ringspot viruses and tobraviruses. In addition, some tylenchid nematodes such as *Anguina tritici* transmit *Dilophospora alopecuri*, *Rathayibacter tritici*, etc. (Khan *et al.*, 2023b).

Nematode–microbe–plant interactions

This is the most complex interaction and represents the intricate relationship between nematodes and other pathogens that leads to the formation of serious diseases commonly referred

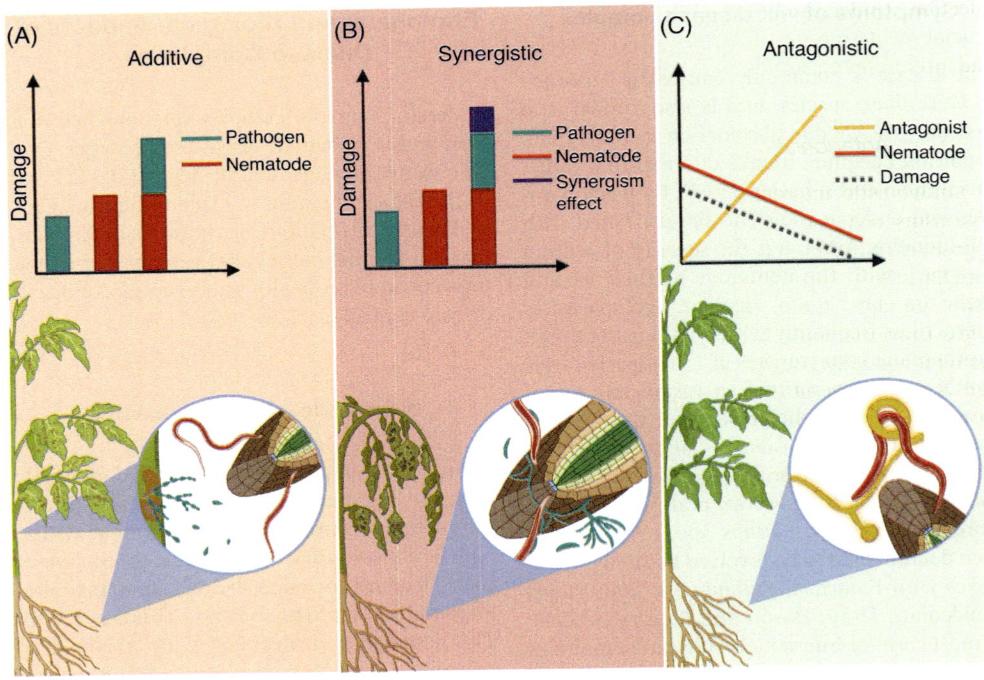

Fig. 2.2. Interaction between nematodes and other microbes. In the context of plant health, plant-parasitic nematodes interact with other microbes through three primary approaches. (A) Additive interactions are formed when a plant-parasitic nematode and a different plant pathogen simultaneously infect a plant, and the resulting damage is the sum of the harm they would have caused if infecting alone. (B) Similarly, synergistic interactions also involve a plant-parasitic nematode and an additional pathogen. However, this type of interaction results in damage beyond the sum of what the individual pathogens could cause alone. (C) In contrast, antagonistic interactions occur when a plant-parasitic nematode serves as prey for a different microbe. In this case, the antagonistic microbe would reduce the abundance of plant-parasitic nematodes resulting in plant disease suppression. [Figure used with permission from Melanie Medina López.]

to as disease complexes. In these interactions, the etiological relationships between individual pathogens and the host as well as their inter-relationships largely contribute to the overall impact on the host which determines the type of interactions, i.e. synergistic, antagonistic and additive.

Synergistic interactions

Synergistic interactions between nematodes and other pathogens can significantly promote the progress and severity of the disease caused by the secondary pathogen (Fig. 2.2B). For instance, nematodes can work synergistically with fungal pathogens such as *Fusarium* spp., leading to rapid and severe of damping-off and root decay. In such interactions, the combined effects of the fungus and nematode can result in the crop damage being greater than the sum of their individual effects. In synergistic interactions, the colonization of fungal and bacterial pathogens is often enhanced; however, nematode population dynamics are negatively impacted in most cases, especially with *Fusarium*, *Verticillium*, *Rhizoctonia*, etc. (Evans and Haydock, 1993; Francl and Wheeler, 1993). In addition, some plant pathogens secrete compounds that can attract or repel nematodes, influencing their movement and distribution in the soil. Conversely, certain plant species can induce systemic acquired resistance, reducing plant susceptibility to nematode infection. Understanding these relationships is crucial for designing sustainable crop rotation and intercropping strategies.

Antagonistic interactions

In antagonistic interactions, the co-occurring organisms are generally non-compatible and oppose each other or at least one becomes antagonistic to the other; as a result, the cumulative damage caused by the two pathogens is less than the sum of the damage caused by them individually. For example, the colonization and development of *Pythium graminicola* was decreased in *M. incognita*-infected sugarcane roots (Valle-Lamboy and Ayala, 1980). The same nematode species was also found providing protection to bean plants against *R. solani* (Costa Manso and Huang, 1986).

The concept of antagonistic interaction has also been widely exploited in the biocontrol of pests and diseases. Antagonistic microbes, such as *Purpureocillium lilacinum*, *Pochonia chlamydosporia*, *Pasteuria penetrans*, etc., can effectively suppress nematode populations in soil (Khan and Mohiddin, 2023). Similarly, nematophagous fungi, such as *Arthrobotrys* spp., have the capability to trap and parasitize nematodes, leading to antagonistic interactions (Khan, 2016).

Impact of Climate Change

Climate change has been a recurring phenomenon since the creation of the Earth. One significant effect of climate change is the rise of global mean temperature. The origin of climate change can be primarily attributed to both natural factors such as radiation from the Sun and volcanic eruptions, as well as anthropogenic sources, especially the burning of fossil fuel which leads to the release of greenhouse gases (Khan and Rizvi, 2020). The Sixth Assessment IPCC Report states that the Earth has been experiencing increasingly warmer temperatures over the past four decades (IPCC, 2021). Global warming can have a substantial impact on pests and pathogens, including nematodes, and their relationship with other organisms and plants.

Climate change can lead to significant impacts at different levels, from individual organisms to entire ecosystems and agricultural systems, influencing reproduction, survival rates and the overall structure of the ecosystem. A recent study on the impact of soil temperature on the survival, virulence and pathogenesis of *Meloidogyne floridensis* and *Rotylenchulus reniformis* suggests that climate change will result in more severe plant damage by these nematode species compared to the current conditions (Khanal and Land, 2023).

The elevated levels of CO_2 alone may influence pathogen performance. Khan and Rizvi (2020) reported significantly greater galling and reproduction of *M. incognita* on cucurbits exposed to 450–500 ppm CO_2. In addition, heatwaves, droughts, fires, excessive rainfall, and increased pressure from pathogens and pests can also cause the expansion of certain pest and pathogen species into new areas. For example, *M. enterolobii* has gained recent attention in European countries. This nematode species can overcome resistance in important crops, such as tomatoes, conferred by genes such as *Mi-1.2*, as the resistance governed by this gene can be lost at temperatures above 28°C (Williamson and Kumar, 2006). However, the effect of climate change on nematodes with regard to their biology, ecology and host plant interactions has not been understood adequately (Dutta and Phani, 2023). It is evident that agricultural practitioners worldwide are primarily concerned about drought stress and water shortages, which may not be guaranteed even for farms with irrigation systems during prolonged drought periods (van Dijk *et al.*, 2013). Climate change can affect the composition, population dynamics, host plant interactions, host reaction, population dynamics of natural enemies, spread of nematodes to exotic areas, etc. Introduction of new species can have catastrophic consequences, especially in the absence of their natural enemies and the non-effective control methods. For instance, the cyst nematode *Heterodera schachtii* has caused significant economic losses due to its rapid spread in sugarbeet cultivation areas. Similarly, climate change is attributed for the spread of soybean cyst nematode *H. glycines* in Canada (Boland *et al.*, 2004). The changes in climatic factors are likely to impact the interactions between hosts, parasites and the environment, which contribute to the development of diseases (Kalinkat *et al.*, 2015). Plant nematodes also contribute to the spread of fungal, bacterial and viral pathogens, leading to more severe damage (Ragozzino and d'Errico, 2012).

Certain nematode species, such as *M. incognita*, *Pratylenchus* spp., *Helicotylenchus* spp., and *Ditylenchus dipsaci*, have already experienced a shift towards northern latitudes since the 1960s (Bebber *et al.*, 2013). However, a study on *H. schachtii* populations indicated that northern populations of this nematode species may not be strongly affected by global warming (Fournet *et al.*, 2018). Global warming can also influence the geographical distribution patterns of virus vectors such as longidorids and trichodorids, along with other nematode groups, leading to the colonization of new areas (Dutta *et al.*, 2012). Changes in temperature and humidity can also cause shifts in the abundance and distribution of nematodes, as observed in the case of the pine wood nematode *Bursaphelenchus xylophilus* and its vector beetle, *Monochamus* spp. (d'Errico *et al.*, 2015; Hirata *et al.*, 2017; Li *et al.*, 2022).

Management of Disease Complexes

Plant nematodes are major pests of agricultural crops and inflict 7–15% yield losses (Khan, 2023b). Chemical pesticides are used globally to suppress the nematodes in soil, particularly depending on the crop and geographical areas. However, the widespread use of chemical products has not yielded satisfactory control of nematodes, rather the damage has remained largely unchanged or worsened over time coupled with pesticide contamination. As a result, a number of efficacious nematicides such as most fumigants, and some organophosphates and dithiocarbamates, have been banned for use in agriculture because of their adverse impacts on almost all segments of the environment.

In protected structures such as polyhouses and greenhouses, soils are commonly co-infested with plant nematodes, particularly *Meloidogyne* and *Pratylenchus* species, and phytopathogenic fungi such as *Fusarium*, *Verticillium*, *Rhizoctonia*, etc. Moreover, the use of synthetic pesticides can lead to the selection of resistant strains among fungi and bacteria due to repeated exposure to the same mode of action in polyhouse cultivation. To deal with such issues, researchers and manufacturing companies are now focusing on alternative strategies that are effective, durable, environmentally safe, selective and economically sustainable. The use of antagonistic microbial agents is gaining significant attention in this context. Biocontrol agents possess multiple mechanisms of action which can prevent or slow down the development of disease complexes, in particular because of the multifaceted action of some microbial antagonists (Khan and Mohiddin, 2018).

In conclusion, based on worldwide research, it is evident that combinations of treatments, including various cultural practices (soil solarization, crop rotation, planting resistant varieties, soil amendment, grafting, etc.), chemicals, plant extracts, trap crops, sanitation and biological agents, may effectively control plant nematode populations and the disease complexes (Jaiti *et al.*, 2008; d'Errico *et al.*, 2022). Innovative technologies such as nanobiotechnology can be exploited in offering effective and economic solutions for plant diseases. Preventive measures are generally the most effective means of controlling disease complexes, although management measures do exist. The increasing demand for agricultural products combined with the restoration of natural agroecosystems as a response to climate change pose significant challenges to global food production systems. Therefore, sustainable management strategies necessitate holistic approaches that consider the entire production system. The management of disease complexes has evolved over time and will continue to evolve in response to specific needs and variables.

Future Prospects and Conclusions

The recent advances in the field of nematology and plant pathology as well as the interdisciplinary approaches have deepened our understanding of nematode diseases and their interactions with other pathogens. However, further concerted efforts and resources are required to be invested for developing novel management strategies and preventive measures. The future is the use of next-generation sequencing techniques, such as metabarcoding and metagenomics, which can enable the identification and characterization of multiple pathogens simultaneously, and may also unravel the intricate interactions between nematodes and other disease-causing agents. This knowledge can lead to the development of risk assessment tools, targeted control methods and more effective management strategies. Additionally, these techniques allow for a comprehensive analysis of the various organisms involved in disease complexes, including potential biocontrol agents. By detecting and identifying these organisms, researchers can develop targeted and effective control measures.

Further, remote sensing and precision agriculture technologies can provide valuable insights into the identification and monitoring of nematode disease complexes, even before they can be observed visually. By using remote sensing, researchers and farmers can swiftly detect risk factors, and will be able to implement timely interventions, thereby reducing the spread and impact of these diseases.

In conclusion, the study of nematode disease complexes is crucial for sustainable agriculture and effective plant protection. Ongoing research and interdisciplinary collaborations in this field are essential for developing effective management strategies and preventive measures. The integration of next-generation sequencing, remote sensing, precision agriculture technologies and histopathology techniques offers great promise for advancing our understanding of the complex disease systems.

References

Abad, P., Gouzy, J., Aury, J.-M., Castagnone-Sereno, P., Danchin, E.G.J. *et al.* (2008) Genome sequence of the metazoan plant-parasitic nematode *Meloidogyne incognita*. *Nature Biotechnology* 26(8), 909–915. DOI: 10.1038/nbt.1482.

Atkinson, G.F. (1892) Some diseases of cotton. *Alabama Agricultural Experiment Station Bulletin* 41, 61–65.

Back, M.A., Haydock, P.P.J. and Jenkinson, P. (2002) Disease complexes involving plant parasitic nematodes and soilborne pathogens. *Plant Pathology* 51(6), 683–697.

Bebber, D., Ramotowski, M. and Gurr, S. (2013) Crop pests and pathogens move polewards in a warming world. *Nature Climate Change* 3(11), 985–988. DOI: 10.1038/nclimate1990.

Bergeson, G.B. (1972) Concepts of nematode–fungus associations in plant disease complexes: A review. *Experimental Parasitology* 32(2), 301–314. DOI: 10.1016/0014-4894(72)90037-9.

Bohlmann, H. (2015) Introductory chapter on the basic biology of cyst nematodes. *Advances in Botanical Research* 73, 33–59. DOI: 10.1016/bs.abr.2014.12.001.

Boland, G.J., Melzer, M.S., Hopkin, A., Higgins, V. and Nassuth, A. (2004) Climate change and plant diseases in Ontario. *Canadian Journal of Plant Pathology* 26(3), 335–350.

Costa Manso, E.S.B.G. and Huang, C.S. (1986) Interacao entre *Meloidogyne incognita* (Race 4) e isoladoes de *Rhizocotonia solani*, em *Phaseolus vulgaris* cv. Rico. *Fitopathologia Brasileria* 23, 857–864.

Coyne, D.L., Cortada, L., Dalzell, J.J., Claudius-Cole, A.O., Haukeland, S. *et al.* (2018) Plant-parasitic nematodes and food security in Sub-Saharan Africa. *Annual Review of Phytopathology* 56(1), 381–403.

d'Errico, G., Carletti, B., Schröder, T., Mota, M., Vieira, P. *et al.* (2015) An update on the occurrence of nematodes belonging to the genus *Bursaphelenchus* in the Mediterranean area. *Forestry: An International Journal of Forest Research* 88(5), 509–520.

d'Errico, G., Greco, N., Vinale, F., Marra, R., Stillittano, V. *et al.* (2022) Synergistic effects of *Trichoderma harzianum*, 1,3 dichloropropene and organic matter in controlling the root-knot nematode *Meloidogyne incognita* on tomato. *Plants* 11(21), 2890.

Dutta, T.K. and Phani, V. (2023) The pervasive impact of global climate change on plant–nematode interaction continuum. *Frontiers in Plant Science* 14, 1143889. DOI: 10.3389/fpls.2023.1143889.

Dutta, T.K., Ganguly, A.K. and Gaur, H.S. (2012) Global status of rice root-knot nematode, *Meloidogyne graminicola*. *African Journal of Microbiology Research* 6(32), 6016–6021.

Evans, K. and Haydock, P.P.J. (1993) Interactions of nematodes with root-rot fungi. In: Khan, M.W. (ed.) *Nematode Interactions*. Chapman & Hall, London, pp. 104–133.

FAO (2005) *The State of Food and Agriculture 2005: Agricultural Trade and Poverty: Can Trade Work for the Poor?* Food and Agriculture Organization of the United Nations, Rome.

Ferraz, S. and Lear, B. (1976) Interaction of four plant parasitic nematodes and *Fusarium oxysporum* f. sp. *dianthi* on carnation. *Experientiae* 22, 272–277.

Fournet, S., Pellan, L., Porte, C., Piriou, C., Grenier, E. *et al.* (2018) Populations of the beet cyst nematode *Heterodera schachtii* exhibit strong differences in their life-history traits across changing thermal conditions. *Frontiers in Microbiology* 9, 2801.

Francl, L.J. and Wheeler, T.A. (1993) Interaction of plant-parasitic nematodes with wilt-inducing fungi. In: Khan, M.W. (ed.) *Nematode Interactions*. Chapman & Hall, London, pp. 79–103.

Haque, Z. and Khan, M.R. (2021) *Handbook of Invasive Plant-Parasitic Nematodes*. CAB International, Wallingford, UK.

Hasan, N. and Bhaskar, R. (2004) Disease complex of berseem involving nematode and two soil inhabiting fungi. *Annals of Plant Protection Sciences* 12, 159–161.

Hassan, M.A., Pham, T.H., Shi, H. and Zheng, J. (2013) Nematodes threats to global food security. *Acta Agriculturae Scandinavica, Section B – Soil & Plant Science* 63(5), 420–425. DOI: 10.1080/09064710.2013.794858.

Hirata, A., Nakamura, K., Nakao, K., Kominami, Y., Tanaka, N. *et al.* (2017) Potential distribution of pine wilt disease under future climate change scenarios. *PLoS ONE* 12(8), e0182837. DOI: 10.1371/journal.pone.0182837.

Holdeman, Q. (1956) The effect of the tobacco stunt nematode on the incidence of Fusarium wilt in flue-cured tobacco. *Phytopathology* 46, 129.

Hussey, R.S. and Grundler, F.M.W. (1998) Nematode parasitism of plants. In: Perry, R.N. and Wright, D.J. (eds) *The Physiology and Biochemistry of Free-Living and Plant-Parasitic Nematodes*. CAB International, Wallingford, UK, pp. 213–243.

IPCC (2021) *Climate Change 2021 – The Physical Science Basis. Working Group I Contribution to the Sixth Assessment Report of the Intergovernmental Panel on Climate Change*. Cambridge University Press, New York.

Jaiti, F., Kassami, M., Meddich, A. and Hadrami, I.E. (2008) Effect of *Arbuscular mycorrhization* on the accumulation of hydroxycinnamic acid derivatives in date palm seedlings challenged with *Fusarium oxysporum* f. sp. *albedinis*. *Journal of Phytopathology* 156, 641–646.

Jones, J.T., Haegeman, A., Danchin, E.G.J., Gaur, H.S., Helder, J. *et al.* (2013) Top 10 plant-parasitic nematodes in molecular plant pathology. *Molecular Plant Pathology* 14(9), 946–961.

Kalinkat, G., Rall, B.C., Björkman, C. and Niemelä, P. (2015) Effects of climate change on the interactions between insect pests and their natural enemies. In: Björkman, C. and Niemelä, P. (eds) *Climate Change and Insect Pests*. CAB International, Wallingford, UK, pp. 74–91.

Khan, M.R. (2008) *Plant Nematodes – Methodology, Morphology, Systematics, Biology and Ecology*. Science Publishers, Enfield, NH. DOI: 10.1201/b10766.

Khan, M.R. (2016) Nematode biocontrol agents: Diversity and effectiveness against phytonematodes in sustainable crop protection. *Indian Phytopathology* 69(4s), 453–463.

Khan, M.R. and Khan, M.W. (1994a) Effects of simulated acid rain and root-knot nematode on tomato. *Plant Pathology* 43, 41–49.

Khan, M.R. and Khan, M.W. (1994b) Single and interactive effects of root-knot nematode and coal-smoke on okra. *New Phytologist* 126, 337–342.

Khan, M.R. and Jairajpuri, M.S. (2010) *Nematode Infestations Part I: Food Crops*. National Academy of Sciences, Allahabad, India.

Khan, M.R. and Rizvi, T.F. (2020) Effect of elevated levels of CO_2 on powdery mildew development in five cucurbit species. *Scientific Reports* 10(1), 4986. DOI: 10.1038/s41598-020-61790-w.

Khan, M.R. and Sharma, R.K. (2020) Fusarium–nematode wilt disease complexes, etiology and mechanism of development. *Indian Phytopathology* 73(4), 615–628. DOI: 10.1007/s42360-020-00240-z.

Khan, M.R. and Quintanilla, M. (2023) *Nematode Diseases of Crops and Their Sustainable Management*. Academic Press, Cambridge, MA.

Khan, M.R. (2023a) Nematode pests of agricultural crops, a global overview. In: Khan, M.R. (ed.) *Novel Biological and Biotechnological Applications in Plant Nematode Management*. Springer, Singapore, pp. 3–45.

Khan, M.R. and Mohiddin, F.A. (2018) Trichoderma: Its multifarious utility in crop improvement. In: Prasad, R., Gill, S.S. and Tuteja, N. (eds) *New and Future Developments in Microbial Biotechnology and Bioengineering: Crop Improvement Through Microbial Biotechnology*. Elsevier, Amsterdam, pp. 263–291.

Khan, M.R., Ahamad, I. and Shah, H. (2021) Emerging important nematode problems in field crops and their management. In: Singh, K.P., Jahagirdar, S. and Sarma, B.K. (eds) *Emerging Trends in Plant Pathology*. Springer, Dordrecht, The Netherlands, pp. 33–62.

Khan, M.R. and Mohiddin, F.A. (2023) Biocontrol strategies for nematode management, an overview. In: Khan, M.R. (ed.) *Novel Biological and Biotechnological Applications in Plant Nematode Management*. Springer, Singapore, pp. 113–131.

Khan, M.R., Ahamad, F., Rizvi, T.F. and Akram, M. (2023a) Root-knot nematodes in vegetables and ornamentals and their management by novel biological and biotechnological tools. In: Khan, M.R. (ed.) *Novel Biological and Biotechnological Applications in Plant Nematode Management*. Springer, Singapore, pp. 261–288.

Khan, M.R. (2023b) Plant nematodes, an underestimated constraint in global crop production. nematode diseases of crops and their sustainable management. In: Khan, M.R. and Quintanilla, M. (eds) *Nematode Diseases of Crops and Their Sustainable Management*. Academic Press, Cambridge, MA, pp. 3–23.

Khan, M.R., Manzoor, S. and Ansari, M.S.A. (2023b) Effectiveness of *Trichoderma* species in controlling the seed-borne infestation of *Anguina tritici* in wheat seed-galls. *Indian Phytopathology* 76(4), 1083–1090.

Khan, M.R., Mohiddin, F.A., Haque, Z. and Sharma, R.K. (2023c) Disease complexes and their sustainable management. In: Khan, M.R. and Quintanilla, M. (eds) *Nematode Diseases of Crops and Their Sustainable Management*. Academic Press, Cambridge, MA, pp. 65–93.

Khan, M.W. and Dasgupta, M.K. (1993) The concept of interaction. In: Khan, M.W. (ed.) *Nematode Interactions*. Chapman & Hall, London, pp. 42–54.

Khan, M.W. (ed.) (1993) *Nematode Interactions*. Springer, Dordrecht, The Netherlands. DOI: 10.1007/978-94-011-1488-2.

Khanal, C. and Land, J. (2023) Study on two nematode species suggests climate change will inflict greater crop damage. *Scientific Reports* 13(1), 14185.

Kumar, V., Khan, M.R. and Walia, R.K. (2020) Crop loss estimations due to plant-parasitic nematodes in major crops in India. *National Academy Science Letters* 43(5), 409–412. DOI: 10.1007/s40009-020-00895-2.

Li, H., Xing, L., Liu, X., Pu, Y., Yang, Y. *et al.* (2022) Potential impact of climate change on the distribution of the pinewood nematode *Bursaphelenchus xylophilus* in Chongqing, China. *Pakistan Journal of Zoology* 54(2), 809–816. DOI: 10.17582/journal.pjz/20190912070900.

Mendoza-de Gives, P. (2022) Soil-borne nematodes: Impact in agriculture and livestock and sustainable strategies of prevention and control with special reference to the use of nematode natural enemies. *Pathogens* 11(6), 640.

Minton, N. and Minton, E. (1966) Effect of root knot and sting nematodes on expression of Fusarium wilt of cotton in three soils. *Phytopathology* 56, 319–322.

Murfin, K.E., Dillman, A.R., Foster, J.M., Bulgheresi, S., Slatko, B.E. *et al.* (2012) Nematode–bacterium symbioses – cooperation and conflict revealed in the "Omics" age. *The Biological Bulletin* 223(1), 85–102. DOI: 10.1086/BBLv223n1p85.

Newsom, L. and Martin, W. (1953) Effects of soil fumigation on populations of parasitic nematodes, incidence of Fusarium wilt, and yield of cotton. *Phytopathology* 43(5), 292–293.

Nicol, J.M., Turner, S.J., Coyne, D.L., Nijs, L.D., Hockland, S. *et al.* (2011) Current nematode threats to world agriculture. In: Jones, J.T., Gheysen, L. and Fenoll, C. (eds) *Genomics and Molecular Genetics of Plant–Nematode Interactions*. Springer, Dordrecht, The Netherlands, pp. 21–43. DOI: 10.1007/978-94-007-0434-3_2.

Perry, R.N., Moens, M. and Jones, J.T. (eds) (2018) *Cyst Nematodes*. CAB International, Wallingford, UK.

Perveen, G., Urooj, F., Moin, S., Farhat, H., Fahim, M.F. *et al.* (2020) Estimation of losses caused by root rotting fungi and root knot nematodes infecting some important crops in lower Sindh and Hub, Balochistan of Pakistan. *Pakistan Journal of Botany* 52(2). DOI: 10.30848/PJB2020-2(15).

Powell, N.T. (1979) Interactions between nematodes and fungi in disease complexes. *Annual Review of Phytopathology* 9(1), 253–274.

Ragozzino, A. and d'Errico, G. (2012) Interactions between nematodes and fungi: A concise review. *Redia* 94, 123–125.

Siddique, S., Radakovic, Z.S., Torre, C.M., Chronis, D. and Grundler, F.M.W. (2019) Parasitic worms stimulate host NADPH oxidases to produce reactive oxygen species that limit plant cell death and promote infection. *Science Signaling* 12(594), 7467.

Siddiqui, Z.A. and Aziz, S. (2024) Plant parasitic nematode–fungus interactions: Recent concepts and mechanisms. *Plant Physiology Reports* 29(1), 37–50. DOI: 10.1007/s40502-023-00762-4.

Sitaramaiah, K. and Pathak, K.N. (1993) Nematode bacterial disease interactions. In: Khan, M.W. (ed.) *Nematode Interactions*. Chapman & Hall, London, pp. 232–250. DOI: 10.1007/978-94-011-1488-2_11.

Topalović, O. and Heuer, H. (2019) Plant–nematode interactions assisted by microbes in the rhizosphere. *Current Issues in Molecular Biology* 30, 75–88. DOI: 10.21775/cimb.030.075.

Topalović, O., Hussain, M. and Heuer, H. (2020) Plants and associated soil microbiota cooperatively suppress plant-parasitic nematodes. *Frontiers in Microbiology* 11, 491484.

Trudgill, D.L., Milliken, G.A. and Webster, J.M. (2003) Nematode vectors of plant viruses. In: Downer, R.G.H., Barker, K.R. and Carter, C.C. (eds) *Vector-Mediated Transmission of Plant Pathogens*. APS Press, St Paul, MN, pp. 85–112.

Valle-Lamboy, S. and Ayala, A. (1980) Pathogenicity of *Meloidogyne incognita* and *Pratylenchus zeae*, and their association with *Pythium graminicola* on roots of sugarcane in Puerto Rico. *The Journal of Agriculture of the University of Puerto Rico* 64(3), 338–347. DOI: 10.46429/jaupr.v64i3.10222.

van Dijk, A.I.J.M., Beck, H.E., Crosbie, R.S., de Jeu, R.A.M., Liu, Y.Y. *et al*. (2013) The millennium drought in Southeast Australia (2001–2009): Natural and human causes and implications for water resources, ecosystems, economy, and society. *Water Resources Research* 49(2), 1040–1057.

Vieira, P. and Gleason, C. (2019) Plant-parasitic nematode effectors – insights into their diversity and new tools for their identification. *Current Opinion in Plant Biology* 50, 37–43. DOI: 10.1016/j.pbi.2019.02.007.

Weischer, B. (1993) Nematode–virus interactions. In: Khan, M.W. (ed.) *Nematode Interactions*. Chapman & Hall, London, pp. 217–231. DOI: 10.1007/978-94-011-1488-2_10.

Williamson, V.M. and Kumar, A. (2006) Nematode resistance in plants: The battle underground. *Trends in Genetics* 22(7), 396–403.

Williamson-Benavides, B.A. and Dhingra, A. (2021) Understanding root rot disease in agricultural crops. *Horticulturae* 7(2), 33. DOI: 10.3390/horticulturae7020033.

3 Understanding the Mechanisms of Nematode Disease Complexes

Gregory C. Bernard[1] **and Mujeebur Rahman Khan**[2*]
[1]*Department of Agricultural and Environmental Sciences, Tuskegee University, Tuskegee, AL, USA;* [2]*Department of Plant Protection, Faculty of Agricultural Sciences, Aligarh Muslim University, Aligarh, India*

Abstract

Plant-parasitic nematodes and their agricultural implications with regard to interactions with other plant pathogens and pests represent a principal point of inquiry within agricultural sciences. These microscopic infectious organisms are responsible for enormous alterations in agricultural productivity globally, overwhelmingly affecting crop yields and economic viability. Largely targeting the root systems of crop plants often results in detrimental effects ranging from decreased plant growth and development to plant disease epidemics. Plant nematodes are often associated with other pathogens, exacerbating their infections. Therefore, understanding the dynamics of nematode infestations and their interactions with host plants and other pests is of paramount significance for devising effective management strategies and safeguarding agricultural sustainability. This chapter elucidates the multifaceted impacts of plant-parasitic nematodes within disease complexes in agricultural systems, highlighting their intricate ecological interactions and the vital need for innovative approaches for mitigation.

Introduction

Plant-parasitic nematodes (PPNs) are aquatic organisms but have adopted terrestrial habitats, requiring adequate moisture for survival and other activities. Plant-parasitic nematodes are present in all kinds of environments and ecosystems, and have a broad host range that leads to significant yield losses to all sorts of crops. On an economic scale, an excess of US$173–200 billion equivalent crop losses are attributed to plant-parasitic nematodes annually (Elling, 2013; Jones *et al.*, 2013; Khan, 2023a). The most significant plant nematode genera affecting agricultural production include *Meloidogyne*, *Heterodera*, *Globodera*, *Pratylenchus*, *Radopholus* and *Rhadinaphelenchus* (= *Bursaphelenchus*) (Jones *et al.*, 2013). However, Khan (2023a) has categorized 14 nematode genera most important with regard to economic significance in crop production, which include *Meloidogyne*, *Pratylenchus*, *Heterodera*, *Ditylenchus*, *Globodera*, *Tylenchulus*, *Xiphinema*, *Radopholus*, *Helicotylenchus*, *Rotylenchulus*, *Tylenchorhynchus*, *Hoplolaimus*, *Aphelenchoides* and *Bursaphelenchus*.

Nematodes, by number, are most ubiquitous in agricultural soils (Khan, 2008), and interact with multiple pests and pathogens, including fungi, bacteria, viruses, insects and other nematodes within the plant rhizosphere (Khan, 1993b). These interactions can be synergistic, where nematodes enrich the pathogenicity of other pathogens, or antagonistic, where competition or antagonism

*Corresponding author: mrkhan777in@yahoo.co.in

 Nematode Disease Complexes in Agricultural Crops (ed. Mujeebur Rahman Khan)
DOI: 10.1079/9781800625228.0003

leads to suppression of either or both of the pathogens. Interestingly, the multifaceted intricacies of host–nematode interactions result from complex evolutionary processes. Environmental conditions such as temperature, soil type, moisture and nutrient levels are crucial in altering nematode populations and pathogenesis. These factors can affect the survival of nematodes and other pathogens, influencing disease severity. When multiple pathogens are present, diverse disease symptoms may occur, constraining the rapid and accurate detection of the causative agents and necessitating additional and comprehensive analyses, including physio-morphological and molecular detections. Furthermore, nematode disease complexes often result in complex symptoms, such as stunted growth, wilting, chlorosis, root necrosis and yield loss, which may mimic symptoms caused by other pathogens or abiotic stresses. Effective management of nematode disease complexes often involves an integrated approach with a broad spectrum of interventions, including chemical pesticides, improved cultural practices, the incorporation of disease-resistant genotypes and consideration of environmental variables (Khan *et al.*, 2021). Elucidating the complex physical and molecular associations of nematodes, other pathogens, hosts and environmental factors is critical in developing novel and efficient management strategies.

Nematode Disease Complexes

Plant-pathogenic fungi or bacteria often co-occur with nematodes in the root zone of plants and develop some relationships. Root invasion by PPNs results in biochemical or physiological alterations in the host, which vary with the pathogenic potential and feeding behavior of the nematode species (Table 3.1). Ectoparasitic nematodes generally induce minor to moderate levels of alterations in the host plants, whereas sedentary endoparasites, specifically *Meloidogyne*, *Heterodera*, *Globodera*, etc., cause massive alterations in the host at biochemical and morphological levels (Khan *et al.*, 2023). Due to the nematode attack,

Table 3.1. Feeding strategies of plant-parasitic nematodes: endoparasitic and ectoparasitic species.

Nematode species	Feeding behavior and structures
Root-knot nematodes (*Meloidogyne*)	Sedentary endoparasitic: giant cells (enlarged and multinucleate plant cells) are induced by cellular reprogramming by nematode secretions (effectors)
Cyst nematodes (*Heterodera* and *Globodera*)	Sedentary endoparasitic: syncytia (enlarged and multinucleate plant cells) are formed by the fusion of multiple adjoining phloem cells induced by the action of the nematode stylet and effectors
Root-lesion nematodes (*Pratylenchus* and *Radopholus*)	Migratory endoparasitic: lesions are created due to migration; the parasite feeds on cortical root cells along its path and causes characteristic lesions
Reniform nematode (*Rotylenchulus reniformis*)	Sedentary semi-endoparasitic: multinucleated syncytia are formed in the pericycle cells by the fusion of adjoining cells, which is mediated by nematode stylets and effectors
Spiral nematodes (*Helicotylenchus*)	Ectoparasitic: these nematodes feed on the outer layers of root tissues, causing localized damage to root cells
Sting nematodes (*Belonolaimus*)	Ectoparasitic: these nematodes use their stylet to puncture and feed on individual root cells, leading to necrosis and damage
Dagger nematodes (*Xiphinema*)	Ectoparasitic: these nematodes feed on the exterior of root tissues, often causing wounds and damage to the terminal galls on the roots

the resulting modifications make the host susceptible or more suitable for the infection of interacting phytopathogenic microorganisms. This leads to the formation of the disease complex. In a disease complex, the progress of infection, incidence and severity of symptoms of fungi and bacteria become aggravated with or without an increase in the nematode symptoms and damage (Khan, 2023b). The advanced symptoms include gall formation, root necrosis, deep cracks in cortical tissue, corky resemblance, root atrophy, accelerated root decay and root rot, stunted growth, wilting, etc. The multiplication of fungi and bacteria is generally promoted, but sometimes nematode reproduction is adversely affected in a fungus–nematode disease complex (Taher *et al.*, 2017; Wolfgang *et al.*, 2019).

The abundant presence of PPNs in all agro-ecosystems has led to a gradual co-adaptation with other microorganisms, especially soil-borne plant-pathogenic fungi and bacteria (Khan and Khan, 1995). The resulting co-occurrence has led to the development of different relationships among the cohabiting nematodes and soil microorganisms (Khan *et al.*, 2009). These relationships may be saprophytic, beneficial, antagonistic, mutualistic, etc. (Khan, 1993b). The plant-pathogenic fungi and bacteria interact comprehensively with the nematodes, developing significant and discernible disease complexes among the soil's vast and varied microbial communities. However, nematodes also develop specialized relationships with plant viruses and some insect pests. Several plant viruses are transmitted by plant nematodes, whereas some insect pests act as vectors for PPNs. In addition, nematodes also interact with other nematodes in cohabitance.

Types of Disease Complexes Involving Nematodes and Other Pests and Pathogens

Nematodes are highly ubiquitous in occurrence and present in every climate under different ecosystems. They have co-adapted with various micro- and macroorganisms, including plant pathogens and pests. The primary disease complexes involving nematodes and other pathogens include nematode–fungus disease complexes, nematode–bacteria disease complexes, nematode–virus disease complexes, nematode–insect disease complexes and nematode–nematode disease complex, which are summarized below and covered in detail in other chapters throughout this book.

Nematode–fungus disease complexes

The PPNs are frequently associated with disease complexes involving plant-pathogenic fungi; such disease complexes are widespread in occurrence (Khan and Sharma, 2020; Zhang *et al.*, 2020). Nematode–fungus disease complexes are highly damaging to the crop production system and very difficult to manage (Khan *et al.*, 2023). Wilt diseases caused by *Fusarium* and *Verticillium* species become serious, with severe crop damage and yield decline in the presence of endoparasitic nematodes (Francl and Wheeler, 1993). The *Fusarium–Meloidogyne*/*Heterodera* wilt disease complexes can be frequently seen in tomatoes, chilli, chickpeas, pigeon peas, etc. (Khan, 2005). The presence of root-knot nematodes or cyst-forming nematodes may not only aggravate the severity of wilt diseases but may also break the tolerance response of vegetable and pulse cultivars to the wilt fungus (Khan and Sharma, 2020). Co-occurrence of *Pratylenchus thornei* may intensify the impact of *Fusarium culmorum* on susceptible wheat lines and resistant lines, contributing to increased disease damage and diminished yield (Laasli *et al.*, 2022). Similarly, root-rot disease complexes resulting from the synergistic interaction of *Rhizoctonia*, *Sclerotinia* and *Macrophomina* species with *Meloidogyne*, *Pratylenchus*, *Radopholus*, *Tylenchorhynchus*, etc., prevalently occur in several oilseed, pulse and vegetable crops (Evans and Haydock, 1993). Wilt (Francl and Wheeler, 1993) and root-rot disease complexes (Evans and Haydock, 1993) often lead to complete crop failure due to massive post-emergence damping-off, extensive root rot and wilting of foliage. Some fungus–nematode disease complexes also develop due to the aid of nematodes in transmitting the fungal pathogens. For example, ear twist disease caused by *Dilophospora alopecuri* develops due to transmission of the fungus by *Anguina tritici* in wheat and barley (Owen *et al.*, 2023).

Nematode–bacteria disease complexes

Several bacteria need help from PPNs in the infection process. The occurrence of bacterial wilt disease complexes caused by *Ralstonia solanacearum* in collaboration with *Meloidogyne* spp. or *Helicotylenchus nannus* is common in solanaceous plants (Libman *et al.*, 1964; Fukudome and Sakasegawa, 1972; Jatala and Martin, 1977b; Sellam *et al.*, 1982; Sitaramaiah and Sinha, 1984). Hairy root disease in roses incited by *Agrobacterium rhizogenes* is minor but becomes severe in the presence of *Pratylenchus vulnus* (Munnecke *et al.*, 1963). Many plant-pathogenic bacteria cannot reach the susceptible tissue without the aid of PPNs. The larvae of *A. tritici* carry the spores of *Rathayibacter tritici* to the shoot meristem of wheat causing yellow slime or 'tundu' disease (Nandal *et al.*, 2010). Similarly, *Rhodococcus fascians* is introduced by *Aphelenchoides fragariae* or *A. ritzemabosi* into the meristematic tissue, causing cauliflower disease of strawberries (Crosse and Pitcher, 1952; Pitcher and Crosse, 1958). The bacteria *Clavibacter michiganensis* s. sp. *insidiosum* and *R. solanacearum* causing wilt in alfalfa and '*Café au lait*' bacteriosis in garlics, respectively, are transmitted by *Ditylenchus dipsaci* (Hawn, 1963; Caubel and Samson, 1984). The nematode larvae take the bacterial cell to or close to the conductive tissue of the host plant.

Nematode–virus disease complexes

Some plant viruses are transmitted by nematodes from one plant to another. Hewitt *et al.* (1958) were the first to demonstrate that a plant virus (grapevine fanleaf virus) can be transmitted by a PPN (*Xiphinema index*). Since then, several viruses have been vectored by nematodes for their introduction into a host. Around 18 viruses from the genera *Nepovirus* (14 species), *Tobravirus* (three species) and *Cherivirus* (one species) are transmitted by some 30 species of ectoparasitic dorylaim nematodes (Bitterlin and Gonsalves, 1987; Georgi, 1988) These nematodes parasitize the roots of annual and perennial plants and belong to two families, Longidoridae and Trichodoridae, of the order Dorylaimida. Within Longidoridae, 11 species of *Xiphinema*, seven species of *Longidorus* and one species of *Pralongidorus* transmit 14 species of *Nepovirus* spp. and one species of *Cheravirus* (Mayo *et al.*, 1994). Similarly, a total of seven species of *Paratrichodorus*, four species of *Trichodorus* and one species of *Nanodorus* are potential vectors of *Tobravirus* spp. (Abdalreda and Mohammed, 2022). Among nepoviruses, grapevine fanleaf virus is transmitted by *X. index* (Hewitt *et al.*, 1958), whereas *L. elongatus* and *P. maximus* transmit tomato black ring virus and raspberry ringspot virus, respectively (Teliz *et al.*, 1966; Yagita, 1977). There are three tobraviruses: pea early browning virus transmitted by *P. anemones* (Loof, 2020), tobacco rattle virus transmitted by *T. similis* (Hunt, 1993) and pepper ringspot virus transmitted by *N. minor* (Siddiqi, 1977). When potential PPN species feed on virus-infected plants, virus particles are ingested along with the cell contents, and the nematode becomes viruliferous. The virus particles are introduced into the susceptible host during the feeding by viruliferous nematodes (Mayo *et al.*, 1994).

Nematode–insect disease complexes

Nematode–insect disease complexes are generally found in perennial host trees. These complexes involve those nematodes that infect the above-ground parts of trees and need a vector to reach and be introduced into the stem or fruit of the susceptible perennial hosts (Giblin-Davis *et al.*, 2003; Pimentel *et al.*, 2023). Unlike other disease complexes, the nematodes draw significant benefits from the interaction. The insect vectors primarily carry the nematodes from the diseased host to the healthy host (Khan, 2020). The insects that co-evolve with nematodes and infest the same part of the tree are involved in the disease complex. Some severe diseases in forest flora, such as pine wilt, red ring and fig gall, are formed due to this nematode–insect interaction (Pimentel *et al.*, 2023). The nematodes and insects in these disease complexes can interact and develop phoretic and mutualistic relationships (Khan, 2008; Ryss *et al.*, 2011). In the phoretic relationship, the nematode dauers are carried internally from a diseased host and introduced into a healthy tree during the oviposition of the adult vector insect, e.g. pine wilt disease (Zheng and Khan, 2023). The wilt in pines is caused by *Bursaphelenchus xylophilus* (pinewood nematode), which is transmitted by sawyer beetles (*Monochamus alternatus* and *M. carolinensis*) (Giblin-Davis *et al.*, 2003; Togashi *et al.*, 2023). The other relationship is mutualistic, where the nematode, in addition to riding over the vector, feeds and reproduces inside the vector. Whereas, the insect feeds on the galls induced by the nematode, e.g. fig gall disease caused by *Schistonchus caprifici* transmitted by the fig wasp, *Blastophaga psenes* (detailed in Chapter 12).

Nematode–nematode disease complexes

In nature, nematodes occur in polyspecific communities because of their wide distribution, polyphagous nature and weak interspecific competition (Manachini *et al.*, 2005). Nematodes from over 24 genera were recorded from a canola field, showing diversity in the community (Manachini *et al.*, 2005). Tens of nematode genera and species are commonly recorded from the root zone of plants from canola fields. Although saprophytes and microbivores share a larger proportion of the nematode community in the soil, a considerable population represents phytophagous nematodes. Due to largely being polyphagous, the phytophagous nematode species become concentrated in the plant's root zone and develop different relationships depending on the relative population densities and behaviors. Despite being weak competitors, the nematode community is very dynamic. Its members interact continuously; the community's interaction may occur at the genera, species or race level (Eisenback, 1993). In interaction, one population of the nematode influences the other. As a result, the reproduction of a population may be enhanced or inhibited.

Mechanism of Formation of Disease Complexes

A series of complex steps are involved in the infection process of nematodes and other pathogens. However, in disease complexes, the physical, morphological or biochemical modifications induced by nematodes may prove critical in promoting the invasion and colonization of pathogens and pests involved in the specific disease complex. Some of the essential and potential mechanisms are presented below.

Plant root exudates and rhizosphere dynamics

Nematode invasion or feeding causes wounds or perforations on the root surface, which might be numerous by ectoparasitic nematodes or relatively fewer by endoparasites. The wounds may cause leakage of cell sap or may lead to greater exudation from the invaded roots (Wang and Bergeson, 1974). A fungal propagule has limited time (a few hours) to germinate in response to passing lateral root branches. Under normal conditions, numerous fungal propagules lose the opportunity, and the root tip passes away when hyphae approach the root. However, root invasion by the stunt nematode (*Tylenchorhynchus* spp.) may slow down the root's linear extension, making the root tip available to the fungal propagules for a longer period (Huisman and Gerik, 1989). Propagules of *Fusarium* spp. germinate in response to root exudates, especially the root tip serving as the center of stimuli (Griffin, 1981). The juveniles of root-knot nematodes or cyst-forming nematodes have an affinity to the root tip area for the invasion. Hence, nematode feeding may cause greater exudation, which may increase the germination of propagules and their attraction to the root tips. The infection of tomato roots by *M. javanica* was found to reduce the population of the actinomycetes but enhanced the infection with a greater number of propagules of *Fusarium* spp. (Bergeson *et al.*, 1970). The rapid decomposition of the roots due to nematode activity may result in the release of greater numbers of fungus propagules in the soil.

The root exudates from host plants containing cellulose and xylan initiate differential gene expression in cell-wall-degrading enzymes encoded by xylanase-encoding genes (*Pc-xyl*) and β−1,4-endoglucanase (*Pc-eng-1*) in the larvae of *P. coffeae* (Bell *et al.*, 2019); the levels of gene expression were governed by the concentration of the exudates (Carlens *et al.*, 2005). The wilt and root-rot disease complex in guava has emerged as a severe problem in India. A study on the impact of *Meloidogyne enterolobii* on the root exudation from guava roots and its impact on the infection of root-rot fungus, *Neocosmospora falciformis*, showed the synthesis of NP-03-F2–3 + 4, in a fraction which enabled the infection of the fungus in the seedlings of guava (Souza *et al.*, 2023). The greater exudation from the nematode-parasitized roots enriched the rhizosphere with amino acids, carbohydrates and sucrose which may favor a shift in the rhizobiome towards the accumulation of 1,5-dinitrobiuret secreted by the microorganisms. The biuret may adversely affect the physiology of guava roots and resistance to fungus. The study provides valuable insights into the complex interactions between parasitic nematodes, pathogenic fungi and host plants, contributing to a better understanding of guava decline and potential strategies for management (Souza *et al.*, 2023). Shifts in the root microbiome often occur in nematode disease complexes, thus supporting the pathobiome concept, where the contribution of several pathogens initiates the onset and progression of diseases (Jakuschkin *et al.*, 2016; Brader *et al.*, 2017). The 16S rDNA gene from bacteria associated with plant nematodes was sequenced to investigate the composition of microbial communities in tomato and coffee infected with *M. enterolobii* and *M. paranaensis*, respectively. The study revealed that Pseudomonadales, Sphingobacteriales and Actinomycetales were the primary beneficiaries of the disease complexes involving root-knot nematodes (Lamelas *et al.*, 2020).

Injuries on the root surface

The majority of plant nematodes cause micropunctures, ruptures, maceration of cells, etc., during feeding by ectoparasitic nematodes or formation of perforations by endoparasitic nematodes for entering into the root (Taylor, 1979; Khan, 2008). The micropunctures or perforations made on the root surface are left behind by both ecto- and endoparasitic nematodes, which may be utilized by the fungal spores and bacterial cells for initial entry into the host. This is an essential mechanism of synergistic interaction and leads to a disease complex (Khan and Sharma, 2020). According to Powell (1979), the injuries caused by migratory nematodes on the root surface appear logical in enhancing the susceptibility of host plants to the invasion of fungi and bacteria. This mechanism was established as early as 1901 when it was demonstrated that tomato plants wilted readily due to *R.* (= *Pseudomonas*) *solanacearum* in nematode-infested soil; they stood healthy (Hunger, 1901). This mechanism was also found operative in the disease complex caused by *P. penetrans* and *F. oxysporum* on alfalfa (Mouza and Webster, 1982). However, in other cases, this obvious mechanism is non-operative. For example, injuries caused by *M. incognita* on cotton roots did not help the propagules of *F. oxysporum* in causing the infection (Perry, 1963; Starr *et al.*, 1989). However, Westerlund *et al.* (1974) reported that *F. oxysporum* f. sp. *ciceri* required wounding for efficient invasion into chickpea roots. The wounding and the host's physiological alterations may work in the synergetic interaction between nematodes and soil-borne pathogens. The wounding has been particularly helpful to the initial invasion of plant-pathogenic bacteria. Munnecke *et al.* (1963) correlated the mechanical injuries incited by *Pratylenchus vulnus*, rendering rose roots susceptible to invasion by *Agrobacterium rhizogenes*. Similarly, when roots were injured mechanically or by *M. hapla*, the alfa wilt caused by *Clavibacter michiganensis* subsp. *insidiosus* became severe (Hunt *et al.*, 1972). Similarly, the injuries made by *Pratylenchus penetrans* and *Helicotylenchus dihystera* helped *Pseudomonas corrugata* and *P. marginalis* to invade the roots and cause the disease (Bookbinder *et al.*, 1982). The tobacco cv. Dixie Bright developed wilt symptoms caused by *R. solanacearum* in the presence of *M. incognita* of almost equal severity when the roots were mechanically wounded before the bacteria inoculation (Lucas *et al.*, 1955).

Alterations in the host substrate

Nematode penetration, movement or feeding induces some alterations in the host as a substrate, which may affect the infection of fungi and bacteria. The modifications caused by endoparasitic nematodes are intensive and may favor the parasitism of plant-pathogenic bacteria and fungi. Plant nematodes may induce two types of alterations in the host: localized and systemic (Khan, 1993a).

Localized modifications

The modifications induced in the host at or close to the nematode infestation site are considered localized. These modifications may benefit fungi and bacteria only at the site of modification. For example, galls (incited by *Meloidogyne*, *Longidorus*, etc.), nurse cells (incited by sedentary nematodes) and root necrosis (incited by migratory endoparasitic nematodes) are distinctly attractive to several soil-borne plant-parasitic fungi (Khan and Sharma, 2020).

Root-knot nematodes and cyst-forming nematodes incite specialized giant cells or syncytia, respectively, to maintain a continuous supply of nutrients during the sedentary lifespan. Nurse cells are multinucleated specialized cells receiving a regular flow of higher amounts of carbohydrates, proteins, minerals, etc. (Huang, 1985). Such morphological and physiological modifications in a few host cells offer an ideal substrate for root-invading bacteria and fungi for initial establishment and multiplication in the root tissue (Khan, 1993a). *Meloidogyne* species cause typical root galls, which are technically a localized modification and offer an enriched substrate for invasion, establishment and colonization of wilt and root-rot-causing pathogens, even in resistant cultivars. The first evidence of this mechanism of synergistic interaction of sedentary nematodes and fungi was presented by Powell and Nusbaum (1960). They observed that hyphae of *Phytophthora parasitica* var. *nicotianae* luxuriantly colonized the galls and giant cells incited by *M. incognita*, followed by invasion of the vascular tissue, in tobacco cultivars. If galls were formed in the fungus-resistant cultivar, the hyphae colonized the galled tissue and caused the infection. The penetration tract had no role in facilitating the fungus infection. Sitaramaiah and Sinha (1984) observed that giant cells induced by *M. javanica* were invaded by *R. solanacearum* with increased wilt in eggplant.

Translocatable substances

The invasion by plant nematodes also results in changes that are translocated in the plant system. Systemic (translocatable) changes are important mechanisms for forming disease complexes (Khan, 1993a). Classical research by Bowman and Bloom (1966) demonstrated that fungus infection may be promoted due to some systemic modifications induced by nematodes at the root part distant from the site of the fungal infection. This study found that some translocatable substances were introduced, or their synthesis was triggered by the nematode parasitism on the host root, and these substances were translocated within the plant system (Webster, 1985). Melendez and Powell (1967) demonstrated this mechanism through a split-root experiment where the root system was carefully split and grown in adjoining pots without breaking and harming the roots. Both of the root parts of tobacco cultivars were inoculated with *M. incognita* and *F. oxysporum* f. sp. *nicotianae*. The split-root experiment demonstrated that when fungus was inoculated on one half of the root system of a wilt-resistant tobacco cultivar, and another half with *M. incognita*, the wilting was as severe as in the plants where both halves of the root system were inoculated with the two pathogens concomitantly.

Nematodes act as a vector for plant pathogens

There are some disease complexes in which nematodes play an essential role as vectors. Without this contribution, the secondary pathogen may be unable to cause the disease in some cases. Several plant viruses, some bacteria and a few fungi are transmitted by nematodes.

However, the vector role in such cases varies greatly depending on the pathogen group. For example, plant viruses are introduced by the vector nematode into the host, whereas fungal or bacterial spores are carried to the susceptible tissue of the plant, as summarized below.

Nematodes introduce plant viruses to the host plant

Hewitt *et al.* (1958) were the first to demonstrate the transmission of plant viruses by PPNs. When a potential nematode species feeds on the virus-infected plants, the particles are ingested along with the cell contents and the nematode becomes viruliferous. Around 18 viruses from the genera *Nepovirus* (14 species), *Tobravirus* (three species) and *Cheravirus* (one species) are transmitted by some 30 species of dorylaim nematodes from two families (Bitterlin and Gonsalves, 1987; Georgi, 1988). The vector nematodes are ectoparasites on the roots of annual and perennial plants. Within Longidoridae, 11 species of *Xiphinema*, seven species of *Longidorus* and one species of *Pralongidorus* transmit *Nepovirus* spp. and *Cheravirus* (Mayo *et al.*, 1994). Similarly, a total of seven species of *Paratrichodorus*, four species of *Trichodorus* and one species of *Nanodorus* are potential vectors of *Tobravirus* spp. (Abdalreda and Mohammed, 2022). For example, among nepoviruses, the grapevine fanleaf virus is transmitted by *X. index* (Gooding and Hewitt, 1962), whereas *L. elongatus* and *P. maximus* transmit tomato black ring virus and raspberry ringspot virus, respectively (Teliz *et al.*, 1966; Yagita and Komuro, 1972). There are three tobraviruses: pea early browning virus is transmitted by *P. anemones* (Loof, 2020), tobacco rattle virus is transmitted by *T. similis* (Hunt, 1993) and pepper ringspot virus is transmitted by *N. minor* (Siddiqi, 1977).

The nematode larvae become viruliferous when they feed on the virus-infected plants and acquire the virus particles. The virus particles are absorbed at specific sites in the lumen of the stylet or esophagus of the nematode. When a viruliferous nematode feeds on a healthy plant, the virus particles retained at the specific sites become free and are introduced into the plant along with the esophageal gland secretions (Khan, 2008) (detailed information is presented in Chapter 9).

Nematodes carry the pathogens to susceptible host tissue:

Nematodes also transmit some plant-pathogenic bacteria and fungi by carrying their inoculum on the external body surface (Sitaramaiah and Pathak, 1993; Asghar *et al.*, 2020). Typical examples are those where the apical shoot tissue is susceptible to the pathogen, and the spores may not reach there because of natural protection by the compact leaf scales, bud scales, etc. In such cases, the spores of certain bacteria and fungi adhere to the body of vector nematodes, which also have an affinity to the same plant tissue or the site close to the pathogen-susceptible tissue. The vector role of the nematode is usually limited to carrying the pathogens only a few centimeters from the soil or plant surface to the meristem. A typical example is yellow ear rot disease, where the infectious cells of the bacterium *Rathayibacter tritici* (= *Clavibacter tritici* = *Corynebacterium tritici*) are carried by the larvae of *A. tritici* to the developing wheat inflorescence. This interaction leads to developing a serious disease in wheat, i.e. yellow slime, ear rot or 'tundu' (Khan, 2023a). Another good example is cauliflower disease of strawberries caused by *R. fascians* (= *Corynebacterium fascians*) (Depuydt *et al.*, 2008; Dhaouadi *et al.*, 2020). The bacterium is unable to reach the meristem of strawberry plants. The nematodes *A. ritzemabosi* or *A. fragariae* take the bacterial cells to the meristematic tissue. Ear twist disease is caused by the fungus *D. alopecuri* under a similar situation, where *A. tritici* larvae carry the fungus to the shoot meristem of wheat (Owen *et al.*, 2023). The bristle-like appendages help the fungal spores adhere to the nematode body.

Insects carry or introduce nematodes into susceptible hosts

Some plant-parasitic nematodes, especially those causing infection in the above-ground parts of trees, need vectors to reach the stem or fruit of the susceptible perennial hosts (Zheng and Khan, 2023). The vector requirement of such nematodes is fulfilled by some vector insects which have co-evolved with nematodes and infest the same part of the tree. Some severe diseases, such as pine wilt, red ring and fig gall, are formed due to nematode–insect interaction (Pimentel *et al.*, 2023). Nematodes and insects can interact differently and develop a synergistic or antagonistic relationship. Antagonistic effects of nematodes, such as entomopathogenic nematodes, have been known for almost a century, and a lot of information is available on this aspect (Gaugler, 2002), although this is beyond the scope of the current chapter and is covered in detail in other chapters throughout this book. However, the relevant information on synergistic interaction is presented here. Unlike other synergistic interactions, in the nematode–insect synergistic interactions, the nematode draws more significant benefits from cohabitance. The nematode rides on the insect and parasitizes the vector (dual parasitism) (Mamiya, 2008). As detailed below, the nematode develops two distinct associations with the insect: mutualistic and phoretic relationships.

Mutualistic relationship

In this relationship, the nematode, besides being transmitted, draws nutrition from the vector insect. The vector larvae feed in the galls, which are primarily induced by the nematode (Zhao *et al.*, 2019). A typical example of this relationship is flower bud gall disease of eucalyptus, *Syzygium*, caused by the nematode *Fergusobia tumifaciens*, which is transmitted by the *Fergusonina* fly (Siddiqi, 1986; Nelson *et al.*, 2014). The adult female fly oviposits into young leaves of *Eucalyptus stuartiana*, *Syzygium cumini* or flower buds of *E. macrorhyncha*, along with nematode dauers (dispersion adapted to stage J_4). Nematode activity causes the formation of galls in the fruit or on the leaf (Giblin-Davis *et al.*, 2004). The fly also injects some gall-inciting chemicals during oviposition. Inside the galls, the nematode juveniles soon develop into parthenogenetic females, which feed on the plant tissue and lay eggs (Fisher and Nickle, 1968). The hatched larvae feed on mesophyll cells and become obese (Giblin-Davis *et al.*, 2004). By this time, *Fergusonina* larvae emerge from the eggs and enter into the galls, feed on the parenchymatous masses, and develop into the third instars. The infective female and male nematodes penetrate the fly larvae, proliferate in the fat body of the pupa and become adults. The nematode female lays eggs in the haemocoel of the fly before it pupates (Giblin-Davis *et al.*, 2001; Khan, 2020). The eggs hatch in the spring after pupation of the insect; the larvae develop to become J_4 (dauer stage) and migrate to the oviduct for transmission during oviposition (Siddiqi, 2000; Giblin-Davis *et al.*, 2001).

The other important example of dual parasitism is the flower gall disease of figs. The adult female of the fig wasp *Blastophaga psenes* carries juveniles and adults of the nematode *Schistonchus caprifici* in the haemocoel (Vovlas *et al.*, 1992; Hunt, 1993). Nematode larvae are introduced during oviposition into male and female florets of the fig, where they develop and reproduce. Endoparasitic feeding of the nematode causes necrosis, and cavities are formed in the cortical parenchyma of florets, leading to the formation of flower galls. Wasp larvae also develop during this time and penetrate the galls which contain nematodes. The nematode larvae enter into the wasp larvae through the mouth and feed on fat bodies to grow further. The nematode-infested wasp pupates and overwinters inside the flower galls (Vovlas *et al.*, 1998).

Phoretic relationship

The role of insects under the phoretic relationship resembles the role of nematodes in nematode–virus interactions. A few diseases are caused by aphelenchid nematodes, where a vector insect is essential for disease development (Hunt, 1993). The nematode larvae in the dauer stage (dispersion-adapted state) are carried internally in the trachea or haemocoel and are introduced into the host plant during feeding or egg laying by the vector insect (Khan, 2020). After introduction, the nematode larvae and insect eggs develop and attack the host independently without any relationship. Pine wilt disease caused by *B. xylophilus* is a classic example representing this relationship (Tomalak and Filipiak, 2021; Zheng and Khan, 2023). The dauer nematodes (J_4) are transmitted by beetles (*Monochamus alternatus* and *M. carolinensis*). The dauers enter into the respiratory system via the mouth of the insect larvae during pupation (Togashi *et al.*, 2023). The adult female beetles emerging from a diseased pine in spring contain 15,000–20,000 J_4 in the trachea. When the insect makes minor wounds in the susceptible conifers to feed or lay eggs, the nematode dauers migrate through the mouth into the wound. A chemical homologous to β-myrcene is believed to be synthesized by the susceptible pines, which incites the out-migration in the nematode dauers (Zhang *et al.*, 2021). Another good example is red ring of coconut, which is caused by *Bursaphelenchus* (= *Rhadinaphelenchus*) *cocophilus*. The nematode is transmitted by the black palm weevil, *Rhynchophorus palmarum* (Oehlschlager *et al.*, 2002), and the sugarcane weevil, *Metamasius* spp. While feeding on the diseased palms, the adult weevil swallows hundreds of dauers (J_3) and redwood. These larvae reach internally at the site of the ovipositor of the weevil and are deposited along with the insect eggs in the leaf or bud axil of healthy palms.

Nematodes as a competitor for other nematodes in cohabitance

The PPNs are generally polyphagous in nature with weak interspecific competition, have a wide distribution and occur in polyspecific communities. Manachini *et al.* (2005) have recorded nematodes belonging to 14 families and 24 genera from a canola field. Despite weak competition, the nematode community is dynamic, and its members interact continuously. The interaction within Nematoda may occur among the genera, species or even races. In interaction, one population of the nematode influences the other. As a result, the reproduction of a population may be enhanced or inhibited.

Plant nematodes exhibit great diversity in feeding habits, survival mechanisms and ecological demands, and the presence of a nematode in an area is determined by its ability to spread, adapt and establish in the area, which depends on its host range and survival ability. Nematodes with similar ecological and physiological needs may usually exhibit the most robust competition. Nematode species co-occur if they have similar life cycle patterns or utilize different niches in an ecosystem. Nematode species with better capabilities to adopt ecological conditions, including the available host plant species, may dominate the locality. Further the prevailing coexisting nematode may interact with other species as competitors or through the host plants, causing different effects on its growth and development. These relationships are interrelated since disease severity often depends on the nematode population in the root zone. Generally, in such interactions, the nematode populations in the root zone and plant tissue are affected due to the weakening of the host, disruption of feeding sites, etc. (detailed information is presented in Chapter 11).

Nematodes altering the host reaction

Modification of the host reaction to pathogens is the most severe impact of nematode interactions in crop production. This has shaken the entire global crop breeding programs involved in developing fungus/bacterial-resistant cultivars. The alteration in host reaction to pathogens is brought about largely by the nematodes. The chickpea cv. Avrodhi, bred for wilt tolerance, became susceptible to fusarium wilt in fields infested with *Meloidogyne* spp. due to non-expression or non-functioning of the gene responsible for imparting resistance against the fungus (Khan *et al.*, 2023). The molecular events leading to successful infection of *Meloidogyne* spp. and responses in the host plants are poorly understood. The cDNA libraries with expressed sequence tags from *Meloidogyne*-challenged root tissues from peanut cvs NemaTAM and Florunner, which were constructed from root-knot nematode-challenged root tissues in the pre-penetration and early infection stages from near-isogenic nematode-resistant and susceptible peanut cvs NemaTAM and Florunner, revealed early onset of compatible host–pathogen interactions in the cv. Florunner roots (Tirumalaraju *et al.*, 2011). The nematode infection may also affect the performance of transgenic cultivars. The induction of specific systemic chemical changes in the host affects the proper expression of the R-gene, and it becomes ineffective through nematode-induced physiological alterations in the host. However, the R-gene does not become non-operational. Resistance against fusarium wilt is generally qualitative and quantitative (Fassuliotis and Rau, 1969). Pitcher reported the qualitative resistance is polygenic and functions on the principle of host–pathogen incompatibility.

The resistance may become affected by the PPN invasion because the nematode induces a greater supply of nutrition to the infected root tissue, which becomes available to the fungus in cohabitance. Root-knot nematodes can break the monogenic resistance in vegetable cultivars against *Fusarium* (Webster, 1985). Mai and Abawi (1987) also emphasized the significant role of *Meloidogyne* spp. in modifying monogenic resistance in plants against the wilt fungus. The reaction of tomato cv. Pusa Ruby to *F. oxysporum* f. sp. *lycopersici* was modified towards greater susceptibility in the presence of *M. incognita* (Khan and Akram, 2000).

Several cultivars resistant to bacteria have also been reported to become susceptible to the pathogen in the presence of plant nematodes. The eggplant cv. Pusa Purple Clusty, a highly resistant bacterial wilt-resistant cultivar, developed severe wilting due to infection by *R. solanacearum* in cohabitance with *M. incognita* (Jatala and Martin, 1977a; Reddy *et al.*, 1979). Similarly, alfalfa cultivars resistant to *C. michiganensis* subsp. *insidiosum* developed wilt symptoms when plants were inoculated with *Ditylenchus dipsaci* (Hawn, 1971), whereas nematode-resistant alfalfa cultivars did not develop wilt (Hawn, 1971).

Effect of Environmental Factors and Climate Change on Disease Complexes

Environmental factors are an essential component of the disease triangle. Their impact assumes greater significance in multipathogenic diseases. Some environmental factors, especially soil type, temperature and moisture, play essential roles in the biology and development of host and cohabiting pathogens and their mutual relationships. The effects of critical environmental factors and climate change are discussed below.

Soil type, pH and fertilization

Light soil or sandy loams are generally found favorable for the activity of plant nematodes. Wilt fungi have also been recorded as prevalent in fields with sandy/acidic soils. The cotton wilt in North America is highly prevalent in areas with sandy, acidic soils. Light soil also favors the activity of root-knot nematodes. Jeffers and Roberts (1993) and Wang and Roberts (2006) observed severe fusarium wilt in cotton in coarse-textured sandy soils infested with *M. incognita*. However, Francl *et al.* (1988) did not record a significant impact of soil textures on the severity of *Verticillium dahlia* and *Pratylenchus penetrans* disease complexes and the resulting yield losses.

Soil pH may also influence the severity of disease complexes. The incorporation of lime was found to significantly reduce the incidence of wilt. The infection of *F. oxysporum* f. sp. *vasinfectum* was promoted in heavy clay soil infested with *Rotylenchulus reniformis*. The reniform nematode synergistically interacted with the fungus, promoting wilt severity (Bell, 1989). Fertilization with ammonium nitrate is reported to enhance wilt severity. In contrast, nitrate may reduce the disease, probably due to increased and decreased soil pH, respectively (Spiegel and Netzer, 1984). Soil pH may also influence the available micronutrients and activity of siderophore-producing bacteria (Louvet, 1989). Oteifa and Elgindi (1976) observed that an Egyptian cotton cultivar resistant to *Fusarium* developed 12% wilt at 50 mg/l K. However, the wilt decreased significantly at 100 mg/l K. The presence of *R. reniformis* significantly reduced the effect of potassium, but the wilt did not decrease until 500 mg/l K. Cotton growers often use potassium fertilizer, lime and resistant cultivars to control fusarium wilt in cotton (Bell, 1989). The mechanism involved in potassium alleviation of wilt is not understood.

Soil moisture and temperature

Basically nematodes are aquatic organisms and have adapted to terrestrial life with some humidity in the habitat (Khan, 2008). Hence, the water available in the soil may significantly affect nematode development and its role as a parasite or a partner of the disease complex. Soil moisture may significantly impact the hatching process of eggs and the development and mobility of nematode larvae (Wilcox-Lee and Loria, 1987). Similarly, fungal and bacterial pathogens require adequate soil moisture for survival and multiplication. The spores of *Verticillium* can germinate in the soil and colonize the roots over varied soil moisture ranges (Besri, 1980). Francl *et al.* (1990) observed that high precipitation and soil moisture in the early season and during the warm weather in the late season may lead to higher disease incidence in potatoes. *V. dahliae*-infested micro plots receiving above-normal irrigation exhibited wilt disease of greater severity with early incidence of symptoms (Cappaert *et al.*, 1990; Wheeler *et al.*, 1991). Faulkner and Bolander (1969) examined the influence of temperature on wilt development in peppermint. A temperature of 24°C was found optimum for developing symptoms incited with *V. dahliae* alone. However, for the disease complex caused by both *P. neglectus* and *V. dahliae*, the best temperature for symptom development was recorded to be 27°C. Overall, the 18–30°C range was most suitable for synergistic interaction between the two pathogens.

Climate change

Climate change is a most severe consequence of energy production/utilization and afforestation of human activities. It has significant counter-consequent effects, including global warming, excessive rain, droughts, sea storms, etc. Climate change-related consequences profoundly impact the biological components of all types of ecosystems. Plant nematodes are highly sensitive to climate change because they may be affected directly by changes in soil temperature and moisture as well as indirectly through the effects on plants.

Climate change may significantly influence host–nematode interactions, impacting the distribution and survival of plant-parasitic and free-living nematodes. Alterations in temperature and precipitation patterns can cause geographical shifts in nematode populations in previously unestablished areas. Empirical evidence suggests that such alterations in environmental conditions can lead to changes in species composition, population densities and sex ratios among nematode communities, serving as indicators of ecological alterations associated with rising temperatures as part of climate change. Warm and humid climates promote faster development, altered life cycles, intensified reproduction rates, prolonged transmission periods, altered sex ratios and spatial-temporal shifts in population densities (Wu *et al.*, 2016; Guo *et al.*, 2021; Dutta and Phani, 2023) Thus, due to their diverse ecological roles and sensitivity to environmental changes, nematodes emerge as valuable biological indicators, reflecting shifts in ecosystem dynamics and highlighting the interplay between climate change and biodiversity. The climate change factors that may affect nematode survival, parasitism and multiplication are discussed below.

Global warming

Global warming (increases in the global mean temperature of the Earth) is the most important consequence of climate change that may profoundly influence nematode activities in soil and parasitism on plants. A fluctuation in the temperature may discernibly influence nematode biology and reproduction. To examine the impact of elevated soil temperature on nematode biology and the crop damage potential, *R. reniformis* and *Meloidogyne floridensis* were subjected to a heating block at 32–34°C for 7 hour, followed by survival analysis after 18 hour (Khanal and Land, 2023). The exposure to these temperature levels for 7 hour during the daytime for 6 weeks revealed species-specific responses. Direct exposure to 34°C increased the survivability of *R. reniformis* by 10% but led to a 12% reduction in *M. floridensis* survival compared to 32°C. Elevating soil temperatures from 32°C to 34°C caused a decrease of 49% and 53% in the reproductive rates of the two nematode species, respectively. Interestingly, an increase from 33°C to 34°C significantly diminished *M. floridensis* reproduction, but *R. reniformis* was not affected, suggesting its superior adaptability to withstand escalating soil temperatures. Furthermore, virulence assays reveal enhanced aggressiveness of reniform nematode at 33°C and 34°C compared to 30°C, highlighting temperature-dependent variations in nematode behavior and pathogenicity. In another experiment, the potato variety Desiree was propagated in pots containing sandy loam soil in a controlled environment maintained at a temperature of 20°C/16-h photoperiod (Jones *et al.*, 2017). After a 3 week growth period, infective juveniles of *Globodera pallida* or *G. rostochiensis* were inoculated into the soil. Thereafter, the pots were placed on heat mats to raise the temperatures to 15°C, 17.5°C, 20°C, 22.5°C and 25°C. The study revealed that *G. pallida* reproduced efficiently at 15°C and less at 20–22.5°C compared to *G. rostochiensis*.

Liu *et al.* (2022) performed a 7year experiment (artificial warming) to assess the impact of warming on nematode communities in alpine meadows in Qinghai, Tibet, using varying temperature gradients. The elimination of Cyperaceae plants was demonstrated with increasing temperatures, promoting alterations in plant community composition across different temperature levels. Soil nematode populations decreased with soil depth as nematodes were shown to inhabit the top soil layer predominantly. The relative abundance of bacterivores increased with elevated temperatures, while plant nematode populations declined under excessive warming conditions. Moreover, nematode community diversity was observed at its highest with a moderate temperature increase, with higher warming gradients leading to a more simplified nematode community. The Wasilewska Index ($P<0.05$), which assesses the mineralization pathway of the soil food web through evaluating the involvement of different groups of soil organisms, highlighted the changes in nematode community diversity towards bacterivores and fungivores. An indirect effect of warming on nematode diversity was found to be governed by changes in the flora and soil properties due to rising temperature (Gendron St-Marseille *et al.*, 2019). The results indicated that soybean cultivation could extend across all cropland in Quebec, Canada, by the year 2050 (2041–2070), driven by the temperature optimum for soybean maturity group I, whereas *H. glycines* may potentially complete 4–5 generations in comparison to a current reproduction rate of 1–3 generations in Quebec regions under optimistic (representative concentration pathway (RCP) 4.5) or pessimistic (RCP 8.5) scenarios, respectively. With the expected drastic increase in *H. glycines*, the disease complexes involving nematodes and plant-pathogenic fungi/bacteria will become significantly greater in occurrence, and the resulting crop losses could be enormous.

Alterations in precipitation

Shifts in rainfall patterns directly influence soil moisture levels, influencing nematode populations. The influence of altered precipitation patterns on plant nematodes in arid, semi-arid and mesic grassland ecosystems was evaluated over 2 years. A precipitation manipulation study evaluated plant nematode abundance, species dynamics and herbivory index in response to varying precipitation levels (Ankrom *et al.*, 2020). Results revealed differential responses of parasitic groups to elevated rainfall. Ectoparasites exhibited minimal response to increased rainfall, with a notable decline at the humid site. In contrast, the abundance of endoparasitic nematodes was unaltered in the arid site, but it increased at the semi-arid site and decreased at the mesic site with elevated precipitation levels. Furthermore, the herbivory index mirrored the trends observed in endoparasite responses, stagnating in arid conditions, increasing in semi-arid environments and decreasing in mesic settings. These findings underscore altered precipitation complex and varied effects on the plant-feeding strategies, increased pathogenicity and the resulting disease complexes, highlighting potential implications for grassland productivity and the broader impacts of climate change on the interaction of nematodes with the host and other pathogens.

Impact of climate change on nematode vectors

There are a notable number of nematodes that are transmitted by insect vectors, importantly pinewood nematode vectored by *Monochamus* spp., red ring nematode by *Rhynchophorus* spp., eucalyptus leaf gall nematode by *Fergusonina* spp., fig flower gall nematode by *Blastophaga* spp., etc. Climate change may significantly affect the distribution of nematode vectors, directly impacting diseases such as pine wilt disease. The global average temperature may increase by 1.1–6.4°C by the end of the 21st century (Stocker *et al.*, 2013; Gao *et al.*, 2019). In Japan, in the 1900s, the pinewood nematode caused mass mortality to *Pinus thunbergii* and *P. densiflora* trees in Nagasaki, Japan (Yano, 1913; Pimentel *et al.*, 2023). The vulnerable regions under existing and future climate conditions were analyzed for four RCPs, with comparative analyses among the four RCPs and two future periods (the 2050s and 2070s) for pine wilt disease. RCPs revealed the expansion of 50% of vulnerable regions in northern Europe, Asia and North America by the 2070s under RCP 8.5. Around 40% of expanded regions fall in the pine wilt disease vulnerable areas, and pine forests are at risk of severe damage due to pine wilt disease or habitat shifts (Taketani *et al.*, 1975; Hirata *et al.*, 2017). The findings revealed the distribution of vulnerable regions in southern parts of North America, Europe and Asia. Incidentally, these regions fall under the natural pine habitat coupled with the occurrence of pine wilt disease. The projections under RCP scenarios have suggested the expansion of vulnerable areas in the northern parts of Asia, Europe and North America. By the 2070s, under RCP 8.5, vulnerable regions may cover almost 50% of the *Pinus* areas. Furthermore, by the 2070s, a considerable number of pine areas in Asia and Europe may become home for *Pinus* species with substantial deemed to pine wilt disease occurrence under RCP 8.5; this implies a dual-threat scenario, where both habitat shifts and the spread of pine wilt disease pose risks to pine forests in these regions. Overall, the impact of climate change on the survival, multiplication and pathogenesis of plant nematodes is complex and multifaceted, with severe consequences for agricultural productivity, ecosystem dynamics and food security.

Impact of climate change on nematode disease complexes

The information presented above has revealed that climate change may significantly increase nematode survival and multiplication and, subsequently, the overall soil population of plant nematodes. An increase in the soil population of nematodes will ultimately affect the extent of modification in the host, which may subsequently help in the infection process of secondary pathogens. Hence, climate change may also influence the intricate interaction between nematodes and other pathogens, and the severity of the resulting disease complex. There may also be indirect effects on PPN prevalence and its interaction. Intensive land use under climate change may increase the populations of opportunistic and plant-parasitic nematode species at the cost of reduced diversity and structure within the nematode community (Siebert *et al.*, 2020). The increase in nematode densities will offer more significant opportunities for the predisposition of the hosts for enhanced invasion and colonization by secondary pathogens. Greater biomass production under elevated CO_2 levels (Khan and Rizvi, 2020) may significantly increase the PPN densities in soil. Besides having a positive impact on nematode reproduction, higher temperatures (Reddy, 2013) and CO_2 levels (Sun *et al.*, 2011) may weaken the host defense and disease resistance that would yield greater susceptibility of the host to both nematodes and other pathogens.

Conclusion and Future Prospects

Several pests and pathogens attack plants in nature. These organisms develop different kinds of relationships when in cohabitance. The relationship where one pathogen may help the other leads to the formation of a disease complex, which is a serious concern with regard to crop health. In addition to directly attacking the plants, nematodes also aid other pathogens, enabling them to cause infection of greater severity or render the resistant/tolerant crop cultivars susceptible to fungi and bacteria. The PPNs play different roles in forming the disease complexes. The available information on this topic has suggested physical, anatomical, morphological and biochemical alterations induced by nematodes contribute to promoting the infection of secondary soil-borne pathogens. The vector role of nematodes is also significant in developing certain viral and other diseases. Although ample information is available, the mechanisms involved in forming disease complexes are doubted and need verification. Further, the molecular aspects with regard to translocatable modification require much more precise research. Similarly, virus specificity, particle retention mechanisms and essential factors regulating particle dissociation require adequate research attention. The overall information on the mechanism of formation of disease complexes will prove vital in developing effective and sustainable management strategies against disease complexes.

References

Abdalreda, E.M. and Mohammed, L.H. (2022) Nematodes that transmit plant pathogenic viruses. *IOP Conference Series: Earth and Environmental Science* 1060, 012125.

Ankrom, K.E., Franco, A.L.C., Fonte, S.J., Gherardi, L.A., Tomasel, C.M. *et al.* (2020) Ecto- and endoparasitic nematodes respond differently across sites to changes in precipitation. *Oecologia* 193(3), 761–771.

Asghar, A., Mukhtar, T., Raja, M.U. and Gulzar, A. (2020) Interaction between *Meloidogyne javanica* and *Ralstonia solanacearum* in chilli. *Pakistan Journal of Zoology* 52(4), 1525–1530.

Bell, A.A. (1989) Role of nutrition in diseases of cotton. In: Englehard, A.W. (ed.) *Soilborne Plant Pathogens: Management of Diseases with Macro- and Microelements*. APS Press, St Paul, MN, pp. 167–204.

Bell, C.A., Lilley, C.J., McCarthy, J., Atkinson, H.J. and Urwin, P.E. (2019) Plant-parasitic nematodes respond to root exudate signals with host-specific gene expression patterns. *PLoS Pathogens* 15(2), e1007503.

Bergeson, G.B., Van Gundy, S.D. and Thomason, I.J. (1970) Effect of *Meloidogyne javanica* on rhizosphere microflora and fusarium wilt of tomato. *Phytopathology* 60(8), 1245–1249.

Besri, M. (1980) Effects of water potential changes on growth of *Fusarium oxysporum* f. sp *lycopersici* and *Verticillium dahliae*. Phytopathologische Zeitschrift 99, 1–9.

Bitterlin, M.W. and Gonsalves, D. (1987) Spatial distribution of *Xiphinema rivesi* and persistence of tomato ringspot virus and its vector in soil. Plant Disease 71(5), 408.

Bookbinder, M.G., Bloom, J.R. and Lukezic, F.L. (1982) Interactions among selected endoparasitic nematodes and three pseudomonads on alfalfa. *Journal of Nematology* 14(1), 105–109.

Bowman, P. and Bloom, J.R. (1966) Breaking the resistance of tomato varieties to fusarium wilt by *Meloidogyne incognita*. Phytopathology 56, 871.

Brader, G., Compant, S., Vescio, K., Mitter, B., Trognitz, F. *et al.* (2017) Ecology and genomic insights into plant-pathogenic and plant-nonpathogenic endophytes. *Annual Review of Phytopathology* 55, 61–83.

Cappaert, M.R., Powelson, M.L. and Christensen, N.W. (1990) Soil moisture, genotype, and *Verticillium dahliae* interactions in potato early dying. *Phytopathology* 80, 971.

Carlens, K., Sági, L., Damme, E.V., Waele, D.D. and Swennen, R. (2005) The use of transgenic *Arabidopsis thaliana* plants to evaluate the nematicidal effect of lectins and lectin-related proteins upon migratory nematodes. *International Journal of Nematology* 15, 1–8.

Caubel, G. and Samson, R. (1984) Effect of the stem nematode *Ditylenchus dipsaci* on the development of "Cafe au lait" bacteriosis in garlic (*Allium sativum*) caused by *Pseudomonas fluorescens*. *Agronomie* 4, 311–313.

Crosse, J.E. and Pitcher, R.S. (1952) Studies in the relationship of eelworms and bacteria to certain plant diseases. The etiology of strawberry cauliflower disease I. *Annals of Applied Biology* 39, 475–484.

Depuydt, S., Dolezal, K., Lijsebettens, M., Moritz, T., Holsters, M. *et al.* (2008) Modulation of the hormone setting by *Rhodococcus fascians* results in ectopic *KNOX* activation in *Arabidopsis*. *Plant Physiology* 146(3), 1267–1281.

Dhaouadi, S., Mougou, A.H. and Rhouma, A. (2020) The plant pathogen *Rhodococcus fascians*. History, disease symptomatology, host range, pathogenesis and plant–pathogen interaction. *Annals of Applied Biology* 177(1), 4–15.

Dutta, T.K. and Phani, V. (2023) The pervasive impact of global climate change on plant–nematode interaction continuum. *Frontiers in Plant Science* 14, 1143889.

Eisenback, J.D. (1993) Innovative multimedia for teaching nematology. *Journal of Nematology* 25(3), 367.

Elling, A.A. (2013) Major emerging problems with minor *Meloidogyne* species. *Phytopathology* 103(11), 1092–1102.

Evans, K. and Haydock, P.P.J. (1993) Interactions of nematodes with root-rot fungi. In: Khan, M.W. (ed.) *Nematode Interactions*. Chapman & Hall, London, pp. 104–133.

Fassuliotis, G. and Rau, G.J. (1969) The relationship of *Meloidogyne incognita* to the incidence of cabbage yellows. *Journal of Nematology* 1, 219–222.

Faulkner, L.R. and Bolander, W.J. (1969) Interaction of *Verticillium dahliae* and *Pratylenchus minyus* in *Verticillium* wilt of peppermint: Effects of soil temperature. *Phytopathology* 59, 68–70.

Fisher, J.M. and Nickle, W.R. (1968) On the classification and life history of *Fergusobia curriei* (Sphaerulariidae: Nematoda). *Proceedings of the Helminthological Society of Washington* 35, 40–46.

Francl, L.J., Rowe, R.E., Riedel, R.M. and Madden, L.V. (1988) Effects of three soil types on potato early dying disease and associated yield reduction. *Phytopathology* 78(2), 159–166.

Francl, L.J., Madden, L.V., Rowe, R.E. and Riedel, R.M. (1990) Correlation of growing season environmental variables and the effect of early dying on potato yield. *Phytopathology* 80(4), 425–432.

Francl, L.J. and Wheeler, T.A. (1993) Interaction of plant-parasitic nematodes with wilt-inducing fungi. In: Khan, M.W. (ed.) *Nematode Interactions*. Chapman & Hall, London, pp. 79–103.

Fukudome, N. and Sakasegawa, Y. (1972) Interaction between root-knot nematode and other diseases. II. Influence of the root-knot nematode (*Meloidogyne incognita*) on the occurrence of granville wilt on tobacco. *Proceedings of the Association for Plant Protection of Kyushu* 18, 100–102.

Gao, R., Wang, Z., Wang, H., Hao, Y. and Shi, J. (2019) Relationship between pine wilt disease outbreaks and climatic variables in the three gorges reservoir region. *Forests* 10(9), 816.

Gaugler, R. (ed.) (2002) *Entomopathogenic Nematology*. CAB International, Wallingford, UK.

Gendron St-Marseille, A.-F., Bourgeois, G., Brodeur, J. and Mimee, B. (2019) Simulating the impacts of climate change on soybean cyst nematode and the distribution of soybean. *Agricultural and Forest Meteorology* 264, 178–187.

Georgi, L.L. (1988) Morphological variation in *Xiphinema* spp. from New York orchards. *Journal of Nematology* 20(1), 47–57.

Giblin-Davis, R.M., Makinson, J., Center, B.J., Davies, K.A., Purcell, M. *et al*. (2001) *Fergusobia/Fergusonina*-induced shoot bud gall development on *Melaleuca quinquenervia*. *Journal of Nematology* 33(4), 239.

Giblin-Davis, R.M., Williams, D.S., Bekal, S., Dickson, D.W., Brito, J.A. *et al*. (2003) "*Candidatus pasteuria usgae*" sp. nov., an obligate endoparasite of the phytoparasitic nematode *Belonolaimus longicaudatus*. *International Journal of Systematic and Evolutionary Microbiology* 53(1), 197–200.

Giblin-Davis, R.M., Davies, K.A., Taylor, G.S. and Thomas, W.K. (2004) Entomophilic nematode models for studying biodiversity and cospeciation. In: Chen, Z.X., Chen, S.Y. and Dickson, D.W. (eds) *Nematology: Advances and Perspectives. Volume 1: Nematode Morphology, Physiology and Ecology*. CAB International, Wallingford, UK, pp. 493–540.

Gooding, G.V. and Hewitt, W.B. (1962) Grape yellow vein: Symptomatology, identification, and the association of a mechanically transmissible virus with the disease. *American Journal of Enology and Viticulture* 13(4), 196–203.

Griffin, G.I. (1981) Physiology of conidium and chlamydospore germination in *Fusarium*. In: Nelson, P.E., Toussoun, T.A. and Cook, R.I. (eds) *Fusarium: Diseases, Biology and Taxonomy*. Pennsylvania State University Press, University Park, pp. 331–339.

Guo, X., Endler, A., Poll, C., Marhan, S. and Ruess, L. (2021) Independent effects of warming and altered precipitation pattern on nematode community structure in an arable field. *Agriculture, Ecosystems & Environment* 316, 107467.

Hawn, E.J. (1963) Transmission of bacterial wilt of alfalfa by *Ditylenchus dipsaci* (Kühn). *Nematologica* 8(1), 65–68.

Hawn, E.J. (1971) Mode of transmission of *Corynebacterium insidiosum* by *Ditylenchus dipsaci*. *Journal of Nematology* 3(4), 420–421.

Hewitt, W.B., Raski, D.J. and Goheen, A.C. (1958) Nematode vector of soil-borne fanleaf virus of grapevines. *Phytopathology* 48, 568–595.

Hirata, A., Nakamura, K., Nakao, K., Kominami, Y., Tanaka, N. *et al*. (2017) Potential distribution of pine wilt disease under future climate change scenarios. *PLoS ONE* 12(8), e0182837.

Huang, S.C. (1985) Formation, anatomy and physiology of giant cells induced by root-knot nematodes. In: Sasser, J.N. and Carter, C.C. (eds) *An Advanced Treatise on Meloidogyne. Volume I. Biology and Control*. North Carolina State University Graphics, Raleigh, NC, pp. 143–154.

Huisman, O.C. and Gerik, J.S. (1989) Dynamics of colonization of plant roots by *Verticillium dahliae* and other fungi. In: Tjamos, E.C. and Beckman, C.H. (eds) *Vascular Wilt Disease of Plants*. Springer, Berlin, pp. 1–18.

Hunger, F.W.T. (1901) Een bacterie-ziekte den tomaat. *Mededlingen Plantentuin Batavia* 48, 4–57.

Hunt, D.J. (1993) *Aphelenchida, Longidoridae, and Trichodoridae: Their Systematics and Bionomics*. CAB International, Wallingford, UK.

Hunt, T., Vanderhoff, G. and London, I.M. (1972) Control of globin synthesis: The role of heme. *Journal of Molecular Biology* 66(3), 471–481.

Jakuschkin, B., Fievet, V., Schwaller, L., Fort, T., Robin, C. *et al*. (2016) Deciphering the pathobiome: Intra- and interkingdom interactions involving the pathogen *Erysiphe alphitoides*. *Microbial Ecology* 72(4), 870–880.

Jatala, J. and Martin, E. (1977a) Interactions of *Meloidogyne incognita acrita* and *Pseudomonas solanacearum* on *Solanum chacoense* and *S. Sparsipilum*. *Proceedings of the American Phytopathological Society* 4, 178.

Jatala, J. and Martin, E. (1977b) Interaction of *Meloidogyne incognita acrita* and *Pseudomonas solanacearum* on field grown potatoes. *Proceedings of the American Phytopathological Society* 4, 177.

Jeffers, D.P. and Roberts, P.A. (1993) Effect of planting date and host genotype on the root-knot nematode–Fusarium wilt disease complex of cotton. *Phytopathology* 83(6), 645.

Jones, J.T., Haegeman, A., Danchin, E.G.J., Gaur, H.S., Helder, J. *et al*. (2013) Top 10 plant-parasitic nematodes in molecular plant pathology. *Molecular Plant Pathology* 14(9), 946–961.

Jones, L.M., Koehler, A.K., Trnka, M., Balek, J., Challinor, A.J. *et al*. (2017) Climate change is predicted to alter the current pest status of *Globodera pallida* and *G. rostochiensis* in the United Kingdom. *Global Change Biology* 23(11), 4497–4507.

Khan, M.R. (2005) *Biological Control of Fusarial Wilt and Root-Knot of Legumes*. Department of Biotechnology, Ministry of Science and Technology, Government of India Publication, New Delhi.

Khan, M.R. (2008) *Plant Nematodes – Methodology, Morphology, Systematics, Biology and Ecology*. Science Publishers, Enfield, NH. DOI: 10.1201/b10766.

Khan, M.R. (2020) Nematode infestations, a potential threat to Indian forests. *Indian Phytopathology* 73(3), 397–414.

Khan, M.R. (2023a) Nematode pests of agricultural crops, a global overview. In: Khan, M.R. (ed.) *Novel Biological and Biotechnological Applications in Plant Nematode Management*. Springer, Singapore, pp. 3–45.

Khan, M.R. (2023b) Plant nematodes, an underestimated constraint in the global food production. In: Khan, M.R. and Quintanilla, M. (eds) *Nematode Diseases of Crops and Their Sustainable Management*. Academic Press, Cambridge, MA, pp. 3–26.

Khan, M.R. and Akram, M. (2000) Effects of certain antagonistic fungi and rhizobacteria on wilt disease complex of tomato caused by *Meloidogyne incognita* and *Fusarium oxysporum* f. *Lycopersici*. *Nematologia Mediterranea* 28, sp.

Khan, M.R. and Rizvi, T.F. (2020) Effect of elevated levels of CO_2 on powdery mildew development in five cucurbit species. *Scientific Reports* 10(1), 4986.

Khan, M.R. and Sharma, R.K. (2020) Fusarium–nematode wilt disease complexes, etiology and mechanism of development. *Indian Phytopathology* 73(4), 615–628.

Khan, M.R., Altaf, S., Mohiddin, F.A., Khan, U. and Anwer, A. (2009) Biological control of plant nematodes with phosphate solubilizing microorganisms. In: Khan, M.S. and Zaidi, A. (eds) *Phosphate Solubilizing Microbes for Crop Improvement*. Nova Science Publishers, Hauppauge, NY, pp. 395–426.

Khan, M.R., Ahamad, I. and Shah, M.H. (2021) Emerging important nematode problems in field crops and their management. In: Singh, K.P., Jahagirdar, S. and Sarma, B.K. (eds) *Emerging Trends in Plant Pathology*. Springer, Dordrecht, The Netherlands, pp. 33–62.

Khan, M.R., Mohiddin, F.A., Haque, Z. and Sharma, R.K. (2023) Disease complexes and their sustainable management. In: Khan, M.R. and Quintanilla, M. (eds) *Nematode Diseases of Crops and Their Sustainable Management*. Academic Press, Cambridge, MA, pp. 65–93.

Khan, M.W. (1993a) Mechanism of interactions between nematodes and other plant pathogens. In: Khan, M.W. (ed.) *Nematode Interactions*. Springer, Dordrecht, The Netherlands. DOI: 10.1007/978-94-011-1488-2.

Khan, M.W. (ed.) (1993b) *Nematode Interactions*. Springer, Dordrecht, The Netherlands. DOI: 10.1007/978-94-011-1488-2.

Khan, M.W. and Khan, M.R. (1995) Fungus–nematode interactions. In: Swarup, G., Dasgupta, D.R. and Gill, J.S. (eds) *Nematode Pest Management – An Appraisal of Eco-Friendly Approaches*. Nematological Society of India, New Delhi, pp. 70–78.

Khanal, C. and Land, J. (2023) Study on two nematode species suggests climate change will inflict greater crop damage. *Scientific Reports* 13(1), 14185.

Laasli, S.E., Imren, M., Özer, G., Mokrini, F., Lahlali, R. *et al.* (2022) Interaction of root-lesion nematode (*Pratylenchus thornei*) and crown rot fungus (*Fusarium culmorum*) associated with spring wheat resistance under simulated field conditions. *Phytoparasitica* 50(4), 789–809.

Lamelas, A., Desgarennes, D., López-Lima, D., Villain, L., Alonso-Sánchez, A. *et al.* (2020) The bacterial microbiome of *Meloidogyne*-based disease complex in coffee and tomato. *Frontiers in Plant Science* 11, 136.

Libman, G., Leach, J.G. and Adams, R.E. (1964) Role of certain plant parasitic nematodes in infection of tomatoes by *Pseudomonas solanacearum*. *Phytopathology* 54, 151–153.

Liu, Y., Wang, W., Liu, P., Zhou, H., Chen, Z. *et al.* (2022) Plant-soil mediated effects of long-term warming on soil nematodes of alpine meadows on the Qinghai–Tibetan plateau. *Biology* 11(11), 1596.

Loof, P.A. (2020) The family Pratylenchidae Thorne, 1949. In: Nickle, W.R. (ed.) *Manual of Agricultural Nematology*. CRC Press, Boca Raton, FL, pp. 363–422.

Louvet, J. (1989) Microbial populations and mechanisms determining soil suppressiveness to fusarium wilts. In: Tjamos, E.C. and Beckman, C.H. (eds) *Vascular Wilt Disease of Plants*. Springer, Berlin, pp. 367–384.

Lucas, G.B., Sasser, J.N. and Kelman, A. (1955) The relationship of root-knot nematodes to Granville wilt resistance in tobacco. *Phytopathology* 45, 537–540.

Mai, W.F. and Abawi, G.S. (1987) Interactions among root-knot nematodes and Fusarium wilt fungi on host plants. *Annual Review of Phytopathology* 25(1), 317–338.

Mamiya, Y. (2008) Histopathological observations of *Bursaphelenchus xylophilus* in symptomatic tissues of pinewood. In: Mota, M.M. and Vieira, P. (eds) *Pine Wilt Disease: A Worldwide Threat to Forest Ecosystem*. Springer, Dordrecht, The Netherlands, pp. 321–334.

Manachini, B., Landi, S. and Tomasini, V. (2005) Biodiversity of nematofauna of oilseed rape (*Brassica napus* L). *Communications in Agricultural and Applied Biological Sciences* 70(4), 927–935.

Mayo, M.A., Robertson, W.M., Legorboru, F.J. and Brierley, K.M. (1994) Molecular approaches to an understanding of the transmission of plant viruses by nematodes. In: Lamberti, F., Giorgi, C. and Bird, D. (eds) *Advances in Molecular Plant Nematology*. Springer, Dordrecht, The Netherlands, pp. 277–293.

Melendez, P.L. and Powell, N.T. (1967) Histological aspects of the Fusarium wilt–root-knot complex in flue-cured tobacco. *Phytopathology* 57, 286–292.

Mouza, B.E. and Webster, J.M. (1982) Suppression of alfalfa growth by concomitant *Pratylenchus penetrans* and two *Fusarium* species populations. *Journal of Nematology* 14, 364–367.

Munnecke, D.E., Chandler, P.A. and Starr, M.P. (1963) Hairy root (*Agrobacterium rhizogenes*) of field roses. *Phytopathology* 53, 788–799.

Nandal, S.N., Paruthi, I.J. and Khan, M.R. (2010) *Seed gall nematode infestation in wheat*. In: Khan, M.R. and Jairajpuri, S. (eds) *Nematode Infestations Part I: Food Crops*. National Academy of Sciences, Allahabad, India, pp. 135–150.

Nelson, L.A., Davies, K.A., Scheffer, S.J., Taylor, G.S., Purcell, M.F. *et al.* (2014) An emerging example of tritrophic coevolution between flies (Diptera: Fergusoninidae) and nematodes (Nematoda: Neotylenchidae) on Myrtaceae host plants. *Biological Journal of the Linnean Society* 111(4), 699–718.

Oehlschlager, A.C., Chinchilla, C., Castillo, G. and Gonzalez, L. (2002) Control of red ring disease by mass trapping of *Rhynchophorus palmarum* (Coleoptera: Curculionidae). *Florida Entomologist* 85(3), 507–513.

Oteifa, B.A. and Elgindi, A.Y. (1976) Potassium nutrition of cotton, *Gossypium barbadense* in relation to nematode infection by *Meloidogyne incognita* and *Rotylenchulus reniformis*. In: *Fertilizer Use And Plant Health*. International Potash Institute, Bern, Switzerland, pp. 301–306.

Owen, K., Walia, R.K., Yan, G. and Khan, M.R. (2023) Nematode problems in wheat and barley and their sustainable management. In: Khan, M.R. and Quintanilla, M. (eds) *Nematode Diseases of Crops and Their Sustainable Management*. Academic Press, Cambridge, MA, pp. 97–131.

Perry, D.A. (1963) Interaction of root-knot and Fusarium wilt of cotton. *Empire Cotton Growing Review* 40, 41–47.

Pimentel, C., Khan, M.R., Zheng, Y. and Quintanilla, M. (2023) Nematode problems in forests and their sustainable management. In: Khan, M.R. and Quintanilla, M. (eds) *Nematode Diseases of Crops and Their Sustainable Management*. Academic Press, Cambridge, MA, pp. 457–493.

Pitcher, R.S. and Crosse, J.E. (1958) Studies in the relationship of eelworms and bacteria to certain plant diseases II. Further analysis of the strawberry cauliflower disease complex. *Nematologica* 3(3), 244–256.

Powell, N.T. and Nusbaum, C.J. (1960) The black shank-root-knot complex in flue-cured tobacco. *Phytopathology* 50, 899–906.

Powell, N.T. (1979) Internal synergisms among organisms inducing disease. In: Horsfall, J.G. and Cowling, E.B. (eds) *Plant Disease: An Advanced Treatise Volume IV. How Pathogens Induce Disease*. Academic Press, New York, pp. 113–133.

Reddy, P.P. (2013) Impact of climate change on insect pests, pathogens and nematodes. *Pest Management in Horticultural Ecosystems* 19(2), 225–233.

Reddy, P.P., Singh, D.B. and Kishnu, R. (1979) Effect of root-knot nematodes on the susceptibility of Pusa Purple Cluster brinjal to bacterial wilt. *Current Science* 48, 915–916.

Ryss, A.Y., Kulinich, O.A. and Sutherland, J.R. (2011) Pine wilt disease: A short review of worldwide research. *Forestry Studies in China* 13(2), 132–138.

Sellam, M.A., Rushdi, M.H. and Gendi, D.M. (1982) Interrelationship of *Meloidogyne incognita* chitwood and *Pseudomonas solanacearum* on tomato. *Egyptian Journal of Phytopathology* 12, 35–42.

Siddiqi, M.R. (1977) *Telotylenchus obtusus* sp. n. and *Histotylenchus sudanensis* sp. n.(Nematoda: Tylenchida: Telotylenchinae). *Nematologia Mediterranea* 5, 281–289.

Siddiqi, M.R. (2000) *Tylenchida: Parasites of Plants and Insects*. CAB International, Wallingford, UK.

Siddiqi, M.R. (1986) A review of the nematode genus *Fergusobia currie* (hexatylina) descriptions of *F. jambophila* n. sp. and *F. magna* n. sp. In: Swarup, G. and Dasgupta, D.R. (eds) *Plant Parasitic Nematodes of India, Problems and Progress*. Indian Agricultural Research Institute, New Delhi, pp. 264–278.

Siebert, J., Ciobanu, M., Schädler, M. and Eisenhauer, N. (2020) Climate change and land use induce functional shifts in soil nematode communities. *Oecologia* 192(1), 281–294.

Sitaramaiah, K. and Sinha, S.K. (1984) Interaction between *Meloidogyne javanica* and *Pseudomonas solanacearum* on brinjal. *Indian Journal of Nematology* 14, 1–5.

Sitaramaiah, K. and Pathak, K.N. (1993) Nematode bacterial disease interactions. In: Khan, M.W. (ed.) *Nematode Interactions*. Chapman & Hall, London, pp. 232–250.

Souza, R.M., Oliveira, D.F., Gomes, V.M., Viana, A.J.S., Silva, G.H. *et al.* (2023) *Meloidogyne enterolobii*-induced changes in guava root exudates are associated with root rotting caused by *Neocosmospora falciformis*. *Journal of Nematology* 55(1), 20230055.

Spiegel, Y. and Netzer, D. (1984) Effect of nitrogen form at various levels of potassium on the *Meloidogyne–Fusarium* wilt complex in muskmelon. *Plant and Soil* 81(1), 85–92.

Starr, J.L., Jeger, M.L., Martyn, R.D. and Schilling, K. (1989) Effects of *Meloidogyne incognita* and *Fusarium oxysporum* f. sp. *vas infectum* on plant mortality and yield of cotton. *Phytopathology* 79, 640–646.

Stocker, T.F., Qin, D., Plattner, G.K., Tignor, M.M.B., Allen, S.K. *et al.* (2013) *The Physical Science Basis: Working Group I Contribution to the 5th Intergovernmental Panel on Climate Change Assessment*. Cambridge University Press, Cambridge, UK.

Sun, Y., Yin, J., Cao, H., Li, C., Kang, L.E. *et al.* (2011) Elevated CO_2 influences nematode-induced defense responses of tomato genotypes differing in the JA pathway. *PloS ONE* 6(5), e19751.

Taher, I.Y., Kassab, A.S., Mahgoob, A.E., Taha, E.H. and Banora, M. (2017) Disease complex in tomato cv. 035 involving *Meloidogyne incognita* and *Fusarium oxysporum* f. *lycopersici*. *Arab Universities Journal of Agricultural Sciences* 25(2), sp.

Taketani, A., Okuda, M. and Hosoda, R. (1975) The meteorological analysis on the epidemic mortality of pine trees, with special reference to the effective accumulated temperature. *Journal of the Japanese Forestry Society* 57(6), 169–175.

Taylor, C.E. (1979) Meloidogyne interrelationships with microorganisms. In: Lamberti, F. and Taylor, C.E. (eds) *Root-Knot Nematodes (Meloidogyne Species) Systematics, Biology and Control*. Academic Press, London, pp. 375–398.

Teliz, D., Grogan, R.G. and Lownsbery, B.F. (1966) Transmission of tomato ringspot, peach yellow bud mosaic, and grape yellow vein viruses by *Xiphinema americanum*. *Phytopathology* 56, 658–663.

Tirumalaraju, S.V., Jain, M. and Gallo, M. (2011) Differential gene expression in roots of nematode-resistant and -susceptible peanut (*Arachis hypogaea*) cultivars in response to early stages of peanut root-knot nematode (*Meloidogyne arenaria*) parasitization. *Journal of Plant Physiology* 168(5), 481–492.

Togashi, K., Kasuga, H., Kasuga, S., Matsunaga, K. and Jikumaru, S. (2023) Pathogenicity and boarding ability of hybrid-derived populations between *Bursaphelenchus xylophilus* and *B. mucronatus* (Nematoda: Aphelenchoididae). *Nematology* 26(1), 1–13.

Tomalak, M. and Filipiak, A. (2021) Effects of inter-specific crossbreeding between the invasive pine wood nematode, *Bursaphelenchus xylophilus* and native *B. mucronatus* on morphology and reproduction of the hybrid offspring. *Forest Pathology* 51(2), e12676.

Vovlas, N., Inserra, R.N. and Greco, N. (1992) *Schistonchus caprifici* parasitizing caprifig (*Ficus carica sylvestris*) florets and the relationship with its fig wasp (*Bastophaga psenes*) vector. *Nematologica* 38, 215–226.

Vovlas, N., Troccoli, A., Noort, S.V. and Berg, E. (1998) *Schistonchus africanus* n. sp. (Aphelenchida:Aphelenchoididae) associated with *Ficus thonningii* (Moraceae) and its pollinator wasp *Elisabethiella stuckenbergi* (Chalcidoidea: Agaonidae). *Journal of Nematology* (4), sp.

Wang, C. and Roberts, P.A. (2006) A fusarium wilt resistance gene in *Gossypium barbadense* and its effect on root-knot nematode–wilt disease complex. *Phytopathology* 96(7), 727–734. DOI: 10.1094/PHYTO-96-0727.

Wang, E.L.H. and Bergeson, G.B. (1974) Biochemical changes in root exudate and xylem sap of tomato plants infected with *Meloidogyne incognita*. *Journal of Nematology* 6, 192–194.

Webster, J.M. (1985) Interaction of *Meloidogyne* with fungi on crop plants. In: Sasser, J.N. and Carter, C.C. (eds) *An Advanced Treatise on Meloidogyne. Volume I. Biology and Control*. North Carolina State University Graphics, Raleigh, NC, pp. 183–192.

Westerlund, F.U., Campbell, R.N. and Simble, K.A. (1974) Fungal root-rot and wilt of chick-pea in California. *Phytopathology* 64, 432–436.

Wheeler, T.A., Rowe, R.C. and Madden, L.V. (1991) Effect of soil moisture on incidence of potato early dying and tuber yield. *Phytopathology* 81, 1156.

Wilcox-Lee, D. and Loria, R. (1987) Effect of nematode parasitism on plant–water relations. In: Veech, J.A. and Dickson, D.W. (eds) Vistas on Nematology. Society of Nematologists, Hyattsville, MD, pp. 260–266.

Wolfgang, A., Taffner, J., Guimarães, R.A., Coyne, D. and Berg, G. (2019) Novel strategies for soil-borne diseases: Exploiting the microbiome and volatile-based mechanisms toward controlling *Meloidogyne*-based disease complexes. Frontiers in Microbiology 10, 1296.

Wu, J., Chen, H. and Zhang, Y. (2016) Latitudinal variation in nematode diversity and ecological roles along the Chinese coast. Ecology and Evolution 6(22), 8018–8027.

Yagita, H. (1977) The life history and biology of the needle nematode, *Longidorus martini* Merny III. Studies on the host range of mulberry ringspot virus and mode of transmission by *L. martini*. Japanese Journal of Nematology 7, 15–20.

Yagita, H. and Komuro, Y. (1972) Transmission of mulberry ringspot virus by *Longidorus martini* Merny. Japanese Journal of Phytopathology 38(4), 275–283.

Yano, S. (1913) Investigation on pine death in Nagasaki prefecture. *Sanrin-Kouhou* 4, 1–14.

Zhang, M., Zhang, H., Tan, J., Huang, S., Chen, X. *et al*. (2021) Transcriptome analysis of eggplant root in response to root-knot nematode infection. *Pathogens* 10(4), 470.

Zhang, X., Peng, H., Zhu, S., Xing, J., Li, X. *et al*. (2020) Nematode-encoded RALF peptide mimics facilitate parasitism of plants through the FERONIA receptor kinase. *Molecular Plant* 13(10), 1434–1454.

Zhao, J., Li, L., Liu, Q., Liu, P., Li, S. *et al*. (2019) A MIF-like effector suppresses plant immunity and facilitates nematode parasitism by interacting with plant annexins. *Journal of Experimental Botany* 70(20), 5943–5958.

Zheng, Y. and Khan, M.R. (2023) Pine wood nematode in coniferous forests and their management by novel biological and biotechnological interventions. In: Khan, M.R. (ed.) *Novel Biological and Biotechnological Applications in Plant Nematode Management*. Springer, Singapore, pp. 489–514.

4 Biochemical and Molecular Basis of Nematode Disease Complexes in Plants

Dominick Padilla and Bala Rathinasabapathi*
Horticultural Sciences Department, University of Florida, Gainesville, FL, USA

Abstract
Plant-parasitic nematodes are microscopic pests that significantly affect a wide array of plants. They are prevalent in various climates and can severely impact crop production. Plant-parasitic nematodes mostly incite non-specific symptoms on plants and as a result their damage is often not recognized by crop growers. Nematode parasitism on plants shows great diversity and complexity. This chapter highlights the biochemical and molecular levels of the interactions between plant-parasitic nematodes and their host plants, as well as with other pathogens.

Introduction

Nematodes from certain genera, particularly *Meloidogyne*, *Pratylenchus*, *Heterodera*, *Ditylenchus*, *Globodera*, *Tylenchulus*, *Xiphinema*, *Radopholus*, *Helicotylenchus*, *Rotylenchulus*, etc., are detrimental to agricultural crops, due to their widespread prevalence and pathogenicity (Jones *et al.*, 2013; Khan, 2023b). Plant-parasitic nematode infestations can lead to substantial economic losses, estimated between US$118 and US$200 billion annually (Sasser and Freckman, 1986; Chitwood, 2003; Abad *et al.*, 2008; Khan, 2023a). Traditionally, chemical nematicides applied during the pre-plant stage have been the primary method of management. However, methyl bromide, a very effective nematicide, has been phased out due to its ozone-depleting properties and toxicity (Park *et al.*, 2020). Consequently, development of crop cultivars for improved resistance against plant-parasitic nematodes has become an important goal in plant breeding programs.

The relationship between plants and nematodes involves complex molecular and biochemical processes, and has been reviewed recently (Rutter *et al.*, 2022; Bali and Gleason, 2023). Most studies that revealed these mechanisms have used cyst nematodes and root-knot nematodes and their interactions with major crops or the model species *Arabidopsis thaliana*. To illustrate the processes in this chapter, we have used selective studies using root-knot nematodes (*Meloidogyne* spp.).

Root-knot Nematodes

Root-knot nematodes, *Meloidogyne* spp. (order Tylenchida), are soil-borne obligate plant sedentary endoparasites. *Meloidogyne* spp. females are pyriform except juvenile stages J_1 and J_2, and males are vermiform shaped except J_3 and J_4 (Khan, 2008). They use their anterior protrusible stylet, a needle-like structure, to inject secretions into host root cells to reprogram

*Corresponding author: brath@ufl.edu

DOI: 10.1079/9781800625228.0004

developmental processes and to feed on the cells (Khan and Khan, 1994a). About five to eight root cells enlarge to form the giant cells at the feeding site (Huang, 1985). *Meloidogyne* females (except *M. graminicola*) lay eggs in a gelatinous case on the surface of the roots (Khan and Ahamad, 2020). Among the four developmental stages, J_2 is the infective stage (Khan and Khan, 1994b). Species of *Meloidogyne* that are parasitic on important crops reproduce parthenogenetically (Castagnone-Sereno, 2006). Polyploidy is common in these species, and they unusually lack canonical telomeric repeats and telomerases (Mota *et al.*, 2024).

Understanding the biochemical dynamics of plant–nematode interactions is essential for developing precise interventions. Detailed knowledge of how plants respond biochemically to nematodes is vital for breeding resistant crop varieties and devising integrated management strategies that combine biological control (Khan, 2016), genetic resistance and safer chemical treatments (Manohar *et al.*, 2020; Eves-van den Akker *et al.*, 2021; Haque and Khan, 2021, 2023; Zhong *et al.*, 2024). Identifying specific molecules involved, such as proteins and metabolites, is key to understanding and disrupting these interactions effectively (Desmedt *et al.*, 2020). The processes involved in disease complex development following infection by plant-parasitic nematodes are illustrated in Fig. 4.1.

Factors Affecting Infection: Plant Root Exudates

Root exudates are a complex mixture of organic and inorganic compounds secreted by plant roots, which play a crucial role in the rhizosphere dynamics and interactions with soil organisms, including plant nematodes. The following substances in root exudates were shown to affect plant and nematode interactions: sugars (Warnock *et al.*, 2016; Oota *et al.*, 2022), amino acids (Escudero *et al.*, 2014; Jiang *et al.*, 2022; Kihika-Opanda *et al.*, 2022), organic acids (Escudero *et al.*, 2014; Kirwa *et al.*, 2018; Wang *et al.*, 2021; Kihika-Opanda *et al.*, 2022), fatty acids (Dong *et al.*, 2018) and secondary metabolites (Kirwa *et al.*, 2018; Shivakumara *et al.*, 2018; Desmedt *et al.*, 2022; Kihika-Opanda *et al.*, 2022). Specifically, in the context of plant–nematode interactions, root exudates can alter nematode behavior, either attracting (Oota *et al.*, 2020; Wang *et al.*, 2021) or repelling them (Wang *et al.*, 2021; Lu *et al.*, 2023), thereby impacting nematode infection rates and subsequent plant damage (Sharma *et al.*, 2023). Also, some root exudates were reported to inhibit egg hatching (Lu *et al.*, 2023). Exudates can also influence nematode behavior by attracting beneficial nematophagous microorganisms (Topalović *et al.*, 2020). For example, root exudates can attract *Pasteuria penetrans* that antagonizes plant-parasitic nematodes by degrading their cuticle, thereby providing a form of indirect defense (Mohan *et al.*, 2020). However, the identity of the particular compound in the Mohan *et al.* (2020) study, which could be among the 88 compounds identified, causing this attraction of *P. penetrans* remains unclear. Nonetheless, understanding the specific components of root exudates and their effects on plant nematodes is crucial for exploiting these interactions to develop crop varieties with enhanced plant nematode resistance. Conversely, specific secondary metabolites can directly deter nematode entry or establishment

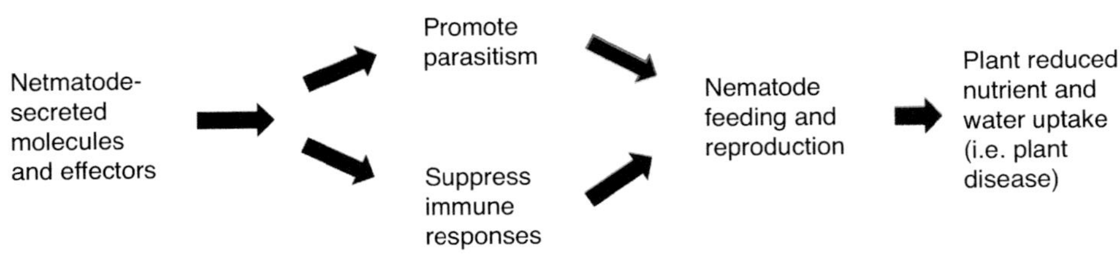

Fig. 4.1. Processes involved in nematode–plant interactions leading to development of a disease complex.

within the root system, and such compounds can be starting points for designing synthetic compounds with nematicidal activities.

The composition of root exudates that impacts nematode behavior varies widely among plant species and is influenced by genetic, environmental and soil health factors (Sharma *et al.*, 2023). Certain compounds in the classes of flavonoids and terpenoids have been shown to possess *Meloidogyne*-repellent properties (Desmedt *et al.*, 2020). Flavonoids may interfere with the nematode hormone systems (Koch *et al.*, 2014), whereas terpenoids are indicated to negatively affect nematode physiology (Desmedt *et al.*, 2020; Mwamba *et al.*, 2021). Flavonoids might mimic or block the effects of natural nematode hormones, potentially by binding to their hormone receptors, thereby preventing normal hormone functions (Koch *et al.*, 2014). This interaction may affect root-knot nematode chemosensory functions, leading to altered perception of host suitability or detection. Moreover, flavonoids can also impact nematode behavior and development by interfering with their signaling pathways, which are crucial for their hormonal regulation (Chin *et al.*, 2018).

The mechanism by which plant terpenoids affect chemosensory systems of nematodes such *Meloidogyne* has not been studied. However, in the case of *Caenorhabditis elegans*, the interaction involves interfering with the function of specific G-protein-coupled receptors (GPCRs) that are essential for chemotaxis and host detection processes (Lei *et al.*, 2010). Research has highlighted that terpenoids may bind to or modulate these GPCRs, thereby altering the ability of the nematode to effectively perceive and respond to environmental cues.

Certain chemical compounds in root exudates have been identified as repellent to root-knot nematodes, thus preventing their larvae from infecting the roots. However, the function and mechanism of chemical compounds in root exudates enacting repellency remains unclear. These include specific secondary metabolites such as saponins (D'Addabbo *et al.*, 2011), which have been shown to disrupt the cellular membranes of microorganisms, and volatile organic compounds that can alter nematode behavior from a distance. The repellent chemicals may interfere with nematode sensory perception or mimic signaling molecules that mislead the larvae about the suitability of the plant as a host. The development and deployment of these compounds can be a critical first line of defense in plant resistance strategies, offering a means to reduce nematode infestation rates.

Nutrient Flow to Nurse Cells

After invading host root tissue, the infective-stage juveniles (J_2) of sedentary nematodes first navigate towards the root tip, where they circulate within the meristem before penetrating the central vascular tissue. Using their stylet, J_2s inject secretions from their dorsal and subventral pharyngeal glands into selected vascular cells (Khan *et al.*, 2023). This action initiates the transformation of these cells into specialized feeding structures called giant cells, which support the nematode throughout its sedentary life phase (Huang, 1985). The initial feeding cell undergoes hyperplasia (an increase in cell number due to changes in the cell cycle) and hypertrophy (enlargement of nuclei, organelle proliferation, metabolic activation and cell wall modifications) (Gheysen and Mitchum, 2011). These processes, in just a few cells in the vascular parenchyma, lead to the development of giant cells, typically observed at the root tips in root-knot nematode-infected roots. Table 4.1 lists examples of multiple physiological processes that have been experimentally linked to plant–root-knot nematode interaction. Formation of nurse cells is a very important event in the process of parasitism of sedentary nematodes because without the formation of nurse cells the sedentary nematode cannot feed and starves to death. Formation of specialized nurse cells followed by greater continuous flow of nutrients to the giant cells provides the required amount of nutrients to the developing female.

Understanding the host mechanisms that inhibit giant cell formation is crucial for developing targeted strategies for *Meloidogyne* resistance, particularly in relation to nutrient exchange. Adaptations to cell-wall-modifying effectors are also critical for preventing giant cell formation. For instance, the *fra* mutant, a katanin mutant with defective microtubules, is affected for cell division and intracellular transport (Escobar and Fenoll, 2015). *Arabidopsis* mutants for katanin are more susceptible to root-knot nematodes than wild-type, suggesting the need for microtubules

Table 4.1. Processes experimentally linked to root-knot nematode interactions with plant roots (only selected studies are listed for illustration purposes).

	Processes	Reference(s)
1	Synthesis and secretion of effector molecules by the root-knot nematode	Bellafiore *et al.*, 2008; Yu *et al.*, 2024
2	Synthesis of cellulose-binding proteins by the root-knot nematode	Adam *et al.*, 2008
3	Reactive oxygen species burst in the plant	Song *et al.*, 2018
4	NADPH oxidase and pattern-triggered immunity	Bakhetia *et al.*, 2005
5	Cell wall modifications in the plant and pectin synthesis	Veronico *et al.*, 2022
6	Antioxidant enzymes in the plant	Oliveira *et al.*, 2012
7	Polyamine synthesis	Oota *et al.*, 2020
8	Defence-related callose formation	Zhang *et al.*, 2015
9	Phenolic compounds	Pegard *et al.*, 2005
10	Salicylic acid-mediated plant defense	Branch *et al.*, 2004
11	Jasmonic acid synthesis, jasmonic acid-mediated plant defense and oxylipins	Ozalvo *et al.*, 2014
12	Regulation of auxin levels in the plant	Hutangura *et al.*, 1999
13	Synthesis of pathogenesis-related proteins in the plant	Hamamouch *et al.*, 2011
14	Alternative splicing	Verma *et al.*, 2018
15	Trafficking of immune proteins via vesicle formation via SNAREs	Bekal *et al.*, 2015
16	Proteosome-dependent ubiquitination of proteins	Shi *et al.*, 2018
17	Histone modifications in plant nuclei	Atighi *et al.*, 2021

for resistance (Meidani *et al.*, 2019). A network of microtubules aligned with the plasma membrane has been observed along the cell wall ingrowths in giant cells, believed to enhance and facilitate solute transport from neighboring xylem and phloem cells (Escobar and Fenoll, 2015).

Plant resistance to root-knot nematode infection could be by disrupting the development or function of these feeding sites through changes in cellular architecture, such as the deposition of callose (Huang *et al.*, 2023b; Yu *et al.*, 2024) and other polysaccharides, which physically block root-knot nematode access to plant cells. Additionally, the plant resistance may interfere with nutrient transport to these sites, effectively hindering nematode development (Gheysen and Mitchum, 2011; Kyndt *et al.*, 2013). At the molecular level, plants regulate genes involved in nutrient allocation. These genes control the synthesis and transport of nutrients such as sugars, amino acids and other essential metabolites (Grundler and Hofmann, 2011). Thus, it would be reasonable to hypothesize that one mechanism of resistance is downregulation of these genes in response to the nematode attack to reduce the availability of critical nutrients at the feeding site. However, in the *Meloidogyne*–host interaction, there is insufficient evidence, albeit one study showed that total protein content, sugar concentration and total phenol content variations were not correlated to *M. javanica* resistance levels in tomato (Lobna *et al.*, 2017). Furthermore, genes that promote the synthesis of defense compounds are upregulated, enhancing the plant's ability to resist nematode penetration and establishment. This genetic reprogramming is a sophisticated method of resource allocation that prioritizes plant survival over the growth of the nematode.

When a parasitic organism such as a nematode invades a plant, the plant's immune system is triggered. This involves the recognition of pathogen-associated molecular patterns (PAMPs) by proteins localized in the plasma membrane and the deployment of defense mechanisms. This initial recognition is crucial for activating downstream genetic responses (Holbein *et al.*, 2019). Following nematode detection, signal transduction pathways are activated. These pathways often involve plant hormones such as salicylic acid, jasmonic acid and ethylene, which are pivotal in modulating defense responses. The signaling molecules regulate specific transcription factors that control the expression of genes involved in the defense (preventing feeding site formation) (Rutter *et al.*, 2022; Bali and Gleason, 2023).

Specific genes induced in response to nematode infection include proteins involved in strengthening cell walls (lignin synthesis) (Denness *et al.*, 2011; Veronico *et al.*, 2018; Yang *et al.*, 2023), toxic compound production (e.g. phytoalexins and terpenoids) (Veech, 1982; Ghorbel *et al.*, 2021) and proteins that directly disrupt the nematode metabolism (protease inhibitors) (Ribeiro *et al.*, 2022). One area that is relatively unknown, however, is whether plant hosts are able to minimize nutrient loss through downregulating genes associated with nutrient transport towards the infected sites. On the other hand, there are numerous studies of root-knot nematodes reprogramming host cell machinery to redirect nutrients to the feeding site.

Plant growth regulators such as jasmonic acid and salicylic acid are particularly significant, as they are involved in signaling pathways that activate plant defenses against nematode attack (Rutter *et al.*, 2022). The presence of nematodes can trigger an increase in jasmonic acid levels, which in turn enhances the plant production of defensive compounds and structural reinforcements at the feeding sites. This hormonal response can effectively modify the local environment within the root, making it inhospitable and less conducive to nematode feeding and establishment. These hormonal changes not only contribute to the structural defenses but also regulate other biochemical pathways that can have a detrimental effect on nematode survival.

The role of plant hormones in resistance mechanisms against nematodes, particularly through pathways involving jasmonic acid and salicylic acid, ties closely into the concept of effector-triggered immunity (ETI). This form of immunity represents an effective defense system activated when host receptors, typically nucleotide-binding site leucine-rich repeat (NBS-LRR) proteins (NLRs), recognize and bind pathogen-derived effectors, which are molecules that pathogens use to facilitate infection. Effectors secreted by the nematodes into plant cells typically manipulate host cell processes when these host resistance factors are absent to favor the parasitic process. Many of these effectors in various plant-parasitic nematodes have been identified (Rutter *et al.*, 2022). Thus, a primary goal is to coincidingly adapt host receptor proteins that can recognize these effectors and activate ETI.

When effectors from nematodes are recognized by specific receptor proteins in the plant, it triggers a signal transduction cascade. Once an effector is recognized, the receptor protein undergoes a conformational change that initiates the signal transduction cascade. Common components of these pathways include various kinases (Bali and Gleason, 2023). Jasmonic acid and salicylic acid, in particular, are central to the signaling pathways that are activated upon effector recognition by major R-genes, typically encoding NLRs. Plants defend against nematodes through a systemic signaling loop between roots and shoots, with nematode resistance relying mainly on jasmonic acid synthesis in the shoots. This defense mechanism involves the propagation of electrical and reactive oxygen species (ROS) signals upon root-knot nematode attack, which activates the mitogen-activated protein kinase MPK1/2 pathway, orchestrating a coordinated response throughout the plant (Wang *et al.*, 2019).

The NLR-mediated resistance response typically manifests in a hypersensitive reaction, wherein the host cell being parasitized is programmed for cell death to essentially trap the nematode. The R-genes encoding and their cognate avirulence gene have been identified (Rutter *et al.*, 2022; Bali and Gleason, 2023). For example, the root-knot nematode effector Mc1194 interacts with RD21A, an R-gene that triggers the hypersensitive response. In

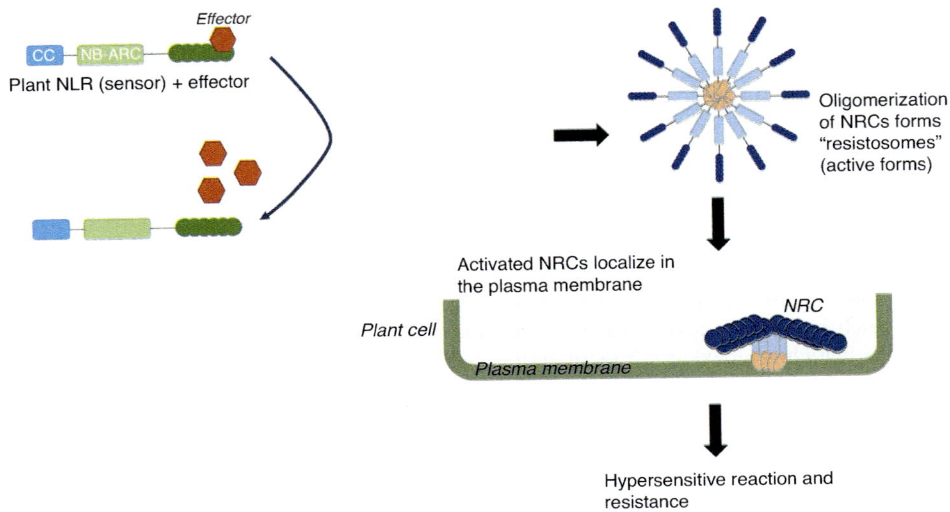

Fig. 4.2. A conceptual model for nucleotide-binding site leucine-rich repeat immune receptor (NLR) immune activation. NRC: NLR required for cell death.

Arabidopsis plants that lack a functional RD21A, there is increased vulnerability to attacks by *Meloidogyne chitwoodi*, highlighting the critical role of RD21A in defending against nematode infections (Yu *et al.*, 2024).

Among the plant disease resistance genes, two types of NLR proteins have been recognized. The 'sensor-type NLR protein' binds the effector of the pathogen first to form a complex. Then, the complex activates the NRC (NLR required for cell death) proteins which multimerize to form a 'resistosome' complex that migrates to the plasma membrane to induce cell death as illustrated in Fig. 4.2.

A genome-wide study revealed 261, 443 and 755 NLR coding genes in the genomes of tomato, potato and pepper, respectively (Seo *et al.*, 2016). The parallel evolution of NLR and NRC protein pairs has generated systems to mediate immunity against diverse plant pathogens including plant-parasitic nematodes (Wu *et al.*, 2017).

Many studies have mapped R-genes in plant genomes without their functional characterization, and the identification of cognate avirulence genes has not been done either. For example, multiple R-genes for root-knot nematode resistance in peppers are known. For example, recently the *Me3* gene in *Capsicum annuum* was cloned and demonstrated to confer resistance against *M. incognita* in transgenic *Arabidopsis* expressing this gene (Liu *et al.*, 2023). The protein coded by the *Me3* gene has two conserved domains: 1) a NB-ARC domain which codes for an ATPase and determines the nucleotide binding state of the protein and 2) a LRR domain which provides structural features facilitating protein–protein interactions. While the identity of the *Me3* gene for root-knot nematode resistance has been resolved, its interaction partners and their role in resistance to *M. incognita* are not known.

Inhibition of Root-knot Nematode Reproduction

There are numerous studies focused on ETI and the resistance genes involved, but only relatively few have identified genes that confer basal immunity. Basal immunity is activated by the detection of general PAMPs, which are broadly conserved across microbial species (Nicaise *et al.*, 2009). These PAMPs are recognized by pattern recognition receptors (PRRs) on the surface of plant cells, initiating pattern-triggered immunity (PTI). Typically, PTI also involves a co-receptor which relies on the signals provided by an activated PRR. A conceptual drawing of PTI induced in *Arabidopsis* plants upon infection by *M. incognita* is shown in Fig. 4.3. The nematode-derived pheromone effector Ascaroside #18 can activate PTI in *Arabidopsis*, relying on the immune co-receptor BAK1 (brassinosteroid-associated kinase 1) (Huang *et al.*, 2023a). A LRR receptor-like kinase, named NILR1, is specifically activated by nematode infection. NILR1 is crucial for launching PTI responses against nematodes, and plants lacking NILR1 are highly susceptible to various nematodes (Mendy *et al.*, 2017). Through BAK1, other cofactors such as brassinosteroids can modulate plant immune responses. For plants infected with *Meloidogyne* spp., the effective activation of PTI through BAK1 can lead to an enhanced immune response, potentially increasing resistance against the nematode (Belkhadir *et al.*, 2012).

In *Arabidopsis*, research indicates that DSC_1 and $WRKY_{19}$, functioning together as an immune receptor pair, significantly influenced reproductive success by reducing the number of egg masses per root tip. However, they do not trigger the typical localized cell death response often associated with resistance (Warmerdam *et al.*, 2020).

There are many studies indicating various phytochemicals associated with inhibition of the root-knot nematode reproduction. These include phytochemicals such as lignin (Wuyts *et al.*, 2006a, 2006b; Basso *et al.*, 2023), flavonoids (Chin *et al.*, 2018; Kirwa *et al.*, 2018; Zhang *et al.*, 2023) and saponins (D'Addabbo *et al.*, 2011). However, fewer studies have detailed the functions of these compounds and their mechanisms of action. Syringyl lignin has been shown to reduce nematode fecundity and development in transgenic *Arabidopsis* mutants overexpressing ferulate 5-hydroxylase (Wuyts *et al.*, 2006a). However, the number of egg masses in these transgenic plants was equal to wild-type plants. Instead, the resistance phenotype manifested as reduced numbers of eggs per egg mass, decreased egg hatch rate and hindrance of juvenile development into females.

Syringyl lignin is one of the three primary types of lignin found in plants, alongside guaiacyl and *p*-hydroxyphenyl lignin, synthesized via the phenylpropanoid pathway. Lignin pathway enzymes may serve as potential targets to modulate lignin biosynthesis, which in turn reduces nematode fecundity and development. For example, orthologues of cinnamyl alcohol dehydrogenase (CAD), a key enzyme in lignin biosynthesis, are suggested to influence resistance (Lee *et al.*, 2022). Results in the study indicate that the resistance

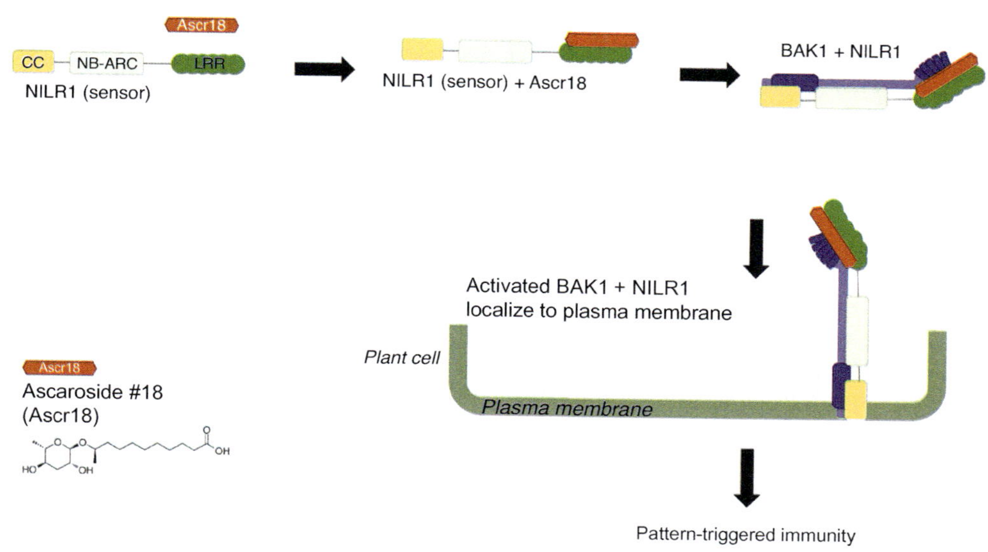

Fig. 4.3. Illustration of pattern-triggered immunity in response to cell surface interaction between ascaroside #18 from the nematode and NILR1. CC: coiled-coil domain; NB-ARC: nucleotide-binding domain shared with APAF-1, various R-proteins and CED-4; LRR: leucine-rich repeat. Based on Mendy *et al.* (2017) in a study of *Arabidopsis* and root-knot nematodes.

in the sweet potato cultivar was due to elevated constitutive *CAD* expression, possibly linking to previous studies where lignin as the final product acts as the compound influencing root-knot nematode reproduction. Paralleling these results are two studies highlighting the role of *GmEXPA1*, a gene encoding an expansin in soybean (Basso *et al.*, 2023; Freitas-Alves *et al.*, 2024). Transgenic lines of *Arabidopsis thaliana* and *Nicotiana tabacum* overexpressing *GmEXPA1* subsequently showed upregulation of genes involved in lignin biosynthesis and moderate resistance to *M. incognita*. Interestingly, although the transgenic lines enabled root-knot nematode infection, giant cell size and the number of neighboring cells were reduced. The relationship between expansins and lignin potentially involves several aspects influencing the reproduction of root-knot nematodes. Expansins might indirectly influence the pathways of lignin biosynthesis through their impact on cell wall architecture and signaling pathways. For example, changes in cell wall integrity, potentially mediated by expansins, can trigger defense signaling pathways such as those involving ROS and jasmonic acid, which are known to enhance lignin biosynthesis as part of the defense response (Denness *et al.*, 2011). Lignin is an intricate polymer that serves as the physical barrier in the defense response to pathogen infection, as lignin is undegradable to most microorganisms. Nonetheless, it is uncertain how lignin affects root-knot nematode colonization. Other studies have shown that when pathogens invade a cell, they induce lignin deposition in the cell wall which provides a physical barrier to resist pathogen infection by limiting the entry of pathogen toxins and cell-wall-degrading enzymes into plants and preventing nutrient transmission from the host to the pathogen.

Another family of compounds involved in the host–nematode interaction are the flavonoids. The biosynthesis of flavonoids begins with the phenylpropanoid pathway. This pathway is also critical for the production of other phenolic compounds. The first committed step in flavonoid biosynthesis involves the enzyme chalcone synthase (CHS). CHS catalyzes the condensation of three molecules of malonyl-CoA with one molecule of 4-coumaroyl-CoA to produce chalcone, the precursor to all flavonoids. From chalcone, a variety of flavonoid subclasses are synthesized through various enzymatic modifications, including flavones, flavonols, flavanones and anthocyanins. These modifications involve hydroxylation, methylation, glycosylation and oxidation, facilitated by specific enzymes (Chin *et al.*, 2018).

Flavonoids are the largest family, comprising of phenolic secondary metabolites, some of which are correlated to root-knot nematode resistance. For example, quercetin is a flavonol shown to be upregulated in resistant plants. Additionally, it has been shown in chemotaxis studies to demonstrate repellent activity. Studies indicate quercetin as a signaling molecule that triggers plant defense mechanisms and induces expression of various genes involved in the plant immune response instead of directly interacting with root-knot nematodes (Wuyts *et al.*, 2006a, 2006b).

In the case of attenuation of root-knot nematode reproduction, flavonoids are more highly induced in resistant cultivars compared to susceptible cultivars. Phenylpropanoid pathway enzymes may serve as potential targets to ultimately regulate root-knot nematode resistance (Chin *et al.*, 2018). One study showed differential expression of CHS between slightly resistant and highly resistant cowpea cultivars. However, CHS expression was unique to the slightly resistant cultivar only, suggesting a basal level of immunity conferred by flavonoids. Another caveat to the utility of CHS as a resistance marker is that some flavonoids are correlated with susceptibility rather than resistance. For instance, higher content of anthocyanidin is positively correlated with root-knot nematode reproduction (Wuyts *et al.*, 2006a; Wuyts *et al.*, 2006b). Overall, the utility of flavonoids and synthetic enzymes in the pathway for resistance in plants should be carefully determined, especially since the flavonoids comprise a variety of compounds with different effects on root-knot nematode reproduction.

Susceptibility Factors

Although pyramiding resistance genes with different mechanisms is an effective strategy to enhance resistance durability, a potential complementary approach is to ‘breed out’ genes conferring susceptibility. Susceptibility genes encode products which confer susceptibility, in contrast to resistance genes which confer resistance (Van Schie and Takken, 2014). Most targets of effectors are positive regulators of host immunity, which become suppressed as a consequence of the effector–target interaction. However, effectors can also indirectly suppress host immunity by interacting with susceptibility factors. Mutant lines for the gene encoding ethylene response factor 6 (ERF6) enabled greater numbers of egg masses per root system by *M. incognita*. Breeding out susceptibility factors seems to be an effective strategy for developing cultivars with root-knot nematode resistance, but there can be trade-offs. For example, ERF6 is assumed to mediate abiotic stress tolerance with ROS and 1-aminocyclopropane-1-carboxy lic acid as cofactors. However, mutants exhibit downregulation of the gene encoding 1-amin ocyclopropane-1-carboxylate oxidase 3, a key enzyme in the ethylene biosynthesis pathway.

Conclusion and Future Prospects

The role of susceptibility factors highlights the nuanced and sometimes antagonistic interactions between abiotic and biotic stress responses in plants. It indicates that pathways activated under abiotic stress may facilitate root-knot nematode reproduction by making the plant more amenable to nematode-induced modifications of plant tissues. Based on the mechanisms of nematode and host plant interactions as discussed here, we have multiple targets available for developing nematode-resistant new crop varieties by using both biotechnology and plant breeding. Also, new strategies can be developed for designing chemicals to manage plant-parasitic nematodes in agriculture.

References

Abad, P., Gouzy, J., Aury, J.-M., Castagnone-Sereno, P., Danchin, E.G.J. *et al.* (2008) Genome sequence of the metazoan plant-parasitic nematode *Meloidogyne incognita*. *Nature Biotechnology* 26(8), 909–915. DOI: 10.1038/nbt.1482.

Adam, M.A.M., Phillips, M.S., Jones, J.T. and Blok, V.C. (2008) Characterisation of the cellulose-binding protein *Mj-cbp-1* of the root knot nematode, *Meloidogyne javanica*. *Physiological and Molecular Plant Pathology* 72(1–3), 21–28. DOI: 10.1016/j.pmpp.2008.05.002.

Atighi, M.R., Verstraeten, B., De Meyer, T. and Kyndt, T. (2021) Genome-wide shifts in histone modifications at early stage of rice infection with *Meloidogyne graminicola*. *Molecular Plant Pathology* 22(4), 440–455. DOI: 10.1111/mpp.13037.

Bakhetia, M., Charlton, W., Atkinson, H.J. and McPherson, M.J. (2005) RNA interference of dual oxidase in the plant nematode *Meloidogyne incognita*. *MPMI* 18, 1099–1106. DOI: 10.1094/MPMI-18-1099.

Bali, S. and Gleason, C. (2023) Unveiling the diversity: Plant parasitic nematode effectors and their plant interaction partners. *MPMI* 37, 179–189. DOI: 10.1094/mpmi-09-23-0124-fi.

Basso, M.F., Lourenço-Tessutti, I.T., Moreira-Pinto, C.E., Mendes, R.A.G., Pereira, D.G. *et al.* (2023) Overexpression of the *GmEXPA1* gene reduces plant susceptibility to *Meloidogyne incognita*. *Plant Cell Reports* 42(1), 137–152. DOI: 10.1007/s00299-022-02941-3.

Bekal, S., Domier, L.L., Gonfa, B., Lakhssassi, N., Meksem, K. *et al.* (2015) A SNARE-like protein and biotin are implicated in soybean cyst nematode virulence. *PLoS ONE* 10(12), e0145601. DOI: 10.1371/journal.pone.0145601.

Belkhadir, Y., Jaillais, Y., Epple, P., Balsemão-Pires, E., Dangl, J.L. *et al.* (2012) Brassinosteroids modulate the efficiency of plant immune responses to microbe-associated molecular patterns. *Proceedings of the National Academy of Sciences of the United States of America* 109(1), 297–302. DOI: 10.1073/pnas.1112840108.

Bellafiore, S., Shen, Z., Rosso, M.N., Abad, P., Shih, P. *et al.* (2008) Direct identification of the *Meloidogyne incognita* secretome reveals proteins with host cell reprogramming potential. *PLoS Pathogogens* 4(10), e1000192.

Branch, C., Hwang, C.-F., Navarre, D.A. and Williamson, V.M. (2004) Salicylic acid is part of the mi-1-mediated defense response to root-knot nematode in tomato. *MPMI* 17(4), 351–356. DOI: 10.1094/MPMI.2004.17.4.351.

Castagnone-Sereno, P. (2006) Genetic variability and adaptive evolution in parthenogenetic root-knot nematodes. *Heredity* 96, 282–289. DOI: 10.1038/sj.hdy.6800794.

Chin, S., Behm, C.A. and Mathesius, U. (2018) Functions of flavonoids in plant–nematode interactions. *Plants* 7(4), 85. DOI: 10.3390/plants7040085.

Chitwood, D.J. (2003) Research on plant-parasitic nematode biology conducted by the United States Department of Agriculture-Agricultural Research Service. *Pest Management Science* 59(6–7), 748–753. DOI: 10.1002/ps.684.

D'Addabbo, T., Carbonara, T., Leonetti, P., Radicci, V., Tava, A. *et al.* (2011) Control of plant parasitic nematodes with active saponins and biomass from *Medicago sativa*. *Phytochemistry Reviews* 10(4), 503–519. DOI: 10.1007/s11101-010-9180-2.

Denness, L., McKenna, J.F., Segonzac, C., Wormit, A., Madhou, P. *et al.* (2011) Cell wall damage-induced lignin biosynthesis is regulated by a reactive oxygen species- and jasmonic acid-dependent process in *Arabidopsis*. *Plant Physiology* 156(3), 1364–1374. DOI: 10.1104/pp.111.175737.

Desmedt, W., Mangelinckx, S., Kyndt, T. and Vanholme, B. (2020) A phytochemical perspective on plant defense against nematodes. *Frontiers in Plant Science* 11, 602079. DOI: 10.3389/fpls.2020.602079.

Desmedt, W., Kudjordjie, E.N., Chavan, S.N., Zhang, J., Li, R. *et al.* (2022) Rice diterpenoid phytoalexins are involved in defence against parasitic nematodes and shape rhizosphere nematode communities. *The New Phytologist* 235(3), 1231–1245. DOI: 10.1111/nph.18152.

Dong, L., Li, X., Huang, C., Lu, Q., Li, B. *et al.* (2018) Reduced *Meloidogyne incognita* infection of tomato in the presence of castor and the involvement of fatty acids. *Scientia Horticulturae* 237, 169–175. DOI: 10.1016/j.scienta.2018.03.066.

Escobar, C. and Fenoll, C. (2015) *Plant Nematode Interactions: A View on Compatible Interrelationships*. Academic Press, New York.

Escudero, N., Marhuenda-Egea, F.C., Ibanco-Canete, R., Zavala-Gonzalez, E.A. and Lopez-Llorca, L.V. (2014) A metabolomic approach to study the rhizodeposition in the tritrophic interaction: Tomato, *Pochonia Chlamydosporia* and *Meloidogyne Javanica*. *Metabolomics* 10, 788–804. DOI: 10.1007/s11306-014-0632-3.

Eves-van den Akker, S., Stojilković, B. and Gheysen, G. (2021) Recent applications of biotechnological approaches to elucidate the biology of plant–nematode interactions. *Current Opinion in Biotechnology* 70, 122–130. DOI: 10.1016/j.copbio.2021.03.008.

Freitas-Alves, N.S., Moreira-Pinto, C.E., Arraes, F.B.M., Costa, L.S. de L., Abreu, R.A. *et al.* (2024) An *ex vitro* hairy root system from petioles of detached soybean leaves for in planta screening of target genes and CRISPR strategies associated with nematode bioassays. *Planta* 259(1), 23. DOI: 10.1007/s00425-023-04286-x.

Gheysen, G. and Mitchum, M.G. (2011) How nematodes manipulate plant development pathways for infection. *Current Opinion in Plant Biology* 14(4), 415–421. DOI: 10.1016/j.pbi.2011.03.012.

Ghorbel, M., Brini, F., Sharma, A. and Landi, M. (2021) Role of jasmonic acid in plants: The molecular point of view. *Plant Cell Reports* 40(8), 1471–1494. DOI: 10.1007/s00299-021-02687-4.

Grundler, F.M. and Hofmann, J. (2011) Water and nutrient transport in nematode feeding sites. In: Jones, J.T., Gheysen, L. and Fenoll, C. (eds) *Genomics and Molecular Genetics of Plant–Nematode Interactions*. Springer, Dordrecht, The Netherlands, pp. 423–439.

Hamamouch, N., Li, C., Seo, P.J., Park, C.M. and Davis, E.L. (2011) Expression of *Arabidopsis* pathogenesis-related genes during nematode infection. *Molecular Plant Pathology* 12(4), 355–364. DOI: 10.1111/j.1364-3703.2010.00675.x.

Haque, Z. and Khan, M.R. (2021) *Handbook of Invasive Plant-Parasitic Nematodes*. CAB International, Wallingford, UK.

Haque, Z. and Khan, M.R. (2023) Nematicidal effect of fluopyram and fluensulfone against rice root-knot nematode, *Meloidogyne graminicola*. *Russian Journal of Nematology* 31, 39–49.

Holbein, J., Franke, R.B., Marhavý, P., Fujita, S., Górecka, M. *et al.* (2019) Root endodermal barrier system contributes to defence against plant-parasitic cyst and root-knot nematodes. *The Plant Journal* 100(2), 221–236. DOI: 10.1111/tpj.14459.

Huang, L., Yuan, Y., Lewis, C., Kud, J., Kuhl, J.C. *et al.* (2023a) NILR1 perceives a nematode ascaroside triggering immune signaling and resistance. *Current Biology* 33(18), 3992–3997. DOI: 10.1016/j.cub.2023.08.017.

Huang, Q., Lin, B., Cao, Y., Zhang, Y., Song, H. *et al.* (2023b) CRISPR/Cas9-mediated mutagenesis of the susceptibility gene *OsHPP04* in rice confers enhanced resistance to rice root-knot nematode. *Frontiers in Plant Science* 14, 1134653. DOI: 10.3389/fpls.2023.1134653.

Huang, S.C. (1985) Formation, anatomy and physiology of giant cells induced by root-knot nematodes. In: Sasser, J.N. and Carter, C.C. (eds) *An Advanced Treatise on* Meloidogyne *Volume I. Biology and Control*. North Carolina State University Graphics, Raleigh, NC, pp. 143–154.

Hutangura, P., Mathesius, U., Jones, M.G.K. and Rolfe, B.G. (1999) Auxin induction is a trigger for root gall formation caused by root-knot nematodes in white clover and is associated with the activation of the flavonoid pathway. *Functional Plant Biology* 26(3), 221. DOI: 10.1071/PP98157.

Jiang, Y., Huang, M., Li, C., Hua, C., Qin, R. *et al.* (2022) Responses of infective juveniles of the soybean cyst nematode (*Heterodera glycines*) and the root-knot nematodes (*Meloidogyne hapla M. incognita*) to amino acids. *Nematology* 24, 1049–1062. DOI: 10.1163/15685411-bja10190.

Jones, J.T., Haegeman, A., Danchin, E.G.J., Gaur, H.S., Helder, J. *et al.* (2013) Top 10 plant-parasitic nematodes in molecular plant pathology. *Molecular Plant Pathology* 14(9), 946–961. DOI: 10.1111/mpp.12057.

Khan, M.R. (2008) *Plant Nematodes – Methodology, Morphology, Systematics, Biology and Ecology*. Science Publishers, Enfield, NH. DOI: 10.1201/b10766.

Khan, M.R. (2016) Nematode biocontrol agents: diversity and effectiveness against phytonematodes in sustainable crop protection. *Indian Phytopathology* 69, 453–463.

Khan, M.R. (2023a) Nematode pests of agricultural crops, a global overview. In: Khan, M.R. (ed.) *Novel Biological and Biotechnological Applications in Plant Nematode Management*. Springer, Singapore, pp. 3–45.

Khan, M.R. (2023b) Plant nematodes, an underestimated constraint in global crop production. Nematode diseases of crops and their sustainable management. In: Khan, M.R. and Quintanilla, M. (eds) *Nematode Diseases of Crops and Their Sustainable Management*. Academic Press, Cambridge, MA, pp. 3–23. DOI: 10.1016/B978-0-323-91226-6.00009-2.

Khan, M.R. and Ahamad, F. (2020) Incidence of root-knot nematode (*Meloidogyne graminicola*) and resulting crop losses in paddy rice in Northern India. *Plant Disease* 104(1), 186–193. DOI: 10.1094/PDIS-12-18-2154-RE.

Khan, M.R. and Khan, M.W. (1994a) Effects of simulated acid rain and root-knot nematode on tomato. *Plant Pathology* 43, 41–49.

Khan, M.R. and Khan, M.W. (1994b) Single and interactive effects of root-knot nematode and coal-smoke on okra. *New Phytologist* 126, 337–342.

Khan, M.R., Ahamad, F., Rizvi, T.F. and Akram, M. (2023) Root-knot nematodes in vegetables and ornamentals and their management by novel biological and biotechnological tools. In: Khan, M.R. (ed.) *Novel Biological and Biotechnological Applications in Plant Nematode Management*. Springer, Singapore, pp. 261–288.

Kihika-Opanda, R., Tchouassi, D.P., Ng'ang'a, M.M., Beck, J.J. and Torto, B. (2022) Chemo-ecological insights into the use of the non-host plant vegetable blackjack to protect two susceptible solanaceous crops from root-knot nematode parasitism. *Journal of Agricultural and Food Chemistry* 70, 6658–6669. DOI: 10.1021/acs.jafc.2c01748.

Kirwa, H.K., Murungi, L.K., Beck, J.J. and Torto, B. (2018) Elicitation of differential responses in the root-knot nematode *Meloidogyne incognita* to tomato root exudate cytokinin, flavonoids, and alkaloids. *Journal of Agricultural and Food Chemistry* 66, 11291–11300. DOI: 10.1021/acs.jafc.8b05101.

Koch, K., Havermann, S., Büchter, C. and Wätjen, W. (2014) *Caenorhabditis elegans* as model system in pharmacology and toxicology: Effects of flavonoids on redox-sensitive signalling pathways and ageing. Scientific World Journal2014920398. DOI: 10.1155/2014/920398.

Kyndt, T., Vieira, P., Gheysen, G. and Almeida-Engler, J. (2013) Nematode feeding sites: Unique organs in plant roots. *Planta* 238, 807–818. DOI: 10.1007/s00425-013-1923-z.

Lee, K.H., Lee, K.-L., Nam, K.J., Yang, J.-W., Lee, J.J. *et al.* (2022) Expression analysis of sweetpotato cinnamyl alcohol dehydrogenase genes in response to infection with the root-knot nematode *Meloidogyne incognita*. *Plant Biotechnology Reports* 16(4), 487–492. DOI: 10.1007/s11816-022-00773-x.

Lei, J., Leser, M. and Enan, E. (2010) Nematicidal activity of two monoterpenoids and SER-2 tyramine receptor of *Caenorhabditis elegans*. *Biochemical Pharmacology* 79(7), 1062–1071. DOI: 10.1016/j.bcp.2009.11.002.

Liu, Y., Cao, H., Ling, J., Yang, Y., Li, Y. *et al.* (2023) Molecular cloning and functional analysis of the pepper resistance gene *Me3* to root-knot nematode. *Horticultural Plant Journal* 9(1), 133–144. DOI: 10.1016/j.hpj.2022.12.003.

Lobna, H., Aymen, E.M., Hajer, R., Naima, M.B. and Najet, H.R. (2017) Biochemical and plant nutrient alterations induced by *Meloidogyne javanica* and *Fusarium oxysporum* f. sp. *radicis lycopersici* co-infection on tomato cultivars with differing level of resistance to *M. javanica*. *European Journal of Plant Pathology* 148(2), 463–472. DOI: 10.1007/s10658-016-1104-6.

Lu, Q., Wang, K., Dou, Z., Zhong, L., Yao, Y. *et al*. (2023) Vanillin in resistant tomato plant root exudate suppresses *Meloidogyne incognita* parasitism. *Journal of Agricultural and Food Chemistry* 71(27), 10269–10276. DOI: 10.1021/acs.jafc.3c00661.

Manohar, M., Tenjo-Castano, F., Chen, S., Zhang, Y.K., Kumari, A. *et al*. (2020) Plant metabolism of nematode pheromones mediates plant–nematode interactions. *Nature Communications* 11(1), 208. DOI: 10.1038/s41467-019-14104-2.

Meidani, C., Ntalli, N.G., Giannoutsou, E. and Adamakis, I.D.S. (2019) Cell wall modifications in giant cells induced by the plant parasitic nematode *Meloidogyne incognita* in wild-type (Col-0) and the *fra2 Arabidopsis thaliana* katanin mutant. *International Journal of Molecular Sciences* 20(21), 5465. DOI: 10.3390/ijms20215465.

Mendy, B., Wang'ombe, M.W., Radakovic, Z.S., Holbein, J., Ilyas, M. *et al*. (2017) *Arabidopsis* leucine-rich repeat receptor-like kinase NILR1 is required for induction of innate immunity to parasitic nematodes. *PLoS Pathogens* 13, e1006284. DOI: 10.1371/journal.ppat.1006284.

Mohan, S., Kiran Kumar, K., Sutar, V., Saha, S., Rowe, J. *et al*. (2020) Plant root-exudates recruit hyperparasitic bacteria of phytonematodes by altered cuticle aging: Implications for biological control strategies. *Frontiers in Plant Science* 11, 763. DOI: 10.3389/fpls.2020.00763.

Mota, A.P.Z., Koutsovoulos, G.D., Perfus-Barbeoch, L., Despot-Slade, E., Labadie, K. *et al*. (2024) Unzipped genome assemblies of polyploid root-knot nematodes reveal unusual and clade-specific telomeric repeats. *Nature Communications* 15(1), 773. DOI: 10.1038/s41467-024-44914-y.

Mwamba, S., Kihika-Opanda, R., Murungi, L.K., Losenge, T., Beck, J.J. *et al*. (2021) Identification of repellents from four non-host asteraceae plants for the root knot nematode, *Meloidogyne incognita*. *Journal of Agricultural and Food Chemistry* 69(50), 15145–15156. DOI: 10.1021/acs.jafc.1c06500.

Nicaise, V., Roux, M. and Zipfel, C. (2009) Recent advances in PAMP-triggered immunity against bacteria: Pattern recognition receptors watch over and raise the alarm. *Plant Physiology* 150(4), 1638–1647. DOI: 10.1104/pp.109.139709.

Oliveira, J.T.A., Andrade, N.C., Martins-Miranda, A.S., Soares, A.A., Gondim, D.M.F. *et al*. (2012) Differential expression of antioxidant enzymes and PR-proteins in compatible and incompatible interactions of cowpea (*Vigna unguiculata*) and the root-knot nematode *Meloidogyne incognita*. *Plant Physiology and Biochemistry* 51, 145–152. DOI: VV10.1016/j.plaphy.2011.10.008.

Oota, M., Tsai, A.Y.L., Aoki, D., Matsushita, Y., Toyoda, S. *et al*. (2020) Identification of naturally occurring polyamines as root-knot nematode attractants. *Molecular Plant* 13(4), 658–665. DOI: 10.1016/j.molp.2019.12.010.

Oota, M., Toyoda, S., Kotake, T., Wada, N., Hashiguchi, M. *et al*. (2022) Rhamnogalacturonan-I as a nematode chemoattractant from *Lotus corniculatus L*. super-growing root culture. *Frontiers in Plant Science* 13, 1008725. DOI: 10.3389/fpls.2022.1008725.

Ozalvo, R., Cabrera, J., Escobar, C., Christensen, S.A., Borrego, E.J. *et al*. (2014) Two closely related members of *Arabidopsis* 13-lipoxygenases (13-LOXs), LOX3 and LOX4, reveal distinct functions in response to plant-parasitic nematode infection. *Molecular Plant Pathology* 15, 319–332. DOI: 10.1111/mpp.12094.

Park, M.G., Choi, J., Hong, Y.S., Park, C.G., Kim, B.G. *et al*. (2020) Negative effect of methyl bromide fumigation work on the central nervous system. *PLoS ONE* 15, e0236694. DOI: 10.1371/journal.pone.0236694.

Pegard, A., Brizzard, G., Fazari, A., Soucaze, O., Abad, P. *et al*. (2005) Histological characterization of resistance to different root-knot nematode species related to phenolics accumulation in *Capsicum annuum*. *Phytopathology* 95, 158–165. DOI: 10.1094/PHYTO-95-0158.

Ribeiro, D.G., Mota, A.P.Z., Santos, I.R., Arraes, F.B.M., Grynberg, P. *et al*. (2022) NBS-LRR-WRKY genes and protease inhibitors (PIs) seem essential for cowpea resistance to root-knot nematode. *Journal of Proteomics* 261, 104575. DOI: 10.1016/j.jprot.2022.104575.

Rutter, W.B., Franco, J. and Gleason, C. (2022) Rooting out the mechanisms of root-knot nematode–plant interactions. *Annual Review of Phytopathology* 60, 43–76. DOI: 10.1146/annurev-phyto-021621-120943.

Sasser, J. and Freckman, D. (1986) A world perspective on nematology – the role of the society. *Journal of Nematology* 18, 596.

Seo, E., Kim, S., Yeom, S.I. and Choi, D. (2016) Genome-wide comparative analyses reveal the dynamic evolution of nucleotide-binding leucine-rich repeat gene family among solanaceae plants. *Frontiers in Plant Science* 7, 1205. DOI: 10.3389/fpls.2016.01205.

Sharma, N., Khanna, K., Jasrotia, S., Kumar, D., Bhardwaj, R. *et al*. (2023) Metabolites and chemical agents in the plant roots: An overview of their use in plant-parasitic nematode management. *Nematology* 25(3), 243–257. DOI: 10.1163/15685411-bja10220.

Shi, Q., Mao, Z., Zhang, X., Ling, J., Lin, R. *et al*. (2018) The novel secreted *Meloidogyne incognita* effector MiISE6 targets the host nucleus and facilitates parasitism in *Arabidopsis*. *Frontiers in Plant Science* 9, 252. DOI: 10.3389/fpls.2018.00252.

Shivakumara, T.N., Dutta, T.K. and Rao, U. (2018) A novel in vitro chemotaxis bioassay to assess the response of *Meloidogyne incognita* towards various test compounds. *Journal of Nematology* 50(4), 487–494. DOI: 10.21307/jofnem-2018-047.

Song, L.X., Xu, X.C., Wang, F.N., Wang, Y., Xia, X.J. *et al*. (2018) Brassinosteroids act as a positive regulator for resistance against root-knot nematode involving RESPIRATORY BURST OXIDASE HOMOLOG-dependent activation of MAPKs in tomato. *Plant Cell Environ* 41, 1113–1125. DOI: 10.1111/pce.12952.

Topalović, O., Hussain, M. and Heuer, H. (2020) Plants and associated soil microbiota cooperatively suppress plant-parasitic nematodes. *Frontiers in Microbiology* 11, 313. DOI: 10.3389/fmicb.2020.00313.

Van Schie, C.C.N. and Takken, F.L.W. (2014) Susceptibility genes 101: How to be a good host. *Annual Review of Phytopathology* 52, 551–581. DOI: 10.1146/annurev-phyto-102313-045854.

Veech, J.A. (1982) Phytoalexins and their role in the resistance of plants to nematodes. *Journal of Nematology* 14, 2–9.

Verma, A., Lee, C., Morriss, S., Odu, F., Kenning, C. *et al*. (2018) The novel cyst nematode effector protein 30D08 targets host nuclear functions to alter gene expression in feeding sites. *The New Phytologist* 219(2), 697–713.

Veronico, P., Paciolla, C., Pomar, F., Leonardis, S., Garcia-Ulloa, A. *et al*. (2018) Changes in lignin biosynthesis and monomer composition in response to benzothiadiazole and root-knot nematode *Meloidogyne incognita* infection in tomato. *Journal of Plant Physiology* 230, 40–50. DOI: 10.1016/j.jplph.2018.07.013.

Veronico, P., Rosso, L.C., Melillo, M.T., Fanelli, E., Luca, F. *et al*. (2022) Water stress differentially modulates the expression of tomato cell wall metabolism-related genes in *Meloidogyne incognita* feeding sites. *Frontiers in Plant Science* 13, 817185. DOI: 10.3389/fpls.2022.817185.

Wang, G., Hu, C., Zhou, J., Liu, Y., Cai, J. *et al*. (2019) Systemic root-shoot signaling drives jasmonate-based root defense against nematodes. *Current Biology* 29(20), 3430–3438. DOI: 10.1016/j.cub.2019.08.049.

Wang, J., Ding, Z., Bian, J., Bo, T. and Liu, Y. (2021) Chemotaxis response of *Meloidogyne incognita* to volatiles and organic acids from root exudates. *Rhizosphere* 17, 100320. DOI: 10.1016/j.rhisph.2021.100320.

Warmerdam, S., Sterken, M.G., Sukarta, O.C.A., Schaik, C.C., Oortwijn, M.E.P. *et al*. (2020) The TIR-NB-LRR pair DSC1 and WRKY19 contributes to basal immunity of *Arabidopsis* to the root-knot nematode *Meloidogyne incognita*. *BMC Plant Biology* 20, 73. DOI: 10.1186/s12870-020-2285-x.

Warnock, N.D., Wilson, L., Canet-Perez, J.V., Fleming, T., Fleming, C.C. *et al*. (2016) Exogenous RNA interference exposes contrasting roles for sugar exudation in host-finding by plant pathogens. *International Journal for Parasitology* 46(8), 473–477. DOI: 10.1016/j.ijpara.2016.02.005.

Wu, C.H., Abd-El-Haliem, A., Bozkurt, T.O., Belhaj, K., Terauchi, R. *et al*. (2017) NLR network mediates immunity to diverse plant pathogens. *Proceedings of the National Academy of Sciences of the United States of America* 114(30), 8113–8118. DOI: 10.1073/pnas.1702041114.

Wuyts, N., Lognay, G., Swennen, R. and De Waele, D. (2006a) Nematode infection and reproduction in transgenic and mutant *Arabidopsis* and tobacco with an altered phenylpropanoid metabolism. *Journal of Experimental Botany* 57(11), 2825–2835. DOI: 10.1093/jxb/erl044.

Wuyts, N., Swennen, R. and De Waele, D. (2006b) Effects of plant phenylpropanoid pathway products and selected terpenoids and alkaloids on the behaviour of the plant-parasitic nematodes *Radopholus similis*, *Pratylenchus penetrans* and *Meloidogyne incognita*. *Nematology* 8(1), 89–101. DOI: 10.1163/156854106776179953.

Yang, J.-W., Park, S.-U., Lee, H.-U., Nam, K.J., Lee, K.-L. *et al*. (2023) Differential responses of antioxidant enzymes and lignin metabolism in susceptible and resistant sweet potato cultivars during root-knot nematode infection. *Antioxidants* 12(6), 1164. DOI: 10.3390/antiox12061164.

Yu, J., Yuan, Q., Chen, C., Xu, T., Jiang, Y. *et al*. (2024) A root-knot nematode effector targets the *Arabidopsis* cysteine protease RD21A for degradation to suppress plant defense and promote parasitism. *The Plant Journal* 118(5), 1500–1515. DOI: 10.1111/tpj.16692.

Zhang, B., Yang, Y., Wang, J., Ling, X., Hu, Z. *et al*. (2015) A CC-NBS-LRR type gene *GHNTR1* confers resistance to southern root-knot nematode in *Nicotiana benthamiana* and *Nicotiana tabacum*. *European Journal of Plant Pathology* 142(4), 715–729. DOI: 10.1007/s10658-015-0646-3.

Zhang, L., Li, S., Shan, C., Liu, Y., Zhang, Y. *et al*. (2023) Integrated transcriptome and metabolome analysis revealed that flavonoids enhanced the resistance of *Oryza sativa* against *Meloidogyne graminicola*. *Frontiers in Plant Science* 14, 1137299. DOI: 10.3389/fpls.2023.1137299.

Zhong, X., Su, G., Hao, L., Chen, H., Li, C. *et al*. (2024) Foliar application of glycine-functionalized nanopesticides for effective prevention and control of root-knot nematodes via a targeted delivery strategy. *Pest Management Science* 80(4), 2120–2130. DOI: 10.1002/ps.7948.

5 Role of Effectors and Elicitors in the Formation of Nematode Disease Complexes

Abdou Lahat Mbaye[1], Aliou Moussa Diouf[1], Zineb Rchiad[1], Zineb Belabess[2], Mustapha Barakate[1,3] and Rachid Lahlali[1,4]*

[1]Mohammed VI Polytechnic University, Ben Guerir, Morocco; [2]Plant Protection Laboratory, Regional Center of Agricultural Research of Meknes, Meknes, Morocco; [3]Laboratory of Microbial Biotechnology, Cadi Ayyad University, Marrakech, Morocco; [4]Department of Plant Protection, Ecole Nationale d'Agriculture de Meknès, Meknes, Morocco

Abstract

Nematode disease complexes are the result of dynamic interactions between nematode effectors and plant triggers. These interactions can be affected by environmental conditions including soil quality, plant vitality and the presence of other microorganisms. A better understanding of novel approaches to crop protection and nematode management can be achieved by researching the molecular mechanisms underlying these interactions, and understanding the dynamic interplay between effectors and elicitors is crucial for unravelling the molecular mechanisms underpinning nematode pathogenesis and for developing effective control strategies against phytopathogenic nematodes. This chapter reviews our current understanding of effectors and elicitors, investigates the emerging trends and current challenges in elucidating their roles in the formation of nematode disease complexes, and discusses their implications for plant–nematode interactions and crop protection.

Introduction

Plant-parasitic nematodes are diminutive roundworms that inflict considerable damage to crop plants (Tileubayeva *et al.*, 2021; Khan, 2023a), and the global agricultural industry faces a major challenge from plant-parasitic nematodes which can cause important economic losses by affecting the productivity of all kinds of crops (Khan, 2023b). Plant-parasitic nematodes can establish complex relationships with their host plants, resulting in severe agricultural losses (Khan, 2008; Perry and Moens, 2013). To develop effective strategies for managing these pests and mitigating crop damage, it is crucial to understand the mechanisms underlying nematode pathogenicity and disease complex formation (Khan *et al.*, 2023). To facilitate parasitism, nematodes use a variety of effector molecules. Plant effectors are secreted into plant tissues during feeding, and they play a crucial role in manipulating host plants. A plant's defense responses can be suppressed, root cells can be altered and nutrient uptake can be modulated by

*Corresponding author: rlahlali@enameknes.ac.ma

 Nematode Disease Complexes in Agricultural Crops (ed. Mujeebur Rahman Khan)
DOI: 10.1079/9781800625228.0005

effectors. Numerous novel nematode effectors have been discovered in recent research that act at various stages of infection. Parasite success can be attributed to proteins such as HgSLP-1 in *Heterodera glycines* (Hu *et al.*, 2019) and Mi-CRT in *Meloidogyne incognita* (Truong *et al.*, 2021).

Elicitors, on the other hand, are mainly linked to triggering plant defense mechanisms. They consist of chemicals or molecular patterns that the plant's immune system identifies, leading to a sequence of defense reactions such as the production of antimicrobial compounds, reinforcement of the cell wall and programmed cell death (apoptosis) at the site of nematode infection (Meena *et al.*, 2022). Plant pattern recognition receptors (PRRs) are activated by known triggers, such as pathogen-associated molecular patterns (PAMPs) and damage-associated molecular patterns (DAMPs), which elicit an immune response (Goverse and Smant, 2014). For example, recognition of nematode-derived peptide elicitors such as Gp-RALF (Mamaeva *et al.*, 2023) by plant receptor kinases (RKs) can lead to the activation of defense responses such as reactive oxygen species (ROS) production and wall reinforcement (Goverse and Smant, 2014).

Nematode disease complexes are the result of dynamic interactions between nematode effectors and plant triggers. This interaction determines the outcome of plant–nematode interactions, ranging from effective parasitism to plant resistance. The balance among these factors can be swayed by diverse environmental conditions including soil quality, plant vitality and the presence of other microorganisms. This chapter aims to provide a comprehensive overview of the roles of effectors and elicitors in shaping nematode disease complexes. A better understanding of novel approaches to crop protection and worm management can be achieved by researching the molecular mechanisms underlying these interactions. Further, understanding the dynamic interplay between effectors and elicitors is crucial for unravelling the molecular mechanisms underpinning nematode pathogenesis and for developing effective control strategies against phytopathogenic nematodes. This chapter investigates the latest advances in elucidating the roles of effectors and elicitors in the formation of nematode disease complexes and discusses their implications for plant–nematode interactions and crop protection.

Nematode Effectors: Molecular Weapons of Parasitism

Types of effectors

In the context of nematodes, effectors refer to molecules or proteins secreted by these parasitic nematodes that manipulate host plants to facilitate infection and nutrient extraction. In order for a nematode and its host to develop a parasitic relationship, nematode effectors are essential. These effectors often disrupt host defense mechanisms, alter host cells and create a more favorable environment for nematode survival. Nematode effectors encompass a variety of specialized functions aimed at manipulating host plants for their benefit. Cell wall degraders such as cellulases and pectate lyases break down plant cell walls, aiding the infiltration of substances into plant tissues (Rehman *et al.*, 2016; Chen *et al.*, 2021). Suppression of plant immunity effectors subvert host defense mechanisms by inhibiting responses and interfering with signaling pathways (Jones, 2006). Gland cell effectors are directly secreted into host tissue to modulate host cell structures and function (Mitchum *et al.*, 2013). Effectors imitate host proteins to manipulate host processes, including key signaling pathways (Vieira and Gleason, 2019; Zhang *et al.*, 2020). Effector proteins for gall formation illustrate the complex tactics nematodes use to take advantage of and colonize the host plant. They also facilitate root invasion, obstruct defensive mechanisms in plants and rewire root cells to produce unique nursery cells. These proteins play a role in the development and maintenance of galls, which act as a store of nutrients for nematodes (Mejias *et al.*, 2019).

Mechanisms of effector secretion

The molecular mechanisms of effector secretion in nematode infections involve a complex interaction between the nematode parasite and its host (Gang and Hallem, 2016). Nematodes are a diverse group of worms that include both free-living and parasitic species. Effector proteins are essential in the complex tactics that parasitic nematodes have developed to control their hosts. Effector proteins are molecules secreted by parasites that modulate host cell functions, facilitating infection (Oosterbeek *et al.*, 2021). In nematodes, secretion of effectors is involved in suppressing the host immune response, altering host cell physiology and promoting the establishment of parasitic associations (Quentin *et al.*, 2013).

To initiate infection, nematodes release effectors through various mechanisms, including stylet secretion, glandular secretion, exosome-mediated secretion and mucin-like secretions (Lilley *et al.*, 2018; Sato *et al.*, 2019). The migration patterns of phytonematodes penetrating plant tissues differ, with some, such as the root-knot nematode, employing a stealthy intercellular migration, while most endoparasites exhibit a forceful intracellular migratory pattern that induces significant damage, prompting a robust plant response. Variations in host root gene expression in response to these various infection techniques have begun to be documented in recent research (Kyndt *et al.*, 2012). It is believed that the plant's early defense against invading nematodes is triggered by conserved PAMPs or DAMPs; however, the precise patterns for phytonematodes have not yet been determined. By dissolving parts of the plant cell wall, effectors released during the migratory phase of infection are essential in enabling nematode penetration and migration (Mitchum *et al.*, 2013; Huang, 2022). In addition, these effectors enable nematodes to suppress the plant's immune system (Lin *et al.*, 2016). Sedentary endoparasites eventually establish feeding sites near the vasculature. For feeding cells to grow and remain viable, plant immune responses must be consistently suppressed. The transcriptional patterns of feeding cells consistently show that sedentary endoparasites downregulate hormone signaling and defense genes during the development of feeding cells (Barcala *et al.*, 2010; Damiani *et al.*, 2012; Mitchum *et al.*, 2013). Effectors released during the sedentary stage of parasitism most likely work in tandem with cellular reprogramming to regulate the inhibition of plant defense. This reprogramming is essential for the formation of metabolically active feeding cells, which the nematode depends on for weeks to sustain its growth and development. However, in resistant cultivars, a hypersensitive response is triggered leading to death of the feeding cells, ultimately leading to nematode starvation and death (Williamson and Kumar, 2006; Balint-Kurti, 2019).

Host–effector interaction and functional roles of effectors in nematode pathogenicity

Nematodes and plants interact through a complicated interplay of molecular signals and effectors. Plant–nematode interactions are dynamic processes that involve the recognition and response of plants and nematodes (Jones *et al.*, 2013). Effector molecules are secreted into plant cells by plant-parasitic nematodes in order to influence host processes and promote parasitism (Table 5.1). Like other pathogens, plant-parasitic nematodes generate effectors that target the host hubs to overcome host defenses (Carella *et al.*, 2018). In plants, however, only a few of their direct targets are known and the precise mechanism of action of these chemicals remains largely unknown. Plant-parasitic nematode effectors may scavenge ROS by interacting with host proteins in response to PAMPs. By interacting with the catalytic subunit of thioredoxin reductase (AtFTRc) in the plant, Lin *et al.* (2016) reported that the *M. javanica* MjTTL5 scavenges ROS. The main roles of the effectors of root-knot nematodes and cyst nematodes are diverse host functions, which may be more directly linked to the establishment and functioning of the specialized feeding site. Cytoskeletal networks need to be significantly reorganized in order for feeding cells to be established, which is triggered by root-knot nematodes and cyst nematodes (de Almeida Engler *et al.*, 2005). Bellafiore *et al.* (2008) showed that the root-knot nematode can release related proteins and cytoskeletal components such as actin or tubulin. Hewezi *et al.* (2010) reported that the effector Hs10A06 from *Heterodera schachtii* interacts with spermidine synthase AtSPDS2 to promote spermidine synthesis and started ROS-scavenging activity upon oxidation of spermidine by polyamine

Table 5.1. Important effectors secreted by plant-parasitic nematodes into plant cells to manipulate host processes and facilitate their parasitism.

Nematode	Effector	Target	Function	Reference(s)
Meloidogyne javanica	MjTTL5	Interacts with a thioredoxin reductase catalytic subunit (AtFTRc)	Scavenges reactive oxygen species	Lin *et al.*, 2016
Meloidogyne graminicola	MgMO237	Interacts with several host pathogenesis-related proteins, a pathogenesis-related Bet v I family protein (OsBetvI), a 1,3-β-glucan synthase (OsGSC) and the cysteine-rich repeat secretory protein 55 (OsCRRSP55)	Suppresses pattern-triggered immunity	Chen *et al.*, 2018
Meloidogyne chitwoodi	Mc01194	Targets papain-like cysteine proteases (a key hub in plant immunity)	Promotes parasitism	Davies *et al.*, 2012
Meloidogyne incognita	'Passe-Muraille' peptide effector	Interacts with subunit 5 of the COP9 signalosome (CSN5)	Unknown	Kazan and Lyons, 2014
	MiPFN3	Binds actin	Alters actin filament structure to favor parasitism	de Almeida Engler and Favery, 2011
Heterodera schachtii	Hs19C07	Interacts with the auxin transporter LAX3	Modulates auxin signaling to facilitate feeding site formation	Lee *et al.*, 2011
	Hs10A07	Affects the expression of auxin-responsive factors	Modulates auxin signaling to facilitate feeding site formation	Hewezi *et al.*, 2015
	Hs10A06	Interacts with a spermidine synthase (AtSPDS2)	Enhances spermidine production and induces reactive oxygen species-scavenging activity	Hewezi *et al.*, 2010
	Hs25A01	Interacts with eIF-2bs	Increases susceptibility to *H. schachtii* infection	Pogorelko *et al.*, 2016
	Hs4E02	Targets papain-like cysteine proteases (a key hub in plant immunity)	Promotes parasitism	Pogorelko *et al.*, 2020
Heterodera glycines	Hg30C02	Targets a β–1,3-endoglucanase (AtPR2)	Increases susceptibility to cyst nematode infection	Hamamouch *et al.*, 2012

oxidase. Additionally, it has been found that certain pathogenesis-related proteins, which are crucial for the mechanism by which plants respond to pathogen invasion by producing antimicrobial proteins, are directly targeted by nematode effectors. The β−1,3-endoglucanase (AtPR2) that *H. glycines* Hg30C02 targets is inactivated in a mutant *Arabidopsis* line, which increases vulnerability to cyst nematode infection (Greco *et al.*, 2012). Through interactions with many host pathogenesis-related proteins, such as 1,3-β-glucan synthase (OsGSC), cysteine-rich repeat secretory protein 55 (OsCRRSP55) and a pathogenesis-related Bet v I family protein (OsBetvI), *M. graminicola* MgMO237 has been demonstrated to reduce pattern-triggered immunity (PTI) (Chen *et al.*, 2018). Plant–nematode effector interaction involves a sophisticated molecular dialogue, with nematode effectors playing a central role in modulating host responses.

Plant Elicitors: Plant Responses to Nematode Attack

Types of elicitors

Elicitors in the context of plant–nematode interactions refer to molecules or signals that induce a response in plants, leading to the activation of defense mechanisms against nematode attacks (Fig. 5.1). These elicitors play a crucial role in the molecular dialogue between plants and nematodes, influencing the outcome of the interaction. Different types of elicitors can trigger various defense responses in plants. The plant's capacity to detect the presence of nematodes and initiate defense mechanisms is mostly dependent on nematode-associated molecular patterns (NAMPs), effectors and effector-triggered immunity (ETI), DAMPs, plant hormones and secondary metabolites. Plant receptors identify PAMPs, which are nematode-derived chemicals that set off a defense mechanism (Choi and Klessig, 2016). When these effectors are recognized, a defense reaction known as ETI is triggered (Goode and Mitchum, 2022). DAMPs are plant chemicals that are exposed or secreted as a result of nematode invasion or feeding, alerting the plant to begin defense mechanisms (Sato *et al.*, 2019). Plant hormones are signaling molecules that control defense responses in response to nematode attacks. Examples of these chemicals are ethylene, salicylic acid and jasmonic acid (Sato *et al.*, 2019). Secondary metabolites are substances produced by plants that can either strengthen the plant's defenses against nematodes or have nematicidal effects (Divekar *et al.*, 2022).

Detection, recognition and induction of plant defense responses

Plants are able to identify particular molecular patterns known as PAMPs or microbe-associated molecular patterns (MAMPs), which are linked to pathogens or bacteria. PTI is the term for the range of defensive reactions that are sparked by this recognition. PAMPs are molecules that have remained constant throughout the evolution of particular pathogen classes, enhancing the overall fitness of the life cycle of the infectious agent. PAMPs that have been identified comprise structural elements of pathogens and biomacromolecules, including proteins and nucleic acids (Gust *et al.*, 2017). In order to counteract PTI responses, viral infections release effectors into the host plants. These effectors weaken or eliminate the effectiveness of the plant's defensive mechanisms by changing plant metabolism, enhancing nutrient uptake and blocking PTI response activation. The second stage of recognizing pathogens involves intracellular proteins with nucleotide-binding domains and leucine-rich repeats (LRR), which are produced by R-genes. When these proteins detect effectors, they initiate a robust defensive response known as ETI (Lolle *et al.*, 2020). ETI frequently results in the hypersensitive response, a form of programmed cell death. Plants recognize PAMPs through germline-encoded receptors called PRRs. Membrane-located receptor-like kinases or receptor-like proteins are PRRs linked to PTI activation, with rare exceptions (Couto and Zipfel, 2016). RKs are composed of distinct structural components: an extracellular ligand-binding domain, a single membrane-spanning domain and a cytoplasmic kinase domain. PRRs such as LRR-RKs, lysine-motif RKs, lectin-RKs (LeC-RKs) and epidermal growth factor RKs are categorized into subclasses based on the structural characteristics of their extracellular domains. Some PRRs have the ability to recognize DAMPs, which are released from infected cells or produced by the host itself (Zipfel, 2014). Most, if not all, PRRs function as elements of multiprotein complexes that are found on the plasma membrane. Brassinosteroid-associated kinase 1 (BAK1), a member of the LRR-RK family, is one such PRR collaborator. BAK1 associates with LRR-type PRRs to produce receptor complexes, such as flagellin sensing 2 (FLS2), which recognizes the bacterial flagellin fragment flg22, a pathogen-associated molecular pattern. When FLS2 and BAK1 recognize flg22, they rapidly form heterodimers. The C-terminus of flg22 binds to FLS2 and interacts closely with BAK1, according to crystallographic studies, strengthening the stability of the FLS2–BAK1 relationship (Sun *et al.*, 2013). When PRRs identify PAMPs or DAMPs, they open the door to a variety of downstream signaling events. Among other things, there is activation of calcium-dependent and mitogen-associated protein kinases, rapid production of ROS, strengthening of the cell wall, and expression of genes related to immunity (Lolle *et al.*, 2020). These processes work together to prevent pathogen growth and development. While a lot of work has gone into investigating which PAMPs cause PRR-mediated PTI in various plants, not as much is known about PAMPs and receptors that are involved in interactions between plants and nematodes (Lolle *et al.*, 2020). Nematodes have several difficulties, such as their obligatory biotrophic lifestyle, asynchronous infection and lack of nematode transformation techniques.

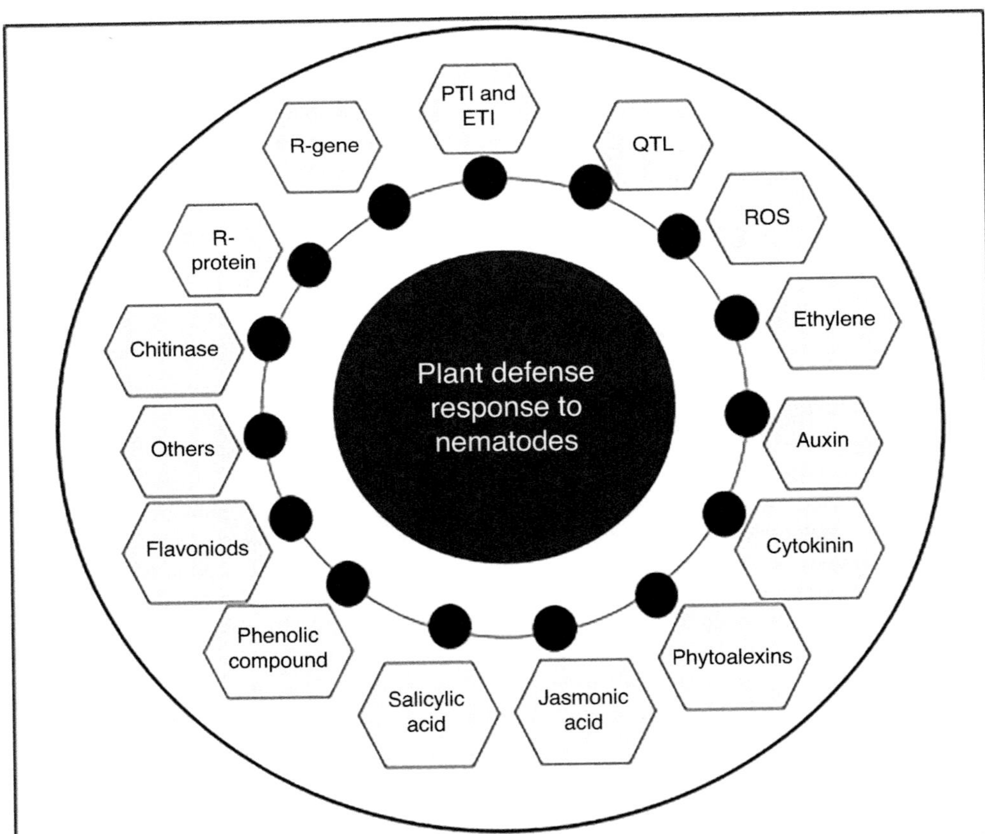

Fig. 5.1. Schematic diagram showing the critical role of several factors in plant defense responses to nematodes. ETI: effector-triggered immunity; PTI: pattern-triggered immunity; QTL: quantitative trait loci; ROS: reactive oxygen species.

Impact on nematode infection

Plant defense responses are influenced by elicitors that trigger defense systems against pathogens, especially nematodes. The production of antimicrobial compounds, the strengthening of cell walls and the induction of systemic acquired resistance are some of these mechanisms. The important point to remember is that chitosan, jasmonic acid, salicylic acid and ROS can all have a substantial impact on nematode infection. Chitin, which is mostly present in nematode cell walls, is the source of chitosan, a well-known elicitor that has been demonstrated in multiple studies to be successful in inducing defense responses in plants against nematodes. In addition to being poisonous and known to impede the growth and development of fungi, chitosan has also been shown to be effective against bacteria, viruses, and other pests (El Hadrami *et al.*, 2010). It was discovered that treating tomato plants with chitosan increased the mortality of the larvae of *M. javanica* and decreased the hatching of eggs (El-Sayed and Mahdy, 2015). Moreover, it has been reported by El Hadrami *et al.* (2010) that chitosan elicits activities in host plants that result in a range of defense responses in response to microbial infections, such as the accumulation of phytoalexins, pathogen-related proteins and proteinase inhibitors, lignin synthesis and callose formation. Plant resistance to nematode infection can be increased by exogenous application of jasmonic acid or its analogues. The plant hormone jasmonic acid controls the plant body's defense mechanisms against a range of illnesses, including nematodes (Bali *et al.*, 2020). Jasmonic acid is a possible signaling molecule that influences various physiological and developmental processes during periods of environmental stress. Previous studies showed that in several economically important crops, jasmonic acid is critical to the basal defense system against root-knot nematodes (Bali *et al.*, 2020). Research has shown that inhibiting strigolactone production led to an increase in jasmonic acid accumulation, which subsequently decreased the level of infection by nematodes in rice (Lahari *et al.*, 2019). The use of a jasmonic acid biosynthesis inhibitor resulted in a reduction in the resistance to soybean cyst nematode provided by *Rhg1*, a gene in soybeans. This suggests that jasmonic acid may have a crucial function in the *Rhg1*-mediated defense response against this pest in soybeans (Guo *et al.*, 2019). Studies suggest that root-knot nematode elevates the expression of genes associated with biosynthesis and signal transduction pathways in tomatoes, leading to an increase in the levels of jasmonic acid and jasmonic acid-isoleucine (Shang *et al.*, 2019). Salicylic acid affects both the development of systemic acquired resistance after an infection by pests that causes necrosis and the hypersensitive reactions of plants to incompatible pathogens. Usually, when plants are in danger from hemi- and biotrophic diseases, salicylic acid is set off (Xie *et al.*, 2022). Furthermore, in order to clarify the functions of salicylic acid pathways in the relationship between rice and *Aphelenchoides besseyi*, it was found that cv. Tetep's defense mechanisms were achieved, at least in part, by triggering the production of defense genes associated to salicylic acid, especially during the flowering stage (Xie *et al.*, 2022). The inhibition of nematode infection, growth and reproduction led to the establishment of Tetep's resistance.

Interaction Between Effectors and Elicitors

Cross-talk between nematode effectors and plant elicitors

The result of the plant–nematode interaction is influenced by the intricate interplay between nematode effectors and plant elicitors (Ali *et al.*, 2018). To promote parasitism, nematodes release effectors, which are proteins or tiny molecules that alter the physiology and defense mechanisms of the host plant (Fig. 5.2). These effectors can modify the immune system of plants in several ways, such as by decreasing defenses or changing hormone signaling pathways (Zhang *et al.*, 2022a). Conversely, chemicals that a plant produces in reaction to nematode infection, or other stresses, are known as plant elicitors. These elicitors have the ability to initiate defense reactions, including the synthesis of antimicrobial substances, the fortification of cell walls and the activation of genes linked to defense (Pršić and Ongena, 2020). Nematode effectors can inhibit

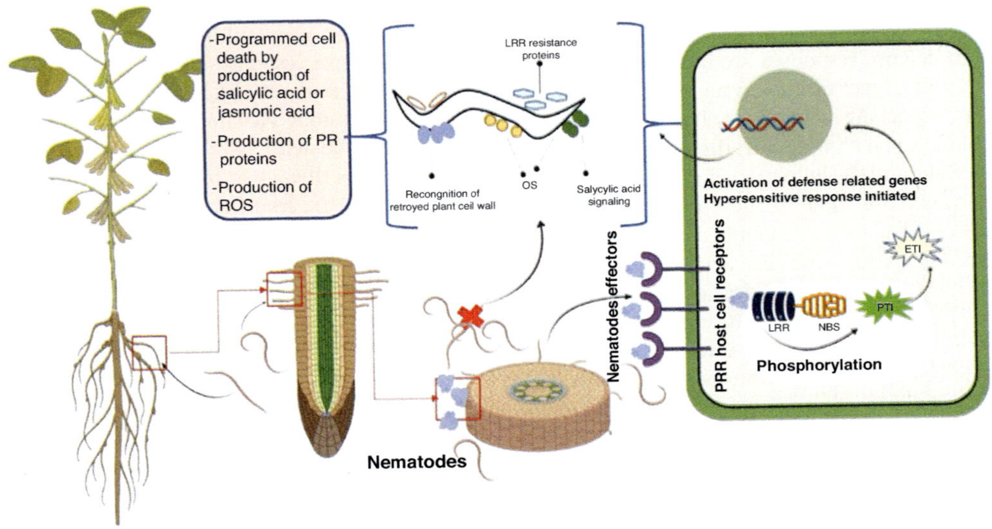

Fig. 5.2. Diagram showing the chemical reaction and interactions between plant and nematodes. The innate immune system of the plant identifies nematode elicitors and mounts a defense process. A hypersensitive reaction known as a localized cell death reaction may occur in the plant at the location of nematode infection. Elicitors have the ability to obstruct a plant's communication channels. ETI: effector-triggered immunity; LRR: leucine-rich repeat; NBS: nucleotide-binding site; PTI: pattern-triggered immunity; ROS: reactive oxygen species; PR: pathogenesis-related; PRR: pattern recognition receptor.

elicitor-triggered reactions in the plant's defense mechanisms, enabling the nematode to evade detection and establish feeding sites. Nematode effectors, for instance, can activate the generation of ROS or obstruct signaling pathways that trigger the activation of defense genes (Jagdale *et al.*, 2021). Additionally, these effectors can control the plant's hormone signaling pathways, which include ethylene, salicylic acid and jasmonic acid. These hormones are essential for controlling the responses of plants to defense. Nematodes have the ability to suppress or manipulate the immunological responses of plants by modifying hormone levels or signals (Pogorelko *et al.*, 2020). Plant defense responses can occasionally be triggered inadvertently by nematode effectors. This may be the result of these effectors being recognized as PAMPs by the plant's immune receptors. By inducing protective reactions, this may stop nematode parasitism. Nematodes and plants at the infection site could compete with one another for nutrients and water. The balance of these resources may have an impact on how nematodes and plants interact as well as the effectiveness of defensive mechanisms (Migunova and Sasanelli, 2021). The development of techniques to increase plant resistance to nematodes requires an understanding of the interactions between nematode effectors and plant elicitors. It might be possible for scientists to create fresh strategies for reducing nematode infestations and raising crop yields by adjusting these relationships.

Co-evolutionary dynamics between nematodes and host plants

Nematodes and their host plants engage in co-evolutionary dynamics that are shaped over time by a complex interaction of genetic, ecological and evolutionary factors. Plants have created defense mechanisms to fend off nematode infections, whereas nematodes have evolved a variety of techniques for parasitizing plants. A co-evolutionary arms race, in which each organism adapts in reaction to the other's adaptations, may result from these interactions.

Nematodes have evolved complex parasitic strategies, such as the ability to avoid or inhibit plant defensive responses, build feeding structures and secrete effectors that alter plant physiology. To defend against the nematode invasion, plants have also evolved a variety of defense mechanisms. These include immunological responses that involve the activation of genes linked to defense, chemical defenses such as the synthesis of poisonous chemicals and physical barriers such as thick cell walls. Nematodes and plants both display genetic variety, which serves as the building block for evolutionary change. Selection for features that increase nematode parasitism and those that make plants resistant to or susceptible to nematode infection is made possible by genetic variation. Gene-for-gene interactions can occasionally occur in plant–nematode interactions when plant-specific R-genes identify nematode effectors and trigger defensive mechanisms. In response, nematodes produce novel effectors to evade plant immune system detection. The dynamics of their co-evolution can also be shaped by the geographic distribution and ecology of both host plants and nematodes. The strength and result of interactions between plants and nematodes can be influenced by variables such as temperature, soil type and diversity of host plants. Populations may diverge into different ecotypes or species as a result of the co-evolutionary dynamics between nematodes and host plants. In addition to genetic drift and other evolutionary processes, the host plant's selection pressure may also be the cause of this divergence.

Formation of Nematode Disease Complexes

Factors involved

Soil is a complex ecosystem where microorganisms and plants form a complex and dynamic relationship that significantly influences plant growth, health and overall ecosystem functioning. In fact, the plant rhizosphere harbors a versatile composition of microorganisms that can impact their development either positively or negatively. Phytonematodes are one of those groups that associate with plants under different environments. While some are considered beneficial to plant growth and development, others feed on plants and are responsible for about 10–25% crop losses worldwide. Inside the group of phytonematodes, endoparasitic nematodes are the most detrimental as they infect, feed and reproduce inside the tissues of their host organisms, causing heavy physical damage to their roots (Khan, 2007; Manzanilla-López *et al.*, 2013). The group of endoparasitic nematodes includes root-knot nematodes, i.e. *Meloidogyne* spp., which invade plant roots, causing the formation of characteristic galls or knots, cyst nematodes, i.e. *Heterodera* and *Globodera* spp., and lesion nematodes, i.e. *Pratylenchus* spp. Despite singularly causing serious threat to plant health, phytonematodes can supplementally facilitate the invasion of plants by other pathogenic microorganisms such as fungi, bacteria or viruses and form what are referred to as disease complexes (Khan, 1993).

A nematode disease complex is the association between a nematode pathogen and another pathogenic organism that results in a synergistic interaction, which may be positive, antagonistic or neutral (Back *et al.*, 2002). A positive association is observed when the damage caused on the plant exceeds the sum of the individual damage caused by the involved pathogenic species, whereas an antagonistic effect would result in less damage than expected from the association of the two pathogens (Khan *et al.*, 2023). The formation of nematode disease complexes involves different factors that facilitate the infection of plants by the second pathogenic organism after the establishment of the nematode infection. One of the factors is the creation of wounds on plant roots during infection, which is very typical of plant-parasitic nematodes (Khan, 2008). Wounding plays a crucial role in the establishment of the fungal infection, serving as an entry point for pathogenic hyphae invasion. Another factor is the modification of the infected plant tissue by nematodes, resulting in the presence of high substrates for fungal growth and reproduction (Khan and Sharma, 2020). Root exudates also play an important role in the formation of disease complexes. Some nematodes can induce changes in root exudates, promoting the germination, growth and reproduction of the pathogenic fungal microbiome in the rhizosphere and in addition altering the microbial community structure by reducing the population of beneficial microorganisms playing a decisive role in disease control (Ravichandra, 2014).

Role of soil environment and host genotype

The development and severity of nematode disease complexes are influenced by the interplay between the environment and the genotype of the host plant. The occurrence and development of plant diseases are affected by a variety of factors that play key roles in disease expression. These factors can be classified into three different variables: pathogen presence, host plant susceptibility and environmental conditions. Each of these factors plays a key role for an infection to occur.

Host plants vary in their susceptibility to nematode infestations because their genetic and physiological characteristics may influence disease resistance or susceptibility. Therefore, some plant genotypes exhibit resistance mechanisms, limiting nematode penetration, reproduction or movement within the root tissues, whereas some others do not possess an innate resistance, exposing them to be disease susceptible. It is estimated that 5–20% of products resulting from photosynthesis are released as root exudates in the rhizosphere (Hütsch *et al.*, 2002). These rhizo-depositions play a crucial role in shaping the structure and composition of microbial communities in the root zone. These exudates, consisting of various organic compounds, serve as a source of nutrients and signaling molecules, influencing the structure and function of the microbial community. Moreover, some root exudates act as chemoattractants and can prime the plant's immune response by influencing the recruitment of specific microbial communities that assist in plant defense against pathogens. They can also act directly on the nematode population by producing nematode-defensive compounds that can be repellent or nematicidal (Sikder and Vestergård, 2020).

On the other hand, root exudates may also elicit some gene expression and favor infection of phytonematodes. For instance, Duarte *et al.* (2015) investigated the potential of tomato root exudates to monitor the expression of several parasitism candidate genes identified in *Meloidogyne hispanica* by exposing 1000 *M. hispanica* second-stage juveniles to exudates from 4-week-old tomato plants for 4 hour. The gene expression levels were further analyzed with quantitative polymerase chain reaction (qPCR), and the authors noticed that several parasitism genes were upregulated by the plant root exudates. From this perspective, we can appreciate the role of the host genotype as being a main element of resistance or predisposition to nematode infection that could later turn into an association with other pathogens.

The surrounding environment must also be considered in the complexity of the disease associations. Soil texture, pH and nutrient levels play a crucial role in nematode distribution and activity. Certain nematode species thrive in specific soil types, and variations in soil conditions can affect their abundance. Throughout the release of organic acids during root exudation, the pH of the rhizosphere changes and modulates the shapes of the microbial communities inhabiting the root zone. Indeed, different nematode species exhibit specific pH preferences. For example, certain plant-parasitic nematodes thrive in acidic soils, while others prefer neutral to alkaline conditions (Perry and Moens, 2006). The populations of *H. glycines* and *M. incognita* in *Glycine max* roots were reported to be reduced in conditions of low pH (Melakeberhan *et al.*, 2004). Similar observations were also reported by another study, which, in addition to different pH conditions, also investigated the effect of different inorganic salt gradients (Hua *et al.*, 2020). Therefore, by controlling several parameters in the root environment, plants may affect the prevalence of phytopathogenic nematode. Soil texture also contributes to the relative abundance of plant-pathogenic nematodes as well as their activity. Jaraba *et al.* (2013) studied the impact of soil texture on *M. incognita–Thielaviopsis basicola* disease complex on cotton using a native silt loam soil (48% sand) and four different artificial soil types (53%, 70%, 74% and 87% sand). They observed an increase of galls on cotton roots with the increase of soil sand; however, they also noticed that soil with 87% sand content affected the nematode–fungus disease complex by reducing the fungal root colonization ability, its reproduction and its survival (Jaraba *et al.*, 2013). *M. incognita* seems to prefer sandy soil for its growth, development and infectivity. A similar observation was also obtained in another study conducted on carrot (Kim *et al.*, 2017).

Tools and Techniques for Studying Effectors and Elicitors

Molecular and biochemical methods

Numerous molecular and biochemical techniques are used in the study of effectors and elicitors in order to recognize, define and work with these molecules. For this study, a few standard tools and techniques such as mass spectrometry, RNA interference (RNAi), protein expression and purification, gene expression analysis, protein–protein interaction studies, and metabolomics and bioinformatics analysis can be used. It is possible to measure the expression of effector genes in nematodes or elicitor-induced genes in plants using methods such as quantitative real-time PCR (qRT-PCR). This may shed light on how effector and elicitor expression are regulated and timed (Chen *et al.*, 2021). The recombinant method can be utilized to create plant elicitors for biochemical research or nematode effectors for protein expression. Protein expression and purification can be examined with methods such as Western blotting and sodium dodecyl sulfate–polyacrylamide gel electrophoresis (SDS-PAGE) (Li *et al.*, 2015). Yeast two-hybrid (Y2H) and co-immunoprecipitation are two techniques that can be used to explore the interaction between effectors and host plant proteins in the study of the protein–protein interaction (Tanowitz and von Zastrow, 2004; Xing *et al.*, 2016). Understanding the molecular processes underlying pathogenicity may be possible through these interactions. Proteins in nematodes and plants, including effectors and elicitors, can also be identified and measured using mass spectrometry. Identification of novel effectors/elicitors and their targets can be aided by mass spectrometry (De Meyer *et al.*, 2020). To investigate the role of nematode genes in plants, RNAi can be utilized to suppress the expression of nematode genes, including effector genes. In a similar vein, RNAi can be employed to investigate the function of certain plant genes in elicitor reactions (Ding *et al.*, 2021). Metabolomics is an additional method that is employed. Plant metabolite profiles can be studied using metabolomics to examine how they vary in response to nematode infection or elicitor treatment. This may shed light on the metabolic processes underlying plant defense (Castro-Moretti *et al.*, 2020). Bioinformatics analysis is a key tool utilized in this investigation. Using sequence similarity to predict and evaluate effector and elicitor sequences, as well as to find conserved motifs and infer their function, are all possible with bioinformatics (Zhao *et al.*, 2024).

Functional assays

Studying the functions and actions of nematode effectors and elicitors in host–parasite interactions requires the use of functional experiments. By using these tests, scientists can determine the mechanisms of action of these compounds and evaluate their impact on host plants. The expression of particular genes in nematodes or host plants can be silenced via RNAi (Banerjee *et al.*, 2017). By reducing the expression of nematode effectors and elicitors and evaluating the ensuing phenotypic alterations in host plants, a technique known as gene silencing can be utilized to investigate the role of these agents. Introducing genes encoding nematode effectors or elicitors into host plants for transient expression is a different strategy known as transient expression testing. This makes it possible to evaluate how these chemicals affect the physiology of plants and their defense mechanisms (Tyurin *et al.*, 2020). In phenotypic assays, different phenotypic characteristics of nematode-infected plants are evaluated, including disease symptoms, growth rates and root shape. These tests can shed light on how nematode effectors and elicitors affect the growth and well-being of plants due to mono- or multipathogenic infections. Furthermore, biochemical assays are employed to investigate specific biochemical pathways or processes in plants infected with nematodes. Enzyme activity assays, for instance, can be used to evaluate how nematode effectors affect plant metabolism (Castro-Moretti *et al.*, 2020). Alongside other tools and methodologies, these functional assays provide crucial insights into the roles and mechanisms of nematode effectors and elicitors in host–parasite interactions and the formation of disease complexes.

Omics approaches

The investigations on biological molecules or processes through omics techniques has transformed the field of nematode effector and elicitor research. These methodologies facilitate the examination of extensive datasets by researchers in order to discern, describe and comprehend the roles of these compounds. Nematode effectors and elicitors are studied using several of the major omics techniques, including transcriptomics, proteomics, metabolomics and genomics. A whole organism's genome is studied in genomics. The field of genomics is utilized to determine the genes that encode nematode effectors and elicitors, forecast their activities and examine their evolutionary connections. Finding distinctive and conserved effectors among various nematode species is another application for comparative genomics (Viney, 2017). The study of every RNA transcript that a cell or organism produces is known as transcriptomics. Analysis of the transcriptomes of nematodes and host plants during infection is frequently done using RNA sequencing (RNA-seq). This method can reveal information on the regulation and expression patterns of differentially expressed genes, including those that encode effectors and elicitors (Lefoulon *et al.*, 2022). The study of all proteins expressed by a cell or organism is known as proteomics. In plant tissues infected with nematodes, proteins, including nematode effectors and elicitors, are identified and quantified using mass spectrometry-based proteomics. This method can shed light on these molecules' composition, quantity and post-translational changes (Walther and Mann, 2010). The study of all of the metabolites found in a cell or organism is the focus of the metabolomics approach. By utilizing metabolomics, it is possible to detect alterations in the metabolite profiles of plants infected with nematodes (Castro-Moretti *et al.*, 2020). This information may be utilized to gain understanding of the metabolic pathways impacted by the infection and the role of effectors and elicitors in regulating plant metabolism.

Current Challenges and Future Directions: Emerging Trends in Effector and Elicitor Research

Nematodes represent a diverse group of organisms with significant ecological and agricultural implications. Among them, plant-parasitic nematodes pose a substantial threat to global crop production by causing extensive damage to a wide variety of plants. In recent years, there has been increasing interest in unravelling the molecular mechanisms underlying the interaction between nematodes and their plant hosts. Central to this research is the exploration of nematode effectors, molecules secreted by these worms to manipulate host cellular processes and establish parasitism.

Nematode effectors play a pivotal role in the parasitic success of these organisms. Researchers have been unravelling the complex interplay between nematode effectors and plant defense with the aim of identifying key molecular players that contribute to the establishment of infection. A key trend in nematode effector research involves the integration of multi-omics approaches, encompassing genomics and transcriptomics analysis. This holistic strategy provides a comprehensive view of the complex interactions between nematodes and their host plants. By deciphering the entire molecular landscape, researchers gain insights into the dynamic changes in gene expression during infection, allowing for a more nuanced understanding of nematode parasitism strategies.

Comparative genomics analyses across different nematode species are uncovering conserved effector repertoires as well as species-specific adaptations. For instance, Danchin *et al.* (2013) conducted a comparative genomic study between phytopathogenic and non-pathogenic nematodes to develop more specific control strategies against root-knot nematodes. They identified 15,952 nematode genes that were conserved in the genomes of phytopathogenic nematodes but absent in the other groups. Moreover, they identified 993 effector-like proteins from this set of unique genes that were later subset to identify compatible genes for RNA silencing experiments. Of the 16 RNA-silenced genes, 12 highly impacted the parasitism of the nematode species (Danchin *et al.*, 2013). This study showed that understanding the genetic diversity and evolution of nematode effectors is crucial for predicting their effectiveness against different host plants and for developing targeted control strategies. The use of comparative genomics has become a milestone in nematode effector research with a particular focus on predicting infection mechanisms that could serve in developing robust technics in nematode disease management (Shinya *et al.*, 2021; Tritten *et al.*, 2021; van Steenbrugge *et al.* 2021, 2023).

Still, in the quest for sustainable and eco-friendly approaches to control nematode infestations, researchers have explored the concept of host-induced resistance. Host-induced resistance has been powered by the introduction of RNAi, which allows access to a powerful and specific tool for controlling plant-pathogenic nematodes. RNAi technology can disrupt essential biological processes in pathogenic nematodes and consequently impair their ability to infect plants. This technology entails introducing double-stranded RNA (dsRNA) molecules, which are complementary to specific nematode genes, into the host plant. These dsRNA molecules undergo processing by the nematode's RNAi machinery, converting them into short interfering RNAs (siRNAs). These siRNAs then act as guides for degrading messenger RNAs (mRNAs) of targeted parasitism genes, ultimately resulting in the silencing of these genes (Davies *et al.*, 2012). The use of transcriptome analysis has allowed the identification of genes with essential roles in plant infection (Zhou *et al.*, 2020; Channale *et al.*, 2021) and these genes have later served as models for RNAi designing.

For instance, the *Mi-msp-1* effector gene has been used in *M. incognita* for RNAi-based host-induced gene silencing on eggplant (Chaudhary *et al.*, 2019a, 2019b). This effector is known to play a crucial role during the infection process and the pathogenicity of plant-pathogenic nematodes (Chaudhary *et al.*, 2019a, 2019b). Therefore, by generating transgenic eggplants carrying a copy of the *Mi-msp-1* RNAi transgene, Chaudhary *et al.* (2019a,2019b) have been able to strengthen their resistance to nematode infection and protect the plant in its early stages. Similar studies with promising results were also reported on soybean infected by cyst nematode where the *Hg-rps23*, *Hg-snb1* and *Hg-cpn1* genes in *H. glycines* have been used to design RNAi transgene expressed by soybean to control nematode infection (Zhang *et al.*, 2022b). RNAi technology for plant-pathogenic nematode control holds immense potential and is being widely adopted in nematode research (Hada *et al.*, 2020; Joshi *et al.*, 2020; Kong *et al.*, 2022). However, continued research is essential to refine and optimize its application to the development of robust and sustainable RNAi-based solutions for managing nematode-induced crop losses.

The emerging trends in nematode effector research represent a dynamic and rapidly evolving field with profound implications for agriculture and crop protection. By unravelling the molecular intricacies of the plant–nematode interaction, researchers are paving the way for innovative and sustainable approaches to manage nematode-induced crop losses. As our understanding of nematode effectors continues to deepen, so too does the potential for developing effective strategies to mitigate the impact of these agricultural pests on global food security.

Conclusion and Future Prospects

Despite advancements in genomics and bioinformatics, there is still a considerable gap in our knowledge concerning the full complement of nematode effectors. Many undiscovered effectors likely exist across different nematode species, and unravelling their identities is a critical challenge (Jones *et al.*, 2013; Mitchum *et al.*, 2013). Furthermore, most of the effectors that have been discovered and studied are implicated in the early stage of infections (Leelarasamee *et al.*, 2018; Mejias *et al.*, 2020; Qin *et al.*, 2022); however, not much is known about secreted effectors during late-stage infections despite some great efforts being made towards this purpose (Rocha *et al.*, 2023).

The molecular details of how nematodes and fungi interact at the cellular and molecular levels are not fully elucidated. Therefore, a deeper understanding of the signaling pathways, gene expression changes and biochemical processes involved in these interactions is essential. Additionally, considering the high pressure exerted on host plants during the establishment of disease complexes, research should explore the specific mechanisms by which the host plant responds and how these interactions influence disease progression?

Another question to address is the impact of climate change on nematode pathogenicity and consequently on disease complex formation and effects on plants. Many studies are continuously being conducted on the impact of climate change on other plant pathogens (Daami-Remadi *et al.*, 2009; Parikka *et al.*, 2012; Liu and Liu, 2016; Alsamir *et al.*, 2017; Maharshi *et al.*, 2021) and the impacts are very alarming. Therefore, investigating how rising temperatures, altered precipitation patterns and other climatic factors influence disease is a critical area of research. Beyond that, research avenues should explore and capitalize on genetic resistance in plants against prevalent nematode disease complexes. Identifying and breeding crop varieties with dual resistance holds promise as a sustainable strategy for disease management.

Nematodes, plants and other microbes interact in a variety of ways to generate nematode disease complexes, and effectors and elicitors are crucial in coordinating the intricate molecular conversation that underlies parasitism. Nematode effectors are proteins or compounds that are released and function as molecular tools to alter host physiology, reduce immunological reactions and aid in the uptake of nutrients. In the end, these effectors promote nematode survival and reproduction by focusing on a variety of biological processes, including hormone signaling pathways, cell wall remodeling and defense gene control. On the other hand, plant elicitors released in reaction to nematode infection set off defense systems meant to fend off nematode parasitism. Nevertheless, complicated interactions between plant elicitors and nematode effectors can result in the development of disease complexes, in which the physiology of plants is altered by nematodes, providing an ideal environment for secondary infections by other pathogens. This underscores the critical importance of understanding the molecular processes that govern the interactions between nematodes and plants, and how these interactions influence the onset of illness. Functional assays, omics technologies and sophisticated imaging methods have all made significant advances possible in analyzing the functions of effectors and elicitors in the development of nematode disease complexes. In the end, creating long-term plans to lessen the negative effects of nematode diseases on agriculture and crop productivity requires a comprehensive understanding of the complex molecular interactions that occur between nematodes, plants and other microbes.

Acknowledgments

We extend our gratitude to the Phytopathology Unit within the Department of Plant Protection and Environment at the Ecole Nationale d'Agriculture de Meknes (ENAM) as well as to the Agrobiosciences Program within the College of Agriculture and Environmental Sciences at Mohammed VI Polytechnic University (UM6P) for their generous financial assistance.

References

Ali, M.A., Anjam, M.S., Nawaz, M.A., Lam, H.M. and Chung, G. (2018) Signal transduction in plant–nematode interactions. *International Journal of Molecular Sciences* 19(6), 1648. DOI: 10.3390/ijms19061648.

Alsamir, M., Ahmad, M., Mahmood, T. and Trethowan, R. (2017) Impact of heat stress on Fusarium wilt (*F. solani*) incidence in cultivated tomato and related species. *Australian Journal of Crop Science* 11(8), 997–1004. DOI: 10.21475/ajcs.17.11.08.pne500.

Back, M.A., Haydock, P.P.J. and Jenkinson, P. (2002) Disease complexes involving plant parasitic nematodes and soilborne pathogens. *Plant Pathology* 51(6), 683–697. DOI: 10.1046/j.1365-3059.2002.00785.x.

Bali, S., Kaur, P., Jamwal, V.L., Gandhi, S.G., Sharma, A. *et al.* (2020) Seed priming with jasmonic acid counteracts root knot nematode infection in tomato by modulating the activity and expression of antioxidative enzymes. *Biomolecules* 10(1), 98. DOI: 10.3390/biom10010098.

Balint-Kurti, P. (2019) The plant hypersensitive response: Concepts, control and consequences. *Molecular Plant Pathology* 20(8), 1163–1178. DOI: 10.1111/mpp.12821.

Banerjee, S., Banerjee, A., Gill, S.S., Gupta, O.P., Dahuja, A. *et al.* (2017) RNA interference: A novel source of resistance to combat plant parasitic nematodes. *Frontiers in Plant Science* 8, 834. DOI: 10.3389/fpls.2017.00834.

Barcala, M., García, A., Cabrera, J., Casson, S., Lindsey, K. *et al.* (2010) Early transcriptomic events in microdissected *Arabidopsis* nematode-induced giant cells. *The Plant Journal* 61(4), 698–712. DOI: 10.1111/j.1365-313X.2009.04098.x.

Bellafiore, S., Shen, Z., Rosso, M.N., Abad, P., Shih, P. *et al.* (2008) Direct identification of the *Meloidogyne incognita* secretome reveals proteins with host cell reprogramming potential. *PLoS Pathogens* 4(10), e1000192. DOI: 10.1371/journal.ppat.1000192.

Carella, P., Evangelisti, E. and Schornack, S. (2018) Sticking to it: Phytopathogen effector molecules may converge on evolutionarily conserved host targets in green plants. *Current Opinion in Plant Biology* 44, 175–180. DOI: 10.1016/j.pbi.2018.04.019.

Castro-Moretti, F.R., Gentzel, I.N., Mackey, D. and Alonso, A.P. (2020) Metabolomics as an emerging tool for the study of plant–pathogen interactions. *Metabolites* 10(2), 52. DOI: 10.3390/metabo10020052.

Channale, S., Kalavikatte, D., Thompson, J.P., Kudapa, H., Bajaj, P. *et al.* (2021) Transcriptome analysis reveals key genes associated with root-lesion nematode *Pratylenchus thornei* resistance in chickpea. *Scientific Reports* 11(1), 17491.

Chaudhary, S., Dutta, T.K., Shivakumara, T.N. and Rao, U. (2019a) RNAi of esophageal gland-specific gene *Mi-msp-1* alters early stage infection behaviour of root-knot nematode, *Meloidogyne incognita*. *Journal of General Plant Pathology* 85(3), 232–242. DOI: 10.1007/s10327-019-00837-x.

Chaudhary, S., Dutta, T.K., Tyagi, N., Shivakumara, T.N., Papolu, P.K. *et al.* (2019b) Host-induced silencing of *Mi-msp-1* confers resistance to root-knot nematode *Meloidogyne incognita* in eggplant. *Transgenic Research* 28(3–4), 327–340. DOI: 10.1007/s11248-019-00126-5.

Chen, J., Hu, L., Sun, L., Lin, B., Huang, K. *et al.* (2018) A novel *Meloidogyne graminicola* effector, MgMO237, interacts with multiple host defence-related proteins to manipulate plant basal immunity and promote parasitism. *Molecular Plant Pathology* 19(8), 1942–1955. DOI: 10.1111/mpp.12671.

Chen, J., Li, Z., Lin, B., Liao, J. and Zhuo, K. (2021) A *Meloidogyne graminicola* pectate lyase is involved in virulence and activation of host defense responses. *Frontiers in Plant Science* 12, 651627. DOI: 10.3389/fpls.2021.651627.

Choi, H.W. and Klessig, D.F. (2016) DAMPs, MAMPs, and NAMPs in plant innate immunity. *BMC Plant Biology* 16(1), 1–10. DOI: 10.1186/s12870-016-0921-2.

Couto, D. and Zipfel, C. (2016) Regulation of pattern recognition receptor signalling in plants. *Nature Reviews. Immunology* 16(9), 537–552. DOI: 10.1038/nri.2016.77.

Daami-Remadi, M., Souissi, A., Oun, H.B., Mansour, M. and Nasraoui, B. (2009) Salinity effects on fusarium wilt severity and tomato growth. *Dynamic Soil, Dynamic Plant* 3, 61–69.

Damiani, I., Baldacci-Cresp, F., Hopkins, J., Andrio, E., Balzergue, S. *et al.* (2012) Plant genes involved in harbouring symbiotic rhizobia or pathogenic nematodes. *The New Phytologist* 194(2), 511–522. DOI: 10.1111/j.1469-8137.2011.04046.x.

Danchin, E.G.J., Arguel, M.J., Campan, A., Perfus-Barbeoch, L., Magliano, M. *et al.* (2013) Identification of novel target genes for safer and more specific control of root-knot nematodes from a pan-genome mining. *PLoS Pathogens* 9(10), e1003745. DOI: 10.1371/journal.ppat.1003745.

Davies, L.J., Lilley, C.J. and Urwin, P.E. (2012) RNA interference in plant parasitic nematodes: A summary of the current status. *Parasitology* 139(5), 630–640. DOI: 10.1017/S0031182011002071.

de Almeida Engler, J., Favery, B., Engler, G. and Abad, P. (2005) Loss of susceptibility as an alternative for nematode resistance. *Current Opinion in Biotechnology* 16(2), 112–117. DOI: 10.1016/j.copbio.2005.01.009.

de Almeida Engler, J. and Favery, B. (2011) The plant cytoskeleton remodelling in nematode induced feeding sites. In: Jones, J.T., Gheysen, L. and Fenoll, C. (eds) *Genomics and Molecular Genetics of Plant–Nematode Interactions*. Springer, Dordrecht, The Netherlands, pp. 369–393.

De Meyer, M., Ryck, J., Goormachtig, S. and Damme, P. (2020) Keeping in touch with type-III secretion system effectors: Mass spectrometry-based proteomics to study effector–host protein–protein interactions. *International Journal of Molecular Sciences* 21(18), 1–24. DOI: 10.3390/ijms21186891.

Ding, S., Wang, D., Xu, C., Yang, S., Cheng, X. *et al.* (2021) A new fungus-mediated RNAi method established and used to study the fatty acid and retinol binding protein function of the plant-parasitic nematode *Aphelenchoides besseyi*. *RNA Biology* 18(10), 1424–1433. DOI: 10.1080/15476286.2020.1852779.

Divekar, P.A., Narayana, S., Divekar, B.A., Kumar, R., Gadratagi, B.G. *et al.* (2022) Plant secondary metabolites as defense tools against herbivores for sustainable crop protection. *International Journal of Molecular Sciences* 23(5), 2690. DOI: 10.3390/ijms23052690.

Duarte, A., Maleita, C., Abrantes, I. and Curtis, R. (2015) Tomato root exudates induce transcriptional changes of *Meloidogyne hispanica* genes. *Phytopathologia Mediterranea* 54, 104–108.

El Hadrami, A., Adam, L.R., El Hadrami, I. and Daayf, F. (2010) Chitosan in plant protection. *Marine Drugs* 8(4), 968–987. DOI: 10.3390/md8040968.

El-Sayed, S.M. and Mahdy, M.E. (2015) Effect of chitosan on root-knot nematode, *Meloidogyne javanica* on tomato plants. *International Journal of ChemTech Research* 7(4), 1985–1992.

Gang, S.S. and Hallem, E.A. (2016) Mechanisms of host seeking by parasitic nematodes. *Molecular and Biochemical Parasitology* 208(1), 23–32. DOI: 10.1016/j.molbiopara.2016.05.007.

Goode, K. and Mitchum, M.G. (2022) Pattern-triggered immunity against root-knot nematode infection: A minireview. *Physiologia Plantarum* 174(2), e13680. DOI: 10.1111/ppl.13680.

Goverse, A. and Smant, G. (2014) The activation and suppression of plant innate immunity by parasitic nematodes. *Annual Review of Phytopathology* 52(1), 243–265. DOI: 10.1146/annurev-phyto-102313-050118.

Greco, M., Chiappetta, A., Bruno, L. and Bitonti, M.B. (2012) In *Posidonia oceanica* cadmium induces changes in DNA methylation and chromatin patterning. *Journal of Experimental Botany* 63(2), 695–709. DOI: 10.1093/jxb/err313.

Guo, W., Zhang, F., Bao, A., You, Q., Li, Z. *et al.* (2019) The soybean *Rhg1* amino acid transporter gene alters glutamate homeostasis and jasmonic acid-induced resistance to soybean cyst nematode. *Molecular Plant Pathology* 20(2), 270–286. DOI: 10.1111/mpp.12753.

Gust, A.A., Pruitt, R. and Nürnberger, T. (2017) Sensing danger: Key to activating plant immunity. *Trends in Plant Science* 22(9), 779–791. DOI: 10.1016/j.tplants.2017.07.005.

Hada, A., Kumari, C., Phani, V., Singh, D., Chinnusamy, V. *et al.* (2020) Host-induced silencing of FMRFamide-like peptide genes, *flp*-1 and *flp*-12, in rice impairs reproductive fitness of the root-knot nematode *Meloidogyne graminicola*. *Frontiers in Plant Science* 11, 894. DOI: 10.3389/fpls.2020.00894.

Hamamouch, N., Li, C., Hewezi, T., Baum, T.J., Mitchum, M.G. *et al.* (2012) The interaction of the novel 30c02 cyst nematode effector protein with a plant β-1,3-endoglucanase may suppress host defence to promote parasitism. *Journal of Experimental Botany* 63(10), 3683–3695. DOI: 10.1093/jxb/ers058.

Hewezi, T., Howe, P.J., Maier, T.R., Hussey, R.S., Mitchum, M.G. *et al.* (2010) *Arabidopsis* spermidine synthase is targeted by an effector protein of the cyst nematode *Heterodera schachtii*. *Plant Physiology* 152(2), 968–984. DOI: 10.1104/pp.109.150557.

Hewezi, T., Juvale, P.S., Piya, S., Maier, T.R., Rambani, A. *et al.* (2015) The cyst nematode effector protein 10A07 targets and recruits host posttranslational machinery to mediate its nuclear trafficking and to promote parasitism in *Arabidopsis*. *The Plant Cell* 27(3), 891–907. DOI: 10.1105/tpc.114.135327.

Hu, Y., You, J., Li, C., Pan, F. and Wang, C. (2019) The *Heterodera glycines* effector Hg16B09 is required for nematode parasitism and suppresses plant defense response. *Plant Science* 289, 110271. DOI: 10.1016/j.plantsci.2019.110271.

Hua, C., Li, C., Jiang, Y., Huang, M., Williamson, V.M. *et al.* (2020) Response of soybean cyst nematode (*Heterodera glycines*) and root-knot nematodes (*Meloidogyne* spp.) to gradients of pH and inorganic salts. *Plant and Soil* 455(1–2), 305–318. DOI: 10.1007/s11104-020-04677-z.

Huang, L. (2022) The NILR1-BAK1 complex mediates resistance to potato cyst nematode and brassinosteroid signaling in potato. Doctoral dissertation, University of Idaho, Moscow, ID.

Hütsch, B.W., Augustin, J. and Merbach, W. (2002) Plant rhizodeposition – an important source for carbon turnover in soils. *Journal of Plant Nutrition and Soil Science* 165(4), 397. DOI: 10.1002/1522-2624(200208)165:43.0.CO;2-C.

Jagdale, S., Rao, U. and Giri, A.P. (2021) Effectors of root-knot nematodes: An arsenal for successful parasitism. *Frontiers in Plant Science* 12, 800030. DOI: 10.3389/fpls.2021.800030.

Jaraba, J., Rothrock, C.S., Kirkpatrick, T.L. and Brye, K.R. (2013) Soil texture influence on *Meloidogyne incognita* and *Thielaviopsis basicola* and their interaction on cotton. *Plant Disease* 98(3), 336–343. DOI: 10.1094/PDIS-04-13-0357-RE.

Jones, J.D.G. (2006) The plant immune system. *Nature* 444(7117), 323–329.

Jones, J.T., Haegeman, A., Danchin, E.G.J., Gaur, H.S., Helder, J. *et al.* (2013) Top 10 plant-parasitic nematodes in molecular plant pathology. *Molecular Plant Pathology* 14(9), 946–961. DOI: 10.1111/mpp.12057.

Joshi, I., Kumar, A., Kohli, D., Singh, A.K., Sirohi, A. *et al.* (2020) Conferring root-knot nematode resistance via host-delivered RNAi-mediated silencing of four *Mi-msp* genes in *Arabidopsis*. *Plant Science* 298, 110592. DOI: 10.1016/j.plantsci.2020.110592.

Kazan, K. and Lyons, R. (2014) Intervention of phytohormone pathways by pathogen effectors. *The Plant Cell* 26(6), 2285–2309. DOI: 10.1105/tpc.114.125419.

Khan, M.R. (2007) Prospects of microbial control of root-knot nematodes infecting vegetable crops. In: Sharma, N. and Singh, H.B. (eds) *Biotechnology: Plant Health Management*. International Book Distributing, Biotechnology, pp. 643–665.

Khan, M.R. (2008) *Plant Nematodes – Methodology, Morphology, Systematics, Biology and Ecology*. Science Publishers, Enfield, NH. DOI: 10.1201/b10766.

Khan, M.R. (2023a) Nematode pests of agricultural crops, a global overview. In: Khan, M.R. (ed.) *Novel Biological and Biotechnological Applications in Plant Nematode Management*. Springer, Singapore, pp. 3–45.

Khan, M.R. (2023b) Plant nematodes, an underestimated constraint in global crop production. Nematode diseases of crops and their sustainable management. In: Khan, M.R. and Quintanilla, M. (eds) *Nematode Diseases of Crops and Their Sustainable Management*. Academic Press, Cambridge, MA, pp. 3–23. DOI: 10.1016/B978-0-323-91226-6.00009-2.

Khan, M.R. and Sharma, R.K. (2020) Fusarium–nematode wilt disease complexes, etiology and mechanism of development. *Indian Phytopathology* 73(4), 615–628. DOI: 10.1007/s42360-020-00240-z.

Khan, M.R., Mohiddin, F.A., Haque, Z. and Sharma, R.K. (2023) Disease complexes and their sustainable management. In: Khan, M.R. and Quintanilla, M. (eds) *Nematode Diseases of Crops and Their Sustainable Management*. Academic Press, Cambridge, MA, pp. 65–93.

Khan, M.W. (1993) Mechanisms of interactions between nematodes and other plant pathogens. In: Khan, M.W. (ed.) *Nematode Interactions*. Chapman & Hall, London, pp. 55–78.

Kim, E., Seo, Y., Kim, Y.S., Park, Y. and Kim, Y.H. (2017) Effects of soil textures on infectivity of root-knot nematodes on carrot. *Plant Pathology Journal* 33(1), 66–74. DOI: 10.5423/PPJ.OA.07.2016.0155.

Kong, L., Shi, X., Chen, D., Yang, N., Yin, C. *et al.* (2022) Host-induced silencing of a nematode chitin synthase gene enhances resistance of soybeans to both pathogenic *Heterodera glycines* and *Fusarium oxysporum*. *Plant Biotechnology Journal* 20(5), 809–811. DOI: 10.1111/pbi.13808.

Kyndt, T., Denil, S., Haegeman, A., Trooskens, G., Bauters, L. *et al.* (2012) Transcriptional reprogramming by root knot and migratory nematode infection in rice. *New Phytologist* 196(3), 887–900. DOI: 10.1111/j.1469-8137.2012.04311.x.

Lahari, Z., Ullah, C., Kyndt, T., Gershenzon, J. and Gheysen, G. (2019) Strigolactones enhance root-knot nematode (*Meloidogyne graminicola*) infection in rice by antagonizing the jasmonate pathway. *New Phytologist* 224(1), 454–465. DOI: 10.1111/nph.15953.

Lee, C., Chronis, D., Kenning, C., Peret, B., Hewezi, T. *et al.* (2011) The novel cyst nematode effector protein 19C07 interacts with the *Arabidopsis* auxin influx transporter LAX3 to control feeding site development. *Plant Physiology* 155(2), 866–880. DOI: 10.1104/pp.110.167197.

Leelarasamee, N., Zhang, L. and Gleason, C. (2018) The root-knot nematode effector MiPFN3 disrupts plant actin filaments and promotes parasitism. *PLoS Pathogens* 14(3), e1006947. DOI: 10.1371/journal.ppat.1006947.

Lefoulon, E., McMullen, J.G. and Stock, S.P. (2022) Transcriptomic analysis of *Steinernema* nematodes highlights metabolic costs associated to *Xenorhabdus* endosymbiont association and rearing conditions. *Frontiers in Physiology* 13, 821845. DOI: 10.3389/fphys.2022.821845.

Li, Y., Wang, K., Xie, H., Wang, D.W., Xu, C.L. *et al.* (2015) Cathepsin B cysteine proteinase is essential for the development and pathogenesis of the plant parasitic nematode *Radopholus similis*. *International Journal of Biological Sciences* 11(9), 1073–1087. DOI: 10.7150/ijbs.12065.

Lilley, C.J., Maqbool, A., Wu, D., Yusup, H.B., Jones, L.M. *et al.* (2018) Effector gene birth in plant parasitic nematodes: Neofunctionalization of a housekeeping glutathione synthetase gene. *PLoS Genetics* 14(4), 1–26. DOI: 10.1371/journal.pgen.1007310.

Lin, B., Zhuo, K., Chen, S., Hu, L., Sun, L. *et al.* (2016) A novel nematode effector suppresses plant immunity by activating host reactive oxygen species-scavenging system. *The New Phytologist* 209(3), 1159–1173. DOI: 10.1111/nph.13701.

Liu, X. and Liu, C. (2016) Effects of drought-stress on fusarium crown rot development in barley. *PLoS ONE* 11(12), e0167304. DOI: 10.1371/journal.pone.0167304.

Lolle, S., Stevens, D. and Coaker, G. (2020) Plant NLR-triggered immunity: From receptor activation to downstream signaling. *Current Opinion in Immunology* 62, 99–105. DOI: 10.1016/j.coi.2019.12.007.

Maharshi, A., Rashid, M.M., Teli, B., Yadav, S.K., Singh, D.P. *et al.* (2021) Salt stress alters pathogenic behaviour of *Fusarium oxysporum* f. sp. *ciceris* and contributes to severity in chickpea wilt incidence. *Physiological and Molecular Plant Pathology* 113, 101602. DOI: 10.1016/j.pmpp.2021.101602.

Mamaeva, A., Lyapina, I., Knyazev, A., Golub, N., Mollaev, T. *et al.* (2023) RALF peptides modulate immune response in the moss *Physcomitrium patens*. *Frontiers in Plant Science* 14, 1077301. DOI: 10.3389/fpls.2023.1077301.

Manzanilla-López, R.H., Esteves, I., Finetti-Sialer, M.M., Hirsch, P.R., Ward, E. *et al.* (2013) *Pochonia chlamydosporia*: Advances and challenges to improve its performance as a biological control agent of sedentary endo-parasitic nematodes. *Journal of Nematology* 45(1), 1–7.

Meena, M., Yadav, G., Sonigra, P., Nagda, A., Mehta, T. *et al.* (2022) Role of elicitors to initiate the induction of systemic resistance in plants to biotic stress. *Plant Stress* 5, 100103. DOI: 10.1016/j.stress.2022.100103.

Mejias, J., Truong, N.M., Abad, P., Favery, B. and Quentin, M. (2019) Plant proteins and processes targeted by parasitic nematode effectors. *Frontiers in Plant Science* 10, 970. DOI: 10.3389/fpls.2019.00970.

Mejias, J., Bazin, J., Truong, N.-M., Chen, Y., Marteu, N. *et al.* (2020) The root-knot nematode effector MiEFF18 interacts with the plant core spliceosomal protein SmD1 required for giant cell formation. *The New Phytologist* 229(6), 3408–3423. DOI: 10.1111/nph.17089.

Melakeberhan, H., Dey, J., Baligar, V.C. and Carter, T.E., Jr (2004) Effect of soil pH on the pathogenesis of *Heterodera glycines* and *Meloidogyne incognita* on *Glycine max* genotypes. *Nematology* 6(4), 585–592. DOI: 10.1163/1568541042665205.

Migunova, V.D. and Sasanelli, N. (2021) Bacteria as biocontrol tool against phytoparasitic nematodes. *Plants* 10(2), 1–15. DOI: 10.3390/plants10020389.

Mitchum, M.G., Hussey, R.S., Baum, T.J., Wang, X., Elling, A.A. *et al.* (2013) Nematode effector proteins: An emerging paradigm of parasitism. *New Phytologist* 199(4), 879–894. DOI: 10.1111/nph.12323.

Oosterbeek, M., Lozano-Torres, J.L., Bakker, J. and Goverse, A. (2021) Sedentary plant-parasitic nematodes alter auxin homeostasis via multiple strategies. *Frontiers in Plant Science* 12, 668548. DOI: 10.3389/fpls.2021.668548.

Parikka, P., Hakala, K. and Tiilikkala, K. (2012) Expected shifts in Fusarium species' composition on cereal grain in Northern Europe due to climatic change. *Food Additives & Contaminants* 29(10), 1543–1555. DOI: 10.1080/19440049.2012.680613.

Perry, R. and Moens, M. (2006) *Plant Nematology*. CAB International, Wallingford, UK.

Perry, R.N. and Moens, M. (2013) Introduction to plant-parasitic nematodes; modes of parasitism. In: Jones, J.T., Gheysen, L. and Fenoll, C. (eds) *Genomics and Molecular Genetics of Plant–Nematode Interactions*. Springer, Dordrecht, The Netherlands, pp. 3–20.

Pogorelko, G., Juvale, P.S., Rutter, W.B., Hewezi, T., Hussey, R. *et al.* (2016) A cyst nematode effector binds to diverse plant proteins, increases nematode susceptibility and affects root morphology. *Molecular Plant Pathology* 17(6), 832–844. DOI: 10.1111/mpp.12330.

Pogorelko, G., Wang, J., Juvale, P.S., Mitchum, M.G. and Baum, T.J. (2020) Screening soybean cyst nematode effectors for their ability to suppress plant immunity. *Molecular Plant Pathology* 21(9), 1240–1247. DOI: 10.1111/mpp.12972.

Pršić, J. and Ongena, M. (2020) Elicitors of plant immunity triggered by beneficial bacteria. *Frontiers in Plant Science* 11, 594530. DOI: 10.3389/fpls.2020.594530.

Qin, X., Xue, B., Tian, H., Fang, C., Yu, J. *et al.* (2022) An unconventionally secreted effector from the root knot nematode *Meloidogyne incognita*, Mi-ISC-1, promotes parasitism by disrupting salicylic acid biosynthesis in host plants. *Molecular Plant Pathology* 23(4), 516–529. DOI: 10.1111/mpp.13175.

Quentin, M., Abad, P. and Favery, B. (2013) Plant parasitic nematode effectors target host defense and nuclear functions to establish feeding cells. *Frontiers in Plant Science* 4, 53. DOI: 10.3389/fpls.2013.00053.

Ravichandra, N.G. (2014) Nematode disease complexes. In: Ravichandra, N.G. (ed.) *Horticultural Nematology*. Springer, New Delhi, pp. 207–238. DOI: 10.1007/978-81-322-1841-8_9.

Rehman, S., Gupta, V.K. and Goyal, A.K. (2016) Identification and functional analysis of secreted effectors from phytoparasitic nematodes. *BMC Microbiology* 16(1), 48. DOI: 10.1186/s12866-016-0632-8.

Rocha, R.O., Hussey, R.S., Pepi, L.E., Azadi, P. and Mitchum, M.G. (2023) Discovery of novel effector protein candidates produced in the dorsal gland of adult female root-knot nematodes. *MPMI* 36(6), 372–380. DOI: 10.1094/MPMI-11-22-0232-R.

Sato, K., Kadota, Y. and Shirasu, K. (2019) Plant immune responses to parasitic nematodes. *Frontiers in Plant Science* 10, 1165. DOI: 10.3389/fpls.2019.01165.

Shang, Y., Wang, K., Sun, S., Zhou, J. and Yu, J.Q. (2019) COP9 signalosome CSN4 and CSN5 subunits are involved in jasmonate-dependent defense against root-knot nematode in tomato. *Frontiers in Plant Science* 10, 1223. DOI: 10.3389/fpls.2019.01223.

Shinya, R., Kirino, H., Morisaka, H., Takeuchi-Kaneko, Y., Futai, K. *et al.* (2021) Comparative secretome and functional analyses reveal glycoside hydrolase family 30 and cysteine peptidase as virulence determinants in the pinewood nematode *Bursaphelenchus xylophilus*. *Frontiers in Plant Science* 12, 640459. DOI: 10.3389/fpls.2021.640459.

Sikder, M.M. and Vestergård, M. (2020) Impacts of root metabolites on soil nematodes. *Frontiers in Plant Science* 10, 1792. DOI: 10.3389/fpls.2019.01792.

Sun, Y., Li, L., Macho, A.P., Han, Z., Hu, Z. *et al.* (2013) Structural basis for flg22-induced activation of the *Arabidopsis* FLS2-BAK1 immune complex. *Science* 342(6158), 624–628. DOI: 10.1126/science.1243825.

Tanowitz, M. and von Zastrow, M. (2004) Identification of protein interactions by yeast two-hybrid screening and coimmunoprecipitation. *Methods in Molecular Biology* 259, 353–369.

Tileubayeva, Z., Avdeenko, A., Avdeenko, S., Stroiteleva, N. and Kondrashev, S. (2021) Plant-parasitic nematodes affecting vegetable crops in greenhouses. *Saudi Journal of Biological Sciences* 28(9), 5428–5433. DOI: 10.1016/j.sjbs.2021.05.075.

Tritten, L., Ballesteros, C., Beech, R., Geary, T.G. and Moreno, Y. (2021) Mining nematode protein secretomes to explain lifestyle and host specificity. *PLoS Neglected Tropical Diseases* 15(9), e0009828. DOI: 10.1371/journal.pntd.0009828.

Truong, N.M., Chen, Y., Mejias, J., Soulé, S., Mulet, K. *et al.* (2021) The *Meloidogyne incognita* nuclear effector MiEFF1 interacts with *Arabidopsis* cytosolic glyceraldehyde-3-phosphate dehydrogenases to promote parasitism. *Frontiers in Plant Science* 12, 641480. DOI: 10.3389/fpls.2021.641480.

Tyurin, A.A., Suhorukova, A.V., Kabardaeva, K.V. and Goldenkova-Pavlova, I.V. (2020) Transient gene expression is an effective experimental tool for the research into the fine mechanisms of plant gene function: Advantages, limitations, and solutions. *Plants* 9(9), 1187. DOI: 10.3390/plants9091187.

van Steenbrugge, J.J.M., Elsen, S., Holterman, M., Sterken, M.G., Thorpe, P. *et al.* (2021) Comparative genomics of two inbred lines of the potato cyst nematode *Globodera rostochiensis* reveals disparate effector family-specific diversification patterns. *BMC Genomics* 22(1), 611. DOI: 10.1186/s12864-021-07914-6.

van Steenbrugge, J.J.M., Elsen, S., Holterman, M., Lozano-Torres, J.L., Putker, V. *et al.* (2023) Comparative genomics among cyst nematodes reveals distinct evolutionary histories among effector families and an irregular distribution of effector-associated promoter motifs. *Molecular Ecology* 32(6), 1515–1529. DOI: 10.1111/mec.16505.

Vieira, P. and Gleason, C. (2019) Plant-parasitic nematode effectors – insights into their diversity and new tools for their identification. *Current Opinion in Plant Biology* 50, 37–43. DOI: 10.1016/j.pbi.2019.02.007.

Viney, M. (2017) How can we understand the genomic basis of nematode parasitism? *Trends in Parasitology* 33(6), 444–452. DOI: 10.1016/j.pt.2017.01.014.

Walther, T.C. and Mann, M. (2010) Mass spectrometry-based proteomics in cell biology. *The Journal of Cell Biology* 190(4), 491–500. DOI: 10.1083/jcb.201004052.

Williamson, V.M. and Kumar, A. (2006) Nematode resistance in plants: The battle underground. *Trends in Genetics* 22(7), 396–403. DOI: 10.1016/j.tig.2006.05.003.

Xie, J., Yang, F., Xu, X., Peng, Y. and Ji, H. (2022) Salicylic acid, jasmonate, and ethylene contribute to rice defense against white tip nematodes *Aphelenchoides besseyi*. Frontiers in Plant Science 12, 755802. DOI: 10.3389/fpls.2021.755802.

Xing, S., Wallmeroth, N., Berendzen, K.W. and Grefen, C. (2016) Techniques for the analysis of protein–protein interactions *in vivo*. Plant Physiology 171(2), 727–758. DOI: 10.1104/pp.16.00470.

Zhang, S., Li, C., Si, J., Han, Z. and Chen, D. (2022a) Action mechanisms of effectors in plant–pathogen interaction. *International Journal of Molecular Sciences* 23(12), 6758. DOI: 10.3390/ijms23126758.

Zhang, X., Peng, H., Zhu, S., Xing, J., Li, X. *et al*. (2020) Nematode-encoded RALF peptide mimics facilitate parasitism of plants through the FERONIA receptor kinase. Molecular Plant 13(10), 1434–1454. DOI: 10.1016/j.molp.2020.08.014.

Zhang, Y., Zhao, Q., Zhang, J., Niu, L., Yang, J. *et al*. (2022b) Enhanced resistance to soybean cyst nematode in transgenic soybean via host-induced silencing of vital *Heterodera glycines* genes. Transgenic Research 31(2), 239–248. DOI: 10.1007/s11248-022-00298-7.

Zhao, M., Lei, C., Zhou, K., Huang, Y., Fu, C. *et al*. (2024) POOE: Predicting oomycete effectors based on a pre-trained large protein language model mSystems 9, e01004-23. DOI: 10.1128/msystems.01004-23.

Zhou, Y., Zhao, D., Shuang, L., Xiao, D., Xuan, Y. *et al*. (2020) Transcriptome analysis of rice roots in response to root-knot nematode infection. *International Journal of Molecular Sciences* 21(3), 848. DOI: 10.3390/ijms21030848.

Zipfel, C. (2014) Plant pattern-recognition receptors. *Trends in Immunology* 35(7), 345–351. DOI: 10.1016/j.it.2014.05.004.

6 Fungus–Nematode Wilt Disease Complex

Mujeebur Rahman Khan[1]*, Awol Seid[2], Yao A. Kolombia[3] and Mohd Shahid Anwar Ansari[1]

[1]Department of Plant Protection, Faculty of Agricultural Sciences, Aligarh Muslim University, Aligarh, India; [2]Department of Plant Sciences, College of Agriculture, Wollo University, Dessie, Ethiopia; [3]International Institute of Tropical Agriculture, Ibadan, Nigeria

Abstract

Nematodes are cosmopolitan in nature, and are considered as the most abundant group of animals on Earth. Of all the known nematode species, only a small proportion are plant parasitic in nature and are considered as economically important pests of agricultural crops. Nematodes attack all kinds of agricultural crops and inflict 7–15% yield decline. The monetary losses to crops due to nematode infestation have been estimated to be US$157–200 billion. In addition to causing damage to plants directly, nematodes also promote the infection of other pathogens, leading to the formation of the diseases of complex etiology, commonly called disease complex. The occurrence of disease complexes in agricultural cropping systems is quite common. A disease complex may inflict yield loss considerably greater than the respective monopathogenic diseases. This chapter offers detailed and up-to-date information on fungus–nematode wilt disease complexes along with the etiology, mechanism of formation and methods of management.

Introduction

The synergistic interaction of plant nematodes and phytopathogenic fungi leads to the formation of disease complexes which are highly damaging to crops in comparison to their individual effects on crop health and yields (Khan and Jairajpuri, 2010b, 2010c). Plant nematodes are known as a biopredisposing agents, and sometimes render resistant or tolerant genotypes susceptible to the fungus while in cohabitance. Several examples of nematode interactions have been reported where the damage caused by fungal or bacterial pathogens is aggravated in the presence of plant nematodes (Back *et al*., 2002; Zhang *et al*., 2020; Khan, 2023a, 2023b). In the literature, three major types of pathogen–nematode interactions have been documented, i.e. synergistic, antagonistic and neutral or additive interactions. In a synergistic interaction, the combined damage by the two pathogens to a plant is higher than the sum of the damage caused by the engaged pathogens individually (Khan, 1993). An antagonistic interaction is just the reverse situation, where the combined damage to a plant is lower than the sum of the individual effects of the pathogens involved in the disease complex. A neutral or additive interaction refers to a situation where the damage by the pathogens together is equal

*Corresponding author: mrkhan777in@yahoo.co.in

 Nematode Disease Complexes in Agricultural Crops (ed. Mujeebur Rahman Khan)
DOI: 10.1079/9781800625228.0006

or close to the sum of their individual effects (Back *et al.*, 2002; Khan, 2005).

Since the first record on synergistic interactions between a plant parasitic nematode and the wilt fungus made by Atkinson (1892), numerous studies have shown that co-occurrence of phytonematodes and wilt fungi leads to aggravation of wilt severity (Francl and Wheeler, 1993). Chindo *et al.* (1991) showed that wilt caused by *Fusarium oxysporum* f. sp. *lycopersici* became severe in the presence of *Meloidogyne incognita* on tomato. Similarly, severity of wilt in cotton (*F. oxysporum* f. sp. *vasinfectum*) was significantly increased in the presence of *M. incognita* (Martin *et al.*, 1956) and *Rotylenchulus reniformis* (Neal, 1954). Nagesh *et al.* (2006) observed that *M. incognita* predisposed a resistant tomato cultivar to *F. oxysporum* f. sp. *lycopersici*. A similar observation was reported by Khan and Akram (2000) on a tomato cultivar. A number of endoparasitic nematodes, *Globodera*, *Heterodera*, *Rotylenchulus*, *Pratylenchus*, *Radopholus*, etc., and ectoparasitic nematodes, *Belonolaimus*, *Tylenchorhynchus*, *Xiphinema*, *Longidorus*, *Trichodorus*, etc., have been found to be involved in wilt disease complexes on different crops. However, the *Fusarium–Meloidogyne* wilt disease complex is the most commonly occurring and economically important disease complex worldwide (Manikandan and Raguchander, 2014; Khan and Sharma, 2020).

Wilt-causing Fungi and Their Impact on Crops

Fusarium oxysporum

Fusarium oxysporum is a soil-borne ascomycetes belonging to the family Nactriaceae. It is responsible for fusarium wilt, which is a devastating disease affecting several agricultural crops, including tomatoes, chickpea, pigeon pea, tobacco, cotton, etc. The fungus enters through the roots and colonizes the plant's vascular system, blocking water and nutrient transport. This leads to symptoms such as wilting, leaf yellowing, growth stunting and often plant death. *Fusarium* spp. produce spores that can survive in soil for many years, making it difficult to eradicate.

Verticillium

Verticillium causes wilt in over 300 plant species, including vegetables, trees and ornamental plants. Key pathogens in this group are *V. dahliae* and *V. albo-atrum*, which invade the host roots and spread through the vascular system, causing blockages that lead to wilting, yellowing and premature leaf drops. The fungus produces microsclerotia, which may survive in the soil for years.

Phytophthora parasitica

Phytophthora parasitica is a highly destructive oomycetes fungus affecting a wide range of agricultural crops, including tomato, pepper, tobacco, citrus and ornamentals. It causes root rot, crown rot and fruit rot, leading to development of symptoms like wilting, yellowing and plant collapse (Okungbowa and Shittu, 2012). The pathogen thrives in warm, moist environments, spreading through water, soil and infected plant material. Regular monitoring and sanitation practices are crucial to prevent the spread and to reduce the impact of this pathogen.

Fungus–Nematode Wilt Disease Complex

The interactions between nematodes and wilt fungi are highly complicated and can result in multiple outcomes, such as the onset of fungal disease, severe growth reduction or increased pathogenicity of the fungus, decreased pathogenicity of nematodes, etc. (Reddy, 2008). The relationship between wilt fungi and nematodes covers multiple roles: it alters the rhizospheric environment, predisposes host plants, modifies host behavior and breaks the resistance (Riedel, 1988; Khan and Khan, 1995). The relationships among the host plants, nematodes and fungi are highly specific and contingent on their precise combination. Differences in the biology and feeding habits of nematode species can influence their ability to predispose hosts to *Fusarium* spp. (Bergeson, 1972).

Numerous investigations have reported on the synergistic interaction between wilt-inducing fungi and plant-parasitic nematodes. The plant-parasitic nematodes interact with fungi in diverse ways to induce plant disease complexes, which vary with the nematode group or genus and the fungal species. Some nematodes carry fungal spores, enhancing their mobility and access to the susceptible host or tissue, whereas other nematode groups or genera create wounds on the root surface (Riedel, 1988) or modify the host as a more favorable substrate for fungal invasion, colonization and reproduction. The important nematode groups and genera instrumental in forming wilt disease complexes are briefly described below (also see Table 6.1).

Root-knot nematodes

Meloidogyne spp., commonly called root-knot nematodes, are widely distributed sedentary endoparasites, having an extremely wide host range and inflicting severe yield losses to agricultural crops worldwide (Khan and Khan, 1997; Karssen *et al.*, 2013). The second-stage juveniles invade the roots and incite the galls (Sasser, 1989; Khan and Khan, 1994a, b). Root-knot nematodes possess an enormous capability

Table 6.1. Disease complexes resulting from synergistic interaction between plant-pathogenic fungi and nematodes.

Fungus	Nematode	Symptoms	Crop/plant	Reference
Fusarium oxysporum f. sp. *lycopersici*	*Radopholus similis*	Yellowing, pseudo stem splitting, drooping of leaves, wilting of the plant and eventually plant death, black head toppling	Banana	Murali *et al.*, 2023
Fusarium oxysporum f. sp. *ciceri*	*Pratylenchus thornei*	Reduced growth leading to wilting	Chickpea	Castillo *et al.*, 1998
Fusarium oxysporum f. sp. *pisi*	*Rotylenchulus reniformis*	Reduced root systems, stunting leading to wilting	Pea	Vats and Dalal, 1997
Rhizoctonia solani	*Meloidogyne hapla*	Root rot, wilting, damping-off	Tomato	Goze Ozdemir *et al.*, 2022
Verticillium dahliae	*Pratylenchus penetrans*	Early dying	Potato	Martin *et al.*, 1982
Rhizoctonia solani	*Trichodoridae* (stubby-root nematodes)	Stem canker	Potato	Karlsson, 2006
Rhizoctonia solani	*Globodera pallida*	Stem canker	Potato	Bhattarai *et al.*, 2009
Verticillium dahliae	*Pratylenchus penetrans*	Wilting	Potato	Bowers, 1996
Verticillium dahliae	*Pratylenchus penetrans*	Wilting	Mint	Johnson and Santo, 2001
Aspergillus niger	*Scutellonema bradys*	Tuber cracking and rotting	Yam	Pepple *et al.*, 2010
Fusarium solani	*Heterodera glycine*	Sudden death syndrome, root rot, crown necrosis, defoliation and abortion of pods	Soyabeans	Gao *et al.*, 2006
Rhizoctonia solani	*Meloidogyne incognita*	Damping-off, root rot	Peanut	Abdel-Momen and Starr, 1998
Verticillium dahliae	*Pratylenchus penetrans*	Wilting	Strawberry	Bélair *et al.*, 2018

to exacerbate the infection of soil-borne phyto-fungi, especially wilt-inducing fungi, and to form disease complexes (Khan and Sharma, 2020). The disease complexes involving root-knot nematodes and wilt fungi are serious and highly detrimental to several agricultural crops (Archana *et al.*, 2023). *Fusarium–Meloidogyne* and *Verticillium–Meloidogyne* disease complexes are often encountered in vegetable and pulse crops (Francl and Wheeler, 1993). The presence of root-knot nematode has been found to greatly increase fusarium wilt symptoms and plant mortality (Mohiddin and Khan, 2014).

The synergism to the wilt fungus typically begins in the root zone, where the *Fusarium* propagules become activated in response to increased root exudation from the host roots invaded by *Meloidogyne* juveniles, as observed on various plant species such as cotton, tomato, tobacco, etc. Eggplants inoculated with root-knot nematodes exhibited higher susceptibility to fusarium wilt (Tharshani and Sivapalan, 2009; Abd-Elgawad, 2021). Concomitant inoculations with *M. incognita* and *F. oxysporum* f. sp. *ciceri* caused severe wilting and greater damage to chickpea plants (Khan *et al.*, 2011; Mohiddin and Khan, 2019). Likewise, simultaneous inoculation of *M. incognita* and various *F. oxysporum* strains inflicted severe wilting in tomato (Khan and Akram, 1997), okra, cotton (Rao *et al.*, 2014, 2017), cauliflower (Rajinikanth *et al.*, 2013), onion (Priti *et al.*, 2018), cucumber (Patil *et al.*, 2020), muskmelon (Dhami *et al.*, 2024), chrysanthemum (Khan *et al.*, 2024), gladiolus (Khan *et al.*, 2017a) and other agricultural crops (Francl and Wheeler, 1993) (Table 6.2).

Cyst-forming nematodes

Cyst-forming nematodes are also important sedentary endoparasites, are worldwide in distribution and have a wide host range (Turner and Subbotin, 2013). They feed on syncytia (specialized nurse cells) in the vascular tissue. The saccate female after maturity turns into a dead cyst (Khan, 2008). The cyst contains numerous embryonated eggs that can remain viable for extended periods (Turner and Subbotin, 2013). The second-stage juveniles are the infective stage and invade the host roots. Cyst nematodes occur globally across temperate, tropical and subtropical areas. These nematodes are included in the genera *Heterodera*, *Cactodera*, *Globodera*, *Punctodera*, *Dolichodera* and *Afenestrata* (Siddiqi, 2000). At least five genera pose significant challenges to crop production in tropical agriculture (Sharma and Nene, 1989). Cyst nematodes do not incite galls, but their rapid enlargement in size causes rupturing of the cortex that leads to extensive damage and rotting in the root (Kolombia and Fabiyi, 2023). Cyst-forming nematodes also possess the capability to interact with wilt fungi causing wilt disease complexes (Barker and McGawley, 1998).

Cyst-forming nematodes induce significant, physical, anatomical and physiological alterations in the host which facilitate the invasion of *Fusarium*, *Macrophomina*, *Rhizoctonia* and *Verticillium* species, leading to the respective disease complexes (Francl and Wheeler, 1993; Lopez-Nicora and Niblack, 2018). Such disease complexes can be seen in *Heterodera*- and *Globodera*-prone areas experiencing continuous cultivation of vegetable and pulse crops (Briar *et al.*, 2023). The common symptoms of the disease complex are wilting, yellowing of leaves, stunted growth and eventually plant death (Barker and McGawley, 1998).

The soybean cyst-forming nematode *H. glycines* causes wilt disease complexes in soybean by synergistically interacting with *F. oxysporum* (Ross, 1965; Stiles *et al.*, 1993), *F. solani*, *F. virguliforme* (McGawley, 1992; Xing and Westphal, 2006) and *Macrophomina* species (Winkler *et al.*, 1994), which are highly damaging and may sometimes cause complete crop failure. Various studies have proven an augmentation in the population of *H. glycines* when it co-occurred with *F. oxysporum* (Ross, 1965; Stiles *et al.*, 1993). In South America, sudden death syndrome in soybean was linked with the interaction between the cyst-forming nematode *H. glycines* and four *Fusarium* species, i.e. *F. brasiliense*, *F. crassistipitatum*, *F. tucumaniae* and *F. virguliforme* (Stiles *et al.*, 1993; Aoki *et al.*, 2003). However, in North America only *F. virguliforme* was found to be associated with the syndrome (Aoki *et al.*, 2003).

On sugarbeet (*Beta vulgaris* subsp. *vulgaris* var. *altissima*), *H. schachtii* was found interacting with *Aphanomyces cochlioides* (Whitney and Doney, 1973), *Pythium aphanidermatum* and *P. ultimum* (Whitney, 1974) and *Rhizoctonia solani*

Table 6.2. Wilt disease complexes caused by *Fusarium* and *Meloidogyne* species on different crops.

Wilt fungus	Root-knot species	Crops	Reference(s)
F. oxysporum	*M. incognita*	Cotton	Cauquil and Shepherd, 1970; Khan *et al.*, 2023a
		Chrysanthemum	Johnson and Littrell, 1969
		Cow pea	Singh and Goswami, 2001
		Okra	Sharma *et al.*, 2007
		Mulberry	Naik *et al.*, 2015
		Crossandra	Ramakrishnan *et al.*, 2014
		Tuberose	Rao *et al.*, 2003; Kavitha and Thirukumaran, 2019
F. oxysporum f. sp. *ciceri*	*M. incognita*	Chickpea	Rao and Krishnappa, 1995; Khan *et al.*, 2011; Mohiddin and Khan, 2019
F. oxysporum f. sp. *lycopersici*		Tomato	Khan and Akram, 1997; Bhagawati *et al.*, 2000; Khan and Akram, 2000; Goswami and Tiwari, 2007; Meena *et al.*, 2021
F. oxysporum f. sp. *vasinfectum*		Okra	Rao *et al.*, 2014; Rao *et al.*, 2017
		Cotton	Yang *et al.*, 1976
F. oxysporum f. sp. *conglutinans*		Cauliflower	Rajinikanth *et al.*, 2013
F. oxysporum f. sp. *cepae*		Onion	Priti *et al.*, 2018
F. oxysporum f. sp. *cucumerinum*		Cucumber	Patil *et al.*, 2020
F. oxysporum f. sp. *lentis*		Lentil	Haseeb *et al.*, 2005
F. oxysporum f. sp. *pisi*		Field pea	Haseeb *et al.*, 2007
F. oxysporum f. sp. *gladioli*		Gladiolus	Sowmya and Rao, 2012; Khan *et al.*, 2017a
F. oxysporum f. sp. *niveum*		Watermelon	Hua *et al.*, 2019
F. oxysporum f. sp. *cucumerinum*	*M. javanica*	Cucumber	Singh *et al.*, 2021
F. oxysporum		Okra	Manisha *et al.*, 2023
F. oxysporum f. sp. *tracheiphilum*		Cowpea	Harris and Ferris, 1991
F. oxysporum	*M. arenaria*	Tomato	Sikora and Fernandez, 2005; Regmi and Desaeger, 2020

Continued

Table 6.2. Continued

Wilt fungus	Root-knot species	Crops	Reference(s)
F. oxysporum f. sp. *asparagi*	*M. hapla*	Asparagus	Abawi *et al.*, 2012
F. oxysporum		Carrot	Hallmann and Meressa, 2018
F. oxysporum f. sp. *lycopersici*	*M. enterolobii*	Tomato	Philbrick *et al.*, 2020
F. oxysporum	*M. arabicida*	Coffee	Bertrand *et al.*, 2000

(Polychronopoulos *et al.*, 1969). On potato (*Solanum tuberosum*), wilt disease complex was caused jointly by the golden cyst nematodes *G. rostochiensis* and *V. dahliae* (Harrison, 1971; Storey and Evans, 1987). LaMondia (1992) reported a synergetic interaction between *G. tabacum* and *F. oxysporum* with enhanced wilt severity on tobacco (LaMondia, 1992). The disease complexes involving interactions between *H. daverti* (dry bean cyst nematode) and *Sclerotium rolfsii* on clover, *H. cajani* and *F. udum* on pigeon pea and *H. oryzicola* and *S. rolfsii* on rice caused wilt with considerably greater severity than that caused by the fungus alone (Nordmeyer and Sikora, 1983; Sharma and Nene, 1989; Correia *et al.*, 2023).

Root-lesion nematode

The root-lesion nematode, *Pratylenchus* spp., is a migratory endoparasite of underground parts of the plant. All stages of the nematode are infective, and upon entering the root tissue of the host, the nematode larvae move inter- and intracellularly in the cortex and feed on the root parenchyma (Khan, 2016). Necrotic lesions are formed on the root due to migration and feeding activities of the nematode. The burrowing nematode, *Radopholus*, and stem and bulb nematode, *Ditylenchus*, are other economically important migratory endoparasites of crops (Khan *et al.*, 2021).

Pratylenchus spp. are sometime referred as exacerbators of wilt infections. The penetration, migration and feeding of the nematode larvae greatly promotes the invasion and infection of *Fusarium* and *Verticillium* species, thereby significantly contributing to the formation of wilt disease complex (Khan and Khan, 1998). The necrotic lesions and biochemical alterations induced by the nematode in the host tissue are attributed to the synergistic interaction with wilt fungi. The root lesions may serve as initial infection sites, enabling the fungus to establish on a few cells before spreading to adjacent healthy cells. Dave (1975) observed increased sensitivity of peppermint to *Verticillium* in the presence of *P. reglectus*, while a similar effect was noted for *V. dahliae* and *R. solani* infections in tuber plants (Taheri *et al.*, 2010). Ferraz and Lear (1976) reported worsening of *Verticillium* wilt in carnation plants in the presence of *P. dianthus*, indicating that the nematode induced morphological and physiological changes in the host enhanced the vulnerability to *Verticillium* infection. Fusarium wilt incidence was significantly exacerbated in alfalfa fields infested with *D. dipsaci* (Griffin, 1990) and *R. similis* in banana (Loos, 1959).

Ectoparasites and other nematodes

Ectoparasitic nematodes exhibit diverse behaviors and feeding site preferences, including surface feeders (e.g. *Belonolaimus* and *Tylenchorhynchus*), subsurface feeders (e.g. *Hoplolaimus* and *Helicotylenchus*) and deep feeders (e.g. *Rotylenchus* and *Helicotylenchus*) (Khan, 2008; Decraemer and Hunt, 2013). Their associations with hosts are generally neither specific nor long-lasting. However, host–parasite interactions of some ectoparasites, such as *Longidorus* and *Xiphinema*, are complex and can induce root galls, resulting to a relatively longer attachment to the feeding site compared to other ectoparasites (Khan *et al.*,

2021). Some ectoparasitic nematodes, such as *Tylenchorhynchus*, *Belonolaimus*, *Xiphinema*, *Helicotylenchus*, etc., have been observed to synergistically interact with *Fusarium* spp. on different crop species. On carnation plants, the infection of *F. oxysporum* f. sp. *vasinfectum* became highly severe due to the co-occurrence of *Criconemella curvata* (Ferraz and Lear, 1976). Similarly, concomitant inoculations with *F. oxysporum* f. sp. *dianthi* and *Belonolaimus longicaudatus* caused highly severe wilting on cotton (Cooper and Brodie, 1963; Da Silva *et al.*, 2019). Likewise, *Tylenchorhynchus claytoni* significantly enhanced the severity of fusarium wilt on cotton (Holdeman, 1956). Another species, *T. vulgaris*, was observed to accelerate the infection and wilt symptoms incited by *F. semitectum* on Egyptian clover (Hasan and Bhaskar, 2004; Singh and Agrawal, 2012). *Tylenchorhynchus* spp. have also been found to aggravate wilt on turfgrasses (Francl and Wheeler, 1993).

Mechanisms of Wilt Disease Complexes

The co-occurrence of nematodes with the fungi in the root zone has been found to be a key factor in inciting the disease complex. On this basis, it can be said that nematodes act as key predisposing agents, making plants more susceptible to the infections of phytofungi. The nematode invasion disrupts the plant's normal physiological processes and weakens its natural defense mechanisms. Soil-borne fungi, particularly the wilt-inducing species, take advantage of the compromised plant defense and invade the host quickly. During the root invasion the nematode causes several injuries on the root surface, which facilitate infection. Once inside the plant, the fungi proliferate and colonize the vascular system, leading to wilt diseases (Palomares-Rius *et al.*, 2017). The extent of increase in disease severity can vary depending on the type of nematodes involved. Ectoparasites, such as *Tylenchorhynchus*, *Belonolaimus*, *Xiphinema*, etc., may facilitate the infection in different dynamics of disease progression compared to their endoparasitic counterparts, such as *Pratylenchus*, *Meloidogyne*, etc., which burrow deeper into the root tissues (Duncan and Moens, 2013; Gheysen and Jones, 2013). Some of the important mechanisms that may operate in the formation of wilt disease complex are summarized below.

Nematodes enhance root exudation

The feeding by nematodes on plant roots may lead to an increase in the amount or alter the constituents of root exudates released by plants into the soil (Wang and Bergeson, 1974). Root exudates are rich in carbon-based compounds, including sugars, amino acids and proteins, specific to each plant species, and may account for 20% of the photosynthetic carbon of the plant. Root exudates are mainly classified into two types of compounds: low-molecular-weight compounds such as amino acids, organic acids, sugars, phenolics and other secondary metabolites which show high diversity, although they represent a relatively smaller share of the root exudates (Narula *et al.*, 2012; More *et al.*, 2020), and high-molecular-weight exudates such as mucilage (polysaccharides) and proteins that are less diverse but often comprise a large proportion of the root exudates (Narula *et al.*, 2012; Ma *et al.*, 2022).

Root exudates can attract and stimulate fungal propagules or may attract to a root part. Therefore, increased root exudation, caused by nematode feeding, increases the likelihood of faster fungal spore germination, thus escalating the chance of root invasion greater than normal. Moreover, if root growth is slowed down by nematodes, it further increases the chance of fungal infection because under normal situations the root comes close to the germinating spores for a shorter time (Huisman and Gerik, 1989). The spores of *Fusarium* spp. are known to germinate in response to exudates from susceptible plants, with the root tip being the primary source of attraction (Griffin, 1981). Nematode juveniles in the soil are also attracted towards root exudates, which also stimulate their egg hatching (Bell *et al.*, 2019). The root exudates from *M. incognita*-infected plants contained an altered composition of sugars and amino acids which may work in predisposing plants to fusarium wilt (Wang and Bergeson, 1974). Galled roots release exudates that stimulate chlamydospore germination, increasing *Fusarium* spore numbers in the rhizosphere of infected tomato plants compared to healthy roots (de Lamo and Takken, 2020).

Nematodes alter root architecture by creating wounds

Nematodes inflict wounds on the surface of host roots, creating distinct types of injuries such as micropunctures, ruptures, cell separation and perforations (Taylor, 1979). Ectoparasitic plant nematodes typically create micropunctures on the root surface, while the juveniles of endoparasites make perforations for entering into the root to start the endoparasitic phase of their life.

These injuries or perforations serve as easy and additional entry points for fungal pathogens, facilitating their invasion into the host tissue. Perry (1963) and Starr *et al.* (1989) showed that the perforations left behind by the root-lesion nematode *P. penetrans* were utilized by *F. oxysporum* and this synergistic interaction resulted in severe wilting in alfalfa (Mouza and Webster, 1982). It was earlier demonstrated by Powell (1979) that mechanical wounding caused by plant nematodes on host roots may facilitate the invasion of soil-borne fungi. Westerlund *et al.* (1974) noted that *F. oxysporum* f. sp. *ciceris* may require wounds for efficient invasion into chickpea roots. Similarly, Khan and Akram (1997) reported higher wilt severity incited by *F. oxysporum* f. sp. *lycopersici* in tomato due to pre-inoculation with *M. incognita*. However, mechanical injuries may not always be involved in all disease complexes, as in some cases the artificial wounding on roots fails to promote wilt disease. It is likely that the wounding along with some physiological alterations in the host operates in the synergistic interactions of *Fusarium* and *Verticillium* species with plant nematodes.

Nematodes induce root necrosis

The migratory endoparasitic nematodes such as *Pratylenchus*, *Radopholus*, *Ditylenchus*, etc., cause necrosis in the host tissue which varies with the nematodes (Khan and Jairajpuri, 2012). Some facultative fungi such as *Verticillium* species switch over from dead organic matter to live tissue faster than directly invading the live tissue (Conroy *et al.*, 1972). The root lesions caused by *P. penetrans* on eggplant and tomato are readily colonized by *V. dahliae*, and thereafter the fungus mycelium spreads from the dead cells to the adjoining healthy cells. A similar interaction is also known to occur between wilt fungi and borrowing nematodes.

Nematodes modify the host substrate

Nematodes, with their remarkable adaptability to host plants, often induce modifications in the substrate they inhabit (Escobar and Fenoll, 2015). These alterations prove critical in making the internal host environment conducive for the invasion and/or infection of the sedentary pathogens, including the wilt-inducing fungi (Topalović and Heuer, 2019). One of the primary ways nematodes modify the host substrate is through the release of various enzymes and metabolites (Sikder and Vestergård, 2019). As nematodes feed on plant tissues, they introduce digestive enzymes through saliva into the host. The enzymes break down complex organic compounds into simpler forms, making them more utilizable for microorganisms, also including plant-pathogenic fungi (Khanna *et al.*, 2021). This enzymatic activity is basically aimed at providing extracorporeally digested food to the feeding nematodes (Hussey and Williamson, 1998).

Nematode feeding induces various modifications in the host plants, such as changes in hormone levels, gene expression and secondary metabolite production (Siddique and Grundler, 2018). These alterations can affect plant defense responses and host susceptibility to other pathogens. Overall, nematode-induced modifications in the host substrate have far-reaching effects on soil health, plant physiology and ecosystem dynamics. The nematode-induced alterations may be localized with regard to local impact or translocatable modifications (systemic in action), impacting secondary pathogens at sites away from the nematode invasion.

Localized modification

Certain nematodes induce remarkable morphological alterations in a specific host part. For example, the root galls incited by endoparasitic nematodes (*Meloidogyne* and *Nacobbus*) and ectoparasitic nematodes (*Xiphinema* and *Longidorus*) are important localized modifications. Some fungi show marked affinity to the galls for initial invasion and colonization. Host substrate modifications have emerged as the primary driving force in enhancing fungal pathogenesis (Pitcher, 1965; Powell, 1979).

In addition to galls, specialized nurse cells are distinctive localized host modifications which are induced by sedentary nematodes such as *Meloidogyne*, *Heterodera*, *Globodera*, *Rotylenchulus*, etc. (Khan, 2008). Powell and Nusbaum (1960) were among the first to show how nematode-induced localized alterations in host tissue can pave the way for secondary pathogen infections. They observed that *Phytophthora parasitica* var. *nicotianae* hyphae thrived well in the enlarged, multinucleated and nutritionally rich nurse cells (giant cells) in the phloem parenchyma created by *M. incognita* in tobacco root galls. The fungal hyphae efficiently invaded giant cells, leading to their destruction, and subsequently penetrated vascular tissue, regardless of cultivar susceptibility. The giant cells provide an enriched environment conducive to pathogen growth, establishment and further invasion, because these cells are metabolically highly active and nutritionally very rich (Huang, 1985). The nurse cells (syncytia) created by *Heterodera cajani* were reported to be invaded first by *F. udum* in pigeon pea roots (Correia *et al.*, 2023). Hyphae of *F. oxysporum* luxuriantly colonized the galls caused by *M. incognita* and *M. javanica*, and the fungus mycelium from the galled tissue spread to adjoining tissues quickly.

Translocatable modifications

The process of inducing translocatable modifications by plant nematodes within the host is a vital mechanism that regulates the synergistic interactions and formation of disease complexes. Bowman and Bloom (1966) were the first to observe the existence of nematode-induced substances or chemicals that can move within the plant system from one part to another. These translocatable host alterations may facilitate *Fusarium* infection at sites distant from the initial nematode invasion site. Webster (1985) further documented the production of translocatable substances by nematodes within roots, which subsequently become systemic and are transported throughout the plant. This phenomenon was demonstrated through physically split root systems of tobacco cultivars engaging *M. incognita* and *F. oxysporum* f. sp. *nicotianae* in an interaction experiment (Melendez and Powell, 1967). The study utilized split-root tests and included both susceptible and resistant cultivars of tobacco. The results demonstrated that when half of the root system was infected with *F. oxysporum* f. sp. *nicotianae*, and the other half with *M. incognita*, the wilt was as severe as in the plants having both parts of the root system inoculated simultaneously with the fungus and nematode. The split-root model when tried on cotton with *F. oxysporum* f. sp. *vasinfectum* and *M. incognita* also proved true (Dyer, 2020). Although the split-root studies have revealed the significance of translocatable substances in promoting wilt severity, a specific mechanism alone may not be attributed to disease enhancement. Hillocks (1986) demonstrated that increased wilt severity was observed in treatments where plants thriving on soil contaminated with *Meloidogyne* were subsequently inoculated with the wilt fungus.

Nematodes alter the rhizospheric microbiome

Nematode-induced soil disturbances may promote the activity and colonization of other microorganisms, maintaining the soil ecosystem. Furthermore, nematodes can influence the composition and activity of the rhizosphere microbiome (Elhady *et al.*, 2021). Through feeding activities, excretion of metabolites and increasing the organic matter (decaying root fragments), nematodes create microenvironments that may favor the growth of certain microbial species over others (Midha *et al.*, 2017). The altered microbiome in the rhizosphere may induce greater pressure on the plant defense, leading to the promotion of fungus pathogenesis on the host.

Altering host resistance

The unique feature of nematode interactions with phytofungi and bacteria is the alteration in the host response which renders resistant cultivars susceptible to the secondary pathogen (Khan and Jairajpuri, 2010a). This consequence of nematode interactions poses a serious challenge to breeding efforts aiming to develop disease-resistant cultivars. The physical and/or chemical modifications induced by nematodes may lead to weakening or dysfunctioning of the genes conferring the resistance. For instance, the chickpea cv. Avrodhi, bred as tolerant to fusarium wilt, succumbed to the wilt fungus in nematode-infested fields, despite possessing functional resistance genes (Ali *et al.*, 2010). This phenomenon highlights the physiological alterations induced by nematodes, which can compromise the effectiveness of resistance genes. Resistance to fusarium wilt is typically categorized as qualitative and quantitative resistance. Qualitative resistance, governed by multiple genes, relies on incompatibility reactions of the host with the pathogen. Nematode infestation can strengthen the competitive capability by making available enhanced nutrition to the fungus, which may be regulated by some genes (Sidhu and Webster, 1974). The first record of interaction was noticed by Atkinson (1892) on the onset of fusarium wilt of cotton cultivars in the presence of root-knot nematode. Similar situations have also been reported in some cultivars of vegetable and pulse crops (Ali *et al.*, 2010; Desaeger *et al.*, 2023).

Role of Environmental Factors on Wilt Disease Complexes

The environment around the host and pathogen can significantly influence the relationship between the interacting wilt fungi and nematodes, consequently impacting the development and severity of the disease complex. Edaphic factors such as soil temperature, moisture, pH, etc., may significantly enhance the formation of disease complexes.

Soil type and texture

Soil type and texture may influence the multiplication and infection of wilt fungi as well as of the phytonematodes. Generally, light soils favor both groups of pathogens. Cotton wilt tended to become prevalent in sandy and acidic soils, which also happen to offer a conducive environment for most plant nematodes. Severe foliar wilt was observed in fields with coarse-textured sandy soils infested with *M. incognita* (Jeffers and Roberts, 1993; Wang and Roberts, 2006). Soil reclamation efforts, particularly liming the fields, have been shown to be significantly effective in reducing wilt incidence. However, other *formae speciales*, such as *F. oxysporum* f. sp. *vasinfectum*, thrive in heavy clay soils.

Soil moisture and temperature

Soil moisture plays a significant role in influencing egg hatching and the movement of nematode larvae (Wilcox-Lee and Loria, 1987). *Verticillium* spores, on the other hand, germinate and thrive well on roots across a wide range of soil moisture levels (Besri, 1980). Studies on potato crops have revealed that relatively high rainfall leads to an increase in soil moisture in the early growing season, coupled with high temperatures later in the season, which were found correlated with the higher yield decline due to *V. dahliae* and *P. penetrans* (Francl *et al.*, 1990). This relationship is presumed to be linked to high soil moisture tension, transiently reduced soil oxygen levels and enhanced mobility of the nematode larvae (Cappaert *et al.*, 1990; Wheeler *et al.*, 1991).

The impact of soil temperature on wilt in peppermint was investigated under greenhouse conditions (Faulkner and Bolander, 1969). Results indicated that a soil temperature of 24°C was found optimum for infestation of wilt caused by *V. dahliae* alone, whereas 27°C was found optimum for the plants inoculated with both the fungus and nematode (*P. neglectus*). The synergistic interaction between the above two pathogens was observed across 18–30°C.

Fertilizers and Soil pH

Inorganic fertilizers such as ammonia nitrogen have been noted to enhance wilt incidence, whereas the application of nitrate nitrogen to the soil induced a decrease in fusarium wilt (Spiegel and Netzer, 1984). This observation is linked to the effects of these fertilizers on soil pH. The addition of nitrate nitrogen elevates the soil pH, whereas ammonium nitrogen lowers the pH. This suggests that soil pH may exert a significant impact on wilt severity, indirectly by affecting the availability of micronutrients (Louvet, 1989). In one study, Francl *et al.* (1988) examined the impact of soil texture on the interaction of *Verticillium* and *Pratylenchus*. Interestingly, no variation in the yield attributable to *V. dahliae* or *P. penetrans* in different textured soils was observed. These findings align with earlier field research supporting the efficacy of potassium fertilizer and lime, along with resistant cultivars, in managing fusarium wilt (Tisdale and Dick, 1939). While the precise mechanism behind potassium mitigation of wilting remains unknown, cotton growers have successfully grown resistant cultivars, applying the combination of lime and potassium fertilizer for years to prevent fusarium wilt (Bell, 1989).

Management of Wilt Disease Complexes

Various management strategies, including cultural, physical, biological, host resistance and chemical, as well as integrated disease management (IDM) approaches, have been devised and developed to manage nematode and fungal disease problems. Each of these approaches has distinctive advantages and disadvantages. For example, chemical controls negatively impact the environment, leading to banning of many such chemicals (Onkendia *et al.*, 2014). Crop rotation and fallowing are often impractical for smallholding farmers because of limited land resources. The presence of various pathotypes, isolates and races within fungi and nematodes also reduces the long-term effectiveness of resistant genotypes (Onkendia *et al.*, 2014). Further, limited treatments are available exclusively to manage disease complexes. These limitations have driven the search for alternative, sustainable and environmentally friendly methods for managing fungus–nematode disease complexes.

Upon recognizing a disease complex in the field, the cause of the disease first needs to be ascertained by soil and root analyses (Nilsson *et al.*, 2012). Thereafter appropriate management modules should be applied to target both the pathogens, with greater emphasis on the organism capable of causing greater damage independently and governing the disease complex. IDM may prove more effective as it may address both nematodes and fungi. Host resistance for both pathogens is the most effective option, but it has limited availability at present.

Disease complex management can be achieved through various tactics, including protective, population-reducing, preventive and supportive measures. Protective measures have been found to effectively prevent the spread of pathogen(s) to new fields or areas. This can be achieved by ensuring the farm inputs are disease-free. In this direction, the use of disease-free planting material and sanitized equipment will prove effective. Other preventive methods include crop rotation, removal of volunteer plants, cultural methods, solarization and flooding. Host resistance is an effective approach against nematodes and has proved economic and successful in many crops (Nilsson *et al.*, 2012).

Biocontrol is an effective and eco-friendly method of management of plant-pathogenic fungi (Stirling, 1991; O'Brien, 2017) and nematodes (Khan, 2016). There are different kinds of biocontrol fungi which can be applied before or at planting, and may give good control of the mono- or multipathogenic diseases (Stirling, 1991; Khan and Anwer, 2011; Khan *et al.*, 2017b, 2021; Sikora and Roberts, 2018). Biocontrol fungi such as *Aspergilus niger*, *Purpureocillium lilacinum*, *Pochonia chlamydosporia*, etc., biocontrol bacteria such as *Pasteuria penetrans*, *Bacillus subtilis*, etc. (Jatala, 1986; Stirling, 1991; Kerry, 2000; Khan, 2007) and phosphate solubilizing microorganisms such as *Pseudomonas*, *Bacillus*, *Penicillium* and *Aspergillus* species/strains (Sikora and Fernandez, 2005; Khan *et al.*, 2009, Khan *et al.*, 2016a, Khan *et al.*, 2016b) may work effectively in suppressing the soil populations of nematodes in agricultural fields. The mycoparasitic fungus *Trichoderma* spp.

has emerged as a most promising and effective biocontrol agents to control plant-pathogenic fungi, especially *Fusarium*, *Verticillium*, *Pythium*, etc. (Khan and Mohiddin, 2018). The fungus has also been found to be quite effective against phytonematodes (Mohiddin *et al.*, 2010). The fungus is recommended to be applied on seeds or in soil through mass production on farmyard manure (Khan *et al.*, 2011). The seed treatment or soil application of *Trichoderma* and *Pseudomonas* biopesticides may significantly suppress the severity of fusarium wilt, root knot and the resulting disease complexes (Sikora and Roberts, 2018; Mohammed and Khan, 2021; Khan *et al.*, 2023a).

Generally, a single method has not been found effective against the disease complexes. In such a situation IDM may work better. The IDM strategies include monitoring of the pathogens (sampling, investigation, and sensing technologies to detect pathogens in the farm or plant tissues), soil management (cleaning farm implements and machinery, field quarantine, stubble management, volunteer plant and weed control, water management, nutrient management, organic matter and tillage), targeted control (biocontrol, solarization, trap crops, granular nematicides and flooding), crop rotation (using non-host crops and cover crops), and cultivar selection and establishment (choosing resistant varieties, adjusting sowing and planting dates, harvest dates, seed rates, and transplanting) (Sikora *et al.*, 2022).

The selection of suitable and specific tools and measures, and their timely application, such as use of disease-free and certified planting materials to prevent the entry of pathogen in the field (Gupta and Kumar, 2020; Haque and Khan, 2021) followed by seed or root-dip treatment with thiophanate-methyl or Metalaxyl targeting fungi and abamectin or thiamethoxam against nematodes, will provide effective protection to the germinating seeds and establishing the nursery (Sardrood and Goltapeh, 2018). However, growing cover crops such as marigold (*Tagetes* spp.) will reduce the infestation load of the nematode in soil (Riascos-Ortiz *et al.*, 2022). Application of biopesticides such as *T. harzianum*, *T. asperellum* and *P. fluorescens* in the soil will suppress pathogens and promote soil and plant health (Hamrouni *et al.*, 2024). Further, at a later stage, if disease is not controlled satisfactorily, fluopyram (suppressive to both the fungus and nematode) or a mixture of appropriate fungicide and nematicide may be applied to control the disease complex (Haque and Khan, 2022).

Conclusion and Future Prospects

Wilt disease complexes pose a serious threat to pulse and vegetable production, and a challenge to plant disease management. The *Fusarium*–*Meloidogyne* disease complex is most destructive and prevalent in important agricultural crops, although disease complex management is challenging but not invincible, and can be achieved with logical and systematic efforts. The management action should be initiated by understanding the nature of the pathogens involved in the disease complex and their population dynamics along with accurate diagnosis followed by using an integrated module targeting both the pathogens involved in the disease. Nematodes play a vital biopredisposing role, and generally make crop cultivars more susceptible by weakening host resistance. Sometimes, even resistant cultivars become vulnerable to *Fusarium* spp. and other fungi. The serious consequences of the biopredisposing role of nematodes warrants the necessity of incorporating nematodes into wilt resistance breeding programs. Identifying novel genes responsible for nematode-induced host modifications and developing their inhibitors through modern molecular techniques may serve as a milestone in developing stable host resistance. Considering the socio-economic conditions of farmers in several developing countries, it is crucial to draft and implement state policies on incentives and subsidies to farmers for implementing disease management programs as well as adopting crop diversification programs. Establishing accessible diagnostic and advisory services to smallholding farmers in particular, to offer vital information on the preventive and curative aspects will greatly help in promoting disease management programs, including those targeting the disease complexes.

References

Abawi, G., Moktan, K., Stewart, C., Hadad, R. and Jones, L. (2012) Northeastern division meeting abstracts. *Phytopathology* 101, S257.

Abd-Elgawad, M.M.M. (2021) Biological control of nematodes infecting eggplant in Egypt. *Bulletin of the National Research Centre* 45(1), 1–9.

Abdel-Momen, S.M. and Starr, J.L. (1998) *Meloidogyne javanica–Rhizoctonia solani* disease complex of peanut. *Fundamentals of Applied Nematology* 21, 611–616.

Ali, S.S., Naimuddin, A.M. and Ali, M. (2010) Nematode infestation in pulses. In: Khan, M.R. and Jairajpuri, M.S. (eds) *Nematode Infestations Part I: Food Crops. National Academy of Sciences*. Allahabad, India, pp. 288–325.

Aoki, T., O'Donnell, K., Homma, Y. and Lattanzi, A.R. (2003) Sudden-death syndrome of soybean is caused by two morphologically and phylogenetically distinct species within the *Fusarium solani* species complex *F. Virguliforme* in North America and *F. Tucumaniae* in South America. *Mycologia* 95(4), 660–684.

Archana, T.S., Kumar, D., Kumar, V. and Shukuru, B.N. (2023) Root-knot disease complex: An interactive perspective with microorganisms. In: Ahmad, F. and Blázquez, G.N. (eds) *Root-Galling Disease of Vegetable Plants*. Springer, Singapore, pp. 237–251.

Atkinson, G.F. (1892) Some diseases of cotton. *Alabama Agricultural Experiment Station Bulletin* 41, 61–65.

Back, M.A., Haydock, P.P.J. and Jenkinson, P. (2002) Disease complexes involving plant parasitic nematodes and soilborne pathogens. *Plant Pathology* 51(6), 683–697.

Barker, R.K. and McGawley, E.C. (1998) Interrelations with other microorganisms and pests. In: Sharma, S.B. (ed.) *The Cyst Nematodes*. Springer, Dordrecht, The Netherlands, pp. 266–293.

Bélair, G., Coulombe, J. and Dauphinais, N. (2018) Management of *Pratylenchus penetrans* and *Verticilllium* symptoms in strawberry. *Phytoprotection* 98(1), 1–3.

Bell, A.A. (1989) Role of nutrition in diseases of cotton. In: Englehard, A.W. (ed.) *Soilborne Plant Pathogens: Management of Diseases with Macro- and Microelements*. APS Press, St Paul, MN, pp. 167–204.

Bell, C.A., Lilley, C.J., McCarthy, J., Atkinson, H.J. and Urwin, P.E. (2019) Plant-parasitic nematodes respond to root exudate signals with host-specific gene expression patterns. *PLoS Pathogens* 15(2), e1007503.

Bergeson, G.B. (1972) Concepts of nematode–fungus associations in plant disease complexes: A review. *Experimental Parasitology* 32, 301–314.

Bertrand, B., Nunez, C. and Sarah, J.L. (2000) Disease complex in coffee involving *Meloidogyne arabicida* and *Fusarium oxysporum*. *Plant Pathology* 49, 383–388.

Besri, M. (1980) Effects of water potential changes on growth of *Fusarium oxysporum* f. sp. *lycopersici* and *Verticillium dahliae*. *Phytopathologische Zeitschrift* 99.

Bhagawati, B., Goswami, B.K. and Singh, C.S. (2000) Management of disease complex of tomato caused by *Meloidogyne incognita* and *Fusarium oxysporum* f. sp. *lycopersici* through Bioagent. *Indian Journal of Nematology* 30(1).

Bhattarai, S., Haydock, P.P.J., Back, M.A., Hare, M.C. and Lankford, W.T. (2009) Interactions between the potato cyst nematodes, *Globodera pallida*, *G. rostochiensis*, and soil-borne fungus, *Rhizoctonia solani* (AG3), diseases of potatoes in the glasshouse and the field. *Nematology* 11(4), 631–640.

Bowers, J. (1996) Infection and colonization of potato roots by *Verticillium dahliae* as affected by *Pratylenchus penetrans* and *P. crenatus*. *Phytopathology* 86, 614–621.

Bowman, P. and Bloom, J.R. (1966) Breaking the resistance of tomato varieties to Fusarium wilt by *Meloidogyne incognita*. *Phytopathology* 56, 871.

Briar, S.S., Khan, M.R. and Zwart, R. (2023) Nematode problems in pulse crops and their sustainable management. In: Khan, M.R. and Quintanilla, M. (eds) *Nematode Diseases of Crops and Their Sustainable Management*. Academic Press, Cambridge, MA, pp. 183–204.

Cappaert, M.R., Powelson, M.L. and Christensen, N.W. (1990) Soil moisture, genotype, and *Verticillium dahliae* interactions in potato early dying. *Phytopathology* 80, 971.

Castillo, P., Mora-Rodriguez, M.P., Navas-Cortes, J.A. and Jimenez-Diaz, R.M. (1998) Interactions of *Pratylenchus thornei* and *Fusarium oxysporum* f. sp. *ciceri* on chickpea. *Phytopathology* 88.

Cauquil, J. and Shepherd, R.L. (1970) Effect of root-knot nematode–fungi combinations on cotton seedling disease. *Phytopathology* 60(3), 448–451.

Chindo, P.S., Khan, F.A. and Erinle, I.D. (1991) Reaction of three tomato cultivars to two vascular diseases in presence of the root-knot nematode, *Meloidogyne incognita* race 1. *Crop Protection* 10(1), 62–64.

Conroy, J.J., Green, R.J., Jr and Ferris, J.M. (1972) Interaction of *Verticillium albo-atrum* and the root lesion nematode, *Pratylenchus penetrans*, in tomato roots at controlled inoculum densities. *Phytopathology* 62(3), 362–366.

Cooper, W. and Brodie, B. (1963) A comparison of fusarium-wilt indices of cotton varieties with root-knot and sting nematodes as predisposing agents. *Phytopathology* 53, 1077–1080.

Correia, V.R., Khan, M.R., Oliveira Lima, F.S. and Ahamad, F. (2023) Nematode problems in vegetable legumes and their sustainable management. In: Khan, M.R. and Quintanilla, M. (eds) *Nematode Diseases of Crops and Their Sustainable Management*. Academic Press, Cambridge, MA, pp. 311–336.

Da Silva, M.B., Richard, F., Doan, H.K., Robert, L., Kemerait, R.C. *et al*. (2019) Fusarium wilt of cotton may commonly result from the interaction of *Fusarium oxysporum* f. sp. *vasinfectum* with *Belonolaimus longicaudatus*. *Journal of Nematology* 51.

Dave, G.S. (1975) Interrelationships of *Rhizoctonia solani* with *Heterodera glycines*, *Pratylenchus scribneri* and *Tylenchoryhnchus martini* on "clark 63" soybeans. Doctoral dissertation, University of Illinois at Urbana-Champaign, Champaign, IL.

Decraemer, W. and Hunt, D.J. (2013) Structure and classification. In: Perry, R.N. and Moens, M. (eds) *Plant Nematology*. CAB International, Wallingford, UK, pp. 3–39.

de Lamo, F.J. and Takken, F.L.W. (2020) Biocontrol by *Fusarium oxysporum* using endophyte-mediated resistance. *Frontiers in Plant Science* 11, 37.

Desaeger, J., Khan, M.R. and Silva, E.H.C. (2023) Nematode problems in tomato, okra, and other common vegetables, and their sustainable management. In: Khan, M.R. and Quintanilla, M. (eds) *Nematode Diseases of Crops and Their Sustainable Management*. Academic Press, Cambridge, MA, pp. 223–250.

Dhami, D.S., Kaur, S., Sharma, A., Sharma, S.P., Dhillon, N.K. *et al*. (2024) Characterization of multiple disease resistance in melons (*Cucumis melo* L.) against *Meloidogyne incognita*, *Fusarium oxysporum* and tomato leaf curl Lalampur virus. *Plant Genetic Resources* 22(1), 27–36.

Duncan, L.W. and Moens, M. (2013) Migratory endoparasitic nematodes. In: Perry, R.N. and Moens, M. (eds) *Plant Nematology*. CAB International, Wallingford, UK, pp. 144–178.

Dyer, D.R. (2020) Evaluation of the genetic diversity, temporal distribution, and management of *Fusarium oxysporum* f. sp. *vasinfectum* in the national cotton fusarium wilt evaluation field. Doctoral dissertation, Auburn University, Auburn, AL.

Elhady, A., Topalović, O. and Heuer, H. (2021) Plants specifically modulate the microbiome of root-lesion nematodes in the rhizosphere, affecting their fitness. *Microorganisms* 9(4), 679.

Escobar, C. and Fenoll, C. (2015) *Plant Nematode Interactions: A View on Compatible Interrelationships*. Academic Press, New York.

Faulkner, L.R. and Bolander, W.J. (1969) Interaction of *Verticillium dahliae* and *Pratylenchus minyusin* verticillium wilt of peppermint: Effects of soil temperature. *Phytopathology* 59, 68–70.

Ferraz, S. and Lear, B. (1976) Interaction of four plant parasitic nematodes and *Fusarium oxysporum* f. sp. *dianthi* on carnation. *Experientiae* 22.

Francl, L.J., Rowe, R., Riedel, R.M. and Madden, L.V. (1988) Effects of three soil types on potato early dying disease and associated yield reduction. *Phytopathology* 78(2), 159–166.

Francl, L.J., Madden, L.V., Rowe, R.C. and Riedel, R.M. (1990) Correlation of growing season environmental variables and the effect of early dying on potato yield. *Phytopathology* 80(4), 425.

Francl, L.J. and Wheeler, T.A. (1993) Interaction of plant-parasitic nematodes with wilt-inducing fungi. In: Khan, M.W. (ed.) *Nematode Interactions*. Chapman & Hall, London, pp. 79–103.

Gao, X., Jackson, T.A., Hartman, G.L. and Niblack, T.L. (2006) Interactions between the soybean cyst nematode and *Fusarium solani* f. sp. *glycines* based on greenhouse factorial experiments. *Phytopathology* 96(12), 1409–1415.

Gheysen, G. and Jones, J.T. (2013) Molecular aspects of plant–nematode interactions. In: Perry, R.N. and Moens, M. (eds) *Plant Nematology*. CAB International, Wallingford, UK, pp. 274–298.

Goswami, J. and Tiwari, D.D. (2007) Management of *Meloidogyne incognita* and *Fusarium oxysporum* f. sp. *lycopersici* disease complex on tomato by *Trichoderma harzianum*, *Tinopsora longifolia* and *Glomus fasciculatum*. *Pesticide Research Journal* 19(1), 51–55.

Goze Ozdemir, F.G., Arici, S.E. and Elekcioglu, I.H. (2022) Disease complex of *Rhizoctonia solani* and *Meloidogyne hapla* chitwood, 1949 (Nemata: Meloidogynidae) on tomato. *Mediterranean Agricultural Sciences* 35(2), 69–74.

Griffin, G.D. (1990) Pathological relationship of *Ditylenchus dipsaci* and *Fusarium oxysporum* f. sp. *medicaginis* on alfalfa. *Journal of Nematology* 22(3).

Griffin, G.I. (1981) Physiology of conidium and chlamydospore germination in Fusarium. In: Nelson, P.E., Toussoun, T.A. and Cook, R.I. (eds) *Fusarium: Diseases, Biology and Taxonomy*. Pennsylvania State University Press, University Park, PA, pp. 331–339.

Gupta, A. and Kumar, R. (2020) Management of seed-borne diseases: An integrated approach. In: Kumar, R. and Gupta, A. (eds) *Seed-Borne Diseases of Agricultural Crops: Detection, Diagnosis & Management*. Springer, Dordrecht, The Netherlands, pp. 717–745.

Hallmann, J. and Meressa, B.H. (2018) Nematode parasites of vegetables. In: Sikora, R.A., Coyne, D., Hallmann, J. and Timper, P. (eds) *Plant Parasitic Nematodes in Subtropical and Tropical Agriculture*, 3rd edn. CAB International, Wallingford, UK, pp. 346–410.

Hamrouni, R., Regus, F., Da Silva, A.-M., Orsiere, T., Boudenne, J.L. *et al*. (2024) Current status and future trends of microbial and nematode-based biopesticides for biocontrol of crop pathogens. *Critical Reviews in Biotechnology* 45, 333–352.

Haque, Z. and Khan, M.R. (2021) *Handbook of Invasive Plant-Parasitic Nematodes*. CAB International, Wallingford, UK.

Haque, Z. and Khan, M.R. (2022) Evaluation of fluopyram against single and concomitant inoculation of *Rhizoctonia solani* and *Meloidogyne graminicola* infesting rice crop. *Phytoparasitica* 50(4), 811–822.

Harris, A.R. and Ferris, H. (1991) Interactions between *Fusarium oxysporum* f. sp. *tracheiphilum* and *Meloidogyne* spp. in *Vigna unguculata*. 2. specificity of different taxa. *Plant Pathology* 40(3), 457–464.

Harrison, J.A.C. (1971) Association between the potato cyst-nematode, *Heterodera rostochiensis* Woll., and *Verticillium dahliae* Kleb. in the early-dying disease of potatoes. *Annals of Applied Biology* 67(2), 185–193.

Hasan, N. and Bhaskar, R.B. (2004) Disease complex of berseem involving nematode and two soil inhabiting fungi. *Annals of Plant Protection Sciences* 12, 159–161.

Haseeb, A., Sharma, A. and Abuzar, S. (2005) Field screening of lentil cultivars against *Meloidogyne incognita–Fusarium oxysporum* f. sp. *lentis* disease complex. *Indian Journal of Nematology* 35(2).

Haseeb, A., Amin, A. and Sharma, A. (2007) Disease complex in *Pisum sativum* involving *Meloidogyne incognita* and *Fusarium oxysporum* f. sp. *pisi*. *Annals of Plant Protection Sciences* 15(1).

Hillocks, R.J. (1986) Localised and systemic effects of root-knot nematode on incidence and severity of Fusarium wilt in cotton. *Nematologica* 32(2), 202–208.

Holdeman, Q.L. (1956) The effect of the tobacco stunt nematode on the incidence of Fusarium wilt in flue-cured tobacco. *Phytopathology* 46, 129.

Hua, G.K.H., Timper, P. and Ji, P. (2019) *Meloidogyne incognita* intensifies the severity of fusarium wilt on watermelon caused by *Fusarium oxysporum* f. sp. *niveum*. *Canadian Journal of Plant Pathology* 41(2).

Huang, S.C. (1985) Formation, anatomy and physiology of giant cells induced by root-knot nematodes. In: Sasser, J.N. and Carter, C.C. (eds) *An Advanced Treatise on Meloidogyne Volume I. Biology and Control*. North Carolina State University Graphics, Raleigh, NC, pp. 143–154.

Huisman, O.C. and Gerik, J.S. (1989) Dynamics of colonization of plant roots by *Verticillium dahliae* and other fungi. In: Tjamos, E.C. and Beckman, C.H. (eds) *Vascular Wilt Disease of Plants*. Springer, Berlin, pp. 1–18.

Hussey, R.S. and Williamson, V.M. (1998) Physiological and molecular aspects of nematode parasitism. *Plant and Nematode Interactions* 36, 87–108.

Jatala, P. (1986) Biological control of plant-parasitic nematodes. *Annual Review of Phytopathology* 24(1), 453–489.

Jeffers, D.P. and Roberts, P.A. (1993) Effect of planting date and host genotype on the root-knot nematode–fusarium wilt disease complex of cotton. *Phytopathology* 83(6), 645.

Johnson, A.W. and Littrell, R.H. (1969) Effect of *Meloidogyne incognita*, *M. hapla*, and *M. javanica* on the severity of Fusarium wilt of chrysanthemum. *Journal of Nematology* 1(2), 122–125.

Johnson, D.A. and Santo, G.S. (2001) Development of wilt in mint in response to infection by two pathotypes of *Verticillium dahliae* and co-infection by *Pratylenchus penetrans*. Plant Disease 85(11), 1189–1192.

Karlsson, A. (2006) Possible interactions between Rhizoctonia solani and plant-parasitic nematodes (PPN) in Swedish potato fields – a pilot study. MSc thesis, Swedish University of Agricultural Sciences, Uppsala.

Karssen, G., Wesemael, W. and Moens, M. (2013) Root-knot nematodes. In: Perry, R.N. and Moens, M. (eds) *Plant Nematology*, 2nd edition. CAB International, Wallingford, UK, pp. 73–108.

Kavitha, K. and Thirukumaran, K. (2019) Integrated management of nematode disease complex in tuberose (*Polianthes tuberose* L). Journal of Krishi Vigyan 8(1), 170.

Kerry, B.R. (2000) Rhizosphere interactions and the exploitation of microbial agents for the biological control of plant-parasitic nematodes. *Annual Review of Phytopathology* 38, 423–441.

Khan, M.R. (2005) *Biological Control of Fusarial Wilt and Root-Knot of Legumes*. Department of Biotechnology, Ministry of Science and Technology, Government of India Publication, New Delhi.

Khan, M.R. (2008) *Plant Nematodes – Methodology, Morphology, Systematics, Biology and Ecology*. Science Publishers, Enfield, NH. DOI: 10.1201/b10766.

Khan, M.R. (2016) Nematode biocontrol agents: Diversity and effectiveness against phytonematodes in sustainable crop protection. *Indian Phytopathology* 69(4s), 453–463.

Khan, M.R. (2023a) Nematode pests of agricultural crops, a global overview. In: Khan, M.R. (ed.) *Novel Biological and Biotechnological Applications in Plant Nematode Management*. Springer, Singapore, pp. 3–45.

Khan, M.R. (2023b) Plant nematodes, hidden constraints in global crop production. In: Khan, M.R. and Quintanilla, M. (eds) *Nematode Diseases of Crops and Their Sustainable Management*. Academic Press, Cambridge, MA, pp. 3–23.

Khan, M.R. and Akram, M. (1997) Effect of certain bacterial and fungal bioagents on a disease complex caused by *Meloidogyne incognita* and *Fusarium oxysporum* f.sp. *lycopersici* on tomato. In: *Abstracts of the International Conference on Integrated Plant Disease Management, New Delhi*, p. 252.

Khan, M.R. and Akram, M. (2000) Effects of certain antagonistic fungi and rhizobacteria on wilt disease complex of tomato caused by *Meloidogyne incognita* and *Fusarium oxysporum* f.sp. *lycopersici*. *Nematologia Mediterranea* 28(2).

Khan, M.R. and Anwer, A. (2011) Fungal bioinoculants for plant disease management. In: Paul, M., Clinton, M. and Ahmad, I. (eds) *Microbes and Microbial Technology*. Springer, Dordrecht, The Netherlands, pp. 447–488.

Khan, M.R. and Jairajpuri, S. (2010a) Nematode infestation in industrial crops – national scenario. Allahabad, India, pp. 1–19.

Khan, M.R. and Jairajpuri, S. (2010b) *Nematode Infestations Part I: Food Crops*. National Academy of Sciences, Allahabad, India.

Khan, M.R. and Jairajpuri, S. (2010c) *Nematode Infestations Part II: Industrial Crops*. National Academy of Sciences, Allahabad, India.

Khan, M.R. and Jairajpuri, S. (2012) *Nematode Infestations Part III: Horticultural crops*. National Academy of Sciences, Allahabad, India.

Khan, M.R. and Khan, M.W. (1994a) Effects of simulated acid rain and root-knot nematode on tomato. *Plant Pathology* 43, 41–49.

Khan, M.R. and Khan, M.W. (1994b) Single and interactive effects of root-knot nematode and coal-smoke on okra. *New Phytologist* 126, 337–342.

Khan, M.R. and Khan, M.W. (1997) Effects of the root-knot nematode, meloidogyne incognita, on the sensitivity of tomato to sulfur dioxide and ozone. *Environmental and Experimental Botany* 38, 117–130.

Khan, M.R. and Khan, M.W. (1998) Biomanagement of fungus–nematode wilt disease complexes of pulses. *Botanica* 48, 150–156.

Khan, M.R. and Mohiddin, F.A. (2018) Trichoderma: Its multifarious utility in crop improvement. In: Prasad, R., Gill, S.S. and Tuteja, N. (eds) *New and Future Developments in Microbial Biotechnology and Bioengineering: Crop Improvement Through Microbial Biotechnology*. Elsevier, Amsterdam, pp. 263–291.

Khan, M.R. and Sharma, R.K. (2020) Fusarium–nematode wilt disease complexes, etiology and mechanism of development. *Indian Phytopathology* 73(4), 615–628.

Khan, M.R. Sharma, N. and Singh, H.B. (eds). (2007) *Prospects of microbial control of root-knot nematodes infecting vegetable crops*. Plant Health Management. International Book Distributing, Biotechnology, pp. 643–665.

Khan, M.R., Altaf, S., Mohiddin, F.A., Khan, U. and Anwer, A. (2009) Biological control of plant nematodes with phosphate solubilizing microorganisms. In: Khan, M.S. and Zaidi, A. (eds) *Phosphate Solubilizing Microbes for Crop Improvement*. Nova Science Publishers, Hauppauge, NY, pp. 395–426.

Khan, M.R., Mohidin, F.A., Khan, U. and Ahamad, F. (2016a) Inoculant rhizobia suppressed root-knot disease, and enhanced plant productivity and nutrient uptake of some field-grown food legumes. *Acta Agriculturae Scandinavica, Section B* 68(2), 1–9.

Khan, M.R., Mohidin, F.A., Khan, U. and Ahamad, F. (2016b) Native *Pseudomonas* spp. suppressed the root-knot nematode in *in-vitro* and *in-vivo*, and promoted the nodulation and grain yield in the field grown mungbean. *Biological Control* 101, 159–168.

Khan, M.R., Majid, S., Mohidin, F.A. and Khan, N. (2011) A new bioprocess to produce low cost powder formulations of biocontrol bacteria and fungi to control Fusarial wilt and root-knot nematode of pulses. *Biological Control* 59(2), 130–140.

Khan, M.R., Shahid, S., Mohidin, F.A. and Mustafa, U. (2017a) Interaction of *Fusarium oxysporum* f. sp. *gladioli* and *Meloidogyne incognita* on gladiolus cultivars and its management through corm treatment with biopesticides and pesticides. *Biological Control* 115, 95–104.

Khan, M.R., Mohiddin, F.A. and Haque, Z. (2017b) Phosphate solubilizing microorganisms and their effectiveness against phytonematode. In: Chakrabourty, B.N., Singh, D. and Sharma, P. (eds) *Microbial Antagonist: Their Role in Biological Control in Plant Diseases*. Today & Tomorrow's Printers and Publishers, New Delhi, pp. 515–553.

Khan, M.R., Ahamad, I. and Shah, H. (2021) Emerging important nematode problems in field crops and their management. In: Singh, K.P., Jahagirdar, S. and Sarma, B.K. (eds) *Emerging Trends in Plant Pathology*. Springer, Dordrecht, The Netherlands, pp. 33–62.

Khan, M.R., Mohiddin, F.A., Haque, Z. and Sharma, R.K. (2023a) Disease complexes and their sustainable management. In: Khan, M.R. and Quintanilla, M. (eds) *Nematode Diseases of Crops and Their Sustainable Management*. Academic Press, Cambridge, MA, pp. 65–93.

Khan, M.R., Haque, Z., Akram, M. and Ansari, M.S.A. (2024) Nematode problems in ornamental plants. In: Khan, M.R. and Haque, Z. (eds) *Diseases of Ornamental, Aromatic and Medicinal Plants*. Bentham Science, Sharjah, UAE, pp. 182–206.

Khan, M.W. (1993) Mechanism of interactions between nematodes and other plant pathogens. In: Khan, M.W. (ed.) *Nematode Interactions*. Chapman & Hall, London, pp. 55–78.

Khan, M.W. and Khan, M.R. (1995) Fungus–nematode interactions. In: Swarup, G., Dasgupta, D.R. and Gill, J.S. (eds) *Nematode Pest Management– An Appraisal of Eco-Friendly Approaches. Nematological Society of India*. New Delhi, pp. 70–78.

Khanna, T.M., Baiocchi, G., Callaghan, M., Creutzig, F., Guias, H. *et al.* (2021) A multi-country meta-analysis on the role of behavioural change in reducing energy consumption and CO_2 emissions in residential buildings. *Nature Energy* 6(9), 925–932.

Kolombia, Y.A. and Fabiyi, O.A. (2023) Nematode problems in tuber crops and their sustainable management. In: Khan, M.R. and Quintanilla, M. (eds) *Nematode Diseases of Crops and Their Sustainable Management*. Academic Press, Cambridge, MA, pp. 251–278.

LaMondia, J.A. (1992) Predisposition of broadleaf tobacco to Fusarium wilt by early infection with *Globodera tabacum tabacum* or *Meloidogyne hapla*. *Journal of Nematology* 24, 425–431.

Loos, C.A. (1959) Symptom expression of Fusarium wilt disease of the gros michel banana in the presence of *Radopholus similis* (Cobb, 1893) Thorne, 1949 and *Meloidogyne incognita acrita* Chitwood, 1949. *Proceedings of the Helminthological Society of Washington* 26, 103–111.

Lopez-Nicora, H.D. and Niblack, T.L. (2018) Nematode interactions with other pathogens. In: Perry, R.N., Moens, M. and Jones, J.T. (eds) *Cyst Nematodes*. CAB International, Wallingford, UK, pp. 271–304.

Louvet, J. (1989) Microbial populations and mechanisms determining soil suppressiveness to Fusarium wilts. In: Tjamos, E.C. and Beckman, C.H. (eds) *Vascular Wilt Diseases of Plants*. Springer, Berlin, pp. 367–384.

Ma, W., Tang, S., Dengzeng, Z., Zhang, D., Zhang, T. *et al.* (2022) Root exudates contribute to below-ground ecosystem hotspots: A review. *Frontiers in Microbiology* 13, 937940.

Manikandan, R. and Raguchander, T. (2014) PCR amplification of ITS region of *Fusarium oxysporum* f. sp. *lycopersici* isolates of tomato in Tamil Nadu. *Trends in Biosciences* 7(15), 1937–1940.

Manisha, B.S.P., Mamata, M.S.K., Abhi, R., Yadav, S. and Meena, R. (2023) Impact of the interaction between *Meloidogyne javanica* and *Fusarium oxysporum* on the okra plantation. *International Journal of Plant & Soil Science* 35(18), 1872–1879.

Martin, M.J., Riedel, R.M. and Rowe, R.C. (1982) *Verticillium dahliae* and *Pratylenchus penetrans*: Interactions in the early dying of potato in Ohio. *Phytopathology* 72, 640–644.

Martin, W.J., Newsom, L.D. and Jones, J.E. (1956) Relationship of nematodes to the development of Fusarium wilt in cotton. *Phytopathology* 4, 285–289.

McGawley, E.C. (1992) Interactions with other organisms. In: Riggs, R.D. and Wrather, J.A. (eds) *Biology and Management of the Soybean Cyst Nematode*. APS Press, Saint Paul, MN, pp. 97–106.

Meena, S., Archana, S., Jonathan, E.T. and Raguchander, T. (2021) Study of viability and effect of liquid formulation of *Bacillus subtilis*, Bbv 57 against nematode–fungus disease complex in tomato *Lycopersicon Esculentum* Mill. *Indian Journal of Nematology* 51(2), 192–201.

Melendez, P.L. and Powell, N.T. (1967) Histological aspects of the Fusarium wilt-root-knot complex in flue-cured tobacco. *Phytopathology* 57, 286–292.

Midha, A., Schlosser, J. and Hartmann, S. (2017) Reciprocal interactions between nematodes and their microbial environments. *Frontiers in Cellular and Infection Microbiology* 7, 144.

Mohammed, R.K.A. and Khan, M.R. (2021) Management of root-knot nematode in cucumber through seed treatment with multifarious beneficial microbes in polyhouse under protected cultivation. *Indian Phytopathology* 74(4), 1035–1043.

Mohiddin, F.A. and Khan, M.R. (2014) Root-knot nematode: Ignored soil borne plant pathogen causing root diseases of chickpea. *European Journal of Biotechnology and Bioscience* 2(1), 4–10.

Mohiddin, F.A. and Khan, M.R. (2019) Efficacy of newly developed biopesticides for the management of wilt disease complex of chickpea (*Cicer arietinum* L.). *Legume Research* 42(4), 550–556.

Mohiddin, F.A., Khan, M.R., Khan, S.M. and Bhat, B.H. (2010) Why trichoderma is considered super hero (super fungus) against the evil parasites? *Plant Pathology Journal* 9(3), 92–102.

More, S.S., Shinde, S.E. and Kasture, M.C. (2020) Root exudates a key factor for soil and plant: An overview. *Pharma Innovation Journal* 8, 449–459.

Mouza, B.E. and Webster, J.M. (1982) Suppression of alfalfa growth by concomitant populations of *Pratylenchus penetrans* and two Fusarium species. *Journal of Nematology* 14, 364–367.

Murali, R., Priyadarshan, A., Nagaraja, H.C. and Narayanaswamy, H. (2023) Effectiveness of integrated management on *Fusarium*–nematode wilt complex disease of banana cv. ney poovan (AB): Insights from a field study. *International Journal of Plant & Soil Science* 35(23), 626–638.

Nagesh, M., Hussaini, S.S. and Chidanandaswamy, B.S. (2006) Management of *Meloidogyne incognita* and *Fusarium oxysporum* f. sp. *lycopersici* wilt complex using antagonistic fungi in tomato. Nematologia Mediterranea 34, 63–68.

Naik, V.N., Sharma, D.D., Thippeswamy, T. and Sivaprasad, V. (2015) Integrated approach for management of root disease complex in mulberry (*Morus* spp.).). Indian Journal of Nematology 45(2), 207–212.

Narula, N., Kothe, E. and Behl, R.K. (2012) Role of root exudates in plant-microbe interactions. *Journal of Applied Botany and Food Quality* 82(2), 122–130.

Neal, D.C. (1954) The reniform nematode and its relationship to the incidence of Fusarium wilt of cotton at Baton Rouge, Louisiana. Phytopathology 44, 447–450.

Nilsson, I., Rölin, Å. and Schie, A. (2012) Odla Potatis: En Handbok, 1st edn. Hushållningssällskapet Skaraborg, Skara, Sweden.

Nordmeyer, D. and Sikora, R.A. (1983) Studies on the interaction between *Heterodera daverti*, *Fusarium avenaceum* and *F. oxysporum* on *Trifolium Subterraneum*. *Revue de Nématologie* 6, 193–198.

O'Brien, P.A. (2017) Biological control of plant diseases. *Australasian Plant Pathology* 46(4), 293–304.

Okungbowa, F.I. and Shittu, H.O. (2012) Fusarium wilts: An overview. *Environmental Research Journal* 6(2), 83–102.

Onkendia, E.M., Kariuki, G.M., Marais, M. and Moleleki, L.N. (2014) The threat of root-knot nematodes (*Meloidogyne* spp.) in Africa: A review. *Plant Pathology* 63, 727–737. DOI: 10.1111/ppa.12202.

Palomares-Rius, J.E., Escobar, C., Cabrera, J., Vovlas, A. and Castillo, P. (2017) Anatomical alterations in plant tissues induced by plant-parasitic nematodes. *Frontiers in Plant Science* 8.

Patil, J.A., Yadav, S. and Kumar, A. (2020) Evaluation of organic oils for the management of root-knot nematode, *Meloidogyne incognita* and fungus infesting cucumber under polyhouse conditions. Indian Journal of Nematology 50(2), 79–86.

Pepple, G.A., UKoima, H.N. and Chukunda, A.F. (2010) Studies on the physiology and rooting potential of *Fusarium solani* and *Aspergillus niger* isolated from nematode lesion of yam tuber. Acta Agronomica Nigeriana 10(1), 35–41.

Perry, D.A. (1963) Interaction of root-knot and fusarium wilt of cotton. Empire Cotton Growing Review 40, 41–47.

Philbrick, A.N., Adhikari, T.B., Louws, F.J. and Gorny, A.M. (2020) *Meloidogyne enterolobii*, a major threat to tomato production: Current status and future prospects for its management. Frontiers in Plant Science 11, 606395.

Pitcher, R.S. (1965) Interrelationships of nematodes and other plant pathogens of plants. *Helminth Abstracts (Series B* 34, 1–17.

Polychronopoulos, A.G., Houston, B.R. and Lownsbery, B.F. (1969) Penetration and development of *Rhizoctonia solani* in sugar beet seedlings infected with *Heterodera schachtii*. *Phytopathology* 59, 482–485.

Powell, N.T. and Nusbaum, C.J. (1960) The black shank-root-knot complex in flue-cured tobacco. *Phytopathology* 50, 899–906.

Powell, N.T. (1979) Internal synergisms among organisms inducing disease. In: Horsfall, J.G. and Cowling, E.B. (eds) *Plant Disease: An Advanced Treatise. Volume IV. How Pathogens Induce Disease*. Academic Press, New York, pp. 113–133.

Priti, K., Rao, M.S., Pavani, K.V., Umamaheswari, R. and Gopalakrishnan, C. (2018) Combination of biopesticide formulations to manage *Meloidogyne incognita* and *Fusarium oxysporum* f. sp. *cepae* disease complex in onion. *Pest Management in Horticultural Ecosystems* 24(2), 163–172.

Rajinikanth, R., Rao, M.S., Pavani, K.V., Manojkumar, R., Chaya, M.K. *et al.* (2013) Management of nematode induced disease complex in seedlings of cauliflower (*Brsassica oleraceae* var. *botrytis*) using bio-pesticides. *Pest Management in Horticultural Ecosystems* 19(2), 203–210.

Ramakrishnan, S., Devrajan, K., Poornima, K., Deepa, S.P. and Jain, R.K. (2014) Management of root-knot nematode and Fusarium wilt complex in crossandra (*Crossandra undulaefolia*). *Indian Journal of Nematology* 44(2), 123–125.

Rao, V.K. and Krishnappa, K. (1995) Model system for the integrated management of *Meloidogyne–Fusarium* wilt disease complex of chickpea. *Indian Journal of Nematology* 25(2), 219–221.

Rao, M.S., Shylaja, M. and Naik, D. (2003) Management of nematode induced wilt disease complex in tuberose (*Polianthus tuberosa* L.) cultivar prajwal using *Pochonia chlamydosporia* (*Verticillium chlamydosporium*) and *Trichoderma harzianum*. *Journal of Ornamental Horticulture* 6(4), 341–346.

Rao, M.S., Dwivedi, M.K., Kumar, R.M., Chaya, M.K., Rathnamma, K. *et al.* (2014) Evaluation of bioefficacy of *Bacillus subtilis* (NBAIMCC-B-01211) against disease complex caused by *Meloidogyne incognita* and *Fusarium oxysporum* f. sp. *vasinfectum* in okra. *Pest Management in Horticultural Ecosystems* 20(2).

Rao, M.S., Umamaheswari, R., Prabu, P., Priti, K., Chaya, M.K. *et al.* (2017) Field performance of *Pseudomonas putida* (IIHR pp-2) for the management of *Meloidogyne incognita* and *Fusarium oxysporum* f. sp. *vasinfectum* disease complex in okra (*Abelmoschus esculentus*). *Vegetos – An International Journal of Plant Research* 30(4), 33.

Reddy, P.P. (2008) *Diseases of Horticultural Crops: Nematode Problems and their Management*. Scientific Publishers, Jodhpur, India.

Regmi, H. and Desaeger, J. (2020) Integrated management of root-knot nematode (*Meloidogyne* spp.) in florida tomatoes combining host resistance and nematicides. *Crop Protection* 134, 105170.

Riascos-Ortiz, D., Mosquera-Espinosa, A.T., Agudelo, F., Oliveira, C.M.G. and Muñoz Flórez, J.E. (2022) Non-conventional management of plant-parasitic nematodes in Musaceas crops. In: Chaudhary, K.K. and Meghvansi, M.K. (eds) *Sustainable Management of Nematodes in Agriculture*. Springer, Cham, Switzerland, pp. 381–422.

Riedel, R.M. (1988) Interactions of plant-parasitic nematodes with soil-borne plant pathogens. *Agriculture, Ecosystems & Environment* 24(1–3), 281–292.

Ross, J.P. (1965) Predisposition of soybeans to fusarium wilt by *Heterodera glycines* and *Meloidogyne incognita*. *Phytopathology* 55, 361–364.

Sardrood, B.P. and Goltapeh, E.M. (2018) Effect of agricultural chemicals and organic amendments on biological control fungi. *Sustainable Agriculture Reviews* 31, 217–359.

Sasser, J.N. (1989) *Plant-Parasitic Nematodes: The Farmer's Hidden Enemy*. North Carolina State University Graphics, Raleigh, NC.

Sharma, S.B. and Nene, Y.L. (1989) Interrelationship between *Heterodera cajani* and *Fusarium udum* in pigeonpea. *Nematropica* 19, 21–28.

Sharma, H.K., Singh, S. and Meher, H.C. (2007) Management of root-knot nematode, *Meloidogyne incognita* and wilt fungus, *Fusarium oxysporum* complex on okra. *Pesticide Research Journal* 19(2), 176–179.

Siddiqi, M.R. (ed.) (2000) *Tylenchida: Parasites of Plants and Insects*, 2nd edn. CAB International, Wallingford, UK.

Siddique, S. and Grundler, F.M. (2018) Parasitic nematodes manipulate plant development to establish feeding sites. *Current Opinion in Microbiology* 46, 102–108.

Sidhu, G. and Webster, J.M. (1974) Genetics of resistance in the tomato to root-knot nematode–wilt-fungus complex. Journal of Heredity 65(3), 153–156.

Sikder, M.M. and Vestergård, M. (2019) Impacts of root metabolites on soil nematodes. *Frontiers in Plant Science* 10, 1792.

Sikora, R.A. and Fernandez, E. (2005) Nematode parasites of vegetables. In: Luc, M., Sikora, R.A. and Bridge, L. (eds) Plant Parasitic Nematodes in Subtropical and Tropical Agriculture, 2nd edn. CAB International, Wallingford, UK, pp. 319–392.

Sikora, R.A. and Roberts, P.A. (2018) Management practices: An overview of integrated nematode management technology. In: Sikora, R.A., Coyne, D., Hallmann, J. and Timper, P. (eds) Plant Parasitic Nematodes in Subtropical and Tropical Agriculture, 3rd edn. CAB International, Wallingford, UK, pp. 795–838.

Sikora, R.A., Desaeger, J. and Molendijk, L.P.G. (2022) *Integrated Nematode Management: State-of-the-Art and Visions for the Future*. CAB International, Wallingford, UK.

Singh, A. and Agrawal, M. (2012) Effects of waste water irrigation on physical and biochemical characteristics of soil and metal partitioning in *Beta vulgaris* L. *Agricultural Research* 1(4), 379–391.

Singh, S. and Goswami, B.K. (2001) Studies on the management of disease-complex caused by root-knot nematode, *Meloidogyne incognita* and wilt fungus, *Fusarium oxysporum* on cowpea by neem cake and carbofuran. *Indian Journal of Nematology* 31(2), 122–125.

Singh, N., Jat, M.K. and Singh, B. (2021) Outbreak of leaf miner and root knot nematode–wilt disease complex on cucumber – a new threat under protected cultivation in district Gurugram Haryana. *Agricultural Science Digest – A Research Journal* 41(3), 460–463.

Sowmya, D.S. and Rao, M.S. (2012) Combined effect of pseudomonas putida and paecilomyces lilacinus in the management of disease complex in gladiolus grandiflorus L. *Pest Management in Horticultural Ecosystems* 18(2), 204–209.

Spiegel, Y. and Netzer, D. (1984) Effect of nitrogen form at various levels of potassium, on the *Meloidogyne–Fusarium* wilt complex in muskmelon. *Plant and Soil* 81(1), 85–92.

Starr, J.L., Jeger, M.L., Martyn, R.D. and Schilling, K. (1989) Effects of *Meloidogyne incognita* and *Fusarium oxysporum* f. sp. *vasinfectum* on plant mortality and yield of cotton. *Phytopathology* 79(6).

Stiles, C.M., Glawe, D.A., Noel, G.R. and Pataky, J.K. (1993) Reproduction of *Heterodera glycines* on soybean in nonsterile soil infested with cyst-colonizing fungi. *Nematropica* 23, 81–89.

Stirling, G.R. (1991) *Biological Control of Plant Parasitic Nematodes: Progress, Problems and Prospects*. CAB International, Wallingford, UK.

Storey, G.W. and Evans, K. (1987) Interactions between *Globodera pallida* juveniles, *Verticillium dahliae* and three potato cultivars, with descriptions of associated histopathologies. *Plant Pathology* 36(2), 192–200.

Taheri, A., Hollamby, G.J. and Vanstone, V.A. (2010) Interaction between root lesion nematode, *Pratylenchus neglectus*. *New Zealand Journal of Crop and Horticultural Science* 22, 181–185.

Taylor, C.E. Lamberti, F. and Taylor, C.E. (eds). (1979) *Meloidogyne interrelationships with microorganisms*. Academic Press, London, pp. 375–398.

Tharshani, N. and Sivapalan, A. (2009) Eco-friendly management of tomato root-knot nematode (*Meloidogyne* spp.) in Calci-Red Yellow Latasol: Environment Sri Lanka blog. *Phytopathology* 91, 681–693.

Tisdale, H.B. and Dick, J.B. (1939) The development of wilt in a wilt-resistant and in a wilt-susceptible variety of cotton as affected by N-P-K ratio in fertilizer. *Soil Science Society of America Journal* 4(C), 333–334.

Topalović, O. and Heuer, H. (2019) Plant–nematode interactions assisted by microbes in the rhizosphere. *Current Issues in Molecular Biology* 30(1), 75–88.

Turner, S.J. and Subbotin, S.A. (2013) Cyst nematodes. In: Perry, R.N. and Moens, M. (eds) *Plant Nematology*, 2nd edn. CAB International, Wallingford, UK, pp. 109–143.

Vats, R. and Dalal, M.R. (1997) Interaction between *Rotylenchulus reniformis* and *Fusarium oxysporum* f. sp. *pisi* on pea (*Pisum sativum* L.). *Annals of Biology* 13, 239–242.

Wang, C. and Roberts, P.A. (2006) A fusarium wilt resistance gene in *Gossypium barbadense* and its effect on root-knot nematode–wilt disease complex. *Phytopathology* 96(7), 727–734.

Wang, E.L.H. and Bergeson, G.B. (1974) Biochemical changes in root exudate and xylem sap of tomato plants infected with *Meloidogyne incognita*. *Journal of Nematology* 6, 192–194.

Webster, J.M. (1985) Interaction of *Meloidogyne* with fungi on crop plants. In: Sasser, J.N. and Carter, C.C. (eds) *An Advanced Treatise on Meloidogyne Volume I. Biology and Control*. North Carolina State University Graphics, Raleigh, NC, pp. 183–192.

Westerlund, F.U., Campbell, R.N. and Simble, K.A. (1974) Fungal root-rot and wilt of chick-pea in California. Phytopathology 64, 432–436.

Wheeler, T.A., Rowe, R.C. and Madden, L.V. (1991) Effect of soil moisture on incidence of potato early dying and tuber yield. *Phytopathology* 81, 1156.

Whitney, E.D. and Doney, D.L. (1973) The effects of *Heterodera schachtii* and *Aphanomyces cochlioides* on root-rot. *Journal of Sugarbeet Research* 17(3), 240–245.

Whitney, E.D. (1974) Synergistic effect of *Pythium ultimum* and the additive effect of *P. aphanidermatum* with *Heterodera schachtii* on sugarbeet. Phytopathology 64(3), 380.

Wilcox-Lee, D. and Loria, R. (1987) Effect of nematode parasitism on plant–water relations. In: Veech, J.A. and Dickson, D.W. (eds) *Vistas on Nematology*. Society of Nematologists, Hyattsville, MD, pp. 260–266.

Winkler, H.E., Hetrick, B.A.D. and Todd, T.C. (1994) Interactions of *Heterodera glycines*, *Macrophomina phaseolina*, and *Mycorrhizal fungi* on soybean in Kansas. *Journal of Nematology* 26(4 Suppl), 675–682.

Xing, L. and Westphal, A. (2006) Interaction of *Fusarium solani* f. sp. glycines and *Heterodera glycines* in sudden death syndrome of soybean. *Ecology and Epidemiology* 96(7), 763–770.

Yang, H., Powell, N.T. and Barker, K.R. (1976) Interactions of concomitant species of nematodes and *Fusarium oxysporum* f. sp. *vasinfectum* on Cotton. *Journal of Nematology* 8(1).

Zhang, Y., Li, S., Li, H., Wang, R., Zhang, K.Q. *et al*. (2020) Fungi–nematode interactions: Diversity, ecology, and biocontrol prospects in agriculture. *Journal of Fungi* 6(4), 206.

7 Fungus–Nematode Root-rot Disease Complex

Mujeebur Rahman Khan[1†], Abolfazl Hajihassani[2*†], Shabeg S. Briar[3] and Dayani Patuwatha Withanage[3]

[1]Department of Plant Protection, Faculty of Agricultural Sciences, Aligarh Muslim University, Aligarh, India; [2]Entomology and Nematology Department, Fort Lauderdale Research and Education Center, University of Florida, Davie, Florida, USA; [3]Olds College Centre for Innovation, Olds, Alberta, Canada

Abstract

This chapter explores the intricate world of root-rot diseases caused by plant-parasitic nematodes and fungal pathogens in agricultural settings, covering essential topics for understanding and managing these diseases effectively. It examines the far-reaching economic consequences of disease complexes worldwide, characterizing groups of nematode and fungal pathogens that contribute to synergistic interactions and their diverse effects on cultivated crops. Specifically, this chapter explores how these synergistic interactions amplify root-rot infections, culminating in devastating plant damage and substantial yield losses. Advanced and recent management approaches, ranging from cultural practices to chemical applications, biological control methods and integrated pest management strategies are examined. Finally, the chapter discusses the prospects and advancements in addressing the challenges posed by the complex interplay of root-rot fungi and nematodes, highlighting the importance of ongoing research, innovation and sustainable practices to ensure agricultural resilience and global food security.

Introduction

Plant-parasitic nematodes are worldwide in distribution and inflict slow and unrecognizable damage to crops, which ultimately assumes economic significance within 1–4 years, depending on cultivated crops (Khan and Jairajpuri, 2010; Hajihassani *et al.*, 2018). Plant-parasitic nematodes attack and injure all kinds of crops, including cereals, vegetables, oilseed, fruit, fiber and beverage crops (Briar *et al.*, 2018; Khan and Quintanilla, 2023). The yield declines caused by nematodes may account for US$200 billion, considering all major and minor crops (Khan, 2023a). In addition to causing direct damage, plant-parasitic nematodes are found frequently interacting with soil-borne plant-pathogenic fungi and bacteria (Powell, 1979), resulting in disease complexes (Khan, 1993). Hence, understanding nematode disease complexes is crucial for crop protection and food security (Back *et al.*, 2002; Khan, 2023b). Plant nematodes and root-rot fungi coexist in the root zone soil, engaging in complex interactions. However, nematodes

*Corresponding author: ahajihassani@ufl.edu
†Mujeebur Rahman Khan and Abolfazl Hajihassani have contributed equally to the work.

DOI: 10.1079/9781800625228.0007

lead and govern the formation of root-rot disease complexes due to their capability to interact in a multifaceted way. During the penetration and feeding processes, nematodes inflict wounds on the host surface, creating entry points for fungal pathogens (Khan and Khan, 1997). Nematode species can also alter plant tissue architecture to create a conducive environment for fungal proliferation, consequently facilitating fungal growth and reproduction at the expense of the host plant. Nematode feeding leads to the secretion of root exudates that induce changes in the rhizosphere, promoting the germination, growth and reproduction of fungal propagules in the root zone (Khan *et al.*, 2023b). The exudates can negatively impact rhizospheric microbes, such as actinomycetes, known for their antagonistic activity against certain plant pathogens (El-Tarabily and Sivasithamparam, 2006). The outcomes of simultaneous infections involving nematodes and fungi are diverse, with some combinations fostering nematode reproduction (e.g. *Fusarium oxysporum* f. sp. *ciceris-Pratylenchus thornei*) (Castillo *et al.*, 1998), while others may cause inhibition (e.g. *Heterodera filipjevi and Fusarium culmorum*) (Hajihassani *et al.*, 2013b) depending on the specific combination of partners involved in the nematode–fungus disease complex.

Root-rot-Causing Fungi and Their Impact on Crops

Root-rot diseases pose a significant threat to agricultural crops, causing substantial yield loss and jeopardizing crop survival. Root rot is typically a complex disease involving multiple pathogens (Williamson-Benavides and Dhingra, 2021). These pathogens infect plant roots and inhibit root growth, leading to a range of symptoms and ultimately suppressing the growth or killing the affected plant (Agrios, 2005). The severity of the root rot depends on the causal agent(s) involved, host susceptibility and environmental conditions (Khan and Rehman, 1998). The causal agents of root rot may include bacteria, oomycetes and fungi. These pathogens often need a weak point or an avenue for entry into the host plant. To meet this, other microorganisms need to come into play, such as nematodes, which unintentionally provide suitable entry points to root-rot-causing pathogens or help them in some other ways (Khan, 1993; Bhattarai *et al.*, 2009). Such an interaction between fungi and nematodes in the soil generally provides benefits to the former, thus making the disease complex. Several fungi such as *Rhizoctonia*, *Phytophthora*, *Macrophomina*, *Pythium*, *Sclerotinia*, *Thielaviopsis*, and *Fusarium* have been reported as causative of root rot on different crops. These fungi have been found to interact synergistically with several economically important plant nematodes, including *Meloidogyne*, *Heterodera*, *Globodera*, *Pratylenchus*, and *Rotylenchulus*.

Root-rot disease is a major limiting factor in the cultivation of various crops worldwide. For instance, *Fusarium* spp. and *Rhizoctonia* spp. have an estimated yield loss of 20–40% to alfalfa worldwide (Browne *et al.*, 2013). Pre- and post-emergence damping-off and root rot in pulses caused by *Pythium*, *Sclerotinia*, *Phytophthora*, etc., are important factors responsible for crop failures (Khan, 2005). The root-specific symptoms incited by the above pathogens may go unnoticed unless the seed germination phase is affected. Often the recovery is irreversible when the symptoms become visible on the above-ground parts. Browning and softening of root tips, root lesions, yellowing and wilting of the foliage, stunted growth, and reduced crop yield would be noticeable with the root-rot disease; however, in extreme situations, complete loss of the plant is inevitable (da Silva *et al.*, 2019; DuPont *et al.*, 2021; Chilakala *et al.*, 2023). Some root-rot pathogens are economically significant in causing post-harvest rots as well. These pathogens thrive under specific conditions, including moderate to high soil moisture, inadequate drainage, favorable temperatures, monoculture practices and notably other stressors that affect plant health (Forghani and Hajihassani, 2020; Faske *et al.*, 2022). The stressors include nematode infestations or unexpected climatic events such as droughts, floods and storms.

Avenues of Association of Nematodes with Root-rot Fungi

The infection process of root-rot fungi begins when fungal spores or mycelium come in contact with a susceptible host root (Nandris *et al.*, 1987). Dormant structures of these fungi, such as chlamydospores and sclerotia, are soil-borne and can survive in the soil under unfavorable conditions. However, they germinate under moist and moderate temperature conditions (Termorshuizen, 2014). *Pythium* and *Phytophthora* species produce zoospores that swim toward plant roots using flagella, guided by chemical signals released from the susceptible plants through root exudates (van West *et al.*, 2003). Once germinated, the fungal spores produce hyphae that penetrate the plant roots (Van Buyten and Höfte, 2013). This penetration can occur either through natural openings, such as root tips and wounds, or directly through the epidermal cell walls using enzymatic degradation and mechanical pressure. *Phytophthora* spp. employ both enzymatic lysis and mechanical force to penetrate the root epidermis (Erwin and Ribeiro, 1996; Bronkhorst, 2022) and colonize the root cortex. The fungus establishes intercellularly or intracellularly. The colonization involves the formation of specialized structures, such as haustoria, which facilitate nutrient uptake from host cells (Beckman, 1987). After invading the tissue, the root-rot fungi proliferate within the root tissues and disrupt the normal physiological processes of the host. The colonization leads to the decay of cortical cells and the degradation of cell walls, primarily occurring through cell-wall-degrading enzymes such as cellulases, pectinases and ligninases secreted by the fungus (Mendgen *et al.*, 1996). The invasion and proliferation of the pathogen cause significant structural damage to the root system. The roots become soft, water-soaked and discolored. In later stages, the roots may appear blackened and necrotic, indicative of severe tissue degradation and cell death (Agrios, 2005).

Naturally fertile soil can host a large number of fungal propagules, ranging from 10^4 to 10^6 CFU/g soil, along with various other soil-borne organisms as well as the nematodes which may number approximately 1×10^7 per m^2 of fertile soil. Most nematodes in soil are saprophytic, and have no adverse impact on agricultural crops, but a very small portion is parasitic migrating up and down in the soil profile and causing damage to host crops (Khan *et al.*, 2021; Marquez and Hajihassani, 2024). Soil with adequate moisture is conducive to the activities of both the nematodes as well as the fungi especially in root zone soil. Consequently, various interactions between nematodes and fungi develop in the soil and cause significant impacts on plants. Plant-pathogenic fungi have often been found in co-occurrence with plant nematodes, forming various interactions, including synergistic, antagonistic, or neutral. Synergism occurs when the combined damage from both organisms exceeds the sum of their individual effects. Antagonism is the opposite, where the combined damage is less than expected. Neutral interactions occur when the damage from the association equals the sum of individual damages. Additive interactions, where the organisms do not interact, can be difficult to distinguish from neutral interactions, as the resulting damage is similar (Back *et al.*, 2002). Root lesions caused by migratory endoparasitic nematodes such as *Pratylenchus* species may increase the susceptibility of plant tissues to fungal invasion, potentially accelerating root rotting. Root tips invaded by sedentary endoparasitic nematodes, such as *Meloidogyne* and *Heterodera* species, may also be common sites of fungal infections (Back *et al.*, 2002). Root colonization by root rot fungi can also lead to an increased presence of root-knot nematodes (Ma *et al.*, 2014). Studies have highlighted the significance of population levels of both nematodes and fungal pathogens in determining disease severity (Walker *et al.*, 2000; Hajihassani *et al.*, 2013a).

Significance of Root-rot Disease Complexes

Root-rot disease complexes pose a significant challenge to agriculture, causing frequent crop failures or severe crop damage worldwide. Understanding the significance of these disease complexes is crucial for developing and implementing appropriate management strategies to minimize their impact on crops. Yield losses induced by disease complexes vary greatly, ranging from marginally above the economic threshold to complete crop failure, depending on the combination of pathogens in the disease complex, the susceptibility of the host species and the prevailing environmental conditions (Williamson-Benavides and Dhingra, 2021;

Khan, 2023a). According to Baker and Parker (2003), the association between soil-borne pathogens has intensified because of a complex interplay of feedback mechanisms and communication networks involving multiple pathways of stress and environmental responses. Unpredictable climate conditions, including low or high temperatures, droughts and floods, are expected to continuously stress the crops, favoring the escalated activity of root-damaging pathogens (Jacobs *et al.*, 2019; Chilakala *et al.*, 2023).

Soil-borne fungi and nematodes are notoriously difficult to manage and can cause significant crop damage individually. When they interact to form a disease complex, yield losses are further aggravated, complicating management efforts (Khan *et al.*, 2023b). In field conditions, plants are often colonized by multiple pathogens simultaneously, and the significance of each species varies depending on factors such as geographical region, soil and climate conditions, and crop management practices (Šišić *et al.*, 2022; Marquez and Hajihassani, 2024). Plants affected by disease complexes with infected roots display a range of symptoms, including browning and softening of root tips, root necrosis of various sizes and colors, yellowing and wilting of shoots and leaves, arrested plant growth, and reduced crop productivity (Khan *et al.*, 2011; Kumari and Katoch, 2020). Identifying these symptoms and signs is crucial for early detection and effective control of disease complexes. Some root-rot species can persist in soil for extended periods, posing a significant risk to plant health and making their management a challenging task (Papavizas and Ayers, 1974). While some soil-borne pathogens involved in the disease complexes are host-specific, others have a broad host range, limiting the efficiency of crop rotation as a control method. This complexity highlights the need for a comprehensive approach to managing root-rot disease complexes, considering the specific conditions and pathogen interactions in each region and crop system. Root-rot disease complexes caused by combinations such as *Meloidogyne–Phytophthora*, *Meloidogyne–Fusarium* and *Meloidogyne–Sclerotinia* pose a serious issue in crop cultivation. The disease complex may lead to 50% decline in the market value of horticultural crops (Vanegas-Villa *et al.*, 2020). *Fusarium* spp. also cause rot in bulbus crops, e.g., corn rot of gladioli, and the resulting disease complex becomes severe in the presence of the root-knot nematode and leads to a 28–32% decline in flowering (Khan *et al.*, 2017).

Root-rot Disease Complexes Involving Nematodes and Different Fungi

Rhizoctonia

Rhizoctonia is a widespread, soil-borne fungal genus that thrives in diverse climates around the world (Sneh *et al.*, 2013). It is economically significant due to its impact on various crops, causing substantial yield losses up to 70% (Kiptoo *et al.*, 2021). *Rhizoctonia* species, especially *R. solani* and *R. bataticola*, have a wide host range that includes cereals, vegetables, ornamentals and turfgrass (Sneh *et al.*, 2013). These fungal species cause different diseases, such as root rot, damping-off, stem canker and sheath blight, depending on the host and age of the plant. The fungus can survive in soil and plant debris under moist and warm conditions for extended periods, making management efforts more challenging (Sharma and Ghosh, 2017).

R. solani is the root-rot-causing fungus that has been extensively studied in interaction studies with nematodes because of the discernible relationship it forms with the nematodes (Evans and Haydock, 1993). This fungus often develops a synergistic relationship with plant nematodes; as a result, the root rot and other associated symptoms become severe, and the plants suffer greater damage (Khan and Khan, 1995). Severe root rot may develop in okra and tomato due to infection with *R. solani* in the presence of *Meloidogyne incognita* (Golden and Van Gundy, 1975). Synergistic interaction of *R. solani* with *Globodera rostochiensis* on tomato (Dunn and Hughes, 1967) was observed in which the fungal infection became significantly aggravated. Field experiments have also reported synergistic interaction between *G. pallida* and *R. solani* resulting in increased incidence of infected stolons, stolon pruning, and stem canker of potato, with effects varying across different nematode population levels in the soil (Bhattarai et al., 2009). A study using gnotobiotic culture on radish showed that pre-inoculation with *M. hapla* helped *R. solani* to invade and colonize the roots efficiently (Khan and Muller, 1982). Another test revealed that concomitant inoculation of *M. javanica* and *R. solani* at the time of sowing of tomato seeds notably enhanced the damping-off (Nath *et al.*, 1984). Hasan and Khan (1985) observed that inoculation of *R. solani* in two cultivars of chili pepper caused some loss of host resistance against *M. incognita*. However, cowpea seedlings inoculation with *M. incognita* caused a breakdown of resistance to *R. solani* as well as to the nematode (Khan and Husain, 1989). Edin *et al.* (2019) found that combining root lesion nematodes (*P. crenatus* and *P. pentrans*) and *R. solani*, either simultaneously or sequentially, led to reduced potato tuber yields. However, the interaction between the two pathogens was not synergistic.

Synergistic interaction between the root-knot nematode and the root-rot fungus has been frequently observed on different crops where host resistance was affected and plants suffered greater damage (Chahal and Chhabra, 1984; Hasan and Khan, 1985). Goel and Gupta (1986a, b) found that inoculation with *M. javanica* decreased the resistance of chickpea cultivars to *R. bataticola* and exhibited higher root damage. Mishra *et al.* (1988) reported a greater incidence of root rot in a white jute field due to simultaneous inoculation of plants with *M. incognita* and *R. bataticola* over nematode absence. Inoculation of chrysanthemum roots with *Pratylenchus coffeae* significantly enhanced the root rot index incited by *R. solani* (Hasan, 1988). In this study, the reproduction of the nematode was increased on *R. solani*-infected plants; however, nematode reproduction decreased with *P. aphanidermatum*, but the fungus multiplication interestingly remained unaffected (Hasan, 1988). LaMondia and Martin (1989) reported that inoculation with a higher population of *P. penetrans* exacerbated black root rot of strawberry incited by Rhizoctonia sp. Palmateer *et al.* (2004) reported that inoculation with *Rotylenchulus reniformis* also synergized the infection by *R. solani* in cotton leading to severe root rot. Further, the root rot became severe when plants of eggplant (Prusty, 2017), okra (Saikia and Bhagawati, 2014), soybean (Anwar and Khan, 2002), chickpea (Siddiqui and Aziz, 2024) and mung bean (Gogoi and Mahanta, 2013) were inoculated concomitantly with *R. solani* and *M. incognita*. The combined infection of *M. incognita* and *R. bataticola* has also been found to severely affect okra (Singh *et al.*, 2003) and tomato (Mehta *et al.*, 1995).

Pythium

Pythium is a genus of oomycetes, which are water molds and are found worldwide (Haque and Khan, 2013; Ho, 2018). These pathogens are particularly common in soils that are moist and waterlogged (Sharma *et al.*, 2020), causing root rot and damping-off diseases and inflicting substantial yield reductions in susceptible crops such as vegetables, ornamentals, turfgrass and field (row) (Syed *et al.*, 2020). The fungus incites several diseases such as seed rot, seedling damping-off, root rot and stem rot. These infections flourish in wet conditions. *Pythium* spp. Can survive in soil and plant debris. Important *Pythium* species such as *P. aphanidermatum*, *P. ultimum*, *P. debaryanum*, *P. tracheiphilum* and *P. myriotylum* which can interact with plant nematodes and cause significant yield decline to a number of important crops.

Plant nematodes have been found to synergistically interact with *P. debaryanum* on cotton, leading to enhanced host susceptibility to fungal invasion (Brodie and Cooper, 1964). Hasan and Khan (1985) observed that when two cultivars of chili seedlings were inoculated with *M. incognita*, they lost resistance to *P. aphanidermatum*. However, in another study, pre-inoculation with *M. incognita* prevented *P. aphanidermatum* root rot in a fungus-susceptible cultivar, but the same nematode aggravated the infection of *F. solani* (Doshi and Mathur, 1987). Root rot severity caused by *P. aphanidermatum* was markedly enhanced in *P. coffeae*-infected chrysanthemum plants (Hasan, 1988), but nematode reproduction was suppressed in the presence of the fungus. A greater damping-off incidence has been reported when *M. incognita* and *P. aphanidermatum* were inoculated in cucumber (Morris *et al.*, 2016). Additive interactions have also been reported between *P. tracheiphilum* and *M. hapla* in lettuce (Gracia *et al.*, 1991) and between *P. myriotylum* and *M. incognita* in ginger (Lanjewar and Shukla, 1985). Damping-off and root rot in alfalfa seedlings were exacerbated in the presence of *M. hapla* in the *P. ultimum* infested soil (Townshend, 1984). The same nematode species was also reported to aggravate root rot caused by *P. myriotylum* in peanuts (Garcia and Mitchell, 1975). Damping-off of tomato seedlings increased notably when *M. javanica* and *P. debaryanum* were inoculated at the time of seed sowing (Nath *et al.*, 1984).

Macrophomina

Macrophomina is a genus of soil-borne fungi which occurs worldwide, with a particular prevalence in warm and arid environments (Shahid and Khan, 2016a, b; Ghosh *et al.*, 2018). This genus is an economically significant pathogen since it affects a broad range of crops, including soybeans, cotton, sunflowers and legumes. The most prevalent species, *M. phaseolina*, causes charcoal rot and inflicts economic yield losses, especially under drought stress (Basandrai *et al.*, 2021). The fungus infects more than 500 plant species, causing root rot, stem rot and seedling blight. Its ability to persist in soil and plant debris for long periods challenges management strategies.

M. phaseolina has been found to benefit from the presence of plant nematodes. Ahmad and Khan (2021) observed a significant increase in root rot caused by *M. phaseolina* in the presence of *M. incognita* on mung bean. Root galling was suppressed in the presence of the fungus while nodulation was severely affected in the root system which became devoid of nodules (Fig. 7.1). Tiyagi *et al.* (1988) reported synergistic interactions between *M. phaseolina* and *M. javanica* resulting in greater damage in lentil when inoculated concomitantly. When the fungus was pre-inoculated, nematode reproduction was suppressed. *H. galeatus* and *Macrophomina* spp. have been reported to form a disease complex in soybean, which greatly increased root rot severity and decreased plant growth (McGawley *et al.*, 1984). Begum and Kumar (2005) documented positive interactions between *Heterodera cajani* and *M. phaseolina* on mung beans.

Phytophthora

Phytophthora forms a genus of plant-pathogenic oomycetes responsible for a range of devastating diseases in plants, including root rot, leaf blight and stem canker (Kamoun *et al.*, 2015). These fungi interact with nematodes causing significant damage to field and vegetable crops,

Fig. 7.1. Photographs depict the synergistic interaction between root-knot nematode *Meloidogyne incognita* and root-rot fungus *Macrophomina phaseolina* on green gram cv. SML-668. (A) Healthy root system with extensive nodulation (control). (B) Extensive galling caused by the nematode and the root system devoid of nodules. (C) Rotting on the root and girdling of the basal part of the stem caused by the root-rot fungus. (D) Extensive rotting of basal stem, roots and galls due to concomitant inoculation with the fungus and nematode. (Source: I. Ahmad and M. R. Khan, Aligarh Muslim University, India)

ornamental plants and fruit trees. The synergistic interaction between *M. incognita* or *M. javanica* and *P. nicotianae* in tobacco was reported to induce a mutually positive impact and greater decline in tobacco growth (Vanegas-Villa *et al.*, 2020). This interaction between the two pathogens had a profound impact on the quality of tobacco leaves, as well as a significant decline in both physical and chemical properties of tobacco leaves and the market price of the leaves. In soybeans, *H. glycines* and *P. sojae* were found to be highly damaging when inoculated jointly and caused a severe root-rot disease complex in soybeans. However, *P. sojae* adversely impacted *H. glycines* development. The cyst nematode also did not affect the resistance of a soybean cultivar to *P. Sojae* (Audette *et al.*, 2020).

Fusarium

Fusarium species are a group of fungi that generally cause wilting in plants, which frequently leads to root rotting when infection initiates through the root system. The wilt fungus inflicts significant damage on a wide range of crops, including cereals, vegetables, legumes and ornamentals (Khan, 2005; Khan *et al.*, 2017). *Fusarium* species cause various diseases, such as root rot, stem wilt and ear blight (Kamoun *et al.*, 2015). Nematodes interact with *Fusarium* species on different crops such as wheat, cotton, cowpea, tomato, chickpea, pea, cucurbits, etc. (Katsantonis *et al.*, 2003; Palmateer *et al.*, 2004; Hajihassani *et al.*, 2013a, b). Among nematode species, interaction with *Meloidogyne* species is well documented (Khan and Sharma, 2020). Numerous studies have reported that natural co-occurrence of *Meloidogyne* and *Fusarium* species has been recorded in the root zone of several crops which generally leads to a synergistic interaction, resulting in more severe disease symptoms and greater plant damage. Khan *et al.* (2011) observed wilting caused by *F. oxysporum* f. sp. *ciceris* and *F. udum* became severe in chickpea and pigeon pea, respectively, when plants were also inoculated with *M. incognita*. In a greenhouse study, inoculating tomato cultivars carrying root-knot nematode resistance genes with *F. oxysporum* f. sp. *lycopersici* race three at the planting time reduced *M. javanica* reproduction when both pathogens were introduced simultaneously (Regmi *et al.*, 2022). However, the cumulative effect of the co-infection led to more severe plant damage in tomatoes. Khan and Akram (2000) recorded a greater reduction in the yields of tomato cv. Pusa Ruby due to concomitant inoculation with *M. incognita* and *F. oxysporum* f. sp. *Lycopersici*, but there was a decrease in galling. Similar observations have also been observed in watermelon and other crops due to the co-occurrence of *F. oxysporum* f. sp. *lycopersici* and *F. oxysporum* f. sp. *niveum*, respectively, and *Meloidogyne* spp. (Lobna *et al.*, 2016; Hua *et al.*, 2019). The infection of *F. oxysporum* f. sp. *gladioli*in gladiolus cultivars was promoted by *M. incognita*, leading to severe colonization of the fungus and resulting crown rot and internal browning of the bulbs (Khan *et al.*, 2017). The co-infection of soybean plants with both *F. solani* and *H. glycines* caused severe root rot and sudden death syndrome due to the positive relationship between the two pathogens on soybean (McLean and Lawrence, 1993). The decrease in gall formation and nematode reproduction can be attributed to physiological changes induced by fungal infection, creating an unfavorable environment for nematode development and reproduction. However, some studies have shown that *R. solani*, *F. solani* and *F. oxysporum* extensively colonized the galls, including giant cells, leading to destruction of the feeding sites and consequent starvation of females, and subsequent significant declines in nematode populations occurred (Powell and Nusbaum, 1960; Webster, 1985).

Other root-rot fungi

In addition to the fungi mentioned above, nematodes also synergize some other important root-rot-causing fungi such as *Sclerotinia* and *Thielaviopsis*. *Sclerotinia* is a highly virulent pathogen responsible for stem and crown rot, affecting over 500 species of higher plants (Khan *et al.*, 2024). It can also cause root rot in some crops, posing a significant threat to a wide range of host plant species (Purdy, 1979; Willlets and Wrong, 1980). Research has shown that co-infection of *M. incognita* and *S. sclerotiorum* can lead to enhancement of root rot severity in brassica vegetables (Bano and Haseeb, 2012) and medicinal plants (Anisha and Haseeb, 2012). Co-infection of cotton with *M. incognita* and *Thielaviopsis basicola* has been shown to negatively impact plant growth, reducing factors such as stem node count, plant height, root fresh weight, and shoot dry weight (Ma *et al.*, 2014). The reniform nematode *R. reniformis* exacerbated the infection of *T. basicola* and disease severity in cotton seedlings caused by *Fusarium* spp. (Palmateer *et al.*, 2004). Studies have also shown that root-lesion nematodes (*Pratylenchus* spp.) can initiate synergistic interactions with *Verticillium* wilt fungi. In such situations, the severity of disease symptoms in potato was significantly increased (Burpee and Bloom, 1978; Orlando *et al.*, 2020).

Mechanisms Involved in the Root-rot Disease Complex

The wide range of interactions involving nematodes and root-rot fungi have shown that weak pathogenic fungi may become pathogenic to plants while in co-occurrence with plant nematodes (Nath and Kamalwanshi, 1989), whereas the pathogenic fungi assume a high degree of virulence in the presence of nematodes and cause severe damage to the host plants (Roy *et al.*, 1989). All these diverse interactions indicate the involvement of complex mechanisms supporting the intricate relationship between the nematode and root-rot fungi. It has been generally observed that two or more mechanisms may operate in strengthening or supporting the synergistic relationship in a root-rot disease complex. The potential mechanisms involved in root-rot disease complexes are briefly described below.

Activation of dormant spores and facilitation of invasion

Nematode feeding at the root surface or host penetration may cause the formation of injuries on the root surface that may lead to leakage of cell sap or increased root exudation. This may also modify the nutrition status of the rhizospheric soil (Curl and Truelove, 2012). It is well documented that root exudates of susceptible plants activate the dormant spores and may promote germination and hyphal growth of plant pathogenic fungi (Bais *et al.*, 2006). Additionally, nematode infection may alter the chemistry of the root exudates, inducing greater activation of germination of conidia and spores (Vos *et al.*, 2012). For instance, tomato plants with galls due to *M. incognita* infection had significantly higher levels of carbohydrates and reducing sugars in the root exudates, which promoted the conidial germination and colonization of *F. Oxysporum* (Kleineke-Borchers and Wyss, 1982). Sometimes, the fungal infections can also render roots more attractive to nematodes and aid the larvae in locating the host roots. For example, larvae of *Hirschmanniella oryzae* exhibited more active migration towards rice roots infected with *R. solani* compared to non-infected roots (Gokulapalan and Nair, 1983). Similarly, increased movement of juveniles and adults of *Pratylenchus* spp. was observed inside maize roots infected with *F. moniliforme*, indicating enhanced root penetration in the presence of fungal infection (Jordaan *et al.*, 1987).

The wounds or injuries caused by ectoparasitic nematodes on the root surface or the perforations made by endoparasitic nematodes may act as avenues or suitable sites for initial entry of the fungus hyphae into the host cells (Webster, 1985). It was experimentally proved that the wounds made on tobacco roots promoted the invasion of *R. solani* and development of blank shank symptoms due to inoculation with *Pratylenchus brachyurus* (EdinInagaki and Powell, 1969). In another study, it was found that mycelium of *R. solani* grew along the invasion tracks formed by the juveniles of *H. schachtii* in beet roots (Polychronopoulos *et al.*, 1969). In addition to injuries, rupturing of the root cortex due to the rapid increase in size of sedentary nematodes such as *Meloidogyne*, *Heterodera*, *Globodera*, etc., provides avenues for entry of fungal hyphae. The root-rot fungus *R. solani* readily penetrated roots through ruptures formed by the obesity-gaining *Meloidogyne* females in the root, and also extensively colonized the galled tissue in tomato roots (Golden and Van Gundy, 1975).

Morphological and anatomical modifications in the host

Sedentary endoparasitic nematodes induce anatomical (nurse cells) and morphological (galls) modifications in the host which may become attractive to root-rot fungi (Webster, 1985). It has been observed that root-knot nematode galls are readily and extensively colonized by *R. solani* even in fungus-tolerant cultivars (Golden and Van Gundy, 1975). In fact, the root system infected with *Meloidogyne* spp. is rich in photosynthates which become further concentrated in the nurse cells to provide the required food to sedentary females for reproduction. In one study, the infection tract of *M. javanica* did not impede the invasion of *Phytophthora parasitica* (= *P. nicotianae*) in tobacco roots. However, in the nematode-susceptible cultivars, when galls were formed, the fungal hyphae readily colonized the giant cells followed by their spread to the galled tissue (Powell and Nusbaum, 1960). Research has also shown that *Pythium arrhenomanes* can hinder the establishment, development, and reproduction of *M. graminicola* in the rice roots, with the intensification of the inhibitory effect as the infection pressure of *P. arrhenomanes* increases (Verbeek *et al.*, 2016).

Physiological alterations in the host

During pathogenesis on susceptible hosts, plant nematodes induce certain alterations in the host physiology which may help in the infection process of both or either of the pathogens. Brodie and Cooper (1964) were probably the first to demonstrate that mechanical wounding on roots did not increase the susceptibility of cotton seedlings to *Rhizoctonia solani* or *Pythium debaryanum*. However, the extracts from the galls accelerated the production of sporangia by *P. debaryanum* approximately ten times more than the healthy root extracts. The above mechanism was supported by other observations where *Meloidogyne*-infected tomato root was heavily colonized by *R. solani* (Golden and Van Gundy, 1975) and radish root with *R. solani* (Khan and Muller, 1982).

Effectors and elicitors

Plant-parasitic nematodes, particularly endoparasites such as root-knot nematodes, possess a diverse array of effector proteins that enable them to manipulate plant defenses and facilitate fungal pathogen infection. Among these, effector proteins are cell-wall-degrading enzymes, which break down plant cell walls, facilitating easier entry of plant-pathogenic fungi. Notably, some of these enzymes are obtained through horizontal gene transfer from fungal and bacterial pathogens, highlighting the dynamic evolution of nematode virulence strategies (Mitchum *et al.*, 2013). Additionally, other effector proteins can defeat plant immune responses, creating an environment conducive to infection by other microorganisms, and demonstrating the complex interplay between nematodes and fungi developing disease complexes.

Altering the host reaction and root exudates

Due to physiological alterations or other mechanisms, the ultimate sufferer is the host, and its resistance reaction to other pathogens often gets weakened or becomes non-functional. Studies have shown that infection by *M. incognita* and *M. javanica* reduces pigeon pea resistance to *Fusarium udum*, possibly due to lower levels of the cajanol phytoalexin (Marley and Hillocks, 1994). In contrast, the mechanisms behind interactions where the fungus *R. solani* makes the cowpea (Khan and Husain, 1989) and tomato (Hasan and Khan, 1985) plants more susceptible to *M. incognita* remain unclear and have not been well-defined. Hua *et al.* (2019) found that *M. incognita* can increase the susceptibility of resistant watermelon genotypes to specific *F. oxysporum* f. sp. *niveum* races 1 and 2, suggesting that host resistance alone may not be enough to effectively manage Fusarium wilt in watermelon fields infested with root-knot nematodes. The incidence and severity of soilborne diseases are influenced by complex interactions between plants and their rhizobiome, facilitated by root exudates (Souza *et al.*, 2023). alterations in root exudates may play a role in synergistic interactions between nematodes and soilborne fungi (Evans and Haydock, 1993). In tomato, for instance, *M. incognita* parasitism has been shown to increase the concentration of electrolytes, sugars, and nitrogen-rich compounds in root exudates, creating an environment that favors root rot caused by *R. solani* (van Gundy *et al.*, 1977). Research on guava has shown that root exudates from plants parasitized by *Meloidogyne enterolobii*, but not from nematode-free plants, facilitate the fungus *Neocosmospora falciformis* to cause root rot, ultimately leading to plant death (Gomes *et al.*, 2013). Further research has revealed that parasitism by *M. enterolobii* enriches guava root exudates, potentially favoring microorganisms that produce 1,5-dinitrobiuret in the rhizosphere. The resulting accumulation of biuret, a phytotoxic substance, may compromise root physiology and weaken guava's innate immunity to *N. falciformis* (Souza *et al.*, 2023).

Management of Root-Rot disease complexes

Various management options are available for plant diseases, including nematode-fungal disease complexes. These include chemical control, biological control, cultural practices, integrated disease management, and other methods.

Chemical control

Growers often prefer chemical control, particularly with new pesticides. Chemical products provide effective solutions for controlling various soil-borne pathogens and nematodes. These products, such as succinate dehydrogenase inhibitors (SDHIs), exhibit both fungicidal and nematicidal activities, enabling simultaneous control of multiple fungal and nematode diseases. For instance, cyclobutrifluram shows promising efficacy against root-knot nematodes, although caution is advised for its use against fusarium wilt in watermelon and gummy stem blight due to its insensitivity to some fungal isolates (Wong and Quesada-Ocampo, 2024). Another SDHI, fluopyram, has demonstrated effectiveness against root-knot nematodes (Ji *et al.*, 2019) and root-rot/wilt-causing fungi (Ezrari *et al.*, 2022), and may work effectively in the management of fungus–nematode disease complexes (Haque and Khan, 2022b). Other new-generation non-fumigant nematicides, i.e. fluensulfone, fluazaindolizine and tioxazafen, have also shown promise against nematodes in different cropping systems (Hajihassani *et al.*, 2019; Faske *et al.*, 2022; Nnamdi *et al.*, 2022). Their unique mode of action makes them effective and sustainable for managing nematode populations. Bionematicides have also garnered significant attention in recent years, demonstrating promise in controlling nematodes, either as a standalone treatment or in combination with other management strategies, such as nematode- or fungal-suppressive cover crops (Marquez and Hajihassani, 2023; Gitonga *et al.*, 2025). Combining nematicides and fungicides in a single application offers an efficient strategy for comprehensive management. In addition, fumigant pesticides such as Pic-Clor 60, a pre-mixed combination of chloropicrin and 1,3-dichloropropene, continues to be used as an effective control method for various pathogens and nematodes in commercial crop production worldwide (Browne *et al.*, 2013; Karki *et al.*, 2022a). Treating maize seeds with fungicides (fludioxonil +mefenoxam + azoxystrobin and thiabendazole) and nematicides (abamectin) significantly impacted root health, revealing complex interactions between *Fusarium* spp. and *P. penetrans* (Da Silva *et al.*, 2016). Despite growing concerns about synthetic pesticides, these fumigants remain a crucial tool for managing pests and diseases in many regions. Soil fumigation can be combined with nematicides and fungicides to target disease complexes, increasing efficacy against multiple pathogens simultaneously.

Biological control

Despite environmental concerns and issues related to synthetic pesticides, research efforts have intensified towards developing disease management systems based on biological and botanical products. As a result, current research has shifted towards targeted approaches to modify indigenous soil microbial communities, such as introducing beneficial and antagonistic microorganisms (Hoitink and Boehm, 1999; Barratt *et al.*, 2018; Forghani and Hajihassani, 2020;

Khan and Mohiddin, 2023). Several fungi and bacteria have been reported to have antagonistic effects on different plant-pathogenic nematodes (Stirling, 1991; Khan *et al.*, 2019) associated with disease complexes. The biocontrol approach aims to enhance soil health and to promote a balanced microbial ecosystem, which can lead to increased disease suppression, improved plant growth and more resilient agroecosystems (Khan *et al.*, 2012). Field research has demonstrated that *Trichoderma* spp., *Pseudomonas aeruginosa*, *Purpureocillium lilacinum*, etc., can effectively suppress a range of soil-borne pathogens, including *P. aphanidermatum*, *R. solani*, *M. phaseolina* and *Meloidogyne* spp. (Siddiqui *et al.*, 2000; Haque and Khan, 2023). *Trichoderma* is a noteworthy multifaceted biocontrol fungus known for suppressing a wide range of soil-borne fungal and nematode pathogens across various environments (Khan and Mohiddin, 2018; Khan *et al.*, 2020, 2023a; Yao *et al.*, 2023). Individual application of *T. harzianum*, *T. hamatum*, *T. viride*, *T. reesei* and *T. koningii* controlled the infection of *M. phaseolina* (Khan and Gupta, 1998), *R. solani* (Khan and Rehman, 1998), *Fusarium* spp. (Khan *et al.*, 2011, 2017; Khan and Sharma, 2020), *P. aphanidermatum* (Al-Ani, 2018), *Meloidogyne* spp. (Khan *et al.*, 2018; Haque and Khan, 2023); *Heterodera* (Khan, 2016), *Pratylenchus* (Khan and Anwer, 2011), *Radopholus* (Khan *et al.*, 2023c) and *R. reniformis* infection on sunflower roots (Haggag and Amin, 2001).

Hence, *Trichoderma* can offer offers an effective biocontrol solution for sustainable management of root-rot disease complexes. The fungus possesses excellent adaptability and broad-spectrum effectiveness against various pathogens that may cause root-rot disease complexes with low development costs (Khan *et al.*, 2011; Sood *et al.*, 2020). Seed treatment with this biopesticide offers an economical and handy option for application. By applying biopesticide to seeds, farmers can help protect their crops from a range of diseases, including the multipathogenic diseases caused by fungi, bacteria and nematodes. Further, the development of innovative seed treatment products such as Avicta (abamectin) and Votivo (*Bacillus firmus*) has expanded the options for controlling nematode pests in soybean, cotton and maize (Da Silva *et al.*, 2016).

Cultural methods and integrated disease management

Plant-derived bioactive compounds, such as essential oils, *Brassica* seed meals, neem seed powder, dried seed powders and leaf extracts, may also be employed to manage nematode disease complexes (Forghani and Hajihassani, 2020). The effectiveness of plant-derived volatile aromatic essential oils has been well documented, highlighting their potential as a natural and sustainable solution for disease management (DuPont *et al.*, 2021; Dutta *et al.*, 2021; Haque and Khan, 2022a, b). However, these alternative methods generally exhibit lower efficacy in controlling soil-borne diseases compared to chemicals and can only provide partial control. Crop rotation may not be an effective control method (Tilston *et al.*, 2020). Furthermore, there is a lack of availability of commercial cultivars resistant simultaneously to both root-rot fungi and nematodes. This severely limits the effectiveness of cultural practices in disease complex management (Wong and Quesada-Ocampo, 2024). The above observations highlight the need for integrating these methods with other control measures. An integrated approach to managing plant disease complexes needs to be employed, utilizing a combination of cultural, physical, biological and chemical control methods. These strategies include cultural controls through modifying agricultural practices (e.g. crop rotation and soil solarization) to prevent disease spread and promote healthy plant growth and host resistance to limit or prevent the development of pathogens from causing damage. Biological controls are also an important component of integrated pest management with the limited use of chemicals, nematicides and/or fungicides to target the pathogens directly. These multifaceted control approaches will reduce the incidence and severity of disease complexes, promoting sustainable and effective disease management. However, these practices have some limitations and drawbacks that need to be addressed to make the management measures effective and reliable. Further, the management of root-rot disease complexes is challenging due to their widespread distribution, broad host range and multipathogenic nature. Combining biocontrol agents with other management strategies along with seed treatment with chemicals can lead to more comprehensive crop protection and enhanced disease control strategies, resulting in more effective, consistent and sustainable outcomes.

Conclusion and Future Prospects

Nematode infestations cause physical damage, disrupt physiological processes and induce biochemical changes in roots, creating an environment that facilitates fungal invasion or colonization leading to the formation of disease complexes. However, conflicting findings in disease complex studies have been reported, which can be misleading and highlight the need for further research to precisely clarify these inconsistencies. Much of the research on disease complexes has been conducted under controlled conditions, rather than in crop field settings, which may not accurately reflect the actual interaction situations occurring in real agricultural environments. Optimum soil temperature, moisture, soil type and soil microbiomes can create an environment in the field conducive to greater growth and reproduction of both nematode and fungal pathogens, resulting in the formation of complex diseases. Despite substantial significance, environmental factors such as elevated temperature have largely been overlooked in examining the interaction between nematodes and other pathogens; without this knowledge, the complex dynamics at play cannot be understood precisely. Further, understanding the genetic basis of the root-rot disease complex is also crucial for developing effective breeding strategies to produce resistant crops, and this knowledge is essential for improving crop resilience and reducing disease susceptibility. Research suggests that micronutrients like iron, manganese, and zinc in watermelon may increase its susceptibility to infection by *F. oxysporum* f. sp. *niveum* and *M. incognita*. However, infection triggers the upregulation of induced resistance genes, which may help prevent further infection through the salicylic acid and jasmonic acid pathways (Karki *et al.*, 2022b). Although chemical pesticides effectively manage multiple pathogens within disease complexes, concerns arise over the development of resistance in target pathogens, particularly when using pesticides from the same chemical group repeatedly. Consequently, deciphering the genetic mechanisms underlying root-rot resistance holds potential to revolutionize crop management by facilitating the development of resilient crops capable of thriving in challenging environments, including those affected by climate change. Harnessing such technological advancements could lead to the engineering of crops endowed with robust resistance against multiple pathogens, ensuring sustainable agricultural practices and food security in the face of uncertain environmental conditions.

References

Agrios, G.N. (2005) *Plant Pathology*, 5th edn. Elsevier Academic Press, Burlington, MA.

Ahmad, I. and Khan, M.R. (2021) Biocontrol of fungus–nematode disease complex in mungbean. In: *Souvenir & Abstracts, IPS-National e-conference on Plant Health and Food Security: Challenges and Opportunities,* New Delhi, March 25-27, pp. 61–62.

Al-Ani, L.K.T. (2018) *Trichoderma*: Beneficial role in sustainable agriculture by plant disease management. In: Egamberdieva, D. and Ahmad, P. (eds) *Plant Microbiome: Stress Response*. Springer, Dordrecht, The Netherlands, pp. 105–126. DOI: 10.1007/978-981-10-5514-0_5.

Anisha, B. and Haseeb, A. (2012) Interactive effect of *Meloidogyne incognita* and *Sclerotinia sclerotiorum* on disease development, plant growth and seed yield of *Plantago ovate*. *Current Nematology* 23, 23–28.

Anwar, A. and Khan, F.A. (2002) Studies on the interaction between *Meloidogyne incognita* and *Rhizoctonia solani* on soybean. *Annals of Plant Protection Sciences* 10(1), 128–130.

Audette, C., Bélanger, R.R. and Mimee, B. (2020) Coinfection of soybean plants with *Phytophthora sojae* and soybean cyst nematode does not alter the efficacy of resistance genes. *Plant Pathology* 69(8), 1437–1444. DOI: 10.1111/ppa.13227.

Back, M.A., Haydock, P.P.J. and Jenkinson, P. (2002) Disease complexes involving plant parasitic nematodes and soilborne pathogens. *Plant Pathology* 51(6), 683–697. DOI: 10.1046/j.1365-3059.2002.00785.x.

Bais, H.P., Weir, T.L., Perry, L.G., Gilroy, S. and Vivanco, J.M. (2006) The role of root exudates in rhizosphere interactions with plants and other organisms. *Annual Review of Plant Biology* 57, 233–266.

Baker, B. and Parker, J. (2003) Biotic interactions: Signals from the environment: The good, the bad and the ugly. *Current Opinion in Plant Biology* 6, 297–299. DOI: 10.1016/S1369-5266(03)00066-9.

Bano, A. and Haseeb, A. (2012) Interactive effect of *Meloidogyne incognita* and *Sclerotinia sclerotiorum* on disease development, plant growth and seed yield of *Plantago ovata*. *Current Nematology* 23(1/2), 23–28.

Barratt, B.I.P., Moran, V.C., Bigler, F. and Lenteren, J.C. (2018) The status of biological control and recommendations for improving uptake for the future. *BioControl* 63(1), 155–167. DOI: 10.1007/s10526-017-9831-y.

Basandrai, A.K., Pandey, A.K., Somta, P. and Basandrai, D. (2021) *Macrophomina phaseolina*–host interface: insights into an emerging dry root rot pathogen of mungbean and urdbean, and its mitigation strategies. *Plant Pathology* 70(6), 1263–1275. DOI: 10.1111/ppa.13378.

Beckman, C.H. (1987) *The Nature of Wilt Diseases of Plants*. CAB International, Wallingford, UK.

Begum, M.Z. and Kumar, M.S. (2005) Management of disease complex involving *Heterodera cajani* koshy, 1967 and *Macrophomina phaseolina* (Tassi.) Goid. On green gram (*Vigna radiata* L. Wilczek). *Indian Journal of Nematology* 35(2), 192–194.

Bhattarai, S., Haydock, P.P., Back, M.A., Hare, M.C. and Lankford, W.T. (2009) Interactions between the potato cyst nematodes, *Globodera pallida*, *G. rostochiensis*, and soil-borne fungus, *Rhizoctonia solani* (AG3), diseases of potatoes in the glasshouse and the field. *Nematology* 11(4), 631–640. DOI: 10.1163/156854108X399173.

Briar, S.S., Shrestha, G. and Reddy, G.V. (2018) Plant parasitic nematodes of Montana and Wyoming. In: Subbotin, S.A. and Chitambar, J.J. (eds) *Plant Parasitic Nematodes in Sustainable Agriculture of North America. Volume 1 - Canada, Mexico and Western USA*. Springer, Dordrecht, The Netherlands, pp. 69–86. DOI: 10.1007/978-3-319-99585-4_3.

Brodie, B.B. and Cooper, W.E. (1964) Relation of parasitic nematodes to post emergence damping-off of cotton. *Phytopathology* 54, 1023–1027.

Bronkhorst, J. (2022) *Under Pressure: Exploring the Mechanobiology of Phytophthora Invasion*. Doctoral dissertation, Wageningen University and Research, Wageningen, The Netherlands.

Browne, G., Lampinen, B., Holtz, B., Doll, D., Upadhyaya, S. *et al.* (2013) Managing the almond and stone fruit replant disease complex with less soil fumigant. *California Agriculture* 67(3), 128–138. DOI: 10.3733/ca.v067n03p128.

Burpee, L.L. and Bloom, J.R. (1978) The influence of Pratylenchus penetrans on the incidence and severity of verticillium wilt of potato. *Journal of Nematology* 10, 95–99.

Castillo, P., Mora-Rodríguez, M.P., Navas-Cortés, J.A. and Jiménez-Díaz, R.M. (1998) Interactions of *Pratylenchus thornei* and *Fusarium oxysporum* f. sp. *ciceris* on chickpea. *Phytopathology* 8(8), 828–836. DOI: 10.1094/PHYTO.1998.88.8.828.

Chahal, P.P.K. and Chhabra, H.K. (1984) Interaction of *Meloidogyne incognita* with *Rhizoctonia solani* on tomato. *Indian Journal of Nematology* 14, 56–57.

Chilakala, A.R., Pandey, P., Durgadevi, A., Kandpal, M., Patil, B.S. *et al.* (2023) Drought attenuates plant responses to multiple rhizospheric pathogens: A study on a dry root rot-associated disease complex in chickpea fields. *Field Crops Research* 298, 108965. DOI: 10.1016/j.fcr.2023.108965.

Curl, E.A. and Truelove, B. (2012) *The Rhizosphere*. Springer, Dordrecht, The Netherlands.

da Silva, M.B., Davis, R.F., Doan, H.K., Nichols, R.L., Kemerait, R.C. *et al.* (2019) Fusarium wilt of cotton may commonly result from the interaction of *Fusarium oxysporum* f. sp. *Vasinfectum* with *Belonolaimus longicaudatus*. *Journal of Nematology* 51, 1–10. DOI: 10.21307/jofnem-2019-015.

Da Silva, M.P., Tylka, G.L. and Munkvold, G.P. (2016) Seed treatment effects on maize seedlings coinfected with *Fusarium* spp. and *Pratylenchus Penetrans*. *Plant Disease* 100(2), 431–437. DOI: 10.1094/PDIS-03-15-0364-RE.

Doshi, A. and Mathur, S. (1987) Symptomatology, interaction and management of rhizome rot of ginger by Xenobiotics. *Korean Journal of Plant Pathology* 26, 261–265.

Dunn, E. and Hughes, W.A. (1967) Interactions of *Oospora pustulans*, *Rhizoctonia solani* and *Heterodera rostochiensis* on the potato. *European Potato Journal* 10(4), 327–328. DOI: 10.1007/BF02364471.

DuPont, S.T., Hewavitharana, S.S. and Mazzola, M. (2021) Field scale application of brassica seed meal and anaerobic soil disinfestation for the control of apple replant disease. *Applied Soil Ecology* 166, 104076. DOI: 10.1016/j.apsoil.2021.104076.

Dutta, A., Mandal, A., Kundu, A., Malik, M., Chaudhary, A. *et al.* (2021) Deciphering the behavioral response of *Meloidogyne incognita* and *Fusarium oxysporum* toward mustard essential oil. *Frontiers in Plant Science* 12, 714730. DOI: 10.3389/fpls.2021.714730.

Edin, E., Gulsher, M., Andersson Franko, M., Englund, J.-E., Flöhr, A. *et al.* (2019) Temporal interactions between root-lesion nematodes and the fungus *Rhizoctonia solani* lead to reduced potato yield. *Agronomy* 9(7), 361. DOI: 10.3390/agronomy9070361.

EdinInagaki, H. and Powell, N.T. (1969) Influence of root-lesion nematode on black shank symptom development in flue-cured tobacco. *Phytopathology* 59, 1350–1355.

El-Tarabily, K.A. and Sivasithamparam, K. (2006) Non-streptomycete actinomycetes as biocontrol agents of soil-borne fungal plant pathogens and as plant growth promoters. *Soil Biology and Biochemistry* 38(7), 1505–1520. DOI: 10.1016/j.soilbio.2005.12.017.

Erwin, D.C. and Ribeiro, O.K. (1996) *Phytophthora Diseases Worldwide*. APS Press, St Paul, MN.

Evans, K. and Haydock, P.P.J. (1993) Interactions of nematodes with root-rot fungi. In: Khan, M.W. (ed.) *Nematode Interactions*. Chapman & Hall, London, pp. 104–133. DOI: 10.1007/978-94-011-1488-2_6.

Ezrari, S., Lazraq, A., El Housni, Z., Radouane, N., Belabess, Z. *et al.* (2022) Evaluating the sensitivity and efficacy of fungicides with different modes of action against *Neocosmospora solani* and *Fusarium* species, causing agents of citrus dry root rot. *Archives of Phytopathology and Plant Protection* 55(9), 1117–1135. DOI: 10.1080/03235408.2022.2081765.

Faske, T.R., Brown, K. and Kelly, J. (2022) Toxicity of tioxazafen to *Meloidogyne incognita* and *Rotylenchulus reniformis*. *Journal of Nematology* 54, 20220007. DOI: 10.2478/jofnem-2022-0007.

Forghani, F. and Hajihassani, A. (2020) Recent advances in the development of environmentally benign treatments to control root-knot nematodes. *Frontiers in Plant Science* 11, 1125. DOI: 10.3389/fpls.2020.01125.

Garcia, R. and Mitchell, D.J. (1975) Interactions of *Pythium myriotylum* with *Fusarium solani*, *Rhizoctonia solani* and *Meloidogyne arena ria* in pre-emergence damping-off of peanut. *Plant Disease Reporter* 59, 665–669.

Ghosh, T., Biswas, M.K., Guin, C. and Roy, P. (2018) A review on characterization, therapeutic approaches and pathogenesis of *Macrophomina phaseolina*. *Plant Cell Biotechnology and Molecular Biology* 19, 72–84.

Gitonga, D., Desaeger, J., Shapiro-Ilan, D. and Hajihassani, A. (2025) Effect of integrating cover crops and bionematicides on nematode management in organic zucchini. *Nematology* 27. DOI: 10.1163/15685411-bja10395.

Goel, S.R. and Gupta, D.C. (1986a) Interaction of *Meloidogyne javanica* and *Fusarium oxysporum* f. sp. *ciceri* on chickpea. *Indian Phytopathology* 39, 112–114.

Goel, S.R. and Gupta, D.C. (1986b) Interaction of *Meloidogyne javanica* and *Rhizoctonia bataticola* on chickpea (*Cicer arietinum* L.). *Indian Journal of Nematology* 16, 133–134.

Gogoi, D. and Mahanta, B. (2013) Comparative efficacy of *Glomus fasciculatum*, *Trichoderma harzianum*, carbofuran and carbendazim in management of *Meloidogyne incognita* and *Rhizoctonia solani* disease complex on French bean. *Annals of Plant Protection Sciences* 21(1), 172–175.

Gokulapalan, C. and Nair, M.C. (1983) Field screening for sheath blight and rice root nematode resistance. *International Rice Research Newsletter* 8(6), 4.

Golden, J.K. and Van Gundy, S.D. (1975) A disease complex of okra and tomato involving the nematode, *Meloidogyne incognita*, and the soil inhabiting fungus. *Rhizoctonia solani*. *Phytopathology* 65(3), 265–273. DOI: 10.1094/Phyto-65-265.

Gomes, V.M., Souza, R.M., Silveira, S.F. and Almeida, A.M. (2013) Guava decline: Effect of root exudates from *Meloidogyne enterolobii*-parasitized plants on *Fusarium solani in vitro* and on growth and development of guava seedlings under controlled conditions. *European Journal of Plant Pathology* 137, 393–401. DOI: 10.1007/s10658-013-0251-2.

Gracia, J.A., Reeleder, R.D., and Belair, G. (1991) Interactions between *Pythium tracheiphilum*, *Meloidogyne hapla*, and *Pratylenchus penetrans* on lettuce. *Phytoprotection* 72:195–114.

Haggag, W.M. and Amin, A.W. (2001) Efficiency of *Trichoderma* species on control of fusarium-rot, root knot and reniform nematodes disease complex on sunflower. *Pakistan Journal of Biological Sciences* 4, 314–318.

Hajihassani, A., Smiley, R.W. and Jahanshahi Afshar, F. (2013a) Effects of co-inoculation with *Pratylenchus thornei* and *Fusarium culmorum* on growth and yield of winter wheat. *Plant Disease* 97, 1470–1477.

Hajihassani, A., Tanha Maafi, Z. and Hosseininejad, A. (2013b) Interactions between *Heterodera filipjevi* and *Fusarium culmorum*, and between *H. filipjevi* and *Bipolaris sorokiniana* in winter wheat. *Journal of Plant Disease and Protection* 120, 77–84.

Hajihassani, A., Lawrence, K. and Jagdale, G.B. (2018) Plant parasitic nematodes of Georgia and Alabama. In: Subbotin, S.A. and Chitambar, J. (eds) *Plant Parasitic Nematodes in Sustainable Agriculture of North America. Vol. 2, Northeastern, Midwestern and Southern USA*. Springer International Publishing, Switzerland, pp. 357–391.

Hajihassani, A., Davis, R.F. and Timper, P. (2019) Evaluation of selected nonfumigant nematicides on increasing inoculation density of *Meloidogyne incognita* on cucumber. *Plant Disease* 103, 3161–3165.

Haque, Z. and Khan, M.R. (2013) Morphological and biochemical responses of five tobacco cultivars to simultaneous infection with *Pythium aphanidermatum* and *Meloidogyne incognita*. *Phytopathologia Mediterranea* 52(1), 695–706.

Haque, Z. and Khan, M.R. (2022a) Host resistance and bio-management of tobacco root-rot caused by *Pythium aphanidermatum*. *Indian Phytopathology* 75, 703–712. DOI: 10.1007/s42360-022-00491-y.

Haque, Z. and Khan, M.R. (2022b) Evaluation of fluopyram against single and concomitant inoculation of *Rhizoctonia solani* and *Meloidogyne graminicola* infesting rice crop. *Phytoparasitica* 50(4), 811–822. DOI: 10.1007/s12600-022-01014-1.

Haque, Z. and Khan, M.R. (2023) Nematicidal effect of fluopyram and fluensulfone against rice root-knot nematode, *Meloidogyne graminicola*. *Russian Journal of Nematology* 31, 39–49.

Hasan, A. (1988) Interaction of *Pratylenchus coffeae* and *Pythium aphanidermatum* and/or *Rhizoctonia solani* on chrysanthemum. *Phytopathologische Zeitschrift* 123, 227–232.

Hasan, A. and Khan, M.N. (1985) The effect of *Rhizoctonia solani*, *Sclerotium rolfsii*, and *Verticillium dahliae* on the resistance of tomato to *Meloidogyne incognita*. *Nematologia Mediterranea* 13, 133–136.

Ho, H.H. (2018) The taxonomy and biology of *Phytophthora* and *Pythium*. *Journal of Bacteriology & Mycology* 6(1), 40–45. DOI: 10.15406/jbmoa.2018.06.00174.

Hoitink, H.A.J. and Boehm, M.J. (1999) Biocontrol within the context of soil microbial communities: A substrate-dependent phenomenon. *Annual Review of Phytopathology* 37, 427–446. DOI: 10.1146/annurev.phyto.37.1.427.

Hua, G.K.H., Timper, P. and Ji, P. (2019) *Meloidogyne incognita* intensifies the severity of fusarium wilt on watermelon caused by *fusarium oxysporum* f.sp. *niveum*. *Canadian Journal of Plant Pathology* 41, 261–269. DOI: 10.1080/07060661.2018.1564939.

Jacobs, J.L., Kelly, J.D., Wright, E.M., Varner, G. and Chilvers, M.I. (2019) Determining the soilborne pathogens associated with root rot disease complex of dry bean in Michigan. *Plant Health Progress* 20(2), 122–127. DOI: 10.1094/PHP-11-18-0076-S.

Ji, X., Li, J., Dong, B., Zhang, H., Zhang, S. *et al.* (2019) Evaluation of fluopyram for southern root-knot nematode management in tomato production in China. *Crop Protection* 122, 84–89. DOI: 10.1016/j.cropro.2019.04.028.

Jordaan, E.M., Loots, G.C., Jooste, W.J. and Waele, D. (1987) Effects of root-lesion nematodes (*Pratylenchus brachyurus* Godfrey and *P. zeae* Graham) and *Fusarium moniliforme* sheldon alone or in combination, on maize. *Nematologica* 33(2), 213–219. DOI: 10.1163/187529287X00344.

Kamoun, S., Furzer, O., Jones, J.D.G., Judelson, H.S., Ali, G.S. *et al.* (2015) The top 10 oomycete pathogens in molecular plant pathology. *Molecular Plant Pathology* 16(4), 413–434. DOI: 10.1111/mpp.12190.

Karki, K., Grant, J., da Silva, A.L.B.R., Petkar, A., Coolong, T. *et al.* (2022a) Evaluation of efficacy of Pic-Clor 60 [choloropicrin pre-mixed with 1,3 dicholoropropene] and soil-applied fungicides to manage Fusarium wilt in watermelon. *Crop Protection* 154, 105894. DOI: 10.1016/j.cropro.2021.105894.

Karki, K., Negi, V., Coolong, T., Petkar, A., Mandal, M. *et al.* (2022b) Micronutrients affect expression of induced resistance genes in hydroponically-grown watermelon against *Fusarium oxysporum* f. sp. *Niveum* and *Meloidogyne incognita*. *Pathogens* 11, 1136. DOI: 10.20944/preprints202208.

Katsantonis, D., Hillocks, R.J. and Gowen, S. (2003) Comparative effect of root-knot nematode on severity of verticillium and fusarium wilt in cotton. *Phytoparasitica* 31(2), 154–162. DOI: 10.1007/BF02980785.

Khan, M.R. (2005) *Biological Control of Fusarial Wilt and Root-Knot of Legumes*. Department of Biotechnology, Ministry of Science and Technology, Government of India Publication, New Delhi.

Khan, M.R. (2016) Nematode biocontrol agents: Diversity and effectiveness against phytonematodes in sustainable crop protection. *Indian Phytopathology* 69(4s), 453–463.

Khan, M.R. (2023a) Nematode pests of agricultural crops, a global overview. In: Khan, M.R. (ed.) *Novel Biological and Biotechnological Applications in Plant Nematode Management*. Springer, Singapore, pp. 3–45. DOI: 10.1007/978-981-99-2893-4_1.

Khan, M.R. (2023b) Plant nematodes, an underestimated constraint in global crop production. In: Khan, M.R. and Quintanilla, M. (eds) *Nematode Diseases of Crops and Their Sustainable Management*. Academic Press, Cambridge, MA, pp. 3–23. DOI: 10.1016/B978-0-323-91226-6.00009-2.

Khan, M.R. and Akram, M. (2000) Effects of certain antagonistic fungi and rhizobacteria on wilt disease complex of tomato caused by *Meloidogyne incognita* and *Fusarium oxysporum* f. sp. *Lycopersici*. *Nematologia Mediterranea* 28, 139–144.

Khan, M.R. and Anwer, A. (2011) Fungal bioinoculants for plant disease management. In: Paul, M., Clinton, M. and Ahmad, I. (eds) *Microbes and Microbial Technology*. Springer, Dordrecht, The Netherlands, pp. 447–488.

Khan, M.R. and Gupta, J. (1998) Antagonistic efficacy of *Trichoderma* species against *Macrophomina phaseolina* on eggplant. *Zeitschrift Für Pflanzenkrankheiten Und Pflanzenschutz* 105, 387–393.

Khan, M.R. and Jairajpuri, S. (2010) Nematode infestation in food crops - national scenario. In: Khan, M.R. and Jairajpuri, M.S. (eds) *Nematode Infestations Part I: Food Crops*. National Academy of Sciences, Allahabad, India, pp. 1–16.

Khan, M.W. and Khan, M.R. (1995) Fungus–nematode interactions. In: Swarup, G., Dasgupta, D.R. and Gill, J.S. (eds) *Nematode Pest Management – An Appraisal of Eco-Friendly Approaches*. Nematological Society of India, New Delhi, pp. 70–78.

Khan, M.R. and Khan, M.W. (1997) Effects of the root-knot nematode, *Meloidogyne incognita*, on the sensitivity of tomato to sulfur dioxide and ozone. *Environmental and Experimental Botany* 38, 117–130.

Khan, M.R. and Mohiddin, F.A. (2018) *Trichoderma*: Its multifarious utility in crop improvement. In: Prasad, R., Gill, S.S. and Tuteja, N. (eds) *New and Future Developments in Microbial Biotechnology and Bioengineering: Crop Improvement Through Microbial Biotechnology*. Elsevier, Amsterdam, pp. 263–291. DOI: 10.1016/B978-0-444-63987-5.00013-X.

Khan, M.R. and Mohiddin, F.A. (2023) Biocontrol strategies for nematode management, an overview. In: Khan, M.R. (ed.) *Novel Biological and Biotechnological Applications in Plant Nematode Management*. Springer, Singapore, pp. 113–131. DOI: 10.1007/978-981-99-2893-4_5.

Khan, M.W. and Muller, J. (1982) Interaction between *Rhizoctonia solani* and *Meloidogyne hapla* on radish in gnotobiotic culture. *Libyan Journal of Agriculture* 11, 133–140.

Khan, M.R. and Quintanilla, M. (2023) *Nematode Diseases of Crops and Their Sustainable Management*. Academic Press, Cambridge, MA.

Khan, M.R. and Rehman, Z. (1998) Bio-management of root-rot of chickpea caused by *Rhizoctonia solani*. *Vasundhara* 3, 22–26.

Khan, M.R. and Sharma, R.K. (2020) Fusarium–nematode wilt disease complexes, etiology and mechanism of development. *Indian Phytopathology* 73(4), 615–628. DOI: 10.1007/s42360-020-00240-z.

Khan, M.R., Majid, S., Mohidin, F.A. and Khan, N. (2011) A new bioprocess to produce low cost powder formulations of biocontrol bacteria and fungi to control fusarial wilt and root-knot nematode of pulses. *Biological Control* 59(2), 130–140. DOI: 10.1016/j.biocontrol.2011.04.007.

Khan, M.R., Anwer, M.A., Bhattacharya, C. and Goswami, B.K. (2012) Biomanagement of nematodes infesting horticultural crops. In: Khan, M.R. and Jairajpuri, M.S. (eds) *Nematode Infestations Part III: Horticultural Crops*. National Academy of Sciences, Allahabad, India, pp. 552–606.

Khan, M.R., Shahid, S., Mohidin, F.A. and Mustafa, U. (2017) Interaction of *Fusarium oxysporum* f. sp. *Gladioli* and *Meloidogyne incognita* on gladiolus cultivars and its management through corm treatment with biopesticides and pesticides. *Biological Control* 115, 95–104. DOI: 10.1016/j.biocontrol.2017.09.010.

Khan, M.R., Mohiddin, F.A. and Ahamad, F. (2018) Inoculant rhizobia suppressed root-knot disease, and enhanced plant productivity and nutrient uptake of some field-grown food legumes. *Acta Agriculturae Scandinavica, Section B – Soil & Plant Science* 68(2), 166–174. DOI: 10.1080/09064710.2017.1374448.

Khan, M.R., Goswami, B.K., Singh, N., Akram, M., Ahmad, I. *et al.* (2019) Microbial management of plant parasitic nematodes. In: Khan, M.R., Mukhopadhyay, A.N., Pandey, R.N., Thakur, M.P. and Singh, D. (eds) *Bio-Intensive Approaches: Application and Effectiveness in the Management of Plant Nematodes, Insects and Weeds*. Today & Tomorrow's Printers and Publishers, New Delhi, pp. 1–47.

Khan, M.R., Ahamad, I. and Shah, H. (2020). Emerging important nematode problems in field crops and their management. In: *Emerging trends in Plant Pathology*, K. P. Singh, S. Jahagirdar & B. K. Sarma (Ed.) Springer Nature Pp. 33-62.

Khan, M.R., Ahamad, I. and Shah, H. (2021) Emerging important nematode problems in field crops and their management. In: Singh, K.P., Jahagirdar, S. and Sarma, B.K. (eds) *Emerging Trends in Plant Pathology*. Springer, Dordrecht, The Netherlands, pp. 33–62. DOI: 10.1007/978-981-15-6275-4_3.

Khan, M.R., Rizvi, T.F. and Ansari, M.S.A. (2023a) Nematode problems in polyhouse cultivation and their sustainable management in ornamental and vegetable crops. In: Khan, M.R. and Quintanilla, M. (eds) *Nematode Diseases of Crops and Their Sustainable Management*. Academic Press, Cambridge, MA, pp. 685–706. DOI: 10.1016/B978-0-323-91226-6.00002-X.

Khan, M.R., Mohiddin, F.A., Haque, Z. and Sharma, R.K. (2023b) Disease complexes and their sustainable management. In: Khan, M.R. and Quintanilla, M. (eds) *Nematode Diseases of Crops and Their Sustainable Management*. Academic Press, Cambridge, MA, pp. 65–93. DOI: 10.1016/B978-0-323-91226-6.00016-X.

Khan, M.R., Sharma, D.N., Shah, M.H. and Ahmad, I. (2024) Temporal impact of infection with *Sclerotinia sclerotiorum* on important physiological and biochemical variables of tomato plants. *Indian Phytopathology* 78, 1–10.

Khan, M.W. (ed.) (1993) *Nematode Interactions*. Springer, Dordrecht, The Netherlands. DOI: 10.1007/978-94-011-1488-2.

Khan, T.A. and Husain, S.L. (1989) Relative resistance of six cowpea cultivars as affected by the concomitance of two nematodes and a fungus. *Nematologia Mediterranea* 17, 39–41.

Kiptoo, J.J., Abbas, A., Bhatti, A.M., Usman, H.M., Shad, M.A. *et al.* (2021) *Rhizoctonia solani* of potato and its management: A review. *Plant Protection* 5(3), 157–169. DOI: 10.33804/pp.005.03.3925.

Kleineke-Borchers, A. and Wyss, U. (1982) Investigations on changes in susceptibility of tomato plants to *Fusarium oxysporum* f. sp. *Lycopersici* after infection by *Meloidogyne incognita*. *Zeitschrift Fur Pflanzenkrankheiten Und Pflanzenschutz* 89, 67–78.

Kumari, N. and Katoch, S. (2020) Wilt and root rot complex of important pulse crops: Their detection and integrated management. In: Singh, B.P., Singh, G., Kumar, K., Nayak, S.C. and Srinivasa, N. (eds) *Management of Fungal Pathogens in Pulses: Current Status and Future Challenges*. Springer, Dordrecht, The Netherlands, pp. 93–119. DOI: 10.1007/978-3-030-35947-8_6.

LaMondia, J.A. and Martin, S.B. (1989) The influence of *Pratylenchus penetrans* and temperature on black root rot of strawberry by binucleate *Rhizoctonia*. *Plant Disease* 73(2), 107–110. DOI: 10.1094/PD-73-0107.

Lanjewar, R.D. and Shukla, V.N. (1985) Parasitism and interaction between *Pythium myriotylum* and *Meloidogyne incognita* in soft rot complex of ginger. *Indian Journal of Nematology* 15, 170–173.

Lobna, H., Hajer, R., Naima, M. and Najet, H.R. (2016) Studies on disease complex incidence of *Meloidogyne javanica* and *Fusarium oxysporum* f. sp. *Lycopersici* on resistant and susceptible tomato cultivars. *Journal of Agricultural Science and Food Technology* 2, 41–48.

Ma, J., Jaraba, J., Kirkpatrick, T.L. and Rothrock, C.S. (2014) Effects of *Meloidogyne incognita* and *Thielaviopsis basicola* on cotton growth and root morphology. *Phytopathology* 104, 507–512.

Marley, P.S. and Hillocks, R.J. (1994) Effect of root-knot nematodes on cajanol accumulation in the vascular tissues of pigeonpea after stem inoculation with *Fusarium udum*. *Plant Pathology* 43, 172–176. DOI: 10.1111/j.1365-3059.1994.tb00567.x.

Marquez, J. and Hajihassani, A. (2023) Successional effects of cover cropping and deep tillage on suppression of plant-parasitic nematodes and soilborne fungal pathogens. *Pest Management Science* 79, 2737–2747. DOI: 10.1002/ps.7450.

Marquez, J. and Hajihassani, A. (2024) Seasonal fluctuations in plant-parasitic nematode vertical distributions and their interactions with edaphic factors in vegetable fields of South Georgia, USA. *Phytobiomes Journal* 8, 643–655. DOI: 10.1094/PBIOMES-02-24-0014-R.

McGawley, E.C., Winchell, K.L. and Berggren, G.T. (1984) Possible involvement of *Hoplolaimus galeatus* in a disease complex of "Centennial" soybean. *Phytopathology* 74, 831.

McLean, K.S. and Lawrence, G.W. (1993) Interrelationship of *Heterodera glycines* and *Fusarium solani* in sudden-death syndrome of soybean. *Journal of Nematology* 25, 434–439.

Mehta, N., Walia, K.K. and Gupta, D.C. (1995) *Rhizoctonia* and root-knot nematode (*Meloidogyne javanica*) disease complex as influenced by organic amendments. *Indian Journal of Nematology* 25(1), 94–97.

Mendgen, K., Hahn, M. and Deising, H. (1996) Morphogenesis and mechanisms of penetration by plant pathogenic fungi. *Annual Review of Phytopathology* 34(1), 367–386.

Mishra, C., Som, D. and Singh, B. (1988) Interaction of *Meloidogyne incognita* and *Rhizoctonia bataticola* in white jute (*Corchorus capsularis*). *Indian Journal of Agricultural Sciences* 58, 234–235.
Mitchum, M.G., Hussey, R.S., Baum, T.J., Wang, X., Elling, A.A. *et al.* (2013) Nematode effector proteins: An emerging paradigm of parasitism. *The New Phytologist* 199(4), 879–894. DOI: 10.1111/nph.12323.
Morris, K.A., Langston, D.B., Dutta, B., Davis, R.F., Timper, P. *et al.* (2016) Evidence for a disease complex between *Pythium aphanidermatum* and root-knot nematodes in cucumber. *Plant Health Progress* 17, 200–201. DOI: 10.l094/PHP-BR-16-0036.
Nandris, D., Nicole, M. and Geiger, J.P. (1987) Root rot diseases. *Plant Disease* 71(4), 298–306.
Nath, R. and Kamalwanshi, R.S. (1989) Role of non-pathogenic fungi in presence of root-knot nematode in damping-off of tomato seedlings. *Indian Journal of Nematology* 19, 84–85.
Nath, R., Khan, M.N., Kamalwanshi, R.S. and Dwivedi, R.P. (1984) Influence of root-knot nematode, *Meloidogyne javanica* on pre- and post-emergence damping off of tomato. *Indian Journal of Nematology* 14, 135–140.
Nnamdi, C., Grey, T.L. and Hajihassani, A. (2022) Root-knot nematode management for pepper and squash rotations using plasticulture systems with fumigants and non-fumigant nematicides. *Crop Protection* 152, 105844. DOI: 10.1016/j.cropro.2021.105844.
Orlando, V., Grove, I. G., Edwards, S. G., Prior, T., Roberts, D., Neilson, R., and Back, M. (2020) Root-lesion nematodes of potato: Current status of diagnostics, pathogenicity and management. *Plant Pathology* 69:405–417.
Palmateer, A.J., Lawrence, K.S., van Santen, E. and Morgan-Jones, G. (2004) Interaction of *Rotylenchulus reniformis* with seedling disease pathogens of cotton. *Journal of Nematology* 36(2), 160–166.
Papavizas, G.C. and Ayers, W.A. (1974) *Aphanomyces Species and Their Root Diseases in Pea and Sugarbeet*. US Department of Agriculture, Forest Service, Rocky Mountain Research Station, Washington, DC.
Pervez, R., Rajkumar and Olabiyi, T.I. (2023) Nematode problems in major spice crops and their management. In: Khan, M.R. and Quintanilla, M. (eds) *Nematode Diseases of Crops and Their Sustainable Management*. Academic Press, Cambridge, MA, pp. 497–514.
Polychronopoulos, A.G., Houston, B.R. and Lownsbery, B.F. (1969) Penetration and development of *Rhizoctonia solani* in sugarbeet seedlings infected with *Heterodera schachtii*. *Phytopathology* 59, 482–485.
Powell, N.T. and Nusbaum, C.J. (1960) The black shank-root-knot complex in flue-cured tobacco. *Phytopathology* 50(12), 899–906.
Powell, N.T. (1979) Internal synergisms among organisms inducing disease. In: Horsfall, J.G. and Cowling, E.B. (eds) *Plant Disease: An Advanced Treatise. Volume IV. How Pathogens Induce Disease*. Academic Press, New York, pp. 113–133.
Prusty, A. (2017) Screening evaluation of germplasms and biomanagement of *Meloidogyne incognita* in brinjal. Doctoral dissertation, Orissa University of Agriculture and Technology, Bhubaneswar, India.
Purdy, L.H. (1979) *Sclerotinia sclerotiorum*: History, diseases and symptomatology, host range, geographic distribution, and impact. *Phytopathology* 69(8), 875. DOI: 10.1094/Phyto-69-875.
Regmi, H., Vallad, G.E., Hutton, S.F. and Desaeger, J. (2022) Interaction studies between *Meloidogyne javanica* and *Fusarium oxysporum* f. sp. lycopersici (Fol) race 3 on different isolines of tomato cv. Tasti Lee. *Journal of Nematology* 54, 20220018. DOI: 10.2478/jofnem-2022-0018.
Roy, K.W., Lawrence, G.W., Hodges, H.H., McLean, K.S. and Killebrew, J.F. (1989) Sudden death syndrome of soybean: *Fusarium solani* as incitant and relation of *Heterodera glycines* to disease severity. *Phytopathology* 79(2), 191. DOI: 10.1094/Phyto-79-191.
Saikia, D. and Bhagawati, B. (2014) Management of *Meloidogyne incognita* and *Rhizoctonia solani* complex on okra through integration of *Trichoderma viride*, *Glomus fasciculatum* and mustard cake. *Indian Journal of Nematology* 44(2), 139–143.
Shahid, S. and Khan, M.R. (2016a) Biological control of root-rot on mungbean plants incited by *Macrophomina phaseolina* through microbial antagonists. *Plant Pathology Journal* 15(2), 27–39. DOI: 10.3923/ppj.2016.27.39.
Shahid, S. and Khan, M.R. (2016b) Management of root-rot of mungbean caused by *Macrophomina phaseolina* through seed treatment with fungicides. *Indian Phytopathology* 69(2), 128–136.
Sharma, M. and Ghosh, R. (2017) Heat and soil moisture stress differentially impact chickpea plant infection with fungal pathogens. In: Senthil-Kumar, M. (ed.) *Plant Tolerance to Individual and Concurrent Stresses*. Springer, Dordrecht, The Netherlands, pp. 47–57. DOI: 10.1007/978-81-322-3706-8_4.

Sharma, P., Jambhulkar, P.P., Raja, M. and Javeria, S. (2020) *Pythium* spp. on vegetable crops: Research progress and major challenges. *Pythium* 136–161. DOI: 10.1201/9780429296406-10.

Siddiqui, Z.A. and Aziz, S. (2024) Plant parasitic nematode–fungus interactions: Recent concepts and mechanisms. *Plant Physiology Reports* 29(1), 37–50. DOI: 10.1007/s40502-023-00762-4.

Siddiqui, I.A., Qureshi, S.A. and Sultana, V. (2000) Biological control of root rot-root knot disease complex of tomato. *Plant and Soil* 227(1/2), 163–169. DOI: 10.1023/A:1026599532684.

Singh, L., Singh, S. and Goswami, B.K. (2003) Effect of cakes with *Trichoderma viride* for the management of disease-complex caused by *Rhizoctonia bataticola* and *Meloidogyne incognita* on okra. *Annals of Plant Protection Sciences* 11(1), 178–180.

Šišić, A., Baćanović-Šišić, J., Schmidt, H. and Finckh, M.R. (2022) Farming system effects on root rot pathogen complex and yield of faba bean (*Vicia faba*) in Germany. *Frontiers in Plant Science* 13, 1009906. DOI: 10.3389/fpls.2022.1009906.

Sneh, B., Jabaji-Hare, S., Neate, S.M. and Dijst, G. (eds) (2013) *Rhizoctonia Species: Taxonomy, Molecular Biology, Ecology, Pathology and Disease Control*. Springer, Dordrecht, The Netherlands.

Sood, M., Kapoor, D., Kumar, V., Sheteiwy, M.S., Ramakrishnan, M. *et al.* (2020) Trichoderma: The "secrets" of a multitalented biocontrol agent. *Plants* 9(6), 762. DOI: 10.3390/plants9060762.

Souza, R.M., Oliveira, D.F., Gomes, V.M., Viana, A.J.S., Silva, G.H. *et al.* (2023) *Meloidogyne enterolobii*-induced changes in guava root exudates are associated with root rotting caused by *Neocosmospora falciformis*. *Journal of Nematology* 55(1), 20230055. DOI: 10.2478/jofnem-2023-0055.

Stirling, G.R. (1991) *Biological Control of Plant Parasitic Nematodes: Progress, Problems and Prospects*. CAB International, Wallingford, UK.

Syed, R.N., Lodhi, A.M. and Shahzad, S. (2020) Management of *Pythium* diseases. In: Rai, M., Ahmed, K., Abd-Alsalam, A. and Aingle, A.P. (eds) *Pythium*. CRC Press, Boca Raton, FL, pp. 314–343. DOI: 10.1201/9780429296406-22.

Termorshuizen, A.J. (2014) Root pathogens. In: Dighton, J. and Adams Krimins, J. (eds) *Interactions in Soil: Promoting Plant Growth*. Springer, Dordrecht, The Netherlands, pp. 119–137. DOI: 10.1007/978-94-017-8890-8_6.

Tilston, E.L., Deakin, G., Bennett, J., Passey, T., Harrison, N. *et al.* (2020) Effect of fungal, oomycete and nematode interactions on apple root development in replant soil. *CABI Agriculture and Bioscience* 1(1), 14. DOI: 10.1186/s43170-020-00014-7.

Tiyagi, S.A., Zaidi, S.B.I. and Alam, M.M. (1988) Interaction between *Meloidogyne javanica* and *Macrophomina phaseolina* on lentil. *Nematologia Mediterranea* 16, 221–222.

Townshend, J.L. (1984) Inoculum densities of five plant parasitic nematodes in relation to alfalfa seedling growth. *Canadian Journal of Plant Pathology* 6(4), 309–312. DOI: 10.1080/07060668409501534.

Van Buyten, E. and Höfte, M. (2013) *Pythium* species from rice roots differ in virulence, host colonization and nutritional profile. *BMC Plant Biology* 13, 203. DOI: 10.1186/1471-2229-13-203.

van Gundy, S.D., Kirkpatrick, J.D. and Golden, J. (1977) The nature and role of metabolic leakage from root-knot nematode galls and infection by *Rhizoctonia solani*. *Journal of Nematology* 9, 113–121.

van West, P., Appiah, A.A. and Gow, N.A. (2003) Advances in research on oomycete root pathogens. *Physiological and Molecular Plant Pathology* 62(2), 99–113. DOI: 10.1016/S0885-5765(03)00044-4.

Vanegas-Villa, D.M., Navarro-Álzate, R.A., Afanador-Kafuri, L., Gutiérrez-Monsalve, J.A., Morales-Osorio, J.G. *et al.* (2020) Effect of interaction between *Phytophthora nicotianae* and *Meloidogyne* spp. on the productivity and quality of tobacco plants (*Nicotiana tabacum*). *Nematology* 22(10), 1179–1192. DOI: 10.1163/15685411-bja10021.

Verbeek, R.E.M., Banaay, C.G.B., Sikder, M., De Waele, D., Vera Cruz, C.M. *et al.* (2016) Interactions between the oomycete *Pythium arrhenomanes* and the rice root-knot nematode *Meloidogyne graminicola* in aerobic Asian rice varieties. *Rice* 9, 36.

Vos, C., Claerhout, S., Mkandawire, R., Panis, B. and De Waele, D. (2012) Arbuscular mycorrhizal fungi reduce root-knot nematode penetration through altered root exudation of their host. *Plant and Soil* 354, 335–345. DOI: 10.1007/s11104-011-1070-x.

Walker, N.R., Kirkpatrick, T.L. and Rothrock, C.S. (2000) Influence of *Meloidogyne incognita* and *Thielaviopsis basicola* populations on early-season disease development and cotton growth. *Plant Disease* 84, 449–453.

Webster, J.M. (1985) Interaction of Meloidogyne with Fungi on Crop Plants. In: Sasser, J.N. and Carter, C.L., Eds., An Advanced Treatise on Meloidogyne. Vol. I. Biology and Control, North Carolina State University and United States Agency for International Development, pp. 183–192. Raleigh, NC, USA.

Williamson-Benavides, B.A. and Dhingra, A. (2021) Understanding root rot disease in agricultural crops. *Horticulturae* 7(2), 33. DOI: 10.3390/horticulturae7020033.

Willlets, H.J. and Wrong, J.A.L. (1980) The biology of *Sclerotinia sclerotiorum, S. trifolium, S. minor, with emphasis on specific nomenclature*. *The Botanical Review* 46, 101–165. DOI: 10.1007/BF02860868.

Wong, T.W. and Quesada-Ocampo, L.M. (2024) Sensitivity of *Meloidogyne incognita*, *Fusarium oxysporum* f. sp. *niveum*, and *Stagonosporopsis citrulli* to succinate dehydrogenase inhibitors used for control of watermelon diseases. *Plant Diseases* 108(6), 1762–1768. DOI: 10.1094/PDIS-12-22-2922-RE.

Yao, X., Guo, H., Zhang, K., Zhao, M., Ruan, J. *et al*. (2023) *Trichoderma* and its role in biological control of plant fungal and nematode disease. *Frontiers in Microbiology* 14, 1160551. DOI: 10.3389/fmicb.2023.1160551.

8 Nematode–Bacteria Disease Complexes and Their Sustainable Management

Mujeebur Rahman Khan[1]*, Rizwan Ali Ansari[1], Marceline Egnin[2] and Conard Bonsi[2]

[1]Department of Plant Protection, Faculty of Agricultural Sciences, Aligarh Muslim University, Aligarh, India; [2]Department of Plant Science, Tuskegee University, Tuskegee, AL, USA

Abstract

Nematode–bacteria disease complexes represent a critical multipathogenic etiological scenario with significant implications in crop health and agricultural productivity. The disease complexes resulting from the interaction between plant nematodes and phytobacteria exhibit aggravated disease severity leading to substantially greater yield reductions than those caused by either pathogen alone. The interactions between these pathogens are wide in occurrence, and of great economic significance. The nematode-induced physical and physiological alterations in the host lead to degradation of plant tissues and disruption of vital processes such as root exudation, nutrient uptake, etc. These alterations contribute in promoting the bacterial infection in the host in causing stunting and wilting of the plant, and reducing the yield. The commonly occurring and economically significant nematode–bacteria disease complexes involves *Ralstonia*, *Xanthomonas*, *Agrobacterium* and *Pectobacterium*. The management of these disease complexes presents significant challenges to plant protectionists and crop growers. Integrating biocontrol, chemical, cultural and host resistance strategies that target both nematodes and bacteria may offer sustainable solutions, enhancing plant resilience and ensuring better crop yields. This chapter updates the current understanding of nematode–bacteria disease complexes, the probable and actual mechanisms involved, and identifies areas where sincere research efforts are needed. The chapter also explores the sustainable management strategies of disease complexes.

Introduction

Plant-parasitic nematodes are worm-like microscopic organisms that feed on or within plant tissues to obtain nutrients. Plant nematodes are largely soil inhabitants and constitute an important component of the soil ecosystem, participating in various ecological processes that occur in the soil environment (Khan *et al*., 2021). Plant-parasitic nematodes often attack the underground parts of plants, impairing their ability to absorb water and nutrients from the soil, leading to a significant decline in plant growth and yield (Khan and Jairajpuri, 2010b, c, 2012). Plant nematodes attack all kinds of crops (Sikora, 2013; Khan and Quintanilla,

*Corresponding author: mrkhan777in@yahoo.co.in

 Nematode Disease Complexes in Agricultural Crops (ed. Mujeebur Rahman Khan)
DOI: 10.1079/9781800625228.0008

2023) and inflict 7–21% yield loss (Jones *et al.*, 2013; Khan *et al.*, 2023b) that may account to US$125–200 billion annually (Kumar *et al.*, 2020; Mesa-Valle *et al.*, 2020; Pires *et al.*, 2022; Khan, 2023a). In addition to directly inflicting crop damage, plant nematodes also aid plant-pathogenic fungi, bacteria and viruses in causing infection of greater severity (Back *et al.*, 2002). Such etiological situations are called disease complexes (Khan *et al.*, 2023b; Parrado and Quintanilla, 2024).

Plant roots interact with a diverse array of soil microorganisms leading to complex root–microbe interactions (Alabouvette *et al.*, 1996; Khan, 2023b). Nematodes often serve as primary pathogens facilitating the entry and establishment of plant-pathogenic fungi and bacteria that would otherwise struggle to infect plants (Powell, 1979; Khan and Khan, 1995). The nematodes induce modifications in the host plant creating a synergistic environment favorable for the reproduction and multiplication of bacteria (Liang *et al.*, 2019; Khan and Sharma, 2020). The disease complexes involving plant-parasitic nematodes, fungi, bacteria, viruses and other microorganisms are well documented (Back *et al.*, 2002, 2006). In a disease complex, the cumulative damage inflicted on the plants by the concurrent action of multiple pathogens often surpasses the impact caused by individual pathogens acting independently (Dixit *et al.*, 2024). Thus, plant damage frequently intensifies when subjected to simultaneous attacks by two or more pathogens in comparison to the monopathogenic situation.

The first record of nematode–bacteria interaction was made by Hunger in 1901, who noted that tomato plants became more susceptible to infection by *Ralstonia solanacearum* (= *Pseudomonas solanacearum*) in the presence of nematodes (Hunger, 1901; Kundu and Vyshali, 2023; Parrado and Quintanilla, 2024). Building on this foundational work, pioneering studies by Pitcher and Crosse (1958) and Pitcher (1963, 1965) on the interactions between *Aphelenchoides ritzemabosi* and *Corynebacterium fascians* forming cauliflower disease in strawberries established the involvement of synergistic interactions between the two pathogens that led to the manifestation of the bacterial disease. Studies have shown that nematodes breach the plant defenses through continuous feeding, allowing bacterial pathogens to penetrate deeply and modify root exudates, promoting bacterial growth (Siddiqui *et al.*, 2013; Migunova *et al.*, 2021).

Important Nematode–Bacteria Disease Complexes

Nematode–bacteria complexes are widespread, affecting various crops and leading to significant economic impacts with greater yield losses. For instance, the combination of *R. solanacearum* and *Meloidogyne* spp. causes much greater yield decline in potatoes and tomatoes compared to the bacteria alone (Khan, 1993). The interactions between these pathogens can be mostly synergistic, severely affecting plant health and pathogen development (Khan and Jairajpuri, 2010a; Lamichhane and Venturi, 2015). Hence, nematode–bacteria disease complexes pose a significant threat to various economically important crops worldwide. These complex interactions between plant-parasitic nematodes and pathogenic bacteria can lead to severe disease outbreaks, causing substantial yield losses and economic damage to agricultural production (Getu *et al.*, 2024). Exact estimates on the yield losses inflicted by nematode–bacteria disease complexes are not available, but based on cumulative impact assessment, the complexes have great economic and social significance. The most important disease complexes with regard to crop damage include nematode–*Ralstonia*, nematode–*Xanthomonas*, nematode–*Agrobacterium* and nematode–*Pectobacterium*, as detailed below.

Nematode–*Ralstonia solanacearum* disease complexes

Nematode–*R. solanacearum* disease complexes pose a significant threat to various crops, especially solanaceous crops such as tomato, potato, eggplant, chilli, tobacco, etc., due to the synergistic relationship between *M. incognita* and *R. solanacearum* (Furusawa *et al.*, 2019). The root galls provide an ideal environment for *R. solanacearum* invasion and colonization that exacerbates wilt severity. The occurrence of *R.*

solanacearum and associated nematode disease complexes is a major concern for a number of economically important crops across various climatic zones, soil types and farming systems (Hussain and Bora, 2009; Singh and Siddiqui, 2012).

Ralstonia solanacearum

The bacteria *R. solanacearum* (= *P. solanacearum*) forms fluidal, irregularly round, creamy, convex, dull white colonies with a red center on tetrazolium chloride agar medium (Sharma, 2018). It is a non-spore-forming, Gram-negative and non-fluorescent bacterium, and exhibits a positive reaction in the hypersensitivity response test in tobacco leaves (Aslam and Mukhtar, 2023). The bacterium is an aggressive pathogen causing vascular wilt in over 200 crop species and thrives in moist, soil-borne environments (Van der Wolf *et al.*, 1998). The cells of the bacteria can persist for a longer period in infected fields, with reports indicating survival for up to 12 months in potato fields (van Elsas *et al.*, 2000). Potential sources of dissemination of the bacteria include a variety of factors such as contaminated irrigable water, presence of weeds harboring the pathogens and infested soil, crop debris, farm equipment, etc. The bacterium spreads subterraneously from the infected to healthy plants and causes significant crop losses ranging from 35% to 90% under optimum temperature and moisture conditions (Singh *et al.*, 2015). In another report, bacterial wilt disease was shown to be a major global agricultural problem causing yield losses between 20% and 100% on a worldwide basis (Huet, 2014).

R. solanacearum enters the plant through natural openings or wounds, eventually reaching the xylem vessels where it rapidly multiplies and colonizes the vessels. This disrupts the conduction of water and minerals, causing vascular browning, yellowing and stunting of the host plant (Smith, 1920). Additional symptoms include narrow dark streaks under the epidermis, adventitious root growth on the stem, leaf epinasty and the release of a slimy bacterial ooze from the cut mark in the stem of an infected plant (Sarkar and Chaudhuri, 2015). *R. solanacearum* species complex is distributed across multiple countries in Africa, the Americas, Asia and Oceania (EPPO, 2023). This disease significantly impacts crop productivity, with annual economic losses exceeding US$1 billion (Elphinstone, 2005). The pathogen, *R. solanacearum*, infects over 310 plant species across 42 families, with solanaceous crops such as eggplant, tomato, pepper and potato being the major susceptible hosts (Genin, 2010).

Disease complex

The interaction between *R. solanacearum* and plant-parasitic nematodes, especially root-knot nematode, forms a devastating disease complex that causes severe wilting and significant yield losses in a number of crops worldwide. This synergy exacerbates plant damage beyond the capabilities of each of the pathogens individually. The *Meloidogyne* juveniles create injuries on plant roots, providing entry points for *R. solanacearum* through which the spores enter the root tissues and reach the vascular system. This interaction leads to severe wilting, reduced growth and ultimately plant mortality. The wilt disease complex often develops in bacteria-prone areas, and poses a serious threat to the susceptible crops such as tomatoes, potatoes, eggplants, chillies and other members of solanaceous family worldwide (Hayward, 1991; Denny, 2007). The nematode not only facilitates bacterial invasion but also creates a more favorable environment for bacterial colonization by altering root exudation and induces other modifications in the host. Bacterial wilt-resistant tomato cultivars with the *Mi* resistance gene became susceptible to root-knot nematodes in Japan where *Meloidogyne* populations possess the capability to overcome *Mi* gene resistance (Narabu and Momota, 1992; Umekawa *et al.*, 2005; Matsunaga, 2015).

Asghar *et al.* (2020) reported that concomitant inoculation of *R. solanacearum* and *M. javanica* caused 6–31% reduction in plant biomass of chilli plants. Wilt incidence (*R. solanacearum*) was found to be significantly lower in the absence of *M. incognita*. Furusawa *et al.* (2019) concluded that *Meloidogyne*-induced galls create a favorable micro-environment for the proliferation of *R. solanacearum*. A study conducted by Uehara and Nakaho (2018) investigated the impact of *Mi* gene-breaking *M. incognita* populations on the infection of *R. solanacearum*-resistant rootstocks of tomato. The study revealed that wilt resistance in the usual-grafted tomato plants was compromised with the infection of *Mi* gene-breaking populations of *M. incognita*. Furthermore, the presence of the *Meloidogyne* spp. was found to enhance the upward movement of the bacterial cells in the xylem tissue of rootstocks resistant to tomato containing the *Mi* gene. A study on the use of high grafting as a management strategy for controlling bacterial wilt in *M. incognita*-infected plants recorded a substantial decline in the incidence of wilting in *Meloidogyne*-infected plants (Kunwar *et al.*, 2015). Sundaresh *et al.* (2017) observed synergistic interaction between *M. incognita* and *R. solanacearum* that led to a greater reduction in the plant growth of tomato over their individual impacts. The study revealed that inoculating tomato plants with *M. incognita* 10 days before *R. solanacearum* resulted in 100% bacterial wilt incidence.

Das and Swain (2014) explored the interactive effect of *M. incognita* and *R. solanacearum* over a 2 year study on ginger inoculated with either nematodes or bacteria singly, concomitantly or sequentially with a 10 day interval between the inoculations. Sign of wilting began to manifest at different intervals: 27 days after inoculation (DAI) with bacteria alone, 22 DAI with nematodes followed by bacteria, 17 DAI with nematodes followed by a subsequent bacterial inoculation and 31.5 DAI when bacteria were introduced first, followed by the nematode. Concomitant or sequential inoculations of the pathogens led to a marked decrease in leaf water potential, hastening the onset of wilt symptoms. Additionally, there was a substantial decrease in various biomass parameters and rhizome yield. A decrease in the root-knot index was also noted compared to the control. In a controlled condition experiment, Siddiqui *et al.* (2013) investigated the pathogenic impact of *M. incognita*, *R. solanacearum* and *Pectobacterium carotovorum* through different sequential and concomitant inoculation schedules and found that *M. incognita* caused milder growth reductions compared to *R. solanacearum*, but simultaneous inoculation with both pathogens exacerbated growth reduction beyond individual effects. Inoculating *M. incognita* before *R. solanacearum* intensified damage. The individual inoculations of *M. incognita* or *P. carotovorum* caused considerable reductions in plant growth, but pre-inoculating *P. carotovorum* reduced its impact compared to simultaneous inoculation. However, the galling induced by *M. incognita* on potato roots was adversely affected by co-inoculation with either *P. carotovorum* or *R. solanacearum*. Ghosh and Dutta (2014) also observed similar effects of joint inoculations of *M. javanica* and *R. solanacearum* on tomato plants. In another study, joint inoculation of *M. incognita* and *R. solanacearum* increased the susceptibility of potato cvs Spunta and Diamant (Bekhiet *et al.*, 2010). Hazarika *et al.* (2006) conducted a field trial on *M. incognita*–*R. solanacearum* disease complex, and observed 18.3, 26.1 and 35.6% losses in jute fiber at pre-plant populations of 1312 nematode larvae/kg soil and 7.2×10^7 CFU bacteria/g soil.

Nematode–*Xanthomonas* disease complexes

Nematode–*Xanthomonas* disease complexes represent a significant threat to global agriculture due to their wide occurrence and severe impacts on plant health. The synergistic interaction between *Meloidogyne* spp. and *Xanthomonas* spp. leads to the formation of disease complexes that severely affect plant health of a wide range of economically important crops (Alam and Siddiqui, 2013). Nematode–*Xanthomonas* disease complexes are agriculturally important because their widespread distribution and the critical economic implications on crop production worldwide. Crops concomitantly inoculated with these two pathogens often suffer substantial yield reductions, low quality fruits and increased production costs (Gerura *et al.*, 2022).

Xanthomonas

Xanthomonas are yellow-pigmented, aerobic, flagellate bacteria that cause important diseases in many agriculturally important crops (Bansal *et al.*, 2021). Their host specificity, secretion systems and survival mechanisms contribute to their ecological success as plant pathogens. *Xanthomonas* species have straight rod-shaped, single-polar flagellate and motile cells. They are Gram-negative, obligate aerobes that are catalase-positive and oxidase-negative (Dowson, 1939). The yellow, mucoid colony morphology on YDC medium along with biochemical tests and molecular methods are critical for the identification and characterization of *Xanthomonas* species (Timilsina *et al.*, 2015). *Xanthomonas* spp. are widely distributed in Asia, Africa, Australia and America, and may cause substantial yield losses in important agricultural crops such as peppers, tomatoes, chillies and rice (Roach *et al.*, 2018). The *X. oryzae* pathovars are prevalent across Asia, Africa and Australia. In addition, *X. oryzae* pv. *oryzae* has been identified as an important pathogen of rice in the USA and other countries (EFSA, 2019). Furthermore, *X. euvesicatoria*, responsible for bacterial spot disease in pepper production, was found to be prevalent in Bulgaria and Macedonia (Vancheva *et al.*, 2014). *Xanthomonas* spp. causing leaf spot disease have been recorded in solanaceous crops worldwide. Bacterial spot disease caused by *X. euvesicatoria* can cause a yield loss of 23–44% in pepper plants (Bashan *et al.*, 1985). Bacterial leaf streak (*X. campestris* pv. *undulosa*) inflicts significant yield loss in wheat (Duveiller and Maraite, 1993).

Disease complex

Meloidogyne–Xanthomonas disease complexes pose a threat to production of tomato, peppers and beans. The nematode larvae facilitate bacterial invasion by making punctures and micro-perforations on the root surface (Shehabu *et al.*, 2010). Nematode-induced wounds allow *Xanthomonas* to bypass the plant's natural defenses and enable entry into the host without any resistance. This results in accelerated disease progress and leads to greater crop losses than when either pathogen acts alone. *X. campestris* and *M. incognita* can independently cause significant damage on tomato plants. However, their co-occurrence may enhance disease severity substantially (Arain *et al.*, 2020). Siddiqui *et al.* (2013) observed that co-inoculation of *M. incognita* and *X. campestris* reduced the plant growth of chickpea greater than either pathogen alone. Inoculation with *R. solanacearum* or *X. campestris* and *M. javanica* greatly reduced the plant growth of chickpea. The sequence of inoculation may play a critical role; inoculation of *M. javanica* before the bacteria led to the highest reduction in plant growth. This suggests that the nematode aided the bacterial invasion and colonization (Singh and Siddiqui, 2012). The aiding role of plant nematodes in the invasion and colonization of *Xanthomonas* sp. in the onset of banana wilt was also recorded (Shehabu *et al.*, 2010). *Pratylenchus*, *Meloidogyne*, *Helicotylenchus* and *Tylenchorhynchus* species were commonly encountered in infected banana plants, with *Pratylenchus* being the most abundant. The combination of *X. oryzae* pv. *oryzae* and *Meloidogyne* spp. results in a significant increase in disease severity and devastating yield losses in rice (Bui *et al.*, 2020).

Nematode–*Agrobacterium* disease complexes

The interaction between phytonematodes and *Agrobacterium* has been recorded in a number of economically important crops. The wounds created by the nematode on the root surface serve as avenues for entry of *Agrobacterium* cells and facilitate the transfer of T-DNA of the bacteria into the plant cells, promoting tumor formation at the infection sites (Gelvin, 2000). The nematode invasion leads to weakening of the host defense mechanisms, rendering plants susceptible to bacterial infection or accelerating disease severity (Vrain and Copeman, 1987). The interactive effect of *Agrobacterium* and plant-parasitic nematodes results in the formation of crown galls or tumors, disruption of transport of nutrients and water, and overall stunted plant growth (El-Sherif and Elwakil, 1991). The combined stress from both the nematodes and bacteria makes it extremely challenging for the plants to recover, thereby amplifying the overall impact of the disease complex. Crops affected by nematode–*Agrobacterium* disease complex often suffer from severe growth abnormalities, reduced productivity and increased susceptibility to other secondary infections (Karimi *et al.*, 2000). For instance, the combination of *Meloidogyne* spp. and *A. tumefaciens* has been documented to cause substantial losses in grapevines, stone fruits and various ornamental plants (Goumas and Tzortzakakis, 1998; Rubio-Cabetas *et al.*, 2001).

Agrobacterium *spp.*

Agrobacterium spp. are Gram-negative, motile, aerobic, rod-shaped bacteria that induce neoplastic growths in plants by introducing their T-DNA into the host cell, causing crown gall and hairy root diseases (Morton and Fuqua, 2012; Lacroix and Citovsky, 2022). The genus includes various phytopathogenic species, with *A. tumefaciens* being the most studied, known for causing crown gall disease in plants (Nester, 2015). Geographically, *Agrobacterium* spp. thrive in all kinds of climates, demonstrating marked adaptability to diverse environmental conditions. A number of dicot crops of economic significance such as grapevines, stone fruits, tomatoes and ornamental plants are susceptible to the bacteria and develop crown gall disease (Millam, 2004). *Agrobacterium* spp. possess the unique ability to transfer T-DNA to eukaryotic hosts. T-DNA is extensively utilized in genetic engineering, making it a valuable tool for creating transgenic organisms and genetically modified crops (Nester, 2015).

Disease complex

The nematode–*Agrobacterium* disease complex represents a significant challenge in the cultivation of crops from the Rosaceae family, especially *Prunus* species. The bacterium *A. tumefaciens* causes crown gall, induces tumor-like growths on the plant roots and stems, hindering nutrient and water uptake of the plants. Rubio-Cabetas *et al.* (2001) reported that crown gall symptoms caused by *A. tumefaciens* were observed only in *Meloidogyne*-susceptible clones of myrobalan plum (*Prunus cerasifera*), especially those lacking *Ma* resistance genes, in the presence of the nematodes. Conversely, crown galls were absent in plants that were either non-hosts for the root-knot nematode or possessed resistance genes. This indicates that the dominant alleles of the *Ma1* and *Ma2* genes provided protection to *P. cerasifera* against root crown gall, highlighting the role of nematodes as causative agents that facilitate bacterial penetration and subsequent plant transformation via the Ti plasmid. Pak *et al.* (2009) worked on induction of hairy roots in Oriental melon using *A. rhizogenes* and investigated the reproduction of *M. incognita* on the infected host roots. The study demonstrated the successful transformation of Oriental melon and the integration and expression of foreign genes in the transformed hairy roots. The highest reproduction of *M. incognita* was recorded in the transformed host roots. Hairy root disease of roses caused by *A. rhizogenes* is generally of minor significance, but the disease becomes highly severe in the presence of *Pratylenchus vulnus* (Munnecke *et al.*, 1963).

Nematode–*Pectobacterium* disease complexes

Plant nematodes such as *Meloidogyne* spp. and *Pectobacterium* spp. form disease complexes in important vegetable crops. The bacterial cells attach to the outer surface of *Meloidogyne* juveniles to reach the vascular tissues (Mongae *et al.*, 2013). In carrot, co-infection with *M. incognita* and *P. carotovorum* caused greater plant damage than either pathogen alone (Nesha and Siddiqui, 2013). Similarly, in potato and other crops, combined inoculation with *M. incognita* and *Pectobacterium* sp. resulted in a severe disease complex (Siddiqui *et al.*, 2013; Joshi *et al.*, 2015). The bacterial infection disrupts water and nutrient transport within the plant body, leading to poor growth and reduced productivity (Khanna *et al.*, 2021).

Pectobacterium *spp.*

Pectobacterium species are Gram-negative rods belonging to the family Pectobacteriaceae, and produce membrane vesicles, causing soft rot diseases in crops and ornamental plants (Marquez-Villavicencio *et al.*, 2011). Historically, *P. carotovorum* was classified as *Erwinia carotovora* subsp. *carotovora* and *P. atrosepticum* subsp. *atroseptica*. Both were classified under the soft rot group of Erwinias and share a close taxonomic relationship with *E. chrysanthemi*. *P. carotovorum* is widely distributed across various geographic regions, whereas *P. atrosepticum* is found predominantly in cooler climates (Pérombelon, 2002; Toth *et al.*, 2003). Additionally, *Pectobacterium* strains such as *P. carotovorum* subsp. *brasiliense* and *P. carotovorum* subsp. *carotovorum* possess conserved virulence genes encoding plant cell-wall-degrading enzymes, flagella-based mobility and secretion systems contributing to their pathogenicity and adaptive immunity (Arizala and Arif, 2019; Mallick *et al.*, 2022; Soriano, 2023). Furthermore, genomic and phenotypic analyses have revealed distinct features among *Pectobacterium* isolates, leading to the proposal of a novel species, *P. parvum*, characterized by unique virulence traits and biochemical properties (Jonca *et al.*, 2024). On the other hand, *P. atrosepticum* is notably associated with blackleg disease in potatoes, a condition of substantial economic concern in temperate regions (Pérombelon and Kelman, 1980).

Disease complex

The nematode–*Pectobacterium* disease complex incites soft rot in vegetable crops such as potatoes, carrots and other underground vegetables. *Pectobacterium* spp. induce soft rot through the secretion of pectinases, cellulases and proteases (Pérombelon, 2002; Joko *et al.*, 2014). These enzymes facilitate the breakdown of host cell walls resulting in cell maceration and subsequent decay (Davidsson *et al.*, 2013; Kubicek *et al.*, 2014). Such enzymatic activities represent a critical virulence factor in the pathogenesis of *Pectobacterium* spp. Nematodes aid *Pectobacterium* infection entry in the roots by creating wounds which are formed during feeding or penetration. This interaction exacerbates plant damage leading to more severe and widespread soft rot symptoms. In one study, Mohammadi *et al.* (2020) inoculated sugarbeet seedlings with *P. betavasculorum* (10 ml/kg soil, 10^8 CFU/ml) along with 1500 and 2500 juveniles of *H. schachtii* at three distinct infection timings: 7 days prior to nematode inoculation, concurrently and 7 days after nematode inoculation. The pre-inoculation with *H. schachtii* led to substantially higher reductions in the plant growth metrics.

Mechanism of Disease Complex Formation

Understanding the mechanisms involved in nematode–bacteria disease complexes is crucial for developing effective management strategies (Topalović *et al.*, 2019). The synergy between nematodes and bacterial pathogens leads to exacerbation of plant damage, apparently due to physical disruptions to the host tissues caused by nematode larvae during the penetration and feeding activities which provide invasion avenues to the bacterial cells (Escobar *et al.*, 2015). This collaborative assault can lead to enhanced disease severity surpassing the impact of either pathogen alone. Advanced studies have shown that these interactions are not merely additive but can result in synergistic effects that significantly impact plant health. The nematode-induced alterations in root exudation patterns and the subsequent changes in the rhizosphere microbiota create a conducive environment for bacterial proliferation (Meng *et al.*, 2024). Hence, different mechanisms may be involved simultaneously in one or other combination in different disease complexes. Some of the important mechanisms contributing to the formation of nematode–bacteria disease complexes are summarized below (also see Table 8.1).

Nematodes as invasion avenue creators

Plant nematodes, both ecto- and endoparasites, create injuries and wounds on the root surface during feeding or penetration, respectively, which serve as critical avenues for entry of bacterial pathogens (Khan, 2008). These wounds may allow bacteria to bypass the plant's natural defenses and to establish infection in the host tissues (Elhady *et al.*, 2021; Topalović *et al.*, 2022). The mechanical injuries caused by plant-parasitic nematodes serve as a critical factor for the bacterial invasion into the host (Shehabu *et al.*, 2010; Li *et al.*, 2023). Pitcher (1965) observed that injuries made by plant-parasitic nematodes tend to help bacteria to invade the host plant as bacteria are relatively less adapted to penetrate the host's epidermis. It was observed that wilt or root-rot symptoms resembling nematode–bacterium disease complex symptoms can be replicated by inducing mechanical injury in absence of the plant-parasitic nematodes (Libman *et al.*, 1964; Bozbuga *et al.*, 2022). The tomato cvs Venus and Saturn resistant to bacterial wilt developed severe wilting when *P. solanacearum* was inoculated along with *Meloidogyne* spp. (McCarter, 1976). This pattern has been seen in other solanaceous crops such as eggplant, potato and

Table 8.1. Nematode–bacteria disease complexes and the probable mechanisms involved.

Bacterium	Nematode	Host	Types of mechanisms (reference)	Effects
Agrobacterium vitis	*Meloidogyne hapla*	Grapevine	Root injuries, rhizosphere manipulations (Sule and Lehoczky, 1993)	The simultaneous inoculation of *M. hapla* and *A. vitis* resulted in elevated levels of root infestation, while a high population of *A. vitis* alone demonstrated the ability to invade root tissues. The presence of *M. hapla* at lower bacterial densities considerably augmented the severity of *A. vitis* infection.
Agrobacterium tumefaciens	*Meloidogyne* spp.	*Prunus* spp.	Root wounding (Rubio-Cabetas *et al*., 2001)	Under conditions of high inoculum pressure, the formation of galls induced by *Meloidogyne* spp. and *A. tumefaciens* was both extensive and uniform. The increased inoculum pressure likely accelerated the infection process leading to more pronounced gall development and a higher density of affected tissue.
Ralstonia (= *Pseudomonas*) *solanacearum*	*Meloidogyne incognita* race 1	Tomato	Root wounding, and predisposition of roots by nematodes (Chindo *et al*., 1991)	The presence of the nematode considerably exacerbated wilt severity, accelerated disease progression, increased pathogen load and more severe wilting, stunting and overall decline in the plant growth.
Ralstonia (= *Pseudomonas*) *solanacearum*	*Meloidogyne incognita*	Banana	Root wounding (Pathak *et al*., 1999)	Following 10 days prior inoculation of the nematode, the bacteria significantly accelerated the development of wilt disease to its maximum severity. The interaction led to synergistic effects and disease progression dramatically increased.
Ralstonia (= *Pseudomonas*) *mendocina*	*Meloidogyne incognita*	Tomato	Toxic metabolites, rhizosphere engineering, competition for essential nutrients and rhizosphere space (Siddiqui and Husain, 1991)	The inoculation of bacteria notably suppressed nematode multiplication. Collectively, these mechanisms diminished nematode reproductive capacity and survival rates.

Continued

Table 8.1. Continued

Bacterium	Nematode	Host	Types of mechanisms (reference)	Effects
Ralstonia solanacearum	*Meloidogyne incognita*, *Rotylenchulus reniformis*	Tomato	Resistance breaking (Deberdt *et al.*, 1999)	Under elevated temperatures, the root-knot nematode markedly intensified bacterial wilt in resistant and susceptible cultivars. In contrast, the reniform nematode did not demonstrate a similar effect on wilt severity under the same conditions. This differential impact indicates a species-specific interaction with the host plant.
Ralstonia solanacearum	*Meloidogyne hapla*	Alfalfa	Modification in the host physiology and altering of the host immune system (Partridge, 2008)	The simultaneous infection by the bacteria and nematode significantly amplified both the frequency and severity of the wilt and root-knot nematode diseases. This synergistic interaction exacerbates the overall negative impact on the crops. The nematode activity may have also altered the host physiology and immune responses, making it more susceptible to bacterial infections.
Ralstonia solanacearum, *Fusarium* spp.	*Meloidogyne incognita*	Coleus and Withania	Rhizosphere alteration, root wounding (Mallesh *et al.*, 2009)	The concurrent inoculation with *R. solanacearum*, *M. incognita* and *Fusarium* spp. accelerated the disease development compared to monopathogenic situations. The triple-pathogen interaction led to rapid manifestation of wilting, chlorosis and necrosis.
Ralstonia solanacearum	*Meloidogyne incognita*	Eggplant	Puncturing of root by nematodes (Hussain and Bora, 2009)	The interaction between *R. solanacearum* and *M. incognita* significantly influenced the population dynamics of the soil and within plant roots. The presence of *R. solanacearum* can effectively mitigate the proliferation of the nematode.

tobacco, etc., where nematode-induced injuries on the roots facilitated the proliferation and aggressiveness of the bacteria.

Stewart and Schindler (1956) demonstrated that both ectoparasitic and endoparasitic nematodes can exacerbate the wilt severity by damaging the roots and facilitating bacterial invasion. Wilt-inducing bacteria rely primarily on such wounds to penetrate and establish in the host (Goodman *et al.*, 1967). Numerous bacterial pathogens inhabiting the soil can become active and pathogenic when they encounter with nematode-induced wounds on the root surface. The mechanical damage caused by *Rotylenchulus reniformis* promoted the invasion and colonization of *A. tumifaciens*, thereby contributing to the progression of the disease. Studies have shown that the injuries or wounds caused by *M. hapla*, *P. penetrans* and *H. dihystera* in the roots of alfalfa were subsequently utilized by *Pseudomonas* species for invasion (Bookbinder *et al.*, 1982). However, at low population densities of nematodes, crown gall symptoms did not exacerbate beyond those induced by the bacteria alone following wounding (Nigh, 1966).

Nematodes as vectors of plant-pathogenic bacteria

Some plant nematodes also act as vectors to transmit plant-pathogenic bacteria. This vector relationship enables bacterial cells in accessing the susceptible tissue. In most such cases, the bacterial cells are carried by the nematode on the body surface from the soil or from the plant surface to the susceptible host tissue inaccessible to the bacteria (Mongae, 2013). This relationship can be seen in the diseases such as yellow slime or 'tundu' disease of wheat, caused by *Rathayibacter tritici* (= *Clavibacter tritici*) and *Anguina tritici* (Khan *et al.*, 2023a; Khan, 2024). Likewise, 'Café au lait' bacteriosis (*Ditylenchus dipsaci* +*Pseudomonas fluorescens*) (Caubel and Samson, 1984), annual rye grass toxicity (*Anguina agrostis* +*R. tritici*) and cauliflower disease of strawberry (*Aphelenchoides fragariae* +*Rhodococcus fascians*) are formed in a similar way. This topic is comprehensively discussed in Chapter 10.

Nematodes as host resistance breakers

A notable aspect of nematode–bacteria interactions is the breakdown of resistance in many crop cultivars to bacterial pathogens which often act as secondary invaders. The plant nematodes facilitate bacterial invasion by providing specific conditions absent in resistant cultivars when nematodes are not present (Ayala-Doñas *et al.*, 2020). *Meloidogyne* spp. and other plant nematodes have been reported to render resistant cultivars susceptible to the plant-pathogenic bacteria (Kunwar *et al.*, 2015). This indicates that nematodes, in addition to causing physical wounding, also induce physicochemical changes in plants that render the R-genes non-functional, partially or fully (Zhuo *et al.*, 2015). In a study examining the interaction between nematodes and bacterial wilt, it was found that the wilt-resistant tobacco cv. Dixie Bright 101 became susceptible when plants were inoculated with both *P. solanacearum* and *M. incognita acrita* (Lucas *et al.*, 1955). Nematode infection triggers changes in host gene expression leading to the synthesis of defense-related proteins and metabolites (Dasgupta *et al.*, 1981).

The phenomenon of breakdown of resistance in plants to bacterial pathogens by nematodes is well studied (Riggs, 2020). Key pathways affecting the host resistance include systemic acquired resistance, xenobiotic metabolism and primary metabolism (Afzal *et al.*, 2009). Nematodes have evolved mechanisms to overcome plant defenses such as secretion of effector proteins that modify host tissues (Zinovieva, 2014). In the experiment, three variants of pathogen treatments were applied to cultivated tobacco plants: a bacterial suspension, nematode-infested soil and a combination of both. It was observed that within 21 days, 100% of the plants developed wilting in cohabitance with nematodes, whereas 10% wilted when plants were inoculated with bacteria alone. Some nematode–bacteria complexes can overcome plant resistance by the mechanisms as seen in root-knot nematodes breaking *Mi* gene resistance in tomatoes (Kaloshian *et al.*, 1998).

Several factors contribute to resistance breakdown, including long-term use of resistant cultivars, shifts in nematode races or species and multiple nematode species infestations (Davies and Elling, 2015). Strategies to improve resistance durability include rotating susceptible and resistant cultivars using nematicides alongside resistant cultivars and developing multispecies resistant cultivars (Young, 1992). Experimental studies have shown that pyramiding resistance genes in a single genotype or alternating different resistance genes in rotation can effectively suppress virulent nematode populations and improve long-term crop yields (Djian-Caporalino *et al.*, 2014). However, the development and use of resistant cultivars face challenges such as limited gene transfer, specificity of resistance and environmental factors affecting resistance expression (Roberts, 1992)

Nematodes as rhizoengineers

Nematodes may promote the infection ability of bacterial pathogens by altering the composition of root exudates, thereby enhancing microbial proliferation in the rhizosphere. These modifications stimulate the multiplication of pathogenic bacteria as well as potentially inhibit the development of antagonistic microorganisms, creating a favorable environment for the pathogen. Bhuwaneswari *et al.* (1965) recorded an increase in bacterial populations in the rhizosphere following root infection by nematodes. These root exudates terminate the dormancy of fungal propagules. In some studies, it has been recorded that the root exudates from infected plants contained higher amount of carbohydrates, proteins and amino acids compared to those from nematode-free plants (Čepulytė *et al.*, 2018). The feeding sites and the cellular modifications, particularly the nurse cells (giant cells) induced by *Meloidogyne* spp., offer an ideal substrate for bacterial colonization and growth within the plant (Khanna *et al.*, 2021). The influence of nematodes extends beyond their immediate feeding sites. Nematode-induced factors can translocate to various parts of the host plant including above-ground tissues, thereby altering the plant's resistance to bacterial pathogens and directly stimulating bacterial proliferation.

The biochemical composition of root exudates significantly influences the survival and proliferation of root-infecting pathogens in the soil (Costa *et al.*, 2021). The development of complex diseases is regulated by the changes in the soil microflora resulting from greater release of root exudates from the nematode-infected roots. The exudates can also influence the growth of naturally occurring antagonists in soil, thereby impacting disease dynamics (Khan *et al.*, 2016; Wen *et al.*, 2021; Lamichhane *et al.*, 2023). Furthermore, nematode-infected roots release substances that can break down natural defenses against fungi, making it easier for bacteria and nematodes to thrive (Riedel, 1988; Kenfaoui *et al.*, 2023). Through this mechanism, nematodes not only facilitate bacterial infection but also influence their own reproduction on the host plants.

Nematodes as host substrate modifiers

The modification in the host tissue following nematode invasion is an important part of nematode pathogenesis. Sedentary endoparasitic nematodes such as *Meloidogyne* spp. induce major structural and physiological changes in the host plant (Powell, 1979; Webster, 1980; Khan, 1993). For example, major structural changes induced by *Meloidogyne* spp. include root galls and giant cells (nurse cells) in the galled tissue (Huang, 1986; Khan and Khan, 1994a, b). Histochemical analyses of root galls have shown that the wall of the giant cells contain cellulose and pectin, but lacks lignin, suberin, starch and ninhydrin-positive substances. The absence of lignin and suberin, which are typically found in the cell walls of normal plant cells, suggests that the giant cells have a unique cell wall composition compared to the neighboring cells (Manohar *et al.*, 2020; Willig *et al.*, 2022). Infected roots display elevated concentrations of sugars, cellulose, phosphorylated intermediates, carbohydrates, RNA, free amino acids, proteins, nucleic acids and nucleotides compared to normal root cells (Siddique *et al.*, 2022).

The subventral glands of root-knot and cyst nematode juveniles exhibit enhanced activity prior to host penetration, but activity of these glands is reduced with the onset of parasitism (Endo and Wegin, 1988; Vieira *et al.*, 2020). Concurrently, the activity of the dorsal gland increases during this phase of host invasion (Bird, 1983). Phytohormones such as auxin which are crucial in the formation of feeding sites are introduced by root-knot nematodes (Khan and Khan, 1997). Elevated levels of auxin and its precursors were detected in the galls formed by *Meloidogyne* spp. (Jagdale *et al.*, 2021; Noureddine *et al.*, 2022, 2023). The nematode-infected roots also exhibit increased biologically active cytokinin (De Meutter *et al.*, 2003). Research has indicated that roots infected with *Meloidogyne* spp. are metabolically active, thus creating a favorable environment for the multiplication of bacteria and fungi (Jiang *et al.*, 2020; Habteweld *et al.*, 2024). Additionally, nematode-infected roots are found to be nutritionally much richer than healthy roots that facilitate the growth and colonization of bacterial pathogens. The nematode penetration and resulting damage to the host tissue result in the release of substances into the soil as root exudates.

The emerging evidence suggests that nematodes induce alterations in the host substrate in a complex way. Biochemical changes in the host are likely the most significant factors favoring nematode-related disease complexes (Slack, 1963; Ravichandra, 2014). The nematode invasion may lead to production of metabolites in the host that may degrade chemicals conferring resistance to pathogens or favor their pathogenesis. In the same way, *Meloidogyne* spp. modify infested and surrounding tissues or cells towards greater susceptibility to bacterial colonization (Johnson and Powell, 1969). Bird *et al.* (2003) observed that pre-inoculation of nematodes 3–4 weeks before bacterial inoculation resulted in severe wilt. They found that the nematode induced substantial physiological changes in the host plant which favored the infection of the wilt bacteria.

Nematodes as a necrosis inducers

Some migratory endoparasitic nematodes such as *Pratylenchus* and *Radopholus* species induce necrotic lesions on the host roots (Khan, 2008). These lesions, in addition to affecting plant biomass and yield production, also create conditions that favor the establishment of infection of some root-rot-causing secondary pathogens. One of the primary ways nematodes induce necrotic lesions is that they insert the stylets into root cells, secrete enzymes that degrade cell walls and facilitate nutrient extraction (Wyss, 2002; Fosu-Nyarko and Jones, 2016). This mechanical injury, combined with enzymatic degradation, causes cells to die, leading to the formation of necrotic areas. The necrotic tissue creates a favorable environment for opportunistic fungi and bacteria (Back *et al.*, 2002; Darling *et al.*, 2022). Often, the unspecialized and facultative parasites typically thrive on decaying plant matter but can also invade live tissues (Walters, 2011). Sometimes the progress of infection is slow in the direct shifting of the parasitism from the dead organic matter to live plant tissue. *Pratylenchus* spp. cause distinctive necrotic lesions on the surface and throughout the cortex of infected roots (MacGuidwin and Rouse, 1990). Initially, these lesions appear as reddish-brown spots but eventually turn black and can be found scattered along the root surface. As the nematodes continue their migration and feeding within the roots, these lesions can merge into larger necrotic areas that may eventually girdle the root (Loof, 1991). The rot-causing bacteria may utilize the necrotic lesions as infection parts for initial establishment on the host. Once established on the lesions, the bacteria multiply rapidly and spread to healthy parenchymatous cells of the root cortex (Davis and MacGuidwin, 2000).

Effect of Environmental Factors and Climate Change on Nematode–Bacteria Disease Complexes

Nematode–pathogen interactions are non-static, rather they are dynamic and continuously evolve in response to changes in the surrounding environment (Lahlali *et al.*, 2024). Hence, environmental factors and climate change profoundly influence the interactions between bacteria and nematodes, and the resulting disease complexes. Temperature, rainfall, humidity, edaphic factors and climate change may potentially influence the development and severity of nematode–bacteria disease complexes.

Temperature

Temperature is an important environmental factor that may influence both the biology of pathogens and the susceptibility of the host plant. Optimal temperatures for the activity of plant-pathogenic bacteria and nematodes often overlap, creating conditions conducive to the formation of disease complexes (Velásquez *et al.*, 2018). For example, *R. solanacearum*, affecting tomatoes, potatoes and eggplants, thrives in warm temperatures, whereas *R. solanacearum* race 3 biovar 2 shows high virulence at 19–28°C (Floyd and Michalak, 2007). The majority of *R. solanacearum* strains exhibit pathogenic activity at 25–30°C (Bocsanczy *et al.*, 2014). Warm temperatures (26–29°C) also favor the reproduction and multiplication activity of *Meloidogyne* spp. (Velloso *et al.*, 2022; Pradhan *et al.*, 2023). Such temperatures accelerate the life cycles of the participating pathogens, leading to more rapid disease progression and greater crop damage (Ploeg and Maris, 1999; Wang *et al.*, 2023).

Rainfall and soil moisture

Rainfall may substantially influence nematode and bacteria interactions, and the resulting disease complexes, by directly influencing the soil moisture as well as the activity and pathogenicity of both the pathogens (Furusawa *et al.*, 2019). High rainfall typically enhances mobility, infection rates and root invasion by nematodes that would ultimately promote bacterial invasion and proliferation, leading to more severe disease complexes. During the monsoon seasons, the *R. solanacearum* population increases dramatically, and may rise from 2.4×10^6 to 3.7×10^7 CFU/g in response to increased moisture levels (Chakrabarti, 2011). This enhances the likelihood of bacterial infection in the host plants, in cohabitance with plant nematodes. The relationship between these two pathogens is particularly pronounced in waterlogged conditions which are often a result of excessive rainfall. Studies have shown that when *M. incognita* and *R. solanacearum* infect plants together, the severity of wilt symptoms and reduced growth are significantly higher compared to either pathogen alone. This is particularly evident in conditions where soil moisture is optimal for both pathogens, facilitating their interaction and enhancing disease complex formation (Khan and Siddiqui, 2017). Plant nematodes require adequate soil moisture for movement and infection (Dropkin, 1976). For instance, in rice crop, the combination of high soil moisture and warm temperatures creates ideal conditions for the proliferation of *M. graminicola* (Mantelin *et al.*, 2017) and *X. oryzae* (Ou, 1985). These conditions lead to the development of the disease complex in a severe form (Khan *et al.*, 2014; Narasimhamurthy *et al.*, 2017).

Humidity

Humidity is also assumed to be a critical factor in the establishment and formation of disease complexes. High humidity facilitates the survival and dispersal of bacterial pathogens on the plant surfaces and within soil (Beattie, 2011; Jones and Barbetti, 2012; Velásquez *et al.*, 2018). The combined stress of *Meloidogyne* spp. and *R. solanacearum* can incite more severe disease complexes as both pathogens exploit similar environmental conditions for survival and propagation. High humidity levels favor nematode-parasitic activity as well as bacterial infections (Asghar *et al.*, 2020; Wang *et al.*, 2023). The humid condition facilitates the movement of *Anguina*, *Ditylenchus* and *Aphelenchoides* larvae on the plant surface leading to their increased efficiency of transmitting the *Rathayibacter*, *Pseudomonas*, *Rhodococcus*, etc. (Khan, 2008).

Edaphic factors

In addition, soil properties such as pH, texture and organic content are key features in the dynamics of multiple pathogen interactions. Acidic soils often favor the activity of certain bacterial pathogens, while nematodes can thrive in a range of soil types. The presence of organic matter can either suppress (Bonanomi *et al.*, 2007) or enhance disease severity (Van

Bruggen, 1995), depending on the specific interactions between soil microbes and plant pathogens. For example, sandy soils with low organic matter content are prone to high nematode populations, which can facilitate bacterial infection (Rimé *et al.*, 2003).

Climate change

Climate change poses a significant challenge to agriculture by modifying the environmental conditions that potentially influence the pathogenesis of phytobacteria. Rising global temperature, changes in precipitation patterns and increased CO_2 levels can affect the distribution and activity of plant-pathogenic nematodes and bacteria, and the resulting disease complexes (Lake and Wade, 2009; Wang *et al.*, 2021). For example, higher temperatures can expand the geographical range of both bacteria and nematodes into regions that were previously unsuitable for their survival (Reddy, 2013; Gullino *et al.*, 2018). This can lead to new outbreaks of disease complexes in temperate regions, threatening crops that were previously unaffected. Additionally, increased atmospheric CO_2 can enhance plant growth, potentially providing more biomass for pathogens to exploit and thereby increasing disease severity (Grover *et al.*, 2015; Khan and Rizvi, 2020).

Tomato crops are severely affected by the disease complex involving *Meloidogyne* spp. and *R. solanacearum* throughout the world (Uehara and Nakaho, 2018). In important agroecological regions (tropical and subtropical) where temperatures are comparatively high and humidity is elevated, diseases in severe form may appear leading to substantial crop losses (Brown and Allen, 2004). In rice cultivation, the disease caused by *M. graminicola* and *X. oryzae* becomes severe due to high soil moisture and warm temperatures (Thri Murty and Devadath, 1984; Rusinque *et al.*, 2021). Thus, such problems are more pronounced in flooded rice systems where soil moisture is consistently high (Hsieh and Buddenhagen, 1975). Climate change with its associated warming trends, and altered precipitation patterns may help increase the incidence of such disease complexes (Bekhiet *et al.*, 2010). Ginger plants are susceptible to a disease complex involving *M. incognita* and *R. solanacearum*. In regions with high temperatures and humidity, this disease complex can devastate ginger cultivations (Das and Swain, 2014).

Management of Nematode–Bacteria Disease Complexes

Effective and sustainable management strategies for nematode–bacteria disease complexes are crucial for safeguarding crop health, maximizing agricultural productivity and attaining food security. Nematode–bacteria disease complex management presents a challenge to crop protectionists and crop growers worldwide. However, these diseases can be effectively managed by thoroughly evaluating the efficiency of conventional methods and integrating them with both existing and emerging technologies, taking into account socio-economic needs and ecological constraints.

Cultural and physical methods

Cultural and physical methods can be considered as basic strategies for managing the complex diseases resulting from nematode and bacteria interactions. These approaches offer sustainable and environmentally friendly alternatives to pesticide treatments (Viaene *et al.*, 2013). Cultural methods including summer ploughing, crop rotation (Viaene *et al.*, 2013; Getu *et al.*, 2023), and physical methods such as soil solarization, steaming and hot water treatments have proven effective against wide spectrum of pathogens (Khan *et al.*, 2012; Abd-Elgawad and Askary, 2018; Getu *et al.*, 2023). Summer ploughing involves the deep turning of soil during the hottest part of the year, and serves to disrupt the life cycle of many soil-borne pathogens including plant-pathogenic nematodes and bacteria by exposing them to desiccation and high temperatures (Karimipour Fard *et al.*, 2019; Chinnaswamy, 2022). The impact of summer ploughing on both nematodes and bacteria, and subsequently on their interactions resulting in a disease complex, is profound and multifaceted, and this practice alone can provide a considerable decline in disease severity. Crop rotation and the use of antagonistic and trap crops may prove as effective as other cultural and physical methods in suppressing the nematode populations in particular.

Host resistance

In soil-based protected cultivation, bell pepper cultivars frequently developed nematode–bacteria wilt disease complex due to successive cropping (Ragassi *et al.*, 2022; Khan *et al.*, 2023c). To combat such challenges and to boost productivity, grafting with robust rootstocks proves beneficial. In a recently conducted study, the potential of grafting as a sustainable strategy for enhancing resistance to *Meloidogyne* spp. and *R. solanacearum* in bell pepper cultivation was explored (Naik *et al.*, 2024). The rootstocks of CRS-8 and CRS-1 grafted with Massilia demonstrated the highest resistance to bacterial wilt, with disease incidence as low as 22.2% and 27.7%, respectively. The resistant grafted plants also exhibited elevated levels of antioxidant enzymes and leaf chlorophylls. Getu *et al.* (2021) examined the response of 13 cultivars of potato to concomitant inoculation with *R. solanacearum* and *M. incognita*. They observed that the cvs Gudenie and Belete demonstrated resistance to *M. incognita* when pre-inoculated with *R. solanacearum*, whereas the cvs Mara Charre, Guassa, Araarsaa and Dagim expressed moderate resistance to *R. solanacearum*.

Biocontrol methods

Biocontrol methods are increasingly recognized as viable and sustainable strategies for managing disease complexes caused by nematode–bacteria interactions. These methods utilize natural antagonists to suppress pathogen populations, thereby reducing disease incidence and severity without relying on chemical pesticides (Meyer and Roberts, 2002; Khan, 2016). Biological control agents offer promising avenues for the management of plant diseases, including disease complexes (Siddiqui and Ehteshamul-Haque, 2001; Khan *et al.*, 2012, 2020). Bhagawati *et al.*, 2018 observed that seed treatment with *Pseudomonas fluorescens*, *Trichoderma viride* and *Purpureocillium lilacinum* significantly improved plant growth and yield of jute cv. Navin and suppressed the severity of diseases caused by *M. incognita* race two and *R. solanacearum*. Application of *Rhizobium* sp. in combination with *Ascochyta rabiei* and *Xanthomonas campestris* pv. *cassiae* had an adverse effect on galling and multiplication by *M. incognita* (Alam and Siddiqui, 2013). In addition, Elsharkawy *et al.* (2015) evaluated *B. thuringiensis* CR-371 and *Streptomyces avermectinius* NBRC14893 in combating the bacterial wilt complex affecting tomatoes caused by *M. incognita* and *R. solanacearum*. The application of both biocontrol agents resulted in a significant reduction in disease severity. Seed or nursery treatment with the biocontrol agents before planting may offer effective control of the soil-borne diseases (Khan, 2007; Khan *et al.*, 2009; Khan and Mohiddin, 2018; Khan and Mohiddin, 2023). The pretreatment of tomato roots with the biocontrol agents significantly curtailed the proliferation of *R. solanacearum* both in the presence and absence of the nematode.

Chemical control

Chemical control methods are a critical component of disease management systems and may also prove effective against nematode–bacteria disease complexes. Chemical pesticides provide immediate and effective control of both nematodes and bacterial pathogens. Jones *et al.* (2017) reported that among fluensulfone, fluopyram and spirotetramat applied under different concentrations to evaluate their impact on root-knot of lima bean, fluensulfone at low and high rates significantly reduced root galls in greenhouse trials. Mohsin *et al.* (2016) evaluated five chemicals, i.e. bleaching powder, streptomycin, neomycin, Bactrol and copper oxychloride, against *R. solanacearum*. Among these, bleaching powder demonstrated the most significant inhibition of *R. solanacearum*, and provided the highest disease suppression and yield in field conditions. Streptomycin, neomycin and Bactrol showed moderate efficacy, while copper oxychloride was found ineffective. Khan and Siddiqui (2018) evaluated two concentrations of ZnO nanoparticles in managing the disease complex in eggplants caused by *R. solanacearum*, *Phomopsis vexans* and *M. incognita*. ZnO nanoparticles were administered using three distinct methods: foliar spray, seed treatment and soil application. The results indicated that spraying with 200 ppm ZnO nanoparticles solutions yielded the most significant improvements in biomass, chlorophyll and carotenoid contents followed by seed treatment.

Integrated disease management

Integrated disease management is a holistic approach crucial for managing the disease complexes induced by nematodes and other plant pathogens, including bacteria. This method uses realistic and compatible combinations of suitable cultural practices, biocontrol agents, chemical treatments and host resistance, each playing a vital role in mitigating the adverse effects of the pathogens involved in complex diseases. Hussain and Bora (2008) demonstrated that neem cake, *Trichoderma harzianum*, carbofuran 3G and streptomycin applications at different timings relative to transplanting showed good results in managing the disease complex in pearl millet caused by *M. incognita* and *R. solanacearum*. The most notable improvement was observed in the treatment that combined summer ploughing with half of the recommended doses of carbofuran 3G, neem cake and streptomycin, along with the full dose of *T. harzianum*. This combination yielded effective control in terms of enhanced plant growth and yield. The second-best treatment involved summer ploughing with half doses of carbofuran 3G, neem cake and streptomycin, which also showed substantial improvements but was slightly less effective than the optimal treatment. Pavithra and Khatib (2014) evaluated the effectiveness of integrated pest management modules in managing the *M. incognita–R. solanacearum* wilt complex in eggplant. The study demonstrated that the treatment involving mustard as an intercrop, streptomycin (0.5 g/l) and copper oxychloride 50% WP at 1 g/l significantly reduced the incidence of bacterial wilt to 23.35%. Additionally, this treatment resulted in 73% reduction in the galling over control. Another effective treatment was the combination of mustard intercrop with *P. fluorescens* at 20 g/m^2. This approach decreased wilt severity by 28.3% and root galling by 77.6%, achieving a yield of 1.78 kg per 25 m^2 plot. In a field experiment, integrated treatment with streptomycin sulfate and tetracycline hydrochloride, carbofuran, poultry manure and *T. harzianum* formulation effectively managed the wilt disease complex caused by *Fusarium oxysporum*, *R. solanacearum* and *Meloidogyne* spp. in eggplant cultivation (Akter *et al.*, 2021).

Conclusion and Future Prospects

Plant nematodes have been recognized as important promoters of bacterial infections in plants, developing synergistic interactions and causing serious disease complexes. The synergy between the nematodes and bacteria often leads to exacerbation of disease severity as well as crop damage. The nematode induces physical, morphological, biochemical and physiological modulations in the host which make it conducive to bacterial proliferation and colonization. Various factors, particularly edaphic factors, play a crucial role in the formation and progression of nematode–bacteria disease complexes. Effective management strategy against nematode–bacteria disease complexes are not available. However, given the complex nature of nematode–bacterium interactions, tailored management approaches based on ecological considerations are essential, and both bacteria and nematodes should be targeted in the management module. This necessitates a multidisciplinary approach that combines advanced research with practical applications aiming to discover and implement a single agent capable of managing both nematodes and bacteria in an integrated manner or the use of a consortium of measures. Further, molecular tools should be evolved to precisely characterize the pathogens as well as the antagonistic microbes present in the rhizosphere. This will greatly help in developing effective and sustainable integrated modules against nematode–bacteria disease complexes.

References

Abd-Elgawad, M.M.M. and Askary, T.H. (2018) Fungal and bacterial nematicides in integrated nematode management strategies. *Egyptian Journal of Biological Pest Control* 28(1), 1–24. DOI: 10.1186/s41938-018-0080-x.

Afzal, A.J., Natarajan, A., Saini, N., Iqbal, M.J., Geisler, M. *et al.* (2009) The nematode resistance allele at the *rhg1* locus alters the proteome and primary metabolism of soybean roots. *Plant Physiology* 151, 1264–1280.

Akter, N., Islam, M.R., Hossain, M.B., Islam, M.N., Chowdhury, S.R. *et al.* (2021) Management of wilt complex of eggplant (*Solanum melongena* L.) caused by *Fusarium oxysporum*, *Ralstonia solanacearum* and *Meloidogyne* spp. *American Journal of Plant Sciences* 12(7), 1155–1171. DOI: 10.4236/ajps.2021.127080.

Alabouvette, C., Hoeper, H., Lemanceau, P. and Steinberg, C. (1996) Soil suppressiveness to diseases induced by soilborne plant pathogens. *Soil Biochemistry* 9(1), 371–413.

Alam, S. and Siddiqui, Z. (2013) Interactions of *Meloidogyne incognita*, *Ascochyta rabiei*, *Xanthomonas campestris* pv. *cassiae* and *Rhizobium* sp. on the disease complex of chickpea. *Acta Phytopathologica et Entomologica Hungarica* 48(2), 227–236.

Arain, M.R., Aslam, M.N., Jabeen, S., Akram, S., Shakeel, Q. *et al.* (2020) Effect of combined inoculation of *Xanthomonas* and *Meloidogyne* pathogens on the development of tomato root knot disease. *Journal of Plant Pathology and Microbiology* 11(8), 510. DOI: 10.35248/2157-7471.20.11.510.

Arizala, D. and Arif, M. (2019) Genome-wide analyses revealed remarkable heterogeneity in pathogenicity determinants, antimicrobial compounds, and CRISPR-Cas systems of complex phytopathogenic genus *Pectobacterium*. *Pathogens* 8(4), 247. DOI: 10.3390/pathogens8040247.

Asghar, A., Mukhtar, T., Raja, M.U. and Gulzar, A. (2020) Interaction between *Meloidogyne javanica* and *Ralstonia solanacearum* in Chili. *Pakistan Journal of Zoology* 52(4), 1525–1530. DOI: 10.17582/journal.pjz/20190501030529.

Aslam, M.N. and Mukhtar, T. (2023) Characterization of *Ralstonia solanacearum* causing bacterial wilt from major chili growing areas of Pakistan. *Bragantia* 82, e20230001. DOI: 10.1590/1678-4499.20230001.

Ayala-Doñas, A., Cara-García, M. de, Talavera-Rubia, M. and Verdejo-Lucas, S. (2020) Management of soil-borne fungi and root-knot nematodes in cucurbits through breeding for resistance and grafting. *Agronomy* 10(11), 1641. DOI: 10.3390/agronomy10111641.

Back, M., Haydock, P. and Jenkinson, P. (2006) Interactions between the potato cyst nematode *Globodera rostochiensis* and diseases caused by *Rhizoctonia solani* AG3 in potatoes under field conditions. *European Journal of Plant Pathology* 114(2), 215–223. DOI: 10.1007/s10658-005-5281-y.

Back, M.A., Haydock, P.P.J. and Jenkinson, P. (2002) Disease complexes involving plant parasitic nematodes and soilborne pathogens. *Plant Pathology* 51(6), 683–697. DOI: 10.1046/j.1365-3059.2002.00785.x.

Bansal, K., Kaur, A., Midha, S., Kumar, S., Korpole, S. *et al.* (2021) *Xanthomonas sontii* sp. nov., a non-pathogenic bacterium isolated from healthy basmati rice (*Oryza sativa*) seeds from India. *Antonie van Leeuwenhoek* 114, 1935–1947.

Bashan, Y., Azaizeh, M., Diab, S., Yunis, H. and Okon, Y. (1985) Crop loss of pepper plants artificially infected with *Xanthomonas campestris* pv. *vesicatoria* in relation to symptom expression. *Crop Protection* 4(1), 77–84. DOI: 10.1016/0261-2194(85)90007-9.

Beattie, G.A. (2011) Water relations in the interaction of foliar bacterial pathogens with plants. *Annual Review of Phytopathology* 49(1), 533–555.

Bekhiet, M.A., Kella, A.M., Khalil, A.E. and Tohamy, A.A. (2010) Interaction between root-knot nematode, *Meloidogyne incognita* and the bacterium, *Ralstonia solanacearum* on potato. *Journal of Plant Protection and Pathology* 1(7), 505–518. DOI: 10.21608/jppp.2010.86773.

Bhagawati, B., Choudhury, B.N. and Singh, S. (2018) Bio-management of disease-complex caused by *Meloidogyne incognita* race-2 and *Ralstonia solanacearum* in jute, *Corchoru solitorius* L. *Indian Journal of Nematology* 48(2), 227–229.

Bhuwaneswari, K., Sadasivan, T.S. and Sulochana, C.S. (1965) Rhizosphere microflora in relation to soil borne diseases of plants, advances in agricultural sciences and their applications. *Madras Agricultural Journal* 538–548.

Bird, A.F. (1983) Changes in the dimensions of the oesophageal glands in root-knot nematodes during the onset of parasitism. *International Journal for Parasitology* 13(4), 343–348. DOI: 10.1016/S0020-7519(83)80038-1.

Bird, D.M., Opperman, C.H. and Davies, K.G. (2003) Interactions between bacteria and plant-parasitic nematodes: Now and then. *International Journal for Parasitology* 33(11), 1269–1276. DOI: 10.1016/s0020-7519(03)00160-7.

Bocsanczy, A.M., Achenbach, U.C.M., Mangravita-Novo, A., Chow, M. and Norman, D.J. (2014) Proteomic comparison of *Ralstonia solanacearum* strains reveals temperature dependent virulence factors. *BMC Genomics* 15, 1–14. DOI: 10.1186/1471-2164-15-280.

Bonanomi, G., Antignani, V., Pane, C. and Scala, F. (2007) Suppression of soilborne fungal diseases with organic amendments. *Journal of Plant Pathology* 89, 311–324.

Bookbinder, M.G., Bloom, J.R. and Lukezic, F.L. (1982) Interactions among selected endoparasitic nematodes and three *Pseudomonads* on alfalfa. *Journal of Nematology* 14(1), 105–109.

Bozbuga, R., Ates, S.Y., Guler, P.G., Yildiz, H.N., Kara, P.A. *et al.* (2022) Host–pathogen and pest interactions: Virus, nematode, viroid, bacteria, and pests in tomato cultivation. In: Viškelis, P., Urbonavičienė, D. and Viškelis, J. (eds) *Tomato – From Cultivation to Processing Technology*. InTechOpen, London. DOI: 10.5772/intechopen.106064.

Brown, D.G. and Allen, C. (2004) *Ralstonia solanacearum* genes induced during growth in tomato: An inside view of bacterial wilt. *Molecular Microbiology* 53(6), 1641–1660. DOI: 10.1111/j.1365-2958.2004.04237.x.

Bui, H.X., Hadi, B.A.R., Oliva, R. and Schroeder, N.E. (2020) Beneficial bacterial volatile compounds for the control of root-knot nematode and bacterial leaf blight on rice. *Crop Protection* 135, 104792. DOI: 10.1016/j.cropro.2019.04.016.

Caubel, G. and Samson, R. (1984) Effect of the stem nematode, *Ditylenchus dipsaci*, on the development of “café au lait” bacteriosis in garlic (*Allium sativum*) caused by *Pseudomonas fluorescens*. cabidigitallibrary.org 311–313.

Čepulytė, R., Danquah, W.B., Bruening, G. and Williamson, V.M. (2018) Potent attractant for root-knot nematodes in exudates from seedling root tips of two host species. *Scientific Reports* 8(1), 10847.

Chakrabarti, S.K. (2011) *Ralstonia solanacearum* – an enigmatic bacterial plant pathogen. *Indian Phytopathology* 64(4), 313–321.

Chindo, P.S., Khan, F.A. and Erinle, I.D. (1991) Reaction of three tomato cultivars to two vascular diseases in presence of the root-knot nematode, *Meloidogyne incognita* race 1. *Crop Protection* 10(1), 62–64. DOI: 10.1016/0261-2194(91)90028-P.

Chinnaswamy, U. (2022) Soil borne diseases management in groundnut – an integrated approach. *International Journal of Modern Developments in Engineering and Science* 1(10), 8–11.

Costa, S.R., Ng, J.L.P. and Mathesius, U. (2021) Interaction of symbiotic rhizobia and parasitic root-knot nematodes in legume roots: From molecular regulation to field application. *MPMI* 34(5), 470–490. DOI: 10.1094/MPMI-12-20-0350-FI.

Darling, E., Quintanilla-Tornel, M. and Chung, H. (2022) Plant-eating nematodes and the key to fighting them. *Frontiers for Young Minds* 9, 604175. DOI: 10.3389/frym.2021.604175.

Das, S. and Swain, P.K. (2014) Quick wilt complex of ginger caused by *Meloidogyne incognita* and *Ralstonia solanacearum*. *Annals of Plant Protection Sciences* 22(1), 159–162.

Dasgupta, D.R., Farooqi, T.N., Ganguly, A.K., Mote, U.N., Ganguly, S. *et al.* (1981) Pathogenicity studies with *Meloidogyne incognita* and *Rotylenchulus reniformis* on two varieties of grape vine. *Indian Journal of Nematology* 11, 108–108.

Davidsson, P.R., Kariola, T., Niemi, O. and Palva, E.T. (2013) Pathogenicity of and plant immunity to soft rot pectobacteria. *Frontiers in Plant Science* 4, 191. DOI: 10.3389/fpls.2013.00191.

Davies, L.J. and Elling, A.A. (2015) Resistance genes against plant-parasitic nematodes: a durable control strategy. *Nematology* 17(3), 249–263. DOI: 10.1163/15685411-00002877.

Davis, E.L. and MacGuidwin, A.E. (2000) Lesion nematode disease. *Plant Health Instructor* 1. DOI: 10.1094/PHI-I-2000-1030-02.

De Meutter, J., Tytgat, T., Witters, E., Gheysen, G., Van Onckelen, H. *et al.* (2003) Identification of cytokinins produced by the plant parasitic nematodes *Heterodera schachtii* and *Meloidogyne incognita*. *Molecular Plant Pathology* 4(4), 271–277. DOI: 10.1046/j.1364-3703.2003.00176.x.

Deberdt, P., Deberdt, P., Quénéhervé, P., Darrasse, A. and Prior, P. (1999) Increased susceptibility to bacterial wilt in tomatoes by nematode galling and the role of the *Mi* gene in resistance to nematodes and bacterial wilt. *Plant Pathology* 48(3), 408–414. DOI: 10.1046/j.1365-3059.1999.00346.x.

Denny, T.P. (2007) Plant pathogenic *Ralstonia* species. In: Gnanamanickan, S.S. (ed.) *Plant-Associated Bacteria*. Springer, Dordrecht, The Netherlands, pp. 573–644. DOI: 10.1007/1-4020-4538-7_16.

Dixit, S., Sivalingam, P.N., Baskaran, R.K.M., Senthil-Kumar, M. and Ghosh, P.K. (2024) Plant responses to concurrent abiotic and biotic stress: Unravelling physiological and morphological mechanisms. *Plant Physiology Reports* 29(1), 6–17. DOI: 10.1007/s40502-023-00766-0.

Djian-Caporalino, C., Palloix, A., Fazari, A., Marteu, N., Barbary, A. *et al.* (2014) Pyramiding, alternating or mixing: Comparative performances of deployment strategies of nematode resistance genes to promote plant resistance efficiency and durability. *BMC Plant Biology* 14, 53–53. DOI: 10.1186/1471-2229-14-53.

Dowson, W.J. (1939) On the systematic position and generic names of the gram-negative bacterial plant pathogens. *Zentralblatt Fur Bakteriologie, Parasitenkunde, Infektionskrankheiten Und Hygiene* 9–13, 177–193.

Dropkin, V.H. (1976) Nematode parasites of plants, their ecology and the process of infection. In: Heitefuss, R. and Williams, P.H. (eds) *Physiological Plant Pathology*. Springer, Berlin, pp. 222–246. DOI: 10.1007/978-3-642-66279-9_10.

Duveiller, E. and Maraite, H. (1993) Study on yield loss due to *Xanthomonas campestris* pv. *undulosa* in wheat under high rainfall temperate conditions. *Zeitschrift Für Pflanzenkrankheiten Und Pflanzenschutz* 100, 453–459.

Elhady, A., Topalović, O. and Heuer, H. (2021) Plants specifically modulate the microbiome of root-lesion nematodes in the rhizosphere, affecting their fitness. *Microorganisms* 9(4), 679. DOI: 10.3390/microorganisms9040679.

Elphinstone, J.G. (2005) The current bacterial wilt situation: A global overview. In: Allen, C., Prior, P. and Hayward, A.C. (eds) *Bacterial Wilt Disease and the Ralstonia Solanacearum Species Complex*. APS Press, Saint Paul, MN, pp. 9–28.

Elsharkawy, M.M., Nakatani, M., Nishimura, M., Arakawa, T., Shimizu, M. *et al.* (2015) Control of tomato bacterial wilt and root-knot diseases by *Bacillus thuringiensis* CR-371 and *Streptomyces avermectinius* NBRC14893. *Acta Agriculturae Scandinavica, Section B – Soil & Plant Science* 65(6), 575–580. DOI: 10.1080/09064710.2015.1031819.

El-Sherif, A.G. and Elwakil, M.A. (1991) Interaction between *Meloidogyne incognita* and *Agrobacterium tumefaciens* or *Fusarium oxysporum* f. sp. *lycopersici* on tomato. *Journal of Nematology* 23(2), 239.

Endo, B.Y. and Wegin, W.P. (1988) Ultrastructure of the second stage juvenile of the root-knot nematode, *Meloidogyne incognita*. *Proceedings of the Helminthological Society of Washington* 55, 286–316.

EPPO (European and Mediterranean Plant Protection Organization) (2023) *Ralstonia solanacearum* species complex (RALSSO). Available at: https://gd.eppo.int/taxon/RALSSO/distribution (accessed 30 March 2024).

Escobar, C., Barcala, M., Cabrera, J. and Fenoll, C. (2015) Overview of root-knot nematodes and giant cells. *Advances in Botanical Research* 73, 1–32. DOI: 10.1016/bs.abr.2015.01.001.

Floyd, J. and Michalak, P.S. (2007) *New Pest Response Guidelines:* Ralstonia solanacearum *race 3 biovar 2*. USDA APHIS-PPQ Emergency and Domestic Programs, Washington, DC.

Fosu-Nyarko, J. and Jones, M.G.K. (2016) Advances in understanding the molecular mechanisms of root lesion nematode host interactions. *Annual Review of Phytopathology* 54(1), 253–278. DOI: 10.1146/annurev-phyto-080615-100257.

Furusawa, A., Uehara, T., Ikeda, K., Sakai, H., Tateishi, Y. *et al.* (2019) *Ralstonia solanacearum* colonization of tomato roots infected by *Meloidogyne incognita*. *Journal of Phytopathology* 167(6), 338–343.

Gelvin, S.B. (2000) Agrobacterium and plant genes involved in T-DNA transfer and integration. *Annual Review of Plant Biology* 51(1), 223–256.

Genin, S. (2010) Molecular traits controlling host range and adaptation to plants in *Ralstonia solanacearum*. *New Phytologist* 187(4), 920–928. DOI: 10.1111/j.1469-8137.2010.03397.x.

Gerura, F.N., Meressa, B.H., Tesfaye, A. and Olango, T.M. (2022) Enset (*Ensete ventricosum*) landraces associated parasitic nematodes and their relationship with *Xanthomonas* wilt disease in Gurage, Ethiopia. *Heliyon* 8, 12. DOI: 10.1016/j.heliyon.2022.e11956.

Getu, T., Mohammed, W., Seid, A., Mekete, T. and Kassa, B. (2021) Reaction of selected potato cultivars to *Meloidogyne incognita* and *Ralstonia solanacearum* under greenhouse conditions. *Archives of Phytopathology and Plant Protection* 54(17–18), 1456–1470. DOI: 10.1080/03235408.2021.1914357.

Getu, T., Mohammed, W., Seid, A., Mekete, T., Kassa, B. *et al.* (2023) Prevalence, occurrence and characterisation of disease complex involving *Ralstonia solanacearum* and *Meloidogyne* spp. on potato (*Solanum tuberosum* L.) in eastern Ethiopia. *Russian Journal of Nematology* 31(1), 17–28.

Getu, T., Mohammed, W., Seid, A., Mekete, T. and Kassa, B. (2024) Evaluation of the reaction of various potato (*Solanum tuberosum* L.) cultivars to the *Meloidogyne incognita* and *Ralstonia solanacearum* disease complex under field conditions. *Potato Research* 67(1), 73–91. DOI: 10.1007/s11540-023-09624-w.

Ghosh, P.P. and Dutta, S. (2014) Effect of sole and combined inoculation of *Ralstonia solanacearum* and *Meloidogyne javanica* on tomato. *The Bioscan* 9(1), 93–100.

Goodman, R.N., Kiraly, Z. and Zaitlin, M. (1967) *The Biochemistry and Physiology of Infectious Plant Disease*. Van Nostrand, Princeton, NJ.

Goumas, D.E. and Tzortzakakis, E.A. (1998) Reproduction of *Xiphinema index* and *Meloidogyne* species and infection of *Agrobacterium vitis* on grapevine rootstocks. *Phytopathologia Mediterranea* 37, 22–27.

Grover, M., Maheswari, M., Desai, S., Gopinath, K.A. and Venkateswarlu, B. (2015) Elevated CO_2: Plant associated microorganisms and carbon sequestration. *Applied Soil Ecology* 95, 73–85. DOI: 10.1016/j.apsoil.2015.05.006.

Gullino, M.L., Pugliese, M., Gilardi, G. and Garibaldi, A. (2018) Effect of increased CO_2 and temperature on plant diseases: A critical appraisal of results obtained in studies carried out under controlled environment facilities. *Journal of Plant Pathology* 100(3), 371–389. DOI: 10.1007/s42161-018-0125-8.

Habteweld, A., Kantor, M., Kantor, C. and Handoo, Z. (2024) Understanding the dynamic interactions of root-knot nematodes and their host: Role of plant growth promoting bacteria and abiotic factors. *Frontiers in Plant Science* 15, 1377453. DOI: 10.3389/fpls.2024.1377453.

Hayward, A.C. (1991) Biology and epidemiology of bacterial wilt caused by *Pseudomonas solanacearum*. *Annual Review of Phytopathology* 29(1), 65–87. DOI: 10.1146/annurev.py.29.090191.000433.

Hazarika, K., Rahman, F. and Bora, L.C. (2006) Avoidable yield loss of jute by combined infestations of *Meloidogyne incognita* and *Ralstonia solanacearum*. *Annals of Plant Protection Sciences* 14(2), 435–438.

Hsieh, S.P.Y. and Buddenhagen, I.W. (1975) Survival of tropical *Xanthomonas oryzae* in relation to substrate, temperature, and humidity. *Phytopathology* 65(5), 513–519.

Huang, S.P. (1986) Penetration, development, reproduction, and sex ratio of *Meloidogyne javanica* in three carrot cultivars. *Journal of Nematology* 18(3), 408.

Huet, G. (2014) Breeding for resistances to *Ralstonia solanacearum*. *Frontiers in Plant Science* 5, 715. DOI: 10.3389/fpls.2014.00715.

Hunger, F.W.T. (1901) En bacterie-ziekte der tomaat. *Mededeelingen Uit's Lands Plantentuin* 48, 4–57. DOI: 10.5962/bhl.title.127618.

Hussain, Z. and Bora, B.C. (2008) Integrated management of *Meloidogyne incognita* and *Ralstonia solanacearum* complex in brinjal. *Indian Journal of Nematology* 38(2), 159–164.

Hussain, Z. and Bora, B.C. (2009) Interrelationship of *Meloidogyne incognita* and *Ralstonia solanacearum* complex in brinjal. *Indian Journal of Nematology* 39(1), 41–45.

Jagdale, S., Rao, U. and Giri, A.P. (2021) Effectors of root-knot nematodes: An arsenal for successful parasitism. *Frontiers in Plant Science* 12, 800030. DOI: 10.3389/fpls.2021.800030.

Jiang, Y., Wu, Y., Hu, N., Li, H. and Jiao, J. (2020) Interactions of bacterial-feeding nematodes and indole-3-acetic acid (IAA)-producing bacteria promotes growth of *Arabidopsis thaliana* by regulating soil auxin status. *Applied Soil Ecology* 147, 103447. DOI: 10.1016/j.apsoil.2019.103447.

Johnson, H.A. and Powell, N.T. (1969) Influence of root knot nematodes on bacterial wilt development in flue-cured tobacco. *Phytopathology* 59, 486–491.

Joko, T., Subandi, A., Kusumandari, N., Wibowo, A. and Priyatmojo, A. (2014) Activities of plant cell wall-degrading enzymes by bacterial soft rot of orchid. *Archives of Phytopathology and Plant Protection* 47(10), 1239–1250. DOI: 10.1080/03235408.2013.838374.

Jonca, J., Pirhonen, M., Waleron, M.M., Gawor, J., Mrozik, A. *et al*. (2024) Comprehensive phenomic and genomic studies of the species, *Pectobacterium cacticida* and proposal for reclassification as *Alcorniella cacticida* comb. nov. *Frontiers in Plant Science* 15, 1323790. DOI: 10.3389/fpls.2024.1323790.

Jones, R.A.C. and Barbetti, M.J. (2012) Influence of climate change on plant disease infections and epidemics caused by viruses and bacteria. *CABI Reviews* 7, 1–33. DOI: 10.1079/PAVSNNR20127022.

Jones, J.G., Kleczewski, N.M., Desaeger, J., Meyer, S.L.F. and Johnson, G.C. (2017) Evaluation of nematicides for southern root-knot nematode management in lima bean. *Crop Protection* 96, 151–157. DOI: 10.1016/j.cropro.2017.02.015.

Jones, J.T., Haegeman, A., Danchin, E.G.J., Gaur, H.S., Helder, J. *et al*. (2013) Top 10 plant-parasitic nematodes in molecular plant pathology. *Molecular Plant Pathology* 14(9), 946–961. DOI: 10.1111/mpp.12057.

Joshi, J.R., Burdman, S., Lipsky, A. and Yedidia, I. (2015) Effects of plant antimicrobial phenolic compounds on virulence of the genus *Pectobacterium*. *Research in Microbiology* 166(6), 535–545. DOI: 10.1016/j.resmic.2015.04.004.

Kaloshian, I., Yaghoobi, J., Liharska, T., Hontelez, J., Hanson, D. *et al.* (1998) Genetic and physical localization of the root-knot nematode resistance locus *Mi* in tomato. *Molecular & General Genetics* 257(3), 376–385. DOI: 10.1007/s004380050660.

Karimi, M., Van Montagu, M. and Gheysen, G. (2000) Nematodes as vectors to introduce agrobacterium into plant roots. *Molecular Plant Pathology* 1(6), 383–387. DOI: 10.1046/j.1364-3703.2000.00043.x.

Karimipour Fard, H., Saeidi, K. and Doryanizadeh, N. (2019) Effects of chicken manure and summer ploughing on root-knot nematode (*Meloidogyne javanica*) management in cantaloupe (*Cucumis* melo var. *cantalupensis*). *Archives of Phytopathology and Plant Protection* 52(15–16), 1193–1205.

Kenfaoui, J., Goura, K., Legrifi, I., Seddiqi Khalil, N., El Hamss, H. *et al.* (2023) Natural product repertoire for suppressing the immune response of *Meloidogyne* species. In: Ahmad, F. and Blázquez, G.N. (eds) *Root-Galling Disease of Vegetable Plants*. Springer, Singapore, pp. 163–197.

Khan, M.R. (2007) Prospects of microbial control of root-knot nematodesinfecting vegetable crops. In: Sharma, N. and Singh, H.B. (eds) *Biotechnology: Plant Health Management*. International Book Distributing Co, Lucknow, India, pp. 643–665.

Khan, M.R. (2008) *Plant Nematodes – Methodology, Morphology, Systematics, Biology and Ecology*. Science Publishers, Enfield, NH. DOI: 10.1201/b10766.

Khan, M.R. (2016) Nematode biocontrol agents: Diversity and effectiveness against phytonematodes in sustainable crop protection. *Indian Phytopathology* 69(4s), 453–463.

Khan, M.R. (2023a) Nematode pests of agricultural crops, a global overview. In: Khan, M.R. (ed.) *Novel Biological and Biotechnological Applications in Plant Nematode Management*. Springer, Singapore, pp. 1–40.

Khan, M.R. (2023b) Plant nematodes, hidden constraints in the global crop production. In: Khan, M.R. and Quintanilla, M. (eds) *Nematode Diseases of Crops and Their Sustainable Management*. Elsevier, Publishers, USA, pp. 3–23.

Khan, M.R. (2024) Wheat seed-gall nematode: going to archives or have local significance in the developing countries. *Indian Phytopathology* 77(1), 1–11. DOI: 10.1007/s42360-023-00696-9.

Khan, M.R. and Jairajpuri, M.S. (2010a) Nematode infestation in industrial crops-national scenario. In: Khan, M.R. and Jairajpuri, M.S. (eds) *Nematode Infestations Part II: Industrial Crops. National Academy of Sciences*. Allahabad, India, pp. 1–19.

Khan, M.R. and Jairajpuri, M.S. (2010b) *Nematode Infestations Part I: Food Crops*. National Academy of Sciences, Allahabad, India.

Khan, M.R. and Jairajpuri, M.S. (2010c) *Nematode Infestations Part II: Industrial Crops*. National Academy of Sciences, Allahabad, India.

Khan, M.R. and Jairajpuri, M.S. (2012) *Nematode Infestations Part III: Horticultural Crops*. National Academy of Sciences, Allahabad, India.

Khan, M.R. and Khan, M.W. (1994a) Effects of simulated acid rain and root-knot nematode on tomato. *Plant Pathology* 43, 41–49.

Khan, M.R. and Khan, M.W. (1994b) Single and interactive effects of root-knot nematode and coal-smoke on okra. *New Phytologist* 126, 337–342.

Khan, M.R. and Khan, M.W. (1997) Effects of the root-knot nematode, *Meloidogyne incognita*, on the sensitivity of tomato to sulfur dioxide and ozone. *Environmental and Experimental Botany* 38, 117–130.

Khan, M.R. and Mohiddin, F.A. (2018) Trichoderma: its multifarious utility in crop improvement. In: Prasad, R., Gill, S.S. and Tuteja, N. (eds) *New and Future Developments in Microbial Biotechnology and Bioengineering: Crop Improvement Through Microbial Biotechnology*. Elsevier, Amsterdam, pp. 263–291. DOI: 10.1016/B978-0-444-63987-5.00013-X.

Khan, M.R. and Mohiddin, F.A. (2023) Biocontrol strategies for nematode management, an overview. In: Khan, M.R. (ed.) *Novel Biological and Biotechnological Applications in Plant Nematode Management*. Springer, Singapore, pp. 113–131.

Khan, M.R. and Quintanilla, M. (2023) *Nematode Diseases of Crops and Their Sustainable Management*. Academic Press, Cambridge, MA.

Khan, M.R. and Rizvi, T.F. (2020) Effect of elevated levels of CO_2 on powdery mildew development in five cucurbit species. *Scientific Reports* 10(1), 4986. DOI: 10.1038/s41598-020-61790-w.

Khan, M.R. and Sharma, R.K. (2020) Fusarium–nematode wilt disease complexes, etiology and mechanism of development. *Indian Phytopathology* 73(4), 615–628. DOI: 10.1007/s42360-020-00240-z.

Khan, M. and Siddiqui, Z.A. (2017) Interactions of *Meloidogyne incognita*, *Ralstonia solanacearum* and *Phomopsis vexans* on eggplant in sand mix and fly ash mix soils. *Scientia Horticulturae* 225, 177–184. DOI: 10.1016/j.scienta.2017.06.016.

Khan, M. and Siddiqui, Z.A. (2018) Zinc oxide nanoparticles for the management of *Ralstonia solanacearum*, *Phomopsis vexans* and *Meloidogyne incognita* incited disease complex of eggplant. *Indian Phytopathology* 71(3), 355–364. DOI: 10.1007/s42360-018-0064-5.

Khan, M.R., Altaf, S., Mohiddin, F.A., Khan, U. and Anwer, A. (2009) Biological control of plant nematodes with phosphate solubilizing microorganisms. In: Khan, M.S. and Zaidi, A. (eds) *Phosphate Solubilizing Microbes for Crop Improvement*. Nova Science Publishers, Hauppauge, NY, pp. 395–426.

Khan, M.R., Ashraf, T. and Shahid, S. (2012) Management of rice root-knot nematode by soil application of carbofuran at different intervals. *Annals of Plant Protection Sciences* 20(2), 444–448.

Khan, M.R., Ashraf, S., Rasool, F., Salati, K.M., Mohiddin, F.A. *et al.* (2014) Field performance of *Trichoderma* species against wilt disease complex of chickpea caused by *Fusarium oxysporum* f. sp. *ciceri* and *Rhizoctonia solani*. *Turkish Journal of Agriculture and Forestry* 38(4), 447–454. DOI: 10.3906/tar-1209-10.

Khan, M.R., Mohidin, F.A., Khan, U. and Ahamad, F. (2016) Native *Pseudomonas* spp. suppressed the root-knot nematode in *in vitro* and *in vivo*, and promoted the nodulation and grain yield in the field grown mungbean. *Biological Control* 101, 159–168. DOI: 10.1016/j.biocontrol.2016.06.012.

Khan, M.R., Haque, Z., Ahamad, F. and Zaidi, B. (2020) Biomanagement of rice root-knot nematode *Meloidogyne graminicola* using five indigenous microbial isolates under pot and field trials. *Journal of Applied Microbiology* 130(2), 424–438. DOI: 10.1111/jam.14788.

Khan, M.R., Ahamad, I. and Shah, M.H. (2021) Emerging important nematode problems in field crops and their management. In: Singh, K.P., Jahagirdar, S. and Sarma, B.K. (eds) *Emerging Trends in Plant Pathology*. Springer, Dordrecht, The Netherlands, pp. 33–62. DOI: 10.1007/978-981-15-6275-4_3.

Khan, M.R., Manzoor, S. and Ansari, M.S.A. (2023a) Effectiveness of *Trichoderma* species in controlling the seed-borne infestation of *Anguina tritici* in wheat seed-galls. *Indian Phytopathology* 76(4), 1083–1090. DOI: 10.1007/s42360-023-00681-2.

Khan, M.R., Mohiddin, F.A., Haque, Z. and Sharma, R.K. (2023b) Disease complexes and their sustainable management. In: Khan, M.R. and Quintanilla, M. (eds) *Nematode Diseases of Crops and Their Sustainable Management*. Academic Press, Cambridge, MA, pp. 65–93. DOI: 10.1016/B978-0-323-91226-6.00016-X.

Khan, M.R., Rizvi, T.F. and Ansari, M.S.A. (2023c) Nematode problems in polyhouse cultivation and their sustainable management in ornamental and vegetable crops. In: Khan, M.R. and Quintanilla, M. (eds) *Nematode Diseases of Crops and Their Sustainable Management*. Academic Press, Cambridge, MA, pp. 685–706.

Khan, M.W. (ed.) (1993) *Nematode Interactions*. Springer, Dordrecht, The Netherlands. DOI: 10.1007/978-94-011-1488-2.

Khan, M.W. and Khan, M.R. (1995) Fungus–nematode interactions. In: Swarup, G., Dasgupta, D.R. and Gill, J.S. (eds) *Nematode Pest Management – An Appraisal of Eco-Friendly Approaches*. Nematological Society of India, New Delhi, pp. 70–78.

Khanna, K., Kohli, S.K., Ohri, P. and Bhardwaj, R. (2021) Plants–nematodes-microbes crosstalk within soil: A trade-off among friends or foes. *Microbiological Research* 248, 126755. DOI: 10.1016/j.micres.2021.126755.

Kubicek, C.P., Starr, T.L. and Glass, N.L. (2014) Plant cell wall-degrading enzymes and their secretion in plant-pathogenic fungi. *Annual Review of Phytopathology* 52, 427–451. DOI: 10.1146/annurev-phyto-102313-045831.

Kumar, V., Khan, M.R. and Walia, R.K. (2020) Crop loss estimations due to plant-parasitic nematodes in major crops in India. *National Academy Science Letters* 43(5), 409–412. DOI: 10.1007/s40009-020-00895-2.

Kundu, A. and Vyshali, G. (2023) Current status of nematode–bacteria interaction: A mini review. *Agricultural Reviews* 44(4), 429–440. DOI: 10.18805/ag.R-2291.

Kunwar, S., Paret, M.L., Olson, S.M., Ritchie, L., Rich, J.R. *et al.* (2015) Grafting using rootstocks with resistance to *Ralstonia solanacearum* against *Meloidogyne incognita* in tomato production. *Plant Disease* 99(1), 119–124. DOI: 10.1094/PDIS-09-13-0936-RE.

Lacroix, B. and Citovsky, V. (2022) Agrobacterium. In: Maloy, S. and Hughes, K. (eds) *Brenner's Encyclopedia of Genetics*, 2nd edn. Academic Press, New York, pp. 52–54. DOI: 10.1016/B978-0-12-374984-0.00025-5.

Lahlali, R., Taoussi, M., Laasli, S.-E., Gachara, G., Ezzouggari, R. *et al*. (2024) Effects of climate change on plant pathogens and host–pathogen interactions. *Crop and Environment* 3(3), 159–170. DOI: 10.1016/j.crope.2024.05.003.

Lake, J.A. and Wade, R.N. (2009) Plant–pathogen interactions and elevated CO_2: Morphological changes in favour of pathogens. *Journal of Experimental Botany* 60(11), 3123–3131. DOI: 10.1093/jxb/erp147.

Lamichhane, J.R. and Venturi, V. (2015) Synergisms between microbial pathogens in plant disease complexes: A growing trend. *Frontiers in Plant Science* 6, 141048. DOI: 10.3389/fpls.2015.00385.

Lamichhane, J.R., Barbetti, M.J., Chilvers, M.I., Pandey, A.K. and Steinberg, C. (2023) Exploiting exudates to manage soil-borne disease complexes in a changing climate. *Trends in Microbiology* 32, 27–37. DOI: 10.1016/j.tim.2023.07.011.

Li, Y., Lei, S., Cheng, Z., Jin, L., Zhang, T. *et al*. (2023) Microbiota and functional analyses of nitrogen-fixing bacteria in root-knot nematode parasitism of plants. *Microbiome* 11(1), 48. DOI: 10.1186/s40168-023-01484-3.

Liang, L.M., Zou, C.G., Xu, J. and Zhang, K.Q. (2019) Signal pathways involved in microbe–nematode interactions provide new insights into the biocontrol of plant-parasitic nematodes. *Philosophical Transactions of the Royal Society B* 374(1767), 20180317. DOI: 10.1098/rstb.2018.0317.

Libman, G., Leach, J.G. and Adams, R.E. (1964) Role of certain plant-parasitic nematodes in infection of tomatoes by pseudomonas solanacearum. *Phytopathology* 54, 151–153.

Loof, A.P.P. (1991) The family Pratylenchidae Thorne. In: Nickle, W.R. (ed.) *Manual of Agricultural Nematology*. Marcel Dekker, New York, p. 365. DOI: 10.1201/9781003066576-8.

Lucas, G.B., Sasser, J.N. and Kelman, A. (1955) The relationship of root-knot nematodes to Granville wilt resistance in tobacco. *Phytopathology* 45, 537–540.

MacGuidwin, A.E. and Rouse, D.I. (1990) Role of *Pratylenchus penetrans* in the potato early dying disease of Russet Burbank potato. *Phytopathology* 80(10), 1077–1082. DOI: 10.1094/Phyto-80-1077.

Mallesh, S.B., Lingaraju, S., Byadgi, A.S., Hegde, Y.R., Mokashi, A.N. *et al*. (2009) Bioefficacy of rhizobacteria on root knot/wilt complex in coleus and ashwagandha. *Karnataka Journal of Agricultural Sciences* 22(5), 1116–1120.

Mallick, T., Mishra, R., Mohanty, S. and Joshi, R.K. (2022) Genome wide analysis of the potato soft rot pathogen *Pectobacterium carotovorum* strain ICMP 5702 to predict novel insights into its genetic features. *Plant Pathology Journal* 38(2), 102–114. DOI: 10.5423/PPJ.OA.12.2021.0190.

Manohar, M., Tenjo-Castano, F., Chen, S., Zhang, Y.K., Kumari, A. *et al*. (2020) Plant metabolism of nematode pheromones mediates plant–nematode interactions. *Nature Communications* 11(1), 208. DOI: 10.1038/s41467-019-14104-2.

Mantelin, S., Bellafiore, S. and Kyndt, T. (2017) *Meloidogyne graminicola*: A major threat to rice agriculture. *Molecular Plant Pathology* 18(1), 3–15. DOI: 10.1111/mpp.12394.

Marquez-Villavicencio, M.D.P., Groves, R.L. and Charkowski, A.O. (2011) Soft rot disease severity is affected by potato physiology and *Pectobacterium* taxa. *Plant Disease* 95(3), 232–241. DOI: 10.1094/PDIS-07-10-0526.

Matsunaga, T. (2015) Disease and pest resistance breeding of fruit vegetables: their present conditions and future prospects. *Shokubutu Boeki (Plant Protection)* 69, 334–341.

McCarter, S.M. (1976) Persistence of *Pseudomonas solanacearum* in artificially infested soils. *Ecology and Epidemiology* 998–1000. DOI: 10.1094/Phyto-66-998.

Meng, X.-R., Gan, Y., Liao, L.-J., Li, C.-N., Wang, R. *et al*. (2024) How the root bacterial community of *Ficus tikoua* responds to nematode infection: Enrichments of nitrogen-fixing and nematode-antagonistic bacteria in the parasitized organs. *Frontiers in Plant Science* 15, 1374431. DOI: 10.3389/fpls.2024.1374431.

Mesa-Valle, C.M., Garrido-Cardenas, J.A., Cebrian-Carmona, J., Talavera, M. and Manzano-Agugliaro, F. (2020) Global research on plant nematodes. *Agronomy* 10(8), 1148. DOI: 10.3390/agronomy10081148.

Meyer, S.L.F. and Roberts, D.P. (2002) Combinations of biocontrol agents for management of plant-parasitic nematodes and soilborne plant-pathogenic fungi. *Journal of Nematology* 34(1), 1–8.

Migunova, V.D., Tomashevich, N.S., Konrat, A.N., Lychagina, S.V., Dubyaga, V.M. *et al.* (2021) Selection of bacterial strains for control of root-knot disease caused by *Meloidogyne incognita*. *Microorganisms* 9(8), 1698. DOI: 10.3390/microorganisms9081698.

Millam, S. (2004) Agrobacterium-mediated transformation of potato. In: Curtis, I.S. (ed.) *Transgenic Crops of the World: Essential Protocols*. Springer, Dordrecht, The Netherlands, pp. 257–269. DOI: 10.1007/978-1-4020-2333-0_19.

Mohammadi, F., Charehgani, H., Abdollahi, M. and Rezaei, R. (2020) Interaction effect of cyst nematode, *Heterodera schachtii* and soft rot bacterium, *Pectobacterium betavasculorum* on sugar beet. *Archives of Phytopathology and Plant Protection* 53(19–20), 971–982. DOI: 10.1080/03235408.2020.1807692.

Mohsin, S.M., Nayem, S.A. and Hore, P.K. (2016) In-vitro and in-vivo efficiency of some chemicals to manage the bacterial wilt of tomato caused by *Ralstonia solanacearum*. *International Journal of Agricultural Sustainability* 12(9), 10–15.

Mongae, A.O. (2013) The interaction between root knot nematodes (*Meloidogyne* spp.) and soft rot *Enterobacteriaceae* (*Pectobacterium* spp.) and their host *Solanum tuberosum*. MSc dissertation, University of Pretoria, Pretoria, South Africa.

Mongae, A., Kubheka, G.C., Moleleki, N. and Moleleki, L.N. (2013) The use of fluorescent reporter protein tagging to study the interaction between root-knot nematodes and soft rot *Enterobacteriaceae*. *Letters in Applied Microbiology* 56(4), 258–263. DOI: 10.1111/lam.12045.

Morton, E.R. and Fuqua, C. (2012) Phenotypic analyses of *Agrobacterium*. *Current Protocols in Microbiology* 25(1), 3D–3. DOI: 10.1002/9780471729259.mc03d03s25.

Munnecke, D.E., Chandler, P.A. and Starr, M.P. (1963) Hairy root (*Agrobacterium rhizogenes*) of field roses. *Phytopathology* 53, 788–799.

Naik, S.A.T.S., Hongal, S.V., Hanchinamani, C.N., Manjunath, G., Ponnam, N. *et al.* (2024) Grafting bell pepper onto local genotypes of *Capsicum* spp. as rootstocks to alleviate bacterial wilt and root-knot nematodes under protected cultivation. *Agronomy* 14(3), 470. DOI: 10.3390/agronomy14030470.

Narabu, T. and Momota, Y. (1992) Occurrence and analysis of *Meloidogyne incognita* strains that parasitize resistant tomato cultivars. *Annual Report of the Kanto Plant Protection* 39, 297–299.

Narasimhamurthy, H.B., Ravindra, H., Sehgal, M., Ekabote, S.D. and Ganapathi, G. (2017) Biomanagement of rice root-knot nematode (*Meloidogyne graminicola*). *Journal of Entomology and Zoology Studies* 5(4), 1433–1439.

Nesha, R. and Siddiqui, Z.A. (2013) Interactions of *Pectobacterium carotovorum* pv. *carotovorum*, *Xanthomonas campestris* pv. *carotae*, and *Meloidogyne javanica* on the disease complex of carrot. *International Journal of Vegetable Science* 19(4), 403–411.

Nester, E.W. (2015) *Agrobacterium*: Nature genetic engineer. *Frontiers in Plant Science* 5, 730. DOI: 10.3389/fpls.2014.00730.

Nigh, E.L. (1966) Incidence of crown gall infection in peach as affected by Javenese root-knot nematode. *Phytopathology* 56, 150.

Noureddine, Y., da Rocha, M., An, J., Médina, C., Mejias, J. *et al.* (2022) MiR167–ARF8, an auxin-responsive module involved in the formation of root-knot nematode-induced galls in tomato. *BioRxiv* preprint. DOI: 10.1101/2022.07.29.501986.

Noureddine, Y., da Rocha, M., An, J., Médina, C., Mejias, J. *et al.* (2023) AUXIN RESPONSIVE FACTOR8 regulates development of the feeding site induced by root-knot nematodes in tomato. *Journal of Experimental Botany* 74(18), 5752–5766. DOI: 10.1093/jxb/erad208.

Ou, S.H. (1985) *Rice Diseases*. Commonwealth Agricultural Bureau, Kew, Surrey.

Pak, H.-K., Sim, J.-S., Rhee, Y., Ko, H.-R., Ha, S.-H. *et al.* (2009) Hairy root induction in oriental melon (*Cucumis melo*) by *Agrobacterium rhizogenes* and reproduction of the root-knot nematode (*Meloidogyne incognita*). *Plant Cell, Tissue and Organ Culture* 98(2), 219–228. DOI: 10.1007/s11240-009-9556-4.

Parrado, L.M. and Quintanilla, M. (2024) Plant-parasitic nematode disease complexes as overlooked challenges to crop production. *Frontiers in Plant Science* 15, 1439951. DOI: 10.3389/fpls.2024.1439951.

Partridge, J.E. (2008) *Bacterial Wilt of Alfalfa*. Department of Plant Pathology, University of Nebraska, Lincoln, NE.

Pathak, K.N., Roy, S., Ojha, K.L. and Jha, M.M. (1999) Influence of *Meloidogyne incognita* on the fungal and bacterial wilt complex of banana. *Indian Journal of Nematology* 29(1), 39–43.

Pavithra, S. and Khatib, R. (2014) Integrated management of wilt complex involving *Meloidogyne incognita* and *Ralstonia solanacearum* on brinjal (*Solanum melongena* L.). *The Bioscan* 9, 291–294.

Pérombelon, M.C.M. (2002) Potato diseases caused by soft rot erwinias: An overview of pathogenesis. *Plant Pathology* 51(1), 1–12. DOI: 10.1046/j.0032-0862.2001.Shorttitle.doc.x.

Pérombelon, M.C.M. and Kelman, A. (1980) Ecology of the soft rot erwinias. *Annual Review of Phytopathology* 18(1), 361–387. DOI: 10.1146/annurev.py.18.090180.002045.

Pires, D., Vicente, C.S.L., Menéndez, E., Faria, J.M.S., Rusinque, L. *et al.* (2022) The fight against plant-parasitic nematodes: Current status of bacterial and fungal biocontrol agents. *Pathogens* 11(10), 1178. DOI: 10.3390/pathogens11101178.

Pitcher, R.S. (1963) Role of plant parasitic nematodes in bacterial diseases. *Phytopathology* 53, 35–39.

Pitcher, R.S. (1965) Interrelationships of nematodes and other pathogens of plants. *Helminthological Abstracts* 34, 1–17.

Pitcher, R.S. and Crosse, J.E. (1958) Studies in the relationship of eelworms and bacteria to certain plant diseases. II. Further analysis of the strawberry cauliflower disease complex. *Nematologica* 3(3), 244–256. DOI: 10.1163/187529258X00472.

Ploeg, A. and Maris, P. (1999) Effects of temperature on the duration of the life cycle of a *Meloidogyne incognita* population. *Nematology* 1(4), 389–393. DOI: 10.1163/156854199508388.

Powell, N.T. (1979) Internal synergisms among organisms inducing disease. In: Horsfall, J.G. and Cowling, E.B. (eds) *Plant Disease: An Advanced Treatise. Volume IV. How Pathogens Induce Disease*. Academic Press, New York, pp. 113–133. DOI: 10.1016/B978-0-12-356404-7.50013-6.

Pradhan, P., Naresh, P., Barik, S., Acharya, G.C., Bastia, R. *et al.* (2023) Breeding for root-knot nematode resistance in fruiting solanaceous vegetable crops: A review. *Euphytica* 219(7), 71. DOI: 10.1007/s10681-023-03204-2.

Ragassi, C.F., Ribeiro, C.S. da C., Patiño-Torres, A., Lopes, C.A., Pinheiro, J.B. *et al.* (2022) Bell pepper rootstocks with multiple resistance to soilborne diseases. *Revista Ceres* 69(3), 299–307. DOI: 10.1590/0034-737x202269030007.

Ravichandra, N.G. (2014) Nematode disease complexes. In: Ravichandra, N.G. (ed.) *Horticultural Nematology*. Springer, New Delhi, pp. 207–238. DOI: 10.1007/978-81-322-1841-8_9.

Reddy, P.P. (2013) Impact of climate change on insect pests, pathogens and nematodes. *Pest Management in Horticultural Ecosystems* 19(2), 225–233.

Riedel, R.M. (1988) Interactions of plant-parasitic nematodes with soil-borne plant pathogens. *Agriculture, Ecosystems & Environment* 24(1–3), 281–292. DOI: 10.1016/0167-8809(88)90072-2.

Riggs, R.D. (2020) Resistance-breaking races of plant parasitic nematodes. In: Nickle, W.R. (ed.) *Manual of Agricultural Nematology*. CRC Press, Boca Raton, FL, pp. 827–852. DOI: 10.1201/9781003066576-19.

Rimé, D., Nazaret, S., Gourbière, F., Cadet, P. and Moënne-Loccoz, Y. (2003) Comparison of sandy soils suppressive or conducive to ectoparasitic nematode damage on sugarcane. *Phytopathology* 93(11), 1437–1444. DOI: 10.1094/PHYTO.2003.93.11.1437.

Roach, R., Mann, R., Gambley, C.G., Shivas, R.G. and Rodoni, B. (2018) Identification of *Xanthomonas* species associated with bacterial leaf spot of tomato, capsicum and chilli crops in eastern Australia. *European Journal of Plant Pathology* 150(3), 595–608. DOI: 10.1007/s10658-017-1303-9.

Roberts, P.A. (1992) Current status of the availability, development, and use of host plant resistance to nematodes. *Journal of Nematology* 24(2), 213–227.

Rubio-Cabetas, M.J., Minot, J.C., Voisin, R. and Esmenjaud, D. (2001) Interaction of root-knot nematodes (RKN) and the bacterium *Agrobacterium tumefaciens* in roots of *Prunus cerasifera*: Evidence of the protective effect of the *Ma* RKN resistance genes against expression of crown gall symptoms. *European Journal of Plant Pathology* 107(4), 433–441. DOI: 10.1023/A:1011231201779.

Rusinque, L., Maleita, C., Abrantes, I., Palomares-Rius, J.E. and Inácio, M.L. (2021) *Meloidogyne graminicola* – a threat to rice production: Review update on distribution, biology, identification, and management. *Biology* 10(11), 1163. DOI: 10.3390/biology10111163.

Sarkar, S. and Chaudhuri, S. (2015) New report of additional enterobacterial species causing wilt in West Bengal, India. *Canadian Journal of Microbiology* 61(7), 477–486. DOI: 10.1139/cjm-2015-0017.

Sharma, D.K. (2018) Morphological and biochemical characterization of *Ralstonia solanacearum* (Smith) in brinjal (*Solanum melongena* L.) in Rajasthan (India). *Advances in Plants & Agriculture Research* 8(3), 284–288. DOI: 10.15406/apar.2018.08.00328.

Shehabu, M., Addis, T., Mekonen, S., Waele, D. and Blomme, G. (2010) Nematode infection predisposes banana to soil-borne *Xanthomonas campestris* pv. *musacearum* transmission. *Tree and Forestry Science and Biotechnology* 4(2), 63–64.

Siddique, S., Coomer, A., Baum, T. and Williamson, V.M. (2022) Recognition and response in plant–nematode interactions. *Annual Review of Phytopathology* 60(1), 143–162. DOI: 10.1146/annurev-phyto-020620-102355.

Siddiqui, I.A. and Ehteshamul-Haque, S. (2001) Suppression of the root rot–root knot disease complex by *Pseudomonas aeruginosa* in tomato: The influence of inoculum density, nematode populations, moisture and other plant-associated bacteria. *Plant and Soil* 237(1), 81–89. DOI: 10.1023/A:1013313103032.

Siddiqui, Z.A. and Husain, S.I. (1991) Studies on the biological control of root-knot nematode. *Current Nematology* 2, 5–6.

Siddiqui, Z.A., Shehzad, M. and Alam, S. (2013) Interactions of *Ralstonia solanacearum* and *Pectobacterium carotovorum* with *Meloidogyne incognita* on potato. *Archives of Phytopathology and Plant Protection* 47(4), 449–455. DOI: 10.1080/03235408.2013.811810.

Sikora, R.A. (2013) Biological management of plant parasitic nematodes using fungal and bacterial endophytes – past, present, and future. In: Schneider, C., Leifert, C. and Feldmann, F. (eds) *Endophytes for Plant Protection: The State of the Art*. DPG, Braunschweig, pp. 144–148.

Singh, N. and Siddiqui, Z.A. (2012) Inoculation of tomato with *Ralstonia solanacearum*, *Xanthomonas campestris*, and *Meloidogyne javanica*. *International Journal of Vegetable Science* 18(1), 78–86.

Singh, S., Gautam, R.K., Singh, D.R., Sharma, T.V.R.S., Sakthivel, K. *et al.* (2015) Genetic approaches for mitigating losses caused by bacterial wilt of tomato in tropical islands. *European Journal of Plant Pathology* 143(2), 205–221. DOI: 10.1007/s10658-015-0690-z.

Slack, D.A. (1963) Introduction. Symposium of interrelationships between nematodes and other agents causing plant diseases. *Phytopathology* 53, 27–47.

Smith, E.F. (1920) *An Introduction to Bacterial Diseases of Plants*. Saunders, Philadelphia, PA. DOI: 10.5962/bhl.title.13485.

Soriano, F.V. (2023) Chemotaxis in *Pectobacterium atrosepticum* Scri1043: Functional and structural studies on chemotaxis adaptation proteins and chemoreceptors. Doctoral thesis, University of Granada, Granada, Spain.

Stewart, R.N. and Schindler, A.F. (1956) The effect of some ectoparasitic and endoparasitic nematodes on the expression of bacterial wilt in carnations. *Phytopathology* 46, 219–222.

Sule, S. and Lehoczky, J. (1993) Agrobacterium–nematode interactions on grapevine root. *Novenyvedelem* 29, 412–417.

Sundaresh, V.T., Devappa, C.A. and Priya, N. (2017) Interaction of *Ralstonia solanacearum* and *Meloidogyne incognita* on tomato (*Solanum lycopersicon* L.). *International Journal of Current Microbiology and Applied Sciences* 6(9), 156–160. DOI: 10.20546/ijcmas.2017.609.019.

Thri Murty, V.S. and Devadath, S. (1984) Role of seed in survival and transmission of *Xanthomonas campestris* pv. *oryzae* causing bacterial blight of rice. *Journal of Phytopathology* 110(1), 15–19.

Timilsina, S., Jibrin, M.O., Potnis, N., Minsavage, G.V., Kebede, M. *et al.* (2015) Multilocus sequence analysis of xanthomonads causing bacterial spot of tomato and pepper plants reveals strains generated by recombination among species and recent global spread of *Xanthomonas gardneri*. *Applied and Environmental Microbiology* 81(4), 1520–1529. DOI: 10.1128/AEM.03000-14.

Topalović, O., Elhady, A., Hallmann, J., Richert-Pöggeler, K.R. and Heuer, H. (2019) Bacteria isolated from the cuticle of plant-parasitic nematodes attached to and antagonized the root-knot nematode *Meloidogyne hapla*. *Scientific Reports* 9(1), 11477. DOI: 10.1038/s41598-019-47942-7.

Topalović, O., Santos, S.S., Heuer, H., Nesme, J., Kanfra, X. *et al.* (2022) Deciphering bacteria associated with a pre-parasitic stage of the root-knot nematode *Meloidogyne hapla* in nemato-suppressive and nemato-conducive soils. *Applied Soil Ecology* 172, 104344. DOI: 10.1016/j.apsoil.2021.104344.

Toth, I.K., Bell, K.S., Holeva, M.C. and Birch, P.R.J. (2003) Soft rot Erwiniae: From genes to genomes. *Molecular Plant Pathology* 4(1), 17–30. DOI: 10.1046/j.1364-3703.2003.00149.x.

Uehara, T. and Nakaho, K. (2018) Effects of high grafting on tomato plants infected by *Meloidogyne incognita* and *Ralstonia solanacearum*. *Journal of Phytopathology* 166(1), 53–58.

Umekawa, M., Miyai, S., Yano, E. and Takahashi, K. (2005) Manual for IPM: Guide to integrated pest management with minimal environmental stress. *Agriculture Research Service National Agricultural Research Centre* 55, 338–236.

Van Bruggen, A.H.C. (1995) Plant disease severity in high-input compared to reduced-input and organic farm systems. *Plant Disease* 79(10), 976–984. DOI: 10.1094/PD-79-0976.

Van der Wolf, J.M., Bonants, P.J.M., Smith, J.J., Hagenaar, M., Nijhuis, E. *et al.* (1998) Genetic diversity of *Ralstonia solanacearum* race 3 in western Europe determined by AFLP, RC-PFGE and Rep-PCR. In: Prior, P., Allen, C. and Elphinstone, J. (eds) *Bacterial Wilt Disease: Molecular and Ecological Aspects*. Springer, Berlin, pp. 44–49.

van Elsas, J.D., Kastelein, P., van Bekkum, P., van der Wolf, J.M., de Vries, P.M. *et al.* (2000) Survival of *Ralstonia solanacearum* biovar 2, the causative agent of potato brown rot, in field and microcosm soils in temperate climates. *Phytopathology* 90(12), 1358–1366. DOI: 10.1094/PHYTO.2000.90.12.1358.

Vancheva, T., Stoyanova, M., Tatyozova, M., Bogatzevska, N. and Moncheva, P. (2014) Subspecies diversity of *Xanthomonas euvesicatoria* bulgarian and macedonian strains from pepper. *Biotechnology and Biotechnological Equipment* 28(4), 592–601. DOI: 10.1080/13102818.2014.947722.

Velásquez, A.C., Castroverde, C.D.M. and He, S.Y. (2018) Plant-pathogen warfare under changing climate conditions. *Current Biology* 28(10), R619–R634. DOI: 10.1016/j.cub.2018.03.054.

Velloso, J.A., Maquilan, M.A.D., Campos, V.P., Brito, J.A. and Dickson, D.W. (2022) Temperature effects on development of *Meloidogyne enterolobii* and *M. floridensis*. *Journal of Nematology* 54(1), 20220013. DOI: 10.2478/jofnem-2022-0013.

Viaene, N., Coyne, D.L. and Davies, K.G. (2013) Biological and cultural management. In: Perry, R.N. and Moens, M. (eds) *Plant Nematology*, 2nd edn. CAB International, Wallingford, UK, pp. 383–410.

Vieira, P., Shao, J., Vijayapalani, P., Maier, T.R., Pellegrin, C. *et al.* (2020) A new esophageal gland transcriptome reveals signatures of large scale de novo effector birth in the root lesion nematode *Pratylenchus penetrans*. *BMC Genomics* 21(1), 16. DOI: 10.1186/s12864-020-07146-0.

European Food Safety Authority (EFSA), Vos, S., Camilleri, M. and Diakaki, M. (2019) Pest survey card on *Xanthomonas citri* pv citri and pv. *aurantifolii*. *EFSA Supporting Publications* 16(4), 1587E. DOI: 10.2903/sp.efsa.2019.EN-1587.

Vrain, T.C. and Copeman, R.J. (1987) Interactions between *Agrobacterium tumefaciens* and *Pratylenchus penetrans* in the roots of two red raspberry cultivars. *Canadian Journal of Plant Pathology* 9(3), 236–240. DOI: 10.1080/07060668709501880.

Walters, D. (2011) *Plant Defense: Warding Off Attack by Pathogens, Herbivores and Parasitic Plants*. Wiley, New York. DOI: 10.1002/9781444328547.

Wang, J., Shi, X., Li, L. and Zhang, X. (2021) Changes in soil nematodes in rhizosphere and non-rhizosphere soils following combined elevated [CO2] and canopy warming in a winter wheat field. *Geoderma* 386, 114907. DOI: 10.1016/j.geoderma.2020.114907.

Wang, Z., Luo, W., Cheng, S., Zhang, H., Zong, J. *et al.* (2023) *Ralstonia solanacearum* – a soil borne hidden enemy of plants: Research development in management strategies, their action mechanism and challenges. *Frontiers in Plant Science* 14, 1141902. DOI: 10.3389/fpls.2023.1141902.

Webster, J.M. (1980) Nematodes in an overcrowded world. *Revue de Nematologie* 3, 135–143.

Wen, T., Zhao, M., Yuan, J., Kowalchuk, G.A. and Shen, Q. (2021) Root exudates mediate plant defense against foliar pathogens by recruiting beneficial microbes. *Soil Ecology Letters* 3(1), 42–51. DOI: 10.1007/s42832-020-0057-z.

Willig, J.-J., Guarneri, N., van Steenbrugge, J.J.M., de Jong, W., Chen, J. *et al.* (2022) The *Arabidopsis* transcription factor TCP9 modulates root architectural plasticity, reactive oxygen species-mediated processes, and tolerance to cyst nematode infections. *The Plant Journal* 112(4), 1070–1083. DOI: 10.1111/tpj.15996.

Wyss, U. (2002) Feeding behaviour of plant-parasitic nematodes. In: Lee, D. (ed.) *The Biology of Nematodes*. CRC Press, London, pp. 233–259. DOI: 10.1201/b12614-10.

Young, L.D. (1992) Problems and strategies associated with long-term use of nematode resistant cultivars. *Journal of Nematology* 24(2), 228–233.

Zhuo, H., Ni, H., Yun, M., He, M., Liang, Z. *et al.* (2015) The phytochemicals with antagonistic activities toward pathogens of a disease complex caused by *Meloidogyne incognita* and *Ralstonia solanacearum*. *Journal of Pure and Applied Microbiology* 9(1), 209–213.

Zinovieva, S.V. (2014) Co-adaptation mechanisms in plant–nematode systems. *Parazitologiia* 48(2), 110–130.

9 Plant Diseases Caused by Nematode-vectored Viruses

George Bird[1]* and Guiping Yan[2]
[1]Department of Entomology, Michigan State University, East Lansing, MI, USA;
[2]Department of Plant Pathology, North Dakota State University, Fargo, ND, USA

Abstract
Diseases are detrimental physiological processes that result in characteristic symptoms. In plants, the symptoms can consist of abnormal foliage coloration/structure, under-growth and development such as stunting, or over-growth and development such as root or shoot galls. Infectious diseases are caused by biotic pathogens such as bacteria, fungi and nematodes or by chemical messengers such as viruses. Pathogens invade susceptible plants passively, through natural openings or wounds. Some pathogens, especially plant viruses, are vectored by stylet-bearing insects or nematodes. Only a small portion of the described plant viruses are known to be transmitted by plant nematodes. The known nematode vectors of plant viruses are restricted to the Triplonchida and Dorylaimida. This chapter presents an overview on the plant kingdom, viruses, the animal phylum Nematoda, nematode-vectored plant viruses and virus management strategies, followed by comprehensive descriptions and comparative analyses of selected plant virus diseases vectored by nematodes. The chapter also includes a brief discussion of topics recommended for future research and technology development initiatives.

Introduction

Species classified in the kingdom Plantae are multicellular, have a cell wall, true nucleus and central vacuole. As autotrophs, they convert atmospheric carbon dioxide and water into organic matter in the presence of light through photosynthesis (Michaels *et al.*, 2022). The Plantae is subdivided into five groups: Thallophyta, Bryophyta, Pteridophyta, Gymnosperms and Angiosperms. Thallophytes have a simple body structure and include green, red and brown algae. Some Thallophytes are believed to have evolved about 1.6 billion years ago (Pennisi, 2018). Bryophytes such as mosses, hornworts and liverworts lack a vascular system and require water for reproduction. Pteridophytes such as ferns and equisetum have roots, stems, leaves and a vascular system. As vascular plants, Gymnosperms produce seed that is not enclosed in a fruit. Angiosperms are flowering plants with their seeds encapsulated in a fruit. In general, very little is known about virus diseases of non-vascular plants. No nematode-vectored virus diseases have been reported for non-vascular plants. Most of the known nematode-vectored plant diseases are associated with Angiosperms.

Viruses are submicroscopic particles consisting of a nucleic acid core surrounded by a protein coat. As pathogens, viruses function as chemical messengers. While the origin of viruses is unknown, it is possible that they were part of the process of the formation of

*Corresponding author: birdg@msu.edu

 Nematode Disease Complexes in Agricultural Crops (ed. Mujeebur Rahman Khan)
DOI: 10.1079/9781800625228.0009

Earth about 4.6 billion years ago (Koonin *et al.*, 2006). It is known, however, that bacteria, as the first living organisms on Earth, were present 3.5 billion years ago and that bacteriophages, associations between bacteria and viruses, are common. Numerous viral diseases of humans were important aspects of the environment that kept the human population below 1 billion until less than 300 years ago (Bird and Maredia, 2021). Modern medicine, including the science of virology, was a key aspect of the exponential human population growth during the past two centuries, leading to a current global population of more than 8 billion.

Botany, mycology and bacteriology contributed in important ways to modern medicine and the science of virology. In the late 1840s, Anton de Bary demonstrated that the fungus-like organism *Phytophthora infestans* was the cause of late blight of potato. In 1884, Koch and Loffler's research demonstrated that *Bacillus anthracis* was the cause of anthrax, resulting in the procedures known as Koch's postulates used as the scientific methodology to demonstrate pathogenicity. Pasteur developed a vaccine for rabies in 1885, prior to the pathogen being known to be a virus. In 1892, Ivanovski demonstrated that the cause of tobacco mosaic disease was filterable, making it smaller than a bacterium (Ivanowski, 1892; Lecoq, 2001). In 1898, Beijerinck (1898) was the first individual to coin the word 'virus' to describe the infectious nature of the filterable liquid fluids.

It was not until 1952 that Hershey used radiolabeled components of bacteriophages to show that nucleic acids were the basis of genetics. Five years later, Frankel-Conrat and Singer demonstrated that nuclei acid was the genetic component of tobacco mosaic virus. Virus taxonomy began in 1966 when the International Code of Virus Classification and Nomenclature was formulated. Accordingly, viruses were divided into seven classes: double-stranded (ds) DNA, single-stranded (ss) DNA, dsRNA, positive-sense ssRNA, negative-sense ssRNA, RNA viruses that reserve transcribe and DNA viruses that reverse transcribe. About 12,000 virus species, including satellite viruses and viroids, have been described.

Realm is the highest taxonomic category of viruses. It is based on molecular genetics. The six realms are: 1) Adnaviria which are archaeal filamentous viruses with dsDNA that encodes for an α-helical capsid protein, 2) Duplodnaviria which have dsDNA that encodes for the HK97-fold major capsid protein, 3) Monodnaviria which are ssDNA viruses that encode for the HUH endonucleases and their derivatives, 4) Riboviria which are RNA viruses that encode for RNA polymerase and reverse transcription, 5) Ribozyviria which have negative-sense ssRNA genomes and 6) Varidnaviria which are dsDNA viruses that encode for a Swiss-roll capsid protein. The realms are divided into nine kingdoms, 40 classes, 72 orders and more than 250 families. All known plant viruses vectored by nematodes are classified in the polyhedral-shaped Secoviridae family with ssRNA genomes or the Virgaviridae with positive-sense ssRNA rod-shaped virions (single virus particles).

The process of viral pathogenesis is initiated when the virion attaches to suscept receptors on the plasmalemma or is encapsulated into an endosome. After this fusion, the viral nucleic acid passes through a pore and enters the suscept cytoplasm. The virion's nucleic acid takes over the suscept's metabolism and efficiently produces as many as 1 million new virions, initiating the process of pathogenesis. Although plant viruses are not as well understood as animal viruses, they are estimated to cause US$60 billion annual crop yield losses.

Animal Phylum Nematoda

Modern science recognizes nematodes, classified in the animal phylum Nematoda, as the most prominent group of animals on Earth (Service, 1997). In 1914, Cobb (1914a, 1914b) stated that 'if all the matter in the universe except the nematodes was swept away, our world would still be dimly recognizable – we would find mountaintops, valleys, rivers, lakes and oceans recognizable – by a thin film of nematodes.' The Nematoda can be subdivided into two classes: Enoplea and Chromadorea. The Nematoda contains 12 major orders, colonizing widely variable habitats including marine, freshwater and soil ecosystems, in addition to being vectors of plant viruses and pathogens of plants, animals and humans (Table 9.1). The order Enoplea is believed to contain the most primitive taxa. All plant-parasitic (phytopathogenic) nematodes are

Table 9.1. Twelve major orders of the animal phylum Nematoda and their primary habitats.

Major orders	Marine	Freshwater	Soil	Plant parasitic	Animal parasitic
Rhabitida		X	X	X	X
Plectida	X	X	X		
Araeolaimida	X	X	X		
Monhysterida	X	X	X		
Desmodorida	X	X	X		
Chromodorida	X	X	X		
Enoplida	X	X	X		
Triplonchida		X	X	X	
Trichinellida					X
Mononchida			X		X
Mermithida		X			X
Dorylaimida		X	X	X	X

[Table used with permission from George W. Bird.]

stylet bearing (Khan, 2023a, 2023b) and classified in the orders Rhabditida, Triplonchida or Dorylaimida (Decraemer and Hunt, 2013). The described nematode species that serve as vectors of plant viruses are classified in two of the 12 major nematode orders: Triplonchida and Dorylaimida (Table 9.1).

Nematode-Vectored Plant Viruses

Plant viruses vectored by nematodes are classified in three genera: *Nepovirus* and *Cheravirus* of the family Secoviridae and *Tobravirus* of the Virgaviridae. Nepoviruses are around 30 nm in diameter. They are bipartite with positive-sense ssRNA. Most of the more than 40 *Nepovirus* spp. infect temperate region plants. About one-third of them have known nematode vectors. *Cheravirus* is a relatively new taxon. The reported suscept range of nepoviruses with nematode vectors is limited to eight botanical families ranging from the Moraceae to the Asteraceae (Table 9.2). Six are associated with suscepts classified in the Rosaceae. Thirteen of the nepoviruses are vectored by 11 *Xiphinema* spp., ten by 11 *Longidorus* spp. and one by *Paralongidorus maximus* (Weischer, 1993). Nepoviruses that do not have known nematode vectors have a broader suscept range than those with nematode vectors. These include tropical plants, gymnosperms and arthropod/pollen-vectored viruses.

As a positive bipartite ssRNA virion, the linear RNA strand of nepoviruses is replicated in suscept cytoplasm by transcription. The RNA1 component consists of about 8000 nucleotides and the RNA2 component of 4000–7000 nucleotides. They are encapsulated separately. RNA1 encodes for virion replication proteins, while RNA2 encodes for proteins essential for intercellular transmission and intracellular defenses (Hily *et al.*, 2021).

Plant viruses classified in the Virgaviridae are non-encapsulated and tubular with a helical shape. *Tobravirus* spp. long rod types range in length from 180 to 215 nm. Their associated short rods range from 46 to 115 nm in length. Tobra virion widths range from 20 to 23 nm. The linear positive-sense is bipartite with RNA1 having about 6800 nucleotides and RNA2 from 1800 to 4500 nucleotides. *Tobravirus* spp. have a single structural protein coded for by RNA2. RNA1 codes for the non-structural proteins essential for virion replication (MacFarlane and Neilson, 2009).

Nematode Vectors of Plant Viruses

The first report of nematodes as vectors of plant viruses was made by Hewitt, Raski and Gohen in 1958 (Hewitt *et al.*, 1958) when they discovered transmission of grapevine fanleaf

Table 9.2. Nematode-vectored nepovirus genera/species (Fuchs *et al*., 2022).

Genus/species	Common name	Type species suscept family	Acronym
Nepovirus maculanulatum	Mulberry ringspot virus	Moraceae	MRSV
Nepovirus betae	Beet ringspot virus	Amaranthaceae	BRSV
Nepovirus arabis	Arabis mosaic virus	Celastraceae, Oleaceae, Rosaceae	ArMV
Nepovirus foliumflabelli	Grapevine fanleaf virus	Vitaceae	GFLV
Nepovirus GYV	Grape yellow vein virus	Vitaceae	GYVV
Cheravirus avii	Cherry rasp leaf virus	Rosaceae	CRLV
Nepovirus CRV	Cherry rosette virus	Rosaceae	CRV
Nepovirus persicae	Peach rosette mosaic virus	Rosaceae	PRMV
Nepovirus PYBMV	Peach yellow bud mosaic virus	Rosaceae	PYBMV
Nepovirus rubi	Raspberry ringspot virus	Rosaceae	RpRSV
Nepovirus SLRV	Strawberry latent ringspot virus	Rosaceae	SLRV
Nepovirus nicotianae	Tobacco ringspot virus	Solanaceae	TRSV
Nepovirus TBR	Tomato black ring virus	Solanaceae	TBRV
Nepovirus lycopersici	Tomato ringspot virus	Solanaceae	ToRSV
Nepovirus aegaeum	Artichoke Italian latent virus	Asteraceae	AILV

[Table used with permission from George W. Bird.]

virus by *Xiphinema index*. Currently, nematode species classified in the family Longidoridae (order Dorylaimida) and Trichodoridae (order Triplonchida) are known to vector nepoviruses and tobraviruses, respectively. Of the more than 275 described species of the dagger nematodes, *Xiphinema* (Cobb, 1913), relatively few have been shown to vector plant viruses (Table 9.3). The same is true for needle nematodes, *Longidorus* and the other genera of the Longidoridae. Nineteen species of nematode classified in the Dorylaimida have been shown to vector ten viruses classified in the genus *Nepovirus*.

The genus *Tobravirus* contains a small number of plant viruses classified in the Virgaviridae. They have a broad host range in the monocotyledon and dicotyledon taxa. *Tobravirus* spp. are vectored by *Trichodorus* spp., *Paratrichodorus* spp. and *Nanidorus* spp. of the family Trichodoridae in the nematode order Triplonchida. The three known tobraviruses are pea early browning virus (PEBV), pepper ringspot virus (PepRSV) and tobacco rattle virus (TobRV). While the majority of the more than 90 stubby-root nematode species are not known to vector viruses, 11 are reported to vector TobRV, six vector PEBV and one vectors PepRSV (Table 9.4).

Various nematode species of Longidoridae and Trichodoridae acquire plant virus particles by feeding on viruliferous suscept tissue. This can be achieved by all active vermiform life cycle stages. Virus retention in the nematode is obtained through a selective binding process between the nematode cuticle and protein coat of the virus. In *Longidorus* spp., the virus is retained in the odontostyle. In *Xiphinema*, it is retained in the odontophore and esophageal regions. In the Trichodoridae, the virus is retained in the onchiostyle and esophagus. Plant viruses are not known to replicate in side nematode vectors. They are not retained after molting and do not pass through the egg stage. In some cases, an extended period is required for the vector to become viruliferous and retention has been observed for as long as 1 year. Under optimal soil temperature conditions, the life cycle of Trichodoridae species can be 30 days or less, whereas some Longidoridae have life cycles that are more than 1 year in length (Mayo and Jones, 1999).

Table 9.3. Plant-parasitic nematode vectors of nepoviruses.

Nematode	Virus(es) vectored
Xiphinema americanum (Cobb, 1913)	Tobacco ringspot virus Tomato ringspot virus Peach rosette mosaic virus Cherry rasp leaf virus Yellow bud mosaic of grape Yellow vein virus Strawberry latent ringspot virus
Xiphinema bricolense (Ebsary *et al*., 1989)	Tomato rings pot virus Tobacco ringspot virus Cherry rasp leaf virus Peach rosette mosaic virus
Xiphinema californicum (Lamberti and Bleve-Zacheo, 1979)	Cherry rasp leaf virus Tobacco ringspot virus Tomato ringspot virus Peach rosette mosaic virus
Xiphinema diversicaudatum	Arabis mosaic virus Strawberry latent ringspot virus
Xiphinema inaequale (Khan and Ahmad, 1975)	Tomato ringspot virus Tobacco ringspot virus Cherry rasp leaf virus Peach rosette mosaic virus
Xiphinema incognitum (Lamberti and Bleve-Zacheo, 1979)	Tomato ringspot virus
Xiphinema index	Grapevine fanleaf virus Arabis mosaic virus
Xiphinema italiae	Grapevine fanleaf virus
Xiphinema pachtaicum	Tomato ringspot virus Tobacco ringspot virus
Xiphinema rivesi (Dalmasso, 1969)	Tomato ringspot virus Tobacco ringspot virus Cherry rasp leaf virus Peach rosette mosaic virus
X. tarjanense (Lamberti and Bleve-Zacheo, 1979)	Tomato ringspot virus Tobacco ringspot virus Cherry rasp leaf virus Peach rosette mosaic virus
Longidorus apulus Lamberti and Bleve-Zacheo, 1979)	Artichoke Italian latent virus
Longidorus arthensis (Brown *et al*., 1994)	Cherry rosette nepovirus
Longidorus attenuates	Tomato ringspot virus Tomato black ring virus
Longidorus elongatus (Micoletzky, 1922)	Raspberry ringspot virus Beet rosette mosaic virus Tomato black ring virus
Longidorus fasciatus (Roca *et al*., 1982)	Artichoke Italian latent virus
Longidorus macrosoma	Raspberry ringspot virus
Longidorus martini (Merny, 1966)	Mulberry ringspot virus
Panlongidorus maximus	Raspberry ringspot virus

[Table used with permission from George W. Bird.]

Table 9.4. Plant-parasitic nematode vectors of tobraviruses.

Nematode vectors	Virus
Paratrichodorus anemones (Loof, 1975) *Paratrichodorus pachydermus* (Siddiqi, 1974) *Paratrichodorus teres* (Siddiqi, 1974) *Trichodorus primitivus* (de Man, 1880) *Trichodorus viruliferous* *Trichodorus similis*	Pea early browning virus
Nanidorus minor (Siddiqi, 1974)	Pepper ringspot virus
Nanidorus minor (Siddiqi, 1974) *Nanidorus nanus* (Siddiqi, 1974) Paratrichodorus allius Siddiqi, 1974) *Paratrichodorus anemones* (Siddiqi, 1974; Loof, 1975) *Paratrichodorus hispannus* *Paratrichodorus pachydermus* (Siddiqi, 1974) *Paratrichodorus teres* (Siddiqi, 1974) *Trichodorus cylindricus* *Trichodorus primitivus* (de Man, 1880) *Trichodorus similis* *Trichodorus viruliferus*	Tobacco rattle virus

[Table used with permission from George W. Bird.]

Mechanism of Virus Transmission by Nematodes

The transmission of virus particles by the potential nematode species into the host plant is highly complex and based on interrelated events. It consists of virus acquisition, retention, disassociation, oral secretory egestion into a suscept followed by attachment of the particle to the plasmalemma or endosome encapsulation and infection. Virus transmission by nematodes involves three major steps: acquisition, retention and introduction (e.g.estion) of virus particles.

Acquisition of virus particles

Successful transmission of nematode-vectored plant viruses begins with the nematode acquiring virus particles while feeding on a viruliferous susceptible host plant. This is a highly specific and complex process that involves the chemistry, structure and dynamics of the virus, nematode vector and suscept. In some cases, an extended period is required for the nematode vector to become viruliferous and virus particle retention has been observed as for as long as 1 year. Under optimal soil temperature conditions, the life cycle of Trichodoridae species can be 30 days or less, whereas some Longidoridae species have life cycles that are more than 1 year in length (Mayo and Jones, 1999).

Retention of virus particles

When the nematode feeds on the viruliferous host, the food (cell sap and cytoplasm) passes through the lumen of the stylet and esophagus. During this transfer, the virus particles present in the food become attached to specific sites in the stylet and esophagus, varying with the nematode species. All vermiform stages of *Xiphinema* and *Longidorus* spp. can acquire nepovirus particles through feeding with their odontostyles. This usually takes place near the subapical meristems of young plant root tissue. The particles of *Nepovirus* species become attached (retained) to sites on the lining of the *Xiphinema* esophagus, whereas in *Longidorus* spp. virus particles attach in the lumen of anterior portion of the stylet. An electron micrograph of *X. diversicaudatum* by Harrison *et al.* (1974) shows *Nepovirus arabis* particles attached to the wall of the odontophore (Fig. 9.1). Another electron micrograph showed *Nepovirus rubis* particles located between the odontostyle and guide ring area of the stoma of *Longidorus elongatus*. Timely disassociation of virus particles from the retention sites is essential for successful transmission. The retention of virus particles is believed to be governed by pH, ionic or enzymatic secretions. The opposite charges on the virus particles (positive) and retention surface (negative) are believed to help in the attachment of particles with the retention sites.

Trichodorus, *Paratrichodorus* and *Nanidorus* spp. have a wide host range and complex feeding behaviors. Their curved onchiostyles probe epidermal cells in the region of root elongation. This results in some cell death and the formation of feeding tubes. All vermiform stages of trichodorid nematodes can vector *Tobravirus* species. Electron micrographs of *T. pachydermus* by Harrison *et al.* (1974) showed ingested *Tobravirus tobacum* (tobacco rattle virus) particles associated with the onchiostyle and cuticular lining of the esophagus (Fig. 9.2).

The virus particles retained in *Xiphinema*, *Longidorus*, *Trichodorus* or other nematode species do not pass through the processes of molting and embryogenesis and are lost along with the shedding of internal cuticle of the nematode (Taylor and Robertson, 1975). In summary, *Nepovirus* species associated with *Longidorus* and *Paralongidorus* species are

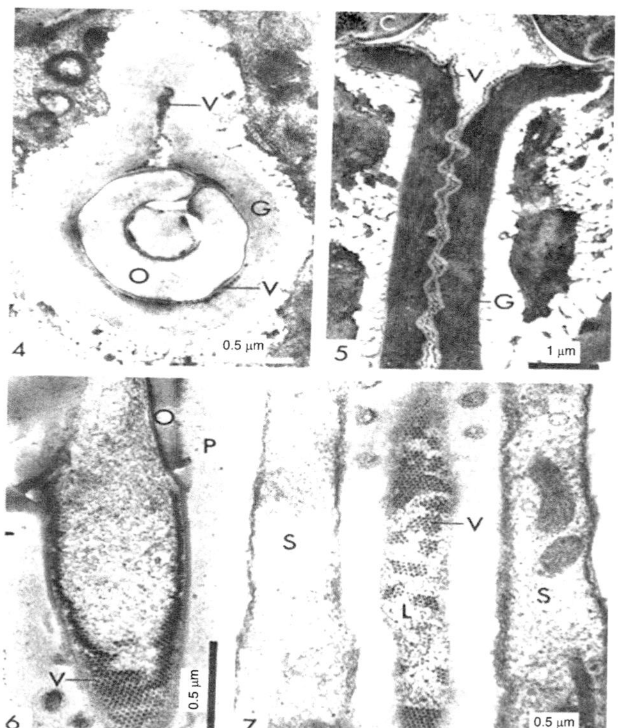

Fig. 9.1. (4–7) Transmission electron micrographs of nematode-transmitted polyhedral viruses. (4) *Nepovirus rubi* (raspberry ringspot virus) associated with the stomodeum of *Longidorus elongatus*, with virus particles (V) between the odontostyle (O) and guide sheath (G), and (5) virus particles lining the guide sheath (G), with none associated with the stoma cuticle (C). (6) Longitudinal section of the odontophore and odontostyle of *Xiphinema diversicaudatum* with *Nepovirus arabis* (Arabis mosaic virus) particles (V) attached to the odontophore wall lumen (P) but not the odontostyle (O), and (7) a longitudinal section of the odontophore with the Arabis virus (V) located where the esophageal lumen (L) cut tangentially and the odontophore sinuses (S) are parallel to the lumen (after Harrison *et al.*, 1974). [Figure used with permission from George W. Bird.]

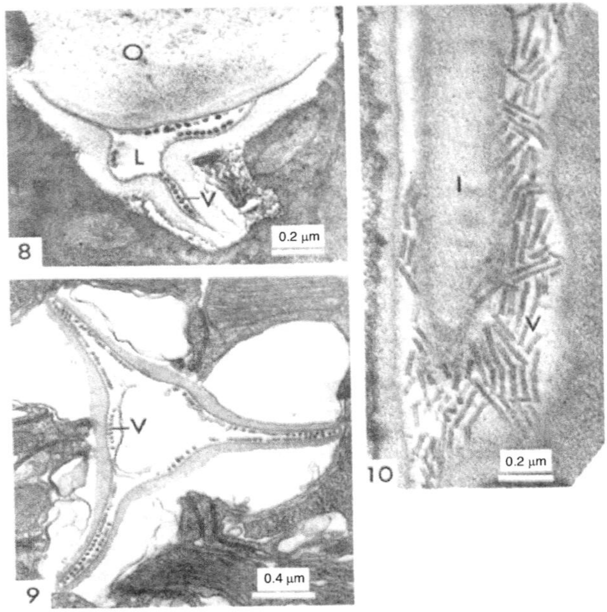

Fig. 9.2. (8–10) Transmission electron micrographs of *Tobravirus tobacum* (tobacco rattle virus (TobRV)) associated with *Trichodorus pachydermus*. (8) Transverse section of the onchiostyle (L) with attached TobRV particles associated with the lumen wall (V). (9) Transverse section of triradiate of esophageal lumen with virus particles (V) in cross-section. (10) Longitudinal section of the esophagus with TobRV particles (V) of two characteristic lengths attached to esophageal wall (after Harrison *et al.*, 1974). [Figure used with permission from George W. Bird.]

retained on the cuticle lining of the odontostye extension and internal surface of the guide ring, whereas *Tobravirus* species associated with *Trichodorus*, *Paratrichodorus* and *Nanidorus* species are retained on the lining of esophagus (Abdalreda and Mohammed, 2022).

Disassociation and egestion of virus particles

When a viruliferous nematodes feeds on the host, the retained virus particles become free at retention sites. Only a small portion of the attached virus particles are disassociated at a time, leaving additional ones available for future transmissions (Taylor and Raski, 1964). The esophageal gland secretions help in disassociation of virus particles. It is assumed that the saliva modifies the pH within the lumen and alters the surface charge on the virus particles (Khan, 2008). In most cases pH becomes 7.0 (neutral). At this pH, charge on the particles becomes neutral and they become free. The dissociation sometimes also depends on the virus strains. Hence, in this final step of virus transmission, a portion of the retained particles disassociates and is made available for oral secretory egestion and introduction along with the esophageal gland secretions into the susceptible host for infection (Weischer, 1993).

While much remains to be learned about the mechanism of the transmission and infection processes, research using molecular technologies has provided new insights into the virus–vector–suscept specificity relationships. One area of significant interest is the function of the virus protein coat in suscept infection, the final stage of successful transmission. An example is the presence of a ligand in the protein coat of the virus (Macho and Lozano-Duran, 2019). It has been shown that in some cases the ligand deactivates the normal plant root cell defense mechanisms designed to prevent invasion of foreign substances. This allows the virus to enter the suscept cell, shed its protein coat and initiate pathogenesis at the infection site level.

Management of Nematode-Vectored Plant Viruses

Food, feed, fiber, timber, ornamental and flower production are biological systems. Biological systems cannot be controlled, but can only be managed (Wessels, 2006). Plant diseases caused by nematode-vectored viruses consist of multiple interactive components and actions that must be monitored for use in the development and implementation of disease management strategies and practices (Fig. 9.3). These include laboratory analyses for accurate identification of the suspected plant pathogen and potential nematode virus vector in relation to overall plant health (Iles *et al.*, 2021). The strategies for management of nematode-vectored plant viruses include pathogen and vector exclusion, pathogen and vector containment, and vector population reduction (Martelli, 1978; Bird and Brewer, 2006; Sokora *et al.*, 2023).

Exclusion and containment

Exclusion can also be referred to as prevention or avoidance. It is by far the best management strategy for plant virus diseases. In some ways, exclusion is like some of the practices used to avoid infection by the COVID-19 virus. Nematode-vectored plant viruses can be excluded from sites that do not have the virus by planting only non-infected seeds or propagative materials. This is extremely important since a significant number of nepovirus- and tobravirus-susceptible plants are usually transplanted. These include tobacco, tomato, pepper, grape, strawberry, raspberry, cherry, peach and mulberry. The viruses associated with the known tomato ringspot and tobacco rattle virus infestation in Michigan are believed to have been introduced through planting virus-infected grapevines, cherry trees and potato seed pieces. In the USA, e.g. in Michigan, all potato seed piece and tree fruit nurseries are inspected for virus diseases by the Michigan Department of Agriculture and Rural Development. This is done prior to certification that they are virus-free. The potato seed piece certification law is the result of interaction between the potato industry and state government. This public/private sector involvement has been highly successful.

The strategy of containment involves preventing the spread of existing virus and nematode vectors from known infested sites to non-infested sites. The management practices can be challenging, and require modern record-keeping technology and actions that prevent virus vectors and infested suscept tissues from being transported to non-infested locations. Power washing all equipment is essential after it

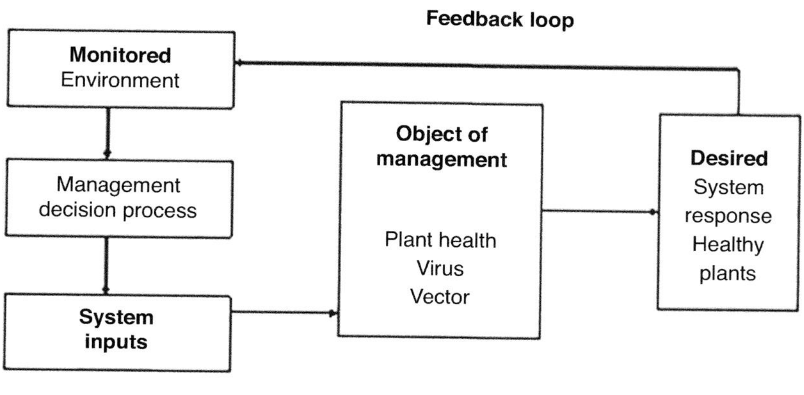

Fig. 9.3. Components and conceptual decision-making process for management of nematode-vectored plant viruses. [Figure used with permission from George W. Bird.]

has been used in an area containing viruliferous nematodes. With perennials, the best approach is to destroy infected vines or trees.

Reduction in the vector nematode population

Vector population reduction is a strategy used in plant virus disease management. Chemical, biological and cultural practices are available for lowering population densities of dagger, needle and stubby-root nematodes. Cultural practices include tillage, crop rotation, non-host plants, biocontrol agents, biofumigation, resistant cultivars and soil health enhancement. While all of these have the potential to lower vector population densities, it must be remembered that the theoretical action threshold is vector detection, since a single vector individual can transmit a virion. The extent of vector mortality from tillage, crop rotation, biological agents and biofumigation is usually inadequate for effective plant virus management. Relatively few non-host plants and suscept plant-resistant cultivars are available for longidoroid and trichodoroid nematodes.

Agricultural plant nematology had its origins with the discovery of the nematicidal properties of soil fumigants. These included 1,3-dichloropropene (1,3-D), ethylene dibromide (EDB), 1,2-dibromo-3-chlororopane (DBCP), chloropicrin, methyldithiocarbamate salts (metam-sodium and metam-potassium) and methyl bromide. Except for DBCP, these chemicals are phytotoxic to plants and most of them have been registered for use only on a pre-plant basis. Methyl bromide is expensive and requires tarping. Chloropicrin is commonly used for management of soil fungi and weeds. 1,3-D is an excellent nematicide. It is sometimes used in combination with chloropicrin. Metham-sodium containing methyl isocyanate has emerged as an effective chemical for the management of nematodes, fungi and weeds. The sites for most vine and tree crops that are susceptible to nematode-vectored viruses are required to be fumigated prior to planting a new location or replanting a site used previously for the crop. In the fall of 2023, most of the north-western US potato acreage was fumigated with 1,3-D, metam or chloropicrin (unpublished information).

The development of organophosphates and organocarbamates provided the opportunity of applying on a wide variety of crops on planting or post-planting stages. However, some of these, such as aldicarb, carbofuran and forate, have been banned for use in agriculture in several countries because of groundwater contamination issues and other environmental issues. In 2023, oxamyl was a commonly used non-fumigant nematicide in vine, tree fruit and potato production systems (unpublished data). During the last decades, a number of new chemistries and products have been developed as nematicides. The amount of information about their efficacy in virus vector management is, however, limited.

Plant Diseases with Nematode Virus Vectors

The following descriptions of plant diseases caused by viruses vectored by nematodes are designed to provide an overview of the nature and management options of each disease. The descriptions include information about the disease, suscepts, symptoms, pathogen, vectors, disease cycle and management. The references are cited as a guide for assessment of the literature.

Virus diseases vectored by dagger nematodes

Tomato ringspot diseases

Tomato ringspot (ToRS) causes at least 44 infectious diseases in angiosperms (CDFA, 2020), These include ToRS of *Buddleja davidii* (butterfly bush), *Capsicum* spp. (peppers), *Capsicum frutescens* (chilli), *Cornus* spp. (dogwood), *Cucumis sativus* (cucumber), *Cydonia oblonga* (quince), *Daphne mezereum* (mezereon), *Dipsacus fullonum* (Fuller's teasel), *Erigeron glaucus* (seaside fleabane), *Fragaria* × *ananassa* (strawberry), *Fragaria* sp. (Chilean strawberry), *Fraxinus americana* (white ash), *Gladiolus* hybrids (sword lily), *Hydrangea* (hydrangeas), *Lilium* spp. (lily),

Lotus corniculatus (bird's foot trefoil), *Malus domestica* (apple), *Nicotiana tabacum* (tobacco), the Orchidaceae (orchids), *Pelargonium* spp. (pelargoniums), *Peperomia* sp. (peperomia), *Picris echioides* (bristly ox tongue), *Plantago* sp. (plantain), *Prunus* spp. (stone fruit), *Prunus armeniaca* (apricot), *Prunus avium* (sweet cherry), *Prunus cerasus* (tart cherry), *Prunus domestica* (plum), *Prunus dulcis* (almond), *Prunus persica* (peach), *Prunus salicina* (Japanese plum), *Punica granatum* (pomegranate), *Rosa* spp. (roses), *Ribes* spp. (currants), *Rubus* spp. (blackberry raspberry), *Rubus idaeus* (raspberry), *Rubus procerus* (Himalayan blackberry), *Sambucus* spp. (elderberry), *Solanum lycopersicum* (tomato), *Solanum melongena* (eggplant), *Taraxacum officinale* (dandelion), *Vaccinium corymbosum* (blueberry), *Vigna* spp. (cowpea) and *Vitis vinifera* (grapevine).

In addition, there are numerous non-cultivated and weed suscepts. Some ToRS diseases have multiple common names such as prunus stem pitting, apple union and grape yellow vein disease. This makes ToRS disease outreach and education challenging.

SYMPTOMS. On herbaceous suscepts, ToRS exhibits a diversity of symptoms (Enelmann and Hamacher, 2008). Some are local and others systemic. Infection results in mosaic and rings-pots, mottling and yellow veins on the foliage. This can be followed by systemic necrosis. Plants become stunted with poor flower and fruit production. Symptoms are often most severe when the plants are young. Most woody perennial suscepts do not exhibit foliar symptoms. On *Musca domestica*, the virus results in an incompatibility between the scion and root system, resulting in poor plant growth and low fruit yield. In some cases, this union necrosis causes the root and shoot system to separate resulting in tree death or at least elimination of the scion cultivar. In a similar manner, ToRSV causes stem pitting of *Prunus persica*. The spongy stem pits are near the soil surface and often result in tree mortality. The virus also causes phloem necrosis on the trunks of *V. vinifera* vines.

PATHOGEN. *Nepovirus lycopersici* is a polyhedral-shaped (isometric) virus classified in the family Secoviridae (Walker *et al.*, 2014). It is about 28 nm in diameter with about 40% RNA and 60% protein. As a positive bipartite ssRNA virion, the linear RNA strand is replicated in suscept cytoplasm by transcription. RNA1 is 8214 nucleotides in length, excluding the 3′ poly(A) tail. RNA2 has between 4000 and 7000 nucleotides. RNA1 and RNA2 are encapsulated separately. RNA1 encodes for virion replication proteins, while RNA2 encodes for proteins essential for intercellular transmission and intracellular defenses. *Nonomuraea lycopersici* has a global distribution with the greatest number of reports from North America. There are, however, important agricultural regions where ToRSV has not been identified.

VECTORS AND TRANSMISSION. ToRSV has been reported to be vectored by nine nematode species classified in the order Dorylaimida. These include *Xiphinema americanum*, *X. bricolense*, *X. californicum*, *X. inaequale*, *X. incognitum*, *X. pachtaicum*, *X. rivesi*, *X. tarjanense* and *Longidorus attenuates*. ToRSV and peach rosette mosaic virus are the only plant viruses reported to be vectored by *Xiphinema* and *Longidorus* spp. (Bitterlin and Gonsalves, 1987; Georgi, 1988). The virion can be transmitted through pollen and seed, in addition to the horticultural processes of grafting and budding. *Xiphinema* and *Longidorus* spp. acquire *N. lycopersici* by feeding on viruliferous suscept tissue. This can take place in all active vermiform life cycle stages. Virion retention in the nematode is achieved through a selective binding process between the nematode cuticle and virus protein coat. In *Xiphinema* spp., the virus is retained in the odontophore and esophageal regions. In *L. attenuates*, the virus particles are retained in the odontostyle. In both cases, the virus is not retained during nematode molting.

DISEASE CYCLE. The ToRS disease cycle involves the virus source, acquisition, retention and transmission resulting in characteristic pathological symptoms (Pinkerton *et al.*, 2008). The first type of disease cycle is where sites already contain viruliferous suscepts or vectors. A second type pertains to sites that do not have viruliferous suscepts or vectors. Following a period of seasonal inactivity caused by temperature or moisture extremes, viable ToRSV exists in suscept root and shoot tissue. This includes plants that are grafted or budded for horticultural purposes. In addition, potential sources of ToRSV include the nine species of *Xiphinema* and one of *Longidorus*, viruliferous seeds and infected pollen of various suscepts including the common weed *Taraxacum officinale* (dandelion). The virus is acquired through the processes of pollination, seedling development, grafting or budding with ToRSV-infected plant material or feeding by viruliferous *Xiphinema* or *Longidorus*. The processes of pollination, grafting and budding transmit ToRSV from infected to healthy tissue. The virus is also transmitted to healthy plants through feeding by viruliferous *Xiphinema* or *Longidorus* spp.

The process of pathogenesis begins when a *N. lycopersici* virion attaches to suscept tissue receptors on the plasmalemma or is encapsulated into an endosome. After the endosome formation, the viral nucleic acid passes through a pore and enters the suscept cytoplasm. The virion's nucleic acid takes over the suscept metabolism and efficiently produces new virions during the process of pathogenesis. This process incurs foliar necrosis resulting in the formation of mosaic and ringspot symptoms including mottling and yellow veins. It can be followed by systemic necrosis. Plants become stunted with reduced flower formation and fruit yield. Additional symptoms vary with different suscepts, sometimes resulting in death of the plant.

MANAGEMENT. ToRS management requires an integrated approach of prevention, containment and nematode vector population reduction (Martelli, 1978; Bird and Brewer, 2006; Sokora *et al.*, 2023). Prevention is the best ToRS management strategy. It requires a combination of practices such as monitoring (scouting), use of virus-free planting propagative stock (rootstock, trees, grafting/budding materials, vines, transplants, pollen, etc.) and clean agricultural equipment. Planting resistant rootstocks reduces the probability of the disease occurrence. Once a site has infected plants or viruliferous nematode vectors, the disease needs to be contained (localized) by destroying infected plants, use of strict sanitation practices and reduction of vector populations. A biological assay is useful for measuring containment success before replanting. The containment process can take more than 1 year.

Most ToRSV vector population reduction must be achieved on a pre-plant basis. This requires determining the presence and population density of *Xiphinema* or *Longidorus* prior to and after implementation of a vector population reduction practice. Site preparation commonly used for perennial woody horticultural and vine crops consists of 1 year or more of appropriate tillage, growing specific cover crops and incorporating their residuals. Relatively little is known about the overall suscept status of cover crops. Halbrendt *et al.* (2008) reported that Dwarf Essex Rape reduced the population of *X. americanum* when it was incorporated as a green manure. This cultivar has been used successfully in tree fruit systems under Michigan growing conditions.

Pre-plant soil fumigation is a common practice in the establishment of perennial woody plants and vines on a replant or new site basis. Historically, in some cases, the cash value of the crop was high enough to warrant use of methyl bromide applied under a plastic tarpaulin. This practice was banned by the Montreal ozone protocol. Metham and 1,3-D are soil fumigants used commonly on a pre-plant basis for perennial woody plant and vine sites. In some cases, 1,3-D is combined with chloropicrin. Regardless of the chemical, it is essential to use the proper dosage. The low rate of metam is usually adequate for nematode population reduction. When chloropicrin is combined with 1,3-D, it is essential to have a high enough dosage of 1,3-D for nematode population reduction. During the past decade there has been concern about a potential negative impact of fumigants on soil health indicators. This is now being questioned. Some important soil health properties may rebound rapidly after fumigation, having a beneficial systemic impact on plant development.

On a post-plant basis, the main options for vector population reduction are tillage and non-fumigant nematicides. Oxamyl has been used on a post-plant basis for decades to reduce population densities of plant-parasitic nematodes, including ToRSV vectors. Recently, new biological and chemical nematicides have become available for management of plant nematodes. However, adequate information under local farm conditions is essential before they should be recommended for management of ToRSV vectors. Tillage is another possibility for vector population reduction. In 1969, the population of *Xiphinema* sp. was affected due to changes in soil temperature and moisture, caused by tillage (Malek, 1969).

Peach yellow bud mosaic

DISEASE, SUSCEPTS AND SYMPTOMS. Peach yellow bud mosaic (PYBM) is a serious disease of major tree fruit crops in California (Karle and Nyland, 1959). It was first described in 1935. PYBM is caused by a strain of *Nepovirus lycopersici* (ToRSV). It is presented here as a separate infectious disease because of its economic significance and an example of the complexity of virus Latin, common and disease names of infectious diseases caused by different strains of plant viruses. The PYBM suscept range includes *Prunus dulcis*, *P. persica* and *P. armeniaca*. It occurs in nature on woody plants. Symptoms of the disease are visible on foliage, tree structure and fruit quality. PYBM foliage is small and distorted forming rosettes or tuffs. It can have holes resulting from the shedding of chlorotic spots. Infected lateral branches are sporadic and have significantly reduced growth, giving trees an open appearance. Fruit of infected trees is small and of poor quality. Infected trees can survive for long periods of time. The rootstock, Marianna 2624 plum, is not a suscept of PYBM.

PATHOGEN AND VECTORS. PYBM is caused by *N. lycopersici*, the virus previously described for ToRS disease. Although it has a global distribution, it is likely an endemic to North America. It has virulent strains that cause symptom specific in the field. The PYBM strain occurs in nature on *P. dulcis*, *P. persica* and *P. armeniaca* in the western USA. Although eight *Xiphinema* spp. and one *Longidorus* sp. have been reported to vector ToRSV, it is likely that *X. americanum* and *X. californicus* are the most common vectors of PYBMV. Since it is known that some nematode vector–suscept relationships are virus strain specific, the PYBMV vector arena needs additional research. Most fruit tree roots are in the upper 1.5m of soil, a zone where dagger nematodes readily feed and reproduce.

DISEASE CYCLE AND MANAGEMENT. In areas where soil temperatures and moisture are continuously appropriate for the root growth of the fruit trees and *Xiphinema* feeding, it is likely that there is a single disease cycle beginning with the introduction of PYBMV and ending with the death and decomposition of the host plant as a source of the virus for the vector. As with most nematode-vectored plant virus diseases, prevention is the best management practice. Once a tree is infected, it cannot be returned to a healthy individual (Adaskaveg *et al.*, 2017). PYBM containment and site restoration are difficult and long-term processes. This is especially true because of the time it takes for tree root systems to decompose at low soil depths.

Peach rosette mosaic

DISEASE. Peach rosette mosaic (PRM) is an infectious disease of tree fruit, small fruit, vines and weed plants (Dias and Cation, 1976). It was first recognized as a problem in peach production in Michigan in 1917. It is caused by *Nepovirus persica* which is vectored by three species of *Xiphinema* and *Longidorus diadecturus*. The known distribution of PRM is limited to Michigan, New York and southwestern Ontario in Canada. The host range of PRM includes *Vitis labrusca* cv. Concord, *Prunus persica*, *Vaccinium corymbosum*, *Taraxacum officinale* and *Solanum carolinense*. Species of Chenopodiaceae, Cucurbitaceae and Leguminosae such as *C. amaranticolor*, *C. quinoa* and *N. tabacum* have been found susceptible under laboratory conditions.

SYMPTOMS. Symptoms of PRM include delayed bud break, delayed leaf development, chlorotic mottling of foliage, severe leaf distortion, short internodes, rosette appearance, poor fruit, stunted plants and plant death. Some susceptible species remain symptomless (Ramsdell and Meyers, 1974).

PATHOGEN. PRM is caused by *N. persica* (PRMV). It is a polyhedral-shaped (isometric) virus classified in the family Secoviridae (Lammers *et al.*, 1999). The virion is believed to be indigenous to North America, where it has been detected in Michigan, southwestern Ontario and New York, but not in California. *N.*

persica is about 28 nm in diameter and has two separate nucleoprotein components consisting of RNA1 or RNA2 and a single protein subunit. The complete 8004 nucleotide sequence of PRMV RNA1, excluding its 3′ polyadenylated tail, has been determined.

VECTORS/TRANSMISSION. *Nepovirus persica* can be transmitted mechanically. It is also known to be vectored by four dorylaimids: *X. americanum*, *X. bricolense*, *X. rivesi* and *Longidorus diadecturus*. *X. americanum* acquires PRMV within 1 hour of feeding and immediately transmits it to a suscept. It is believed that PRMV is retained in the *Xiphinema* odontophore and esophageal regions but does not survive molting (Mayo and Jones, 1999). In *L. diadecturus*, *N. persica* is retained in the odontostyle and lost during molting. The fact that PRMV can be vectored by nematodes in two genera is unique. The vectoring of most nepoviruses is limited to a single genus. In 2017, the European Union Panel on Plant Health recognized *L. diadecturus* as a well-defined vector taxon.

DISEASE CYCLE. As a detrimental physiological process, the PRM disease cycle consists of two types expressed in three phases. The types pertain to previously viruliferous and non-viruliferous sites. The phases include pre-pathogenesis development, pathogenesis and suscept senescence. Pre-pathogenesis development consists of the source of PRMV, acquisition, transport of virions to the suscept, contact with cellular plasmalemma of the infection court and entry of the virion into suscept cells. PRMV sources include viruliferous suscept propagation materials and plants such as rootstocks, shoot system scions, buds, trees, vines and transplants. Viruliferous weeds and nematode vectors are also PRMV sources. Suscept acquisition comes from the plant propagation/planting processes and transmission by nematode vector feeding (Allen and van Schagen, 1998). It is assumed that PRMV plasmalemma attachment and entrance into the cytoplasm is like that described for other nepoviruses. Pathogenesis consists of the virion using the suscept's metabolic processes for virus replication instead of the normal physiology essential for the development and maintenance of a healthy plant. The process results in characteristic necrotic and hypoplastic symptoms. Senescence is initiated with the death of the suscept and terminated when the virus cannot be detected in the site of interest. The disease cycle can be interrupted by temperature and moisture relationships causing suscepts and vectors to become quiescent.

MANAGEMENT. PRM management requires an integrated approach based on the strategies of prevention, containment and system restoration (Bird and Brewer, 2006; Sokora *et al.*, 2023). System restoration involves the reduction of populations of target suscepts, weeds and nematode vectors. Because the key suscepts, peach trees, grapevines and blueberry plants, are propagated perennials, management must begin with the propagation process. The plants must be developed from certified PRMV-free rootstocks, scions and budwood. This must be achieved in nursery soil or media that has no detectable *X. americanum*, *X. bricolense*, *X. rivesi* or *L. diadecturus* or infected weeds. Pre-plant soil fumigation is a common practice in plant nurseries. Prior to being transplanted to orchards, vineyards or blueberry plantings, the plants must be certified as PRMV-free by an appropriate regulatory agency.

When an orchard, vineyard or blueberry planting is viruliferous, it is essential that the disease is contained and not transmitted to healthy plantings. This requires rouging and destroying infected plants. In some cases, removal of the entire orchard or vineyard is necessary. Establishment of PRM-free new or replanted orchards, vineyards or blueberry plantings requires weed-free soil that does not have detectable populations of *X. americanum*, *X. bricolense*, *X. rivesi* or *L. diadecturus*. Most orchard sites are maintained weed-free for at least 1 year prior to planting. If the orchard site preparation process does not involve vector suscepts, there may not be a need for pre-plant soil fumigation. In some cases, post-plant non-fumigant chemicals or biologicals are used to assist in maintaining a PRM-free planting.

Grapevine fanleaf

DISEASE AND SYMPTOMS. Grapevine fanleaf (GFL) is a serious infectious disease of *Vitus* spp., a genus of more than 75 species classified in the family Vitaceae (Link *et al.*, 2004). *Vitus* spp. are vines native to northern temperature regions. The disease was described in Europe in the 1800s (Cazalis-Allut, 1865). In 1958, it became the first plant disease caused by a virus shown to have the pathogen acquired from a viruliferous plant and successfully transferred (vectored) to a healthy plant by a plant-parasitic nematode (Hewitt *et al.*, 1958). Because of the absence of GFL in native North American grapevines, and the long history of grapes in Europe and the Middle East, the disease is assumed to be of European or Middle Eastern origin. Much of its current global distribution is believed to be due to grafting and the distribution of European cultivars on a worldwide basis. In addition to GFL, the disease is known by several names, including grapevine arriciamento, grapevine Bratislava mosaic, grapevine infectious degeneration, grapevine roncet, grapevine vein banding and grapevine yellow mosaic. GFL exhibits necrotic and hypoplastic symptoms (Rupnik-Cigoj *et al.*, 2018). They consist of abnormally fan-shaped leaves with pronounced veins and chlorotic mosaic foliage patterns. Fruit yield and quality is poor. Infected vines become stunted and may die.

PATHOGEN. Hewitt *et al.* (1958) demonstrated that *Nepovirus foliumflabelli* (GFLV) is vectored to suscepts by *X. index*. The suscept range of GFLV is limited to *Vitus* spp. This polyhedral-shaped virus is 30 nm in diameter and classified in the family Secoviridae (Kubina *et al.*, 2023). The GFLV genome encodes proteins involved in virus replication, encapsulation, cell-to-cell movement, host detection, suscept defense activation and transmission (Schellenberger *et al.*, 2017). As a positive bipartite ssRNA virion, the linear RNA strand is replicated in suscept cytoplasm by transcription. In 2013, a South African isolate of GFLV was shown to consist of RNA1, RNA2 and a satellite RNA with 7341, 3816 and 1104 nucleotides, respectively (Lamprecht *et al.*, 2013). RNA1 and RNA2 are encapsulated separately. RNA1 encodes for virion replication proteins, while RNA2 encodes for proteins essential for intercellular transmission and intracellular defenses. Protein prediction software has been used to validate various GFLV protein functions such as RNA silencing, symptom determination, movement, coat structure, transmission and other previously unidentified functions.

VECTORS AND TRANSMISSION. Two nematode species, *X. index* and *X. italiae*, classified in the order Dorylaimida, are the known vectors of GFLV. All mobile life cycle stages of these species are assumed to acquire virions by feeding on viruliferous root tissue. The virus is retained in the odontophore and esophageal regions of the nematode through a selective binding process between the nematode cuticle and protein coat of the virus. Virus retention may be immediate, while in some cases an extended period of feeding is required. The virus, however, is not retained after molting and does not pass through the egg stage. The length of the retention period depends on the length of the life cycle which is mediated by soil temperature. Andret-Link *et al.* (2004) demonstrated that vector transmission is governed by the virus protein coat. As a mechanically transmitted virus, GFLV can be transmitted through rootstocks, scion wood, buds, grafted or budded plants, seeds and pollen. In addition, it can be transmitted by vineyard equipment such as pruning shears.

DISEASE CYCLE. There are at least nine potential sources of the *N. foliumflabelli* virus. These include viruliferous *X. index* or *X. italiae*, grape rootstocks, scion wood, grape buds, grafted or budded vines, seeds and pollen. *X. index* and *X. italiae* acquire *N. foliumflabelli* by feeding on viruliferous grape root tissue. Grapevines acquire GFLV through various horticultural practices including the use of viruliferous pollen, seed, buds, scion wood or rootstocks. Grapevines can also obtain the virus through mechanical practices such as the use of viruliferous pruning equipment. *N. foliumflabelli* is also transmitted to healthy grapevines through the feeding processes of viruliferous *X. index* and *X. italiae*. In addition, vines become infected when viruliferous pollen, seeds, buds, scion wood and root stocks are used in the vine propagation process. The process of pathogenesis begins when a *N. foliumflabelli* virion attaches to grape tissue receptors on the plasmalemma or is encapsulated into an endosome. After this fusion, the viral nucleic acid passes through a pore and enters the cytoplasm. The virion's nucleic acid takes over the suscept metabolism and efficiently produces as many as 1 million new virions, during pathogenesis. This results in abnormally fan-shaped foliage with pronounced veins and chlorotic mosaic foliage patterns.

MANAGEMENT. GFL can be prevented (excluded) from a new vineyard site or an old site that has not had a history by planting vines that have been certified as GFL-free by an appropriate regulatory agency (Raski *et al.*, 1983). Before establishing a new or replanted vineyard, the site should be evaluated for the presence of *X. index* and *X. italiae*. Pre-plant soil fumigation should be considered for use in new or old vineyard site preparation for a new vineyard. Non-fumigant chemical and biological nematode management products should also be considered. Not enough is known at this time about the potential impact of practices such as soil health maintenance, cover crops or composts on *Xiphinema* to make specific recommendations about the potential impacts of these practices regarding GFL management. Laboratory testing is an essential tool for the diagnosis of potential GFLV-infected vines. Diagnostic laboratories should be consulted before collecting samples. Timing of sampling and type of material to be collected and sample maintenance are important aspects of the process. The California University Land Grant System maintains virus-free foundation propagative materials for the most common grape cultivars and clonal selections.

Containment is an essential component of GFL management. If a site in a vineyard is known to have GFL, special precautions need to be taken to ensure that the disease does not spread to other vineyards. It may be necessary to rouge a few vines or possibly the entire vineyard to prevent the spread of the disease. All vineyard equipment used at a problem site needs to be power washed before being used at another site. In most cases, soil fumigation should be considered as part of the disease containment process. Other than fallowing and monitoring a site for a long period of time, soil fumigation with 1,3-D or metam is the best option to assure *X. index* and *X. italiae* population reduction to non-detectible levels. To do nothing is not a viable option for commercial grape production at risk to GFL.

Tobacco ringspot

DISEASE. Tobacco ringspot (TRS) is an infectious disease of *N. tabacum* observed in Virginia by Fromme *et al.* (1927) Its broad suscept range includes *Phaseolus vulgaris*, *Vigna unguiculata*, *Cucumis sativus*, *Vitis* spp., *Prunus persica*, *Glycines max* and numerous weed species. Rush and Gooding (1970) identified 17 plant species indigenous to North Carolina that were suscepts. The first record of TRSV in New Zealand was in 1972 on symptomless horseradish growing in the absence of a nematode vector. It has also been detected in subapical meristem bean root cells (Crowley *et al.*, 1969). The disease can be severe and is distributed globally.

SYMPTOMS. TRS symptoms vary with suscepts. On *N. tabacum*, necrotic rings with green centers are formed. Symptoms are often present as oak-leaf patterns on early-developing leaves, but not on foliage developing later in the growing season. This has been referred to as acquired immunity. Schneider *et al.* (1972) demonstrated that field isolates of TRSV produce different necrotic foliar symptoms. On *G. max*, TRSV causes bud blight.

PATHOGEN. *Nepovirus nicotianae* is classified in the Secoviridae, and is the type species of the genus. It is isometric with a diameter of about 28 nm and has been used as a model for fundamental virus research. Gooding (1970) reported multiple strains of TRSV associated with tobacco in North Carolina. Buzayan et al. (1986a, 1986b) identified a small TRSV RNA satellite of about 359 nucleotides. Hull (2014) reported that helper viruses are involved in the accumulation of satellite RNAs and RNA silencing. Zhao *et al.* (2016) sequenced the cDNA clones of TRSV RNA1 and RNA2.

PATHOGENVECTORS/TRANSMISSION. TRSV is known to be vectored by seven *Xiphinema* spp., thrips and grasshoppers. Fulton (1962) demonstrated that TRSV is vectored by *X. americanum*. The others include *X. bricolense*, *X. californicum*, *X. inaequale*,

X. pachtaicum, *X. rivesi* and *X. tarjanense*. Širca *et al.* (2007) reported that TRSV was present in Slovenia and vectored by *X. rivesi*. It is important to note that in recent years much of TRSV research has been related to molecular aspects of the virus (Fang *et al.*, 2021). In recent fundamental research the term 'vector' has been used widely to describe the process of gene manipulation.

DISEASE CYCLE. The TRS disease cycle begins with the acquisition of the virus from infected suscept cytoplasm or its nematode vector. Seed transmission is important and depends on the time of infection. Early infection can result in no seed production. TRSV is systemic in soybean and moves from infected foliage to root tissue and is readily sap-transmissible. The length of the disease cycle can be less than 1 year with annual suscepts or several decades with perennial species. The disease cycle ends with the death of the suscept or when potential nematode or arthropod vectors are no longer viruliferous.

MANAGEMENT. Successful management requires an integrated approach involving use of only TRSV-free certified seeds and transplants. A limited number of TRSV-resistant cultivars are available for some suscepts. Among these, some soybean cultivars derived from PI 407287, PI 92713 and PI 541194 are resistant to some strains of TRSV. Sites with previous infestations need to be scouted and maintained weed-free through physical, chemical or biological procedures. Populations of *Xiphinema* and insect vectors, such as aphids, grasshoppers, flea beetles and thrips, need to be maintained at non-detectable levels using local management practices. The pathogenicity threshold for a virus vector is zero since a single individual can successfully transmit the virus. With soybeans, it may be desirable to avoid fields infested with *Xiphinema*, use a nematode seed treatment, or treat the site with an appropriate chemical nematicide or biological agent.

Strawberry latent ringspot

DISEASE. Strawberry latent ringspot (SLR) is an infectious disease caused by *Nepovirus* SLRV. It was first identified in 1964 by Lister (1964). The disease has a broad geographical distribution including Europe, New Zealand, California and British Columbia, Canada. The pathogen is vectored by *Xiphinema* spp.

SUSCEPTS AND SYMPTOMS. In 1969, Schmelzer indicated that SLR has a broad suscept range that includes more than 126 species from 27 plant families (Martin *et al.*, 2004). The following fruit crops are SLR suscepts: *Rubus* spp. (blackberries), *Ribes nigrum* (black currents), *Prunus avium* (cherries), *Vitis vinifera* (grapes), *P. prunus* spp. (plums), *P. persica* (peaches), *Rubus idaeobatus* spp. (raspberries) and *Fragaria* × *ananassa* (strawberries). SLR symptoms include yellow flecking of young leaves, chlorotic local lesions, systemic chlorosis and reduced leaf size. Infected plants, including strawberry can be symptomless. SLRV isolated from symptomless strawberry plants caused foliar symptoms on numerous indicator suscepts (Mazyad *et al.*, 2014).

PATHOGEN AND VECTORS. The disease is believed to be caused by a *Nepovirus*. Virus names are approved by the International Committee on the Taxonomy of Viruses (ICTV). The ICTV has not approved an official species name for SLRV. SLR, therefore, is referred to as *Nepovirus* SLRV, tentatively classified in the Secoviridae family. SLRV is a polyhedral with a diameter of about 26 nm. Its RNA1 and RNA2 consist of 7496 and 3842 nucleotides, respectively, and a satellite of 1117 nucleotides. SLRV is known to be vectored by two species of dagger nematodes, *X. americanum* and *X. diversicaudatum*.

DISEASE CYCLE AND MANAGEMENT. SLR management practices are based on the history of the site of interest and disease cycles associated with annual and perennial plant species. The disease cycle begins following an extended period of seasonal inactivity caused by inappropriate temperature or moisture for plant growth. When a suscept site has no history of SLR, the disease can be prevented by planting certified SLRV-free seeds or plants in areas that are vector free and maintained under a strict weed management regime. SLR certified rootstocks, scion tissue or budwood must be used for propagation of fruit suscept species. The disease needs to be prevented from spreading from viruliferous sites to non-viruliferous sites through containment practices such as rouging, removing and destroying infected plants. In addition, adjacent plants should be removed plus weed management and reducing vector populations to non-detectable levels should also be done. Site restoration requires an integrated approach that may require more than 1 year. In addition to molecular and other laboratory assays, a bioassay with a plant such as *Chenopodium quinoa* should be a good species for the assessment technology.

Cherry rasp leaf

DISEASE,SUSCEPTS AND SYMPTOMS. Cheery rasp leaf (CRL) is an infectious disease of cherries. It was first described in Colorado in 1935. CRL is caused by *Cheravirus avii* classified under the Secoviridae family (Le Gall *et al.*, 2007). The suscept range that exhibits symptoms is narrow. It includes *Prunus avium*, *P. mahaleb* (rock cherry), *Malus* spp. and *Solanum lycopersicum*. CRL symptoms include enations on the underside of the leaf midrib. The foliage, however, remains green, distinguishing it from similar virus diseases. In addition, few leaf buds are produced, resulting in barren limbs. Numerous additional cultivated and weed species are relatively symptomless suscepts (Hansen *et al.*, 1974).

PATHOGEN AND VECTORS. The genus *Cheravirus* is a new taxon. It currently has five described species. *Cheravirus avii* (CRLV) is the type species and its genome is bipartite with three capsid proteins. The CRLV genome organization is similar to comoviruses, which are isometric, 28 nm in diameter, non-enveloped particles containing top, middle and bottom components. The virus has positive-strand cytoplasmic RNA replication/transcription. CRLV is seed-borne and vectored by *X. americanum*. It can be transmitted through budding and grafting.

DISEASE CYCLE AND MANAGEMENT. At the non-viruliferous sites, the disease cycle begins when a CRLV-infected plant is used. At viruliferous sites, the cycle begins after a period of seasonal inactivity due to inappropriate temperature or moisture regimes. As with all virus diseases of tree fruit, prevention is the best methodology. Because of the non-cultivated suscept range with plants that are often symptomless, prevention can be difficult. It involves scouting and molecular assessment of known suscepts. New plantings and replants should use certified CRLV-free. This requires all rootstocks, scions and budwood for propagation be free from CRLV. Some cherry rootstocks have been identified as possibly resistant to CRL. Orchard trees expressing symptoms and those adjacent to them should be removed and destroyed. Vector populations should be reduced to non-detectable levels through appropriate physical, chemical or biological measures.

Arabis mosaic

DISEASE. Arabis mosaic (ArM) is an infectious disease of *Arabis* spp. of the plant family Brassicaceae (Zhang *et al.*, 2020). It is caused by *Nepovirus arabis* classified in Subgroup A of the genus *Nepovirus*. ArM was initially reported from England in 1944 by Smith and Markham (1944). Its known global geographical distribution includes 39 European, 11 Asian, six Canadian Providences, seven US states, Chile, Australia and New Zealand.

SUSCEPTS AND PATHOGEN. ArM has a broad suscept range including *Apium graveolems*, *Pelargonium* spp., *Vitis* spp., *Cannabis sativa*, *Humulus* spp., *Armoracia rusticana*, *Lactuca sativa*, *Lamiales* spp., *Prunus persica* and *Beta vulgaris*. *N. arabis* is vectored by *Xiphinema diversicaudatum*. In recent decades, *Arabidopsis thaliana* has become the model plant species for fundamental biological research (Ouibrahim and Caranta, 2013).

Grape yellow vein and grapevine chrome mosaic

DISEASES, SUSCEPT, SYMPTOMS, PATHOGENS AND VECTORS. Grape yellow vein disease (GYVD) was first reported by Gooding in 1962 on *Vitis vinifera* L. at several locations in California (Gooding and Hewitt, 1962). The known suscepts include *V. vinifera* cv. Carignane, Emperor, Valdepenas and Grenache. GYVD symptoms include yellow bands on foliage veins formed from the coalition of chrome yellow flecks. Diseased vine leaves also had chlorotic leaf lines and low fruit yields. GYVV is believed to be a nepovirus vectored by *X. americanum*.

Grapevine chrome mosaic (GCM) was first reported in 1965 from Hungary by Martelli

(1965). It causes grapevine stunting and decline in Europe. GCMV consists of 30 nm diameter isometric two ssRNAs that are separately encapsulated particle strains, one of which can be associated with yellow vein disease of celery (Hollings *et al.*, 1969). Martelli and Sarospataki (1969) reported that GCMV occurred in soil infested with *Xiphinema vuittenezi*.

Although there has been relatively little research on GYVD and GCMV, the diseases are included here as examples of the inadequacies of the literature related to nematode-vectored virus diseases of plants. Although grapes are widely cultivated in China, Japan, Australia and the former USSR, there is distinct absence of literature from these nations related to nematode-vectored plant viruses.

Virus diseases vectored by needle nematodes

Tomato black ring

DISEASE. Tomato black ring (TBR) is an infectious disease caused by *Nepovirus betae*. It was described in 1946 in the UK by Smith (1946). The disease is known to exist in 40 nations throughout Europe, Asia, Africa and the Americas. It is also referred to as celery yellow vein, bean ringspot, beet ringspot, English lettuce ringspot, lettuce ringspot, potato bouquet ringspot and potato pseudo-aucuba. It can be a key limiting factor in strawberry production. TBR is not a common problem in tomato crop. The TBR viral pathogen is seed-borne, transmitted mechanically and vectored by *Longidorus* spp. It is a quarantine disease in many nations.

SUSCEPTS AND SYMPTOMS. TBR has a broad range of suscepts including vegetable, tree fruit, vine and ornamental plants. Weed species such as *Capsella bursa-pastoris*, *Cerastium vulgatum*, *Heracleum sphondylium*, *Polygonum* spp., *Sonchus asper*, *Stellaria media*, *Taraxacum officinale*, *Trifolium repens* and *Urtica urens* are also suscepts. TBR symptoms are necrotic or hypoplastic, and be suscept specific, but are usually expressed as ringspots, chlorotic mottling, vein yellowing, yellow speckling, general yellowing or leaf chlorosis. Leaf curling, reduced plant vigor, stunting and abnormal reproductive morphology are common.

PATHOGEN AND VECTORS. TBR is caused by a strain of *Nepovirus betae* (beet ringspot virus) known as tomato black ring virus (TBRV). This isometric virus is 28 nm in diameter with a protein coat consisting of about 60 subunits arranged icosahedrally. The genome consists of two positive-sense ssRNA molecules. They replicate separately and both are required for pathogenesis. Their lengths are about 7300 and 4600 nucleotides for RNA1 and RNA2, respectively. Subviral satellite RNA particles are also present (Digiaro *et al.*, 2015). Proteins generated from RNA1 are responsible for gene expression and virus replication. Those encoded for by RNA2 are involved in virion formation and movement. *L. elongatus* and *L. attenuates* are vectors of TBRV. In general, these species are limited to coarse-textured soils. TBR is acquired by the ectoparasitic nematode feeding process and retained in the odontostyle. The virion remains viable for around 9 weeks. During this time frame, TBRV can be transferred to non-viruliferous suscepts during subsequent nematode feeding.

DISEASE CYCLE. The primary TBR disease cycle begins after a period of seasonal activity related to temperature or moisture when healthy suscepts begin to grow. The sources of TBRV include virion-infected seed, infected plant propagation materials, infected vines, trees, transplants and viruliferous *L. elongatus* or *L. attenuates*. TBRV-infected plants produce viruliferous seed. Transmission also occurs when infected plants are used for propagation of trees, vines and transplants. Healthy suscepts also acquire TBRV when their root systems are parasitized by viruliferous *L. elongatus* or *L. attenuates*. The process of pathogenesis begins when a *N. betae* virion attaches to suscept tissue receptors on the plasmalemma or is encapsulated into an endosome. After this fusion, the viral nucleic acid passes through a pore and enters the cytoplasm. The virion's nucleic acid takes over the suscept metabolism and efficiently produces as many as 1 million new virions, during the process of pathogenesis. This results in necrotic and hypoplastic symptoms, in addition to serving as the *N. betae* source for the next disease cycle.

MANAGEMENT. TBRV is a regulatory/quarantine pathogen in many nations. In soil that is not infested with viruliferous *L. elongatus* or *L. attenuates*, the disease can be prevented by planting *N. betae*-free seeds, transplants, vines and trees. For this to be achieved, soil samples must be processed for plant-parasitic nematodes. Once a site has viruliferous plants or vectors, containment and vector population reduction are difficult. Infected plant materials must be removed and destroyed, and equipment sanitized. Weed management is also an essential part of the process. Populations of *L. elongatus* and *L. attenuates* need to be reduced to non-detectable levels through available cultural, physical biological and chemical measures.

Beet ringspot

DISEASE. Beet ringspot (BRS) is a soil-borne infectious disease caused by *Nepovirus betae*. It was discovered in Scotland in 1957 and described as a soil-borne virus (Harrison, 1957). BRS is the preferred common name for tomato black ring (TBR). The disease has also been referred to as bean ringspot celery yellow vein, English lettuce ringspot, lettuce ringspot, potato bouquet and potato pseudo-aucuba. BRS is a seed-borne and mechanically transmissible disease. It is vectored by the needle nematode, *Longidorus attenuates*. Although there is significantly limited literature on BRS than on the other plant virus diseases with nematode vectors, recent isolate genomes revealed interspecies recombination inconsistencies (Fowkes *et al.*, 2021). A general overview of BRS is included here because of its history and the need for additional research about its relationship with ToBRS.

SUSCEPTS AND SYMPTOMS. In addition to *Beta vulgaris*, BRS suscepts include *Brassica rapa rapa*, *B. napus*, *Solanum tuberosum*, *Phaseolus vulgaris*, *Lactuca sativa*, *Avena sativa*, *Triticum aestivum* and numerous weeds. BRS symptoms include necrotic ringspots on foliage and fruit, resulting in yield loss and poor quality fruit.

PATHOGEN AND VECTOR. *Nepovirus betae* (BRSV) is a spherical virus of around 30 nm with a single protein coat. RNA1 of its bipartite viral genome has 7394 nucleotides and RNA2 has 4684 nucleotides (Kis *et al.*, 2017). It is vectored by the needle nematode *L. attenuates*. Infestations are generally limited to coarse-textured soils.

Mulberry ringspot

DISEASE AND SUSCEPT. Mulberry ringspot (MRS) is an infectious disease of *Morus alba* classified in the Moraceae. It was first reported in Japan by Tsuchizaki *et al.* (1971). The current suscept range of MRS is restricted to Japan on *M. alba*. The genus *Morus*, however, has at least 64 named species producing mulberry fruit. It occurs in nature and under cultivation. It has been grown for thousands of years in China as a shrub for leaves as food for silkworm larvae. The senior author (G.B.) observed mulberries thriving as ornamentals in Samarkand, Uzbekistan and maintains several highly productive mulberry trees for fruit in his orchard in Michigan. There is very limited recent literature on MRS If MRS is currently a significant factor in mulberry production in Japan, there needs to be a renewed emphasis on its distribution and economic impact.

SUSCEPTS, PATHOGEN AND VECTOR. The symptoms include mosaic patterns and ringspots on foliage. These can easily be confused with those of other diseases caused by viruses. Under laboratory conditions, viruliferous sap is transmissible and replicates in at least ten plant species in five dicotyledon families such as *Vigna sinensis*, *Glycine max*, *Pisum sativum* and *Chenopodium quinoa*. The disease is caused by *Nepovirus maculanulatum* (MRSV) a polyhedral-shaped virus with a diameter of 22–25 nm. Relatively little is known about the molecular attributes of *N. maculanulatum*. It is vectored by *Longidorus martini* and soybean seeds, but not by *Xiphinema* or aphids (Yagita and Komuro, 1972). All life cycle stages of *L. martini* except newly hatched juveniles vector *N. maculanulatum*. In a study of 54 plant species from 20 families, *Morus alba*, *Vitis* sp., *Solanum lycopersicum*, *S. melongena*, *N. tabacum* and *Petunia* sp. were the only suscepts of *L. martini*. It is believed that the virus can be vectored as long as the nematode vector is alive.

DISEASE CYCLE. The MRS disease cycle begins with the source of the virus. This includes infected *M. alba* propagative material, infected trees/shrubs/seed, viruliferous *L. martini*, and viruliferous and infected horticultural instruments. It is assumed that the process of pathogenesis begins when a *N. lycopersici* virion is attached to suscept tissue receptors on the plasmalemma or is encapsulated into an endosome. After endosome formation, the viral nucleic acid passes through a pore and enters the suscept cytoplasm. The virion's nucleic acid takes over the suscept metabolism and efficiently produces new virions, resulting in the process of pathogenesis and symptom expression.

MANAGEMENT. In theory, MRS management is like that of other tree or vine crops grown for fruits or ornamental purposes. The disease can be excluded by planting only MRSV-free trees in soil that does not have viruliferous *L. martini* and using only sanitized horticultural equipment. Once a site has infected trees or viruliferous *L. martini*, it should be isolated from healthy plantings and the diseased trees rouged and destroyed. In new or replant situations, attempts should be made to lower populations of *L. martini* to below detectable levels. Physical, biological and chemical nematode management practices are available to achieve this goal (Sokora *et al.*, 2023).

Cherry rosette

DISEASE,SUSCEPT AND SYMPTOMS. Cherry rosette (CR) is an infectious disease of *Prunus avium*. It is caused by *Nepovirus* CRV. It has been known as a serious problem in Oregon since 1961 and possibly as early as 1953. The symptoms closely resemble boron deficiency and Pfeffingen disease caused by raspberry ringspot virus. They include necrosis of meristematic tissue, bushy or stunted growth, resetting, reduced vigor and inhibition of fruit set. Symptoms are most frequently observed in areas of coarse-textured soil.

PATHOGEN AND VECTOR. *Nepovirus* CRV is classified in the Secoviridae. It is vectored by the needle nematode, *Longidorus arthensis*. CRV particles attach to both the lumen of the odontostyle and the lining of the nematode's stylet guide sheath. The function of the sheath is to allow the odontostyle to move forward for feeding. When the stylet is retracted, the guide sheath forms a double fold. The greatest population densities of *L. arthensis* in cherry orchards are at soil depths greater than 30 cm.

DISEASE CYCLE AND MANAGEMENT. The disease cycle begins when the process of infection is initiated. This can take place during the tree propagation process or when the tree is exposed to viruliferous vectors following orchard establishment. Because *L. arthensis* is active at orchard soil depths greater than the normal frostline, it is probable that the disease cycle is continual and not annual. The best management strategy is prevention. Only certified CRV-free rootstocks, scions and budwood should be used in the cherry tree propagation process. Cherry tree nurseries should not have detectable populations of *L. arthensis*. Only certified CRV-free trees should be planted in new orchards, replanted orchards or as individual tree replacements. The process of containment involves rouging and destroying all trees expressing symptoms and their symptomless neighbors. Extra care should be given to removal of the tree root system. Restoration of infected orchards or individual tree sites involves using physical, chemical or biological nematode management practices to reduce. *L. arthensis* populations to non-detectable levels.

Raspberry ringspot

DISEASE, SUSCEPTS AND SYMPTOMS. Raspberry ringspot (RpRS) is an infectious disease caused by *Nepovirus rubi*. It has numerous suscepts and is present in Europe, Turkey and former USSR Republics. The disease was first reported in 1956 in Scotland and identified as soil-borne in 1957 by Harrison (1957). RpRS suscepts include *Fragaria* × *ananassa*, *Narcissus* spp., *Prunus domestica*, *Rubus* spp. and *Vitis vinifera*. Symptoms of RpRS were described by Harris in 1943 (Harris *et al.*, 1943). Young foliage exhibits yellow lines, ringspots and an oak-leaf pattern. This evolves into net-like chlorosis, necrotic blotches and dieback. Some varieties do not exhibit foliar symptoms. Fruit of diseased plants are small and of poor quality.

PATHOGEN AND VECTORS. *Nepovirus rubi* is classified in the family Secoviridae. It has a wide suscept range. The complete sequence of RpRSV consists of RNA1 and RNA2 with 7935 and 3914 nucleotides, respectively. There are multiple variants including the English, Lloyd George yellow blotch and Scottish RpRSV type strains. RpRSV has many wild and cultivated suscepts. The virion is known to be vectored by *L. elongatus*, *L. macrosoma* and *P. maximus*. As described, the vectors are strain specific. The Scottish strain is vectored mainly by *L. elongatus* and the English strain by *L. macrosoma*. *P. maximus* has a spotty distribution and is the largest species of the Longidoridae in Europe.

DISEASE CYCLE AND MANAGEMENT. The annual RpRS disease cycle at viruliferous sites begins at the end of a period of seasonal activity due to inappropriate temperature or moisture for suscept growth. In non-viruliferous sites it begins with the introduction of RpRSV-infected seeds or plants. Both types of sites should be scouted comprehensively for the disease. This includes both native and cultivated suscept species. Commercial test kits are available for RpRSV strain identification. Since prevention is the best strategy for RpRS management, only RpRSV-certified seed and plants should be used. This requires RpRSV-free rootstocks, scions and budwood for fruit crops. Containment can be achieved through rouging and destroying viruliferous plants. Site restoration requires removal and destruction of all viruliferous plants, weed management and reduction of populations of the *Longidorus* and *Paralongidorus* vectors to non-detectable levels. This is usually a multi-year process.

Artichoke Italian latent

DISEASE, SUSCEPTS, SYMPTOMS AND DISTRIBUTION. Artichoke Italian latent (AIL) is an infectious disease caused by *Nepovirus aegaeum*. It has multiple suscepts including *Cynara cardunculus*, *Gladiolus communis* and wild hosts *Helminthotheca echioides* and *Hypochaeris achyrophorus*. AIL symptoms include yellow chlorotic leaf patterning, vein clearing, mottling and patchy chlorotic stunting. Although patchy chlorotic stunting of artichokes in Greece was not formally reported until 1982 (Roca *et al.*, 1982), it is widespread and believed to have been a problem for many years. The known distribution of AIL is limited to Italy, Tunisia, Greece, Bulgaria, Albania and Turkey. AIL is the most serious disease of artichoke in Greece.

PATHOGEN. *Nepovirus aegaeum* (AILV) is classified in the virus family Secoviridae. It was first identified in 1977. AILV consists of two positive-sense ssRNAs enclosed in a protective protein coat. RNA1 and RNA2 are 7338 and 4630 nucleotides in length, respectively (Gallitelli *et al.*, 2004). In 2023, GenBank information consisted of a single complete sequence from a grape isolate of unknown host origin. In a 2023 molecular study, Elbeaino *et al.* (2023) reported significant RNA2 variation among AILV isolates. While it has been shown that leaf tissue can exhibit symptom recovery, the viroid remains viable in apical meristem tissue.

VECTOR. AILV is vectored by *Longidorus fasciatus*. Viruliferous *L. fasciatus* has a high frequency of AILV transmission within the first 4 weeks after exposure to the virion. AILV particles are retained on the inner surface of the odontostyle. This is followed

DISEASE CYCLE. *Cynara cardunculus* is native to mediterranean climates. Artichokes can be grown from seed or transplants. They grow relatively slowly and take 85–120 days to mature depending on the cultivar and environment. When planting sites have no previous history of AIL, the virus is introduced with infected seed, infected transplants or soil infested with viruliferous *L. fasciatus*. In sites with previous AIL problems, the viroid survives in weed suscepts, volunteer artichokes or the nematode vector. The patchiness associated with the disease is likely due to soil texture. Needle nematodes generally require coarse-textured soils and complete the life cycle in more than 1 year.

MANAGEMENT. AIL disease management requires integrated approaches for sites that have no known history of the disease and those that are viruliferous. Prevention is the key strategy for sites that do not have AIL. This should include using certified AILV-free seeds and transplants, planting into nematode vector-free soil and a comprehensive weed management program. Sites with AIL need to be contained in a way that prevents the spread of the disease. This involves rouging and destroying infected plants. In some cases, the entire planting should be removed and destroyed. In addition, comprehensive weed management and nematode vector population reduction need to be part of the management system. Physical, chemical and biological approaches should be considered for vector population deduction. Because of the potential severity of AIL, overall site restoration is recommended.

Virus diseases vectored by stubby-root nematodes

Tobacco rattle and corky ringspot

DISEASE. Tobacco rattle (TobR) was first reported on *N. tabacum* in 1899 in Germany (Behrens, 1899). Its known distribution includes Europe, North America, South America and Japan. TobR is caused by *Tobravirus* TobRV, the type species of the genus *Tobravirus* classified in the seven-genus virus family Virgaviridae. Corky ringspot (CorRS) is a serious disease of *Solanum tuberosum*. It is caused by a strain of TobRV. CorRS was discovered in Florida and is known to be present in 11 potato-producing states in the USA.

SUSCEPTS AND SYMPTOMS. TobR has a wide suscept range that includes more than 400 herbaceous plant species from 50 families. About 20% are non-cultivated species including weeds. The others are vegetable, ornamental and floriculture plants. Diseased plants exhibit necrotic symptoms, including foliage discoloration, yellow ringspots, streaks, lines or notched leaves. Viruliferous plants can be symptomless, making accurate diagnosis difficult. CorRS-infected *S. tuberosum* tubers exhibit discolored tissue arcs on skin surface and internal necrotic rings and corky tissue, rendering tubers unsuitable for market. While foliar symptoms are rare, leaf tissue can exhibit discolored areas and necrotic ringspots.

PATHOGEN AND VECTORS. In 1931, Bonin demonstrated that the cause of the disease was smaller than a bacterium. It was first referred to as tobacco rattle by Quanjer in 1943 (Robinson, 2003). TobRV was isolated from artichoke in Brazil in 1969 (Chagas and Silberschmidt, 1972). It consists of two 25 nm diameter rod-shaped positive-sense ssRNA strands with separate helically constructed protein coats. RNA1 and RNA2 have 6791 and 3855 nucleotides, respectively. RNA1 and RNA2 are 200 and 55 nm in length, respectively. TobRV RNA1 has four genes. Two of them code for virus replication, one codes for the protein allowing cell-to-cell movement and the function of the fourth gene is unknown. RNA2 has three genes. One codes or the protein coat and the functions of the others are unknown. It is interesting to note that no known human viruses have this type of structure. CorRSV is a strain of TobRV. TobRV is vectored by several nematode species classified in the Trichodoridae. These include *Nanidorus minor*, *N. nanus*, *Paratrichodorus allius*, *P. anemones*, *P. hispannus*, *P. pachydermus*, *P. teres*, *Trichodorus cylindricus*, *T. primitivus*, *T. similis* and *T. viruliferus*. The species appear to be both TobR and geographically specific. *P. allius* is one of the most prevalent stubby-root nematodes in potato-growing areas in the USA and molecular tools are available for detecting and identifying this nematode species (Yan *et al.*, 2016; Huang *et al.*, 2017, 2018; Lawaju *et al.*, 2023; Goraya *et al.*, 2024).

DISEASE CYCLE. TobR has two types of disease cycles, one associated with viruliferous sites and the other with previously non-viruliferous sites. These can be further subdivided into systems with annual and perennial suscepts. At the non-viruliferous sites, the disease cycle begins with the introduction of infected seed, infected plants or soil with viruliferous Trichodoridae species. In viruliferous sites, the annual cycle begins with the end of a period of suscept quiescence due to inappropriate temperature or moisture. In both cases the cycle can be terminated with the death and decomposition of the suscept tissue in the absence of viruliferous nematode vectors. In tropical and subtropical areas, the cycle is continuous.

MANAGEMENT. CorRS of potato disease management provides an example of the strategies and practices required for successful integrated disease management. At the non-viruliferous sites, disease prevention can be achieved by planting TobRV-free seed pieces, eliminating weed suscepts in the surrounding area and making sure nematode vector infested soil is not introduced on agricultural equipment. In viruliferous sites, it is essential to reduce nematode vector populations to non-detectable levels. In the USA and Europe, most potato fields are fumigated for management of *Globodera* spp. (cyst), *Meloidogyne* spp., *Pratylenchus penetrans* (root lesion), *Ditylenchus destructor* (potato rot) or virus vectors. This is done on a pre-plant basis using nematicidal rates of 1,3-D, metam-sodium or metam-potassium. In addition, oxamyl (*N,N*-dimethyl-2-methylcarbamoyloxyimino-2-(methylthio)-acetamide) is often applied pre-plant, at planting or post-emergence for nematode and insect management. Other chemical, biological and physical practices are available for vector population reduction. A biological assay of soil should be used to evaluate the efficacy of the management practices.

Pea early browning

DISEASE. Pea early browning (PEB) is an infectious disease caused by *Tobravirus* PEBV classified in the seven-genus virus family Virgaviridae. PEB was described by Dos and Want in 1962 after being first reported in the Netherlands since 1949. As a serious disease of peas, it is widely distributed throughout Europe and northern Africa.

SUSCEPTS AND SYMPTOMS. PEB has a wide herbaceous suscept range, including *Beta vulgaris*, *Cucumis sativus*, *Eschscholzia californica*, *Lupinus luteus*, *Medicago lupulina*, *Medicago sativa*, *Phaseolus vulgaris*, *Pisum sativum*, *Solanum nigrum*, *Trifolium pratense*, *Trifolium repens*, *Tropaeolum majus* and *Vicia faba*. PEB symptoms on peas include foliar yellowing or browning with green veins or dark colored leaf and pod necrosis. Plants grow slow with delayed maturity and an abnormal amount of plant mortality. Surviving plants produce few pods which possess small peas.

PATHOGEN AND VECTORS. *Tobravirus* PEBV consists of two 21.5 nm diameter rod-shaped positive-sense ssRNA strands with separate helically constructed protein coats having a pitch of 2.5 nm and a 4 nm central core. RNA1 is longer than RNA2. RNA1 can cause PEB systemic infection in the absence of RNA2 and without forming its protein coat. Multiple strains of PEBV have been identified in Europe. PEBV has been reported to be vectored by the following six species of the Trichodoridae: *Paratrichodorus anemones*, *P. pachydermus*, *P. teres*, *Trichodorus primitivus*, *T. viruliferous* and *T. similis*. Vector acquisition time for PEBV can be as short as an hour to as long as several days. The virus does not replicate in the nematode vector and can be retained from weeks to 1 year. For some suscepts, PEBV invades suscept embryos early in their development and is seed-borne in a relatively small portion of seeds produced by individual suscept plants.

DISEASE CYCLE. In non-viruliferous suscept sites, the PEB disease cycle begins with the introduction of infected crop seeds, weed seeds or PEBV-infected nematode vectors in wind-blown soil or from contaminated equipment. In viruliferous sites, the disease cycle begins following a period of seasonal suscept growth inactivity caused by inappropriate temperature or moisture. In these cases, the virus source includes infected perennial crops, weeds and nematode vectors. In the absence of a nematode vector population, a perennial crop suscept or perennial weed suscept, the annual disease cycle is terminated with the death and decomposition of the diseased crop. In the presence of a nematode vector, the virus can survive the next period of seasonal inactivity and initiate a new cycle the following year.

MANAGEMENT. Prevention is the best PEB management strategy for PEBV-free sites. This can be achieved by planting only certified PEBV-free seed and maintaining a rigorous weed management program. Containment can be achieved by destroying infected plants and their immediate neighbors. PEBV-infested site restoration requires reducing vector populations and weed suscept to non-detectable levels. Biological, chemical and physical nematode management practices are available. The best way to determine success is through use of a biological assay with a highly susceptible quickly growing plant that exhibits readily recognizable symptoms.

Pepper ringspot

DISEASE. Pepper ringspot (PepRS) is an infectious disease of several important agricultural crops. It is caused by *Tobravirus* PepRS. PepRS was reported in the 1960s as the Brazilian counterpoint to the tobacco rattle diseases and referred to as CAM tobacco rattle disease by Harrison and Woods (Kitajima and Costa, 1969). PepRS was detected in commercial potato fields in South Africa in 2022 (Uys and Esterhuizen, 2024). It is now recognized as a quarantine issue for adjacent African nations. It has not, however, been detected on *Solanum tuberosum* in Brazil.

SUSCEPTS AND SYMPTOMS. Although no comprehensive surveys have been conducted, PepRS is believed to have a wide suscept range including *Capsicum annuum*, *Cynara cardunculus*, *Solanum lycopersicum* and *S. tuberosum*. PepRS causes yellowing suscept of leaf tissue. Ringspots are formed on *S. tuberosum* tuber skin. Tuber tissue contains discolored arches or sprains making tubers unsuitable for marketing.

PATHOGEN AND VECTOR. *Tobravirus* PepRS is classified in the virus family Virgaviridae. It is believed to be the only *Tobravirus* spp. in Brazil. Although the complete genome of PepRSV has not been reported, it is known that combinations of its RNA1 and RNA2 are infectious on *Nicotiana benthamiana*. Other genomic observations show that PepRSV resembles TobRV. *Nanidorus minor* is the only reported nematode vector of *PepRS*.

DISEASE CYCLE AND MANAGEMENT. There is a lack of published information on the PepRS disease cycle and its management. This may be due to the limited known distribution of the disease or because Brazilian nematology has largely focused on *Heterodera glycines*. It is likely that this will change because of the detection of PepRS associated with potato production in South Africa. It can be assumed that the PepRS disease cycle and effective integrated management are similar to TobR.

Conclusion and Recommendations

The history of plant virus diseases vectored by nematodes can be divided into three phases. The first relates to the discovery of specific infectious diseases based on foliar symptoms and yield loss. This began in the 1800s and was of assistance in the overall understanding of the nature of viruses. The diseases were usually described as soil-borne. The second phase involved demonstrating that species classified in the nematode order Dorylaimida were vectors of phytopathogenic viruses. The third phase focuses on the genomics of the virus and its suscept relationships. The literature on plant diseases caused by viruses vectored by nematodes is spotty. It is often related to the economics of a specific agricultural crop or a unique opportunity to make a significant contribution to science. To achieve a better understanding of the true role of plant nematodes as vectors of viruses in the third decade of the 21st century, there needs to be several new highly funded research laboratories dedicated to the applied and fundamental aspects of this domain on a global basis. In addition, general education at all levels needs to include information about the nature of diseases and food systems in regard to their overall historical, current and potential future impacts on society (Bird and Maredia, 2021).

References

Abdalreda, E.M. and Mohammed, L.H. (2022) Nematodes that transmit plant pathogenic viruses. *IOP Conference Series: Earth and Environmental Science* 1060, 012125. DOI: 10.1088/1755-1315/1060/1/012125.

Adaskaveg, J., Gubler, W., Duncan, R., Stapleton, J., Holtz, B. *et al.* (2017) *UC IPM management guidelines: Almond. UC ANR Publication 3431*. UC IPM, Davis, CA.

Allen, W.R. and van Schagen, J.G.V. (1998) Comparative transmission of the peach rosette mosaic virus by Ontario populations of *Longidorus diadecturus* and *Xiphinema americanum* (Nematoda: Longidoridae). *Canadian Journal of Plant Pathology* 6(1), 29–32. DOI: 10.1080/07060668409501587.

Andret-Link, P., Schmitt-Keichinger, C., Demangeat, G., Komar, V. and Fuchs, M. (2004) The specific transmission of Grapevine fanleaf virus by its nematode vector *Xiphinema* index is solely determined by the viral coat protein. *Virology* 320(1), 12–22. DOI: 10.1016/j.virol.2003.11.022.

Behrens, J. (1899) Die Mauche (Mauke) des Tabaks. *Landwirtsch Versuch* 52, 442.

Beijerinck, M.W. (1898) Concerning a contagium *Vivum fluidium* as a cause of the spot-disease of tobacco leaves. In: *Phytopathology Classics No. 7*. APS Press, St Paul, MN.

Bird, G. and Maredia, K. (2021) Pandemics, famines and global development. *Journal of Medical Arthropodology and Public Health* 2022(2), 51–65.

Bird, G. and Brewer, J. (2006) Integrated pest management, eco-literacy and unexpected consequences. In: Francis, C.A., Poincelot, R. and Bird, G. (eds) *New Social Contract: Developing and Extending Sustainable Agriculture*. Haworth Press, Haworth, UK, pp. 25–50.

Bitterlin, M. and Gonsalves, D. (1987) Spatial distribution of *Xiphinema rivesi* and persistence of tomato ringspot virus and its vector in soil. *Plant Disease* 71(5), 408. DOI: 10.1094/PD-71-0408.

Brown, D., Halbrendt, J., Jones, A., Vrain, T. and Robinson, R. (1994) Transmission of three North American nepoviruses by populations of four distinct species of the *Xiphinema americanum* group. *Phytopathology* 84, 646–649. DOI: 10.1094/Phyto-84-646.

Buzayan, J.M., Gerlach, W.L. and Bruening, G. (1986a) Non-enzymatic cleavage and ligation of RNAs complementary to a plant virus satellite RNA. *Nature* 323(6086), 349–353. DOI: 10.1038/323349a0.

Buzayan, J.M., Gerlach, W.L. and Bruening, G. (1986b) Satellite tobacco ringspot virus RNA: A subset of the RNA sequence is sufficient for autolytic processing. *Proceedings of the National Academy of Sciences of the United States of America* 83, 8859–8862. DOI: 10.1073/pnas.83.23.8859.

Cazalis-Allut, L. (1865) *Oeuvres agricoles de Cazslis-Allut*. Victor Masson et Fils, Paris.

CDFA (2020) *California Pest Rating Proposal for Tomato Ringspot Virus*. California Department of Agriculture, Sacramento, CA.

Chagas, C. and Silberschmidt, K. (1972) Virus da faixa amarela da alcachôfra: Ocorrência, toansmissão mecânica e propriedades físicas. *Biologico* 38, 35.

Cobb, N.A. (1913) New nematode genera found inhabiting fresh water and non-brackish soils. *Journal of the Washington Academy of Sciences* 3(16), 432–444. DOI: 10.5962/bhl.part.20323.

Cobb, N.A. (1914a) *Contributions to a Science of Nematology*. Williams & Wilkins, Baltimore, MD. DOI: 10.5962/bhl.title.20608.

Cobb, N.A. (1914b) *Nematodes and Their Relationships*. US Government Printing Office, Washington, DC.

Crowley, N.C., Davison, E.M., Francki, R.I. and Owusu, G.K. (1969) Infection of bean root meristems by tobacco ringspot virus. *Virology* 39(2), 322–330. DOI: 10.1016/0042-6822(69)90052-x.

Dalmasso, A. (1969) Étudeanatomique et taxonomique des genres *Xiphinema,Longidorus* et *Paralongidorus* (Nemata: Dorylaimida). *Mémoires Du Muséum National d'Histoire Naturelle. Nouvelle Série A. Zoologie* 61, 33–82.

de Man, J.G. (1880) *Die einheimischen, frei in der reinen Erde und im süssen Wasser lebende Nematoden*. E.J. Brill, Leiden, The Netherlands.

Decraemer, W. and Hunt, D. (2013) Structure and classification. In: Perry, R.N. and Moens, M. (eds) *Plant Nematology*. CAB International, Wallingford, UK, pp. 3–39.

Dias, H. and Cation, D. (1976) The characterization of a virus responsible for peach rosette mosaic and grape decline in Michigan. *Canadian Journal of Botany* 54, 1228–1239. DOI: 10.1139/b76-133.

Digiaro, M., Yahyaoui, E., Martelli, G.P. and Elbeaino, T. (2015) The sequencing of the complete genome of a tomato black ring virus (TBRV) and of the RNA2 of three grapevine chrome mosaic virus (GCMV) isolates from grapevine reveals the possible recombinant origin of GCMV. *Virus Genes* 50(1), 165–171. DOI: 10.1007/s11262-014-1094-4.

Ebsary, B.A., Vrain, T.C. and Graham, M.B. (1989) Two new species of *Xiphinema* (Nematoda: Longidoridae) from British Columbia vineyards. *Canadian Journal of Zoology* 67(4), 801–804. DOI: 10.1139/z89-117.

Elbeaino, T., Ben Slimen, A., Belgacem, I., Mnari-Hattab, M., Spanò, R. *et al.* (2023) Identification, sequencing, and molecular analysis of RNA2 of artichoke Italian latent virus isolates from known hosts and a new host plant species. *Viruses* 15(11), 2170. DOI: 10.3390/v15112170.

Enelmann, J. and Hamacher, J. (2008) Plant virus diseases: Ornamental plants. In: *Encyclopedia of Virology*, 3rd edn. Elsevier Academic Press, Amsterdam. DOI: 10.1016/B978-012374410-4.00729-9.

Fang, L., Wei, X.-Y., Liu, L.-Z., Zhou, L.-X., Tian, Y.-P. *et al.* (2021) A tobacco ringspot virus-based vector system for gene and microRNA function studies in cucurbits. *Plant Physiology* 186(2), 853–864. DOI: 10.1093/plphys/kiab146.

Fowkes, A., Adams, I.P., Jones, R.A.C., Fox, A., McGreig, S. *et al.* (2021) Historical and recent tomato black ring virus isolate genomes reveal interspecies recombination and plant health regulation inconsistencies. *Plant Pathology* 71(3), 729–740. DOI: 10.1111/ppa.13507.

Fromme, F., Wingard, S. and Priode, C. (1927) Ringspot of tobacco: An infectious disease of unknown cause. *Phytopathology* 17, 321–328. DOI: 10.1079/cabicompendium.54202.

Fuchs, M.J., Hily, K., Petrzik, H., Sanfaçon, J., Thompson, R. *et al.* (2022) ICTV virus taxonomy profile: Secoviridae 2022. *Journal of General Virology* 103(12), 12. DOI: 10.1099/jgv.0.001807.

Fulton, J. (1962) Transmission of tobacco ringspot virus by *Xiphinema americanum*. *Phytopathology* 52, 375–377.

Gallitelli, D., Rana, G., Vovlas, C. and Martelli, G. (2004) Viruses of globe artichoke: An overview. *Journal of Plant Pathology* 86, 267–281.

Georgi, L.L. (1988) Transmission of tomato ringspot virus by *Xiphinema americanum* and *X. rivesi* from New York apple orchards. *Journal of Nematology* 20(2), 304–308.

Gooding, G.V. (1970) Natural serological strains of tobacco ringspot virus. *Phytopathology* 60(4), 708. DOI: 10.1094/Phyto-60-708.

Gooding, G. and Hewitt, W. (1962) Grape yellow vein: Symptomatology, identification, and the association of a mechanically transmissible virus with the disease. *American Journal of Enology and Viticulture* 13(4), 196–203. DOI: 10.5344/ajev.1962.13.4.196.

Goraya, M., Yan, G., Whitworth, J. and Swisher Grimm, K. (2024) Advancing nematode identification on potato: An isothermal recombinase polymerase amplification assay for stubby root nematode, *Paratrichodorus allius*. *American Journal of Potato Reseaarch* 101, 52–64. DOI: 10.1007/s12230-023-09940-4.

Halbrendt, J., LaMondia, J. and Zasada, I. (2008) Evaluation of millet and rapeseed as rotation or green manure crops to control nematodes in orchard replant. *Abstracts 2008 International Congress of Nemotology* Brisbane, Australia.

Hansen, A., Nyland, G., McElroy, F. and Stace-Smith, R. (1974) Origin, cause, host range and spread of cherry rasp leaf disease in North America. *Phytopathology* 64(5), 721–727. DOI: 10.1094/Phyto-64-721.

Harris, V., Bryce, A. and Foister, C. (1943) Raspberry ringspot virus (ringspot of raspberry). *Report of the East Malling Research Station* Maidstone, UK. DOI: 10.1079/cabicompendium.48120.

Harrison, B.D. (1957) Studies on the host range, properties and mode of transmission of beet ringspot virus. *Annal Applied Biology* 45(3), 462–472. DOI: 10.1111/j.1744-7348.1957.tb05883.x.

Harrison, B., Robertson, W. and Taylor, C. (1974) Specificity of retention of viruses by nematodes. *Journal of Nematology* 6, 1550–164.

Hewitt, W., Raski, D. and Goheen, A. (1958) Nematode vector of soil-borne fan leaf virus of grapevine. *Phytopathology* 48, 586–595. DOI: 10.1079/cabicompendium.57032.

Hily, J., Poulicard, N., Kubina, J., Reynard, J., Spilmont, A. *et al.* (2021) Metaenomic analysis of nepoviruses: Diversity, evolution and identification. *Archives of Virology* 166, 2789–2801. DOI: 10.1007/s00705-021-05111-0.

Hollings, M., Stone, L. and Martelli, G. (1969) Celery yellow vein and Hungarian chrome mosaic viruses. *Report of the Glasshouse Crops Research Institute* 102–103.

Huang, D., Yan, G. and Skantar, A.M. (2017) Development of real-time and conventional PCR assays for identifying stubby root nematode *Paratrichodorus allius*. *Plant Disease* 101(6), 964–972. DOI: 10.1094/PDIS-10-16-1431-RE.

Huang, D., Yan, G.P., Gudmestad, N., Whitworth, J., Frost, K. *et al.* (2018) Molecular characterization and identification of stubby root nematode species from multiple states in the United States. *Plant Disease* 102(11), 2101–2111. DOI: 10.1094/PDIS-10-17-1668-RE.

Hull, R. (2014) *Plant Virology*, 5th edn. Elsevier, Amsterdam. DOI: 10.1016/C2010-0-64974-1.

Iles, L.C., Fulladolsa, A.C., Smart, A., Bonkowski, J., Creswell, T. *et al.* (2021) Everything is faster: How do land-grant university-based plant diagnostic laboratories keep up with a rapidly changing world? *Annual Review of Phytopathology* 59(1), 333–349. DOI: 10.1146/annurev-phyto-020620-102557.

Ivanowski, D. (1892) Concerning the mosaic disease of the tobacco plant. In: *Phytopathology Classics No. 7*. APS Press, St Paul, MN.

Karle, H. and Nyland, G. (1959) Yellow bud mosaic in almond. *Plant Disease Reporter* 43, 5200–5621.

Khan, M.R. (2008) *Plant Nematodes – Methodology, Morphology, Systematics, Biology and Ecology*. Science Publishers, Enfield, NH. DOI: 10.1201/b10766.

Khan, M.R. (2023a) Nematode pests of agricultural crops, a global overview. In: Khan, M.R. (ed.) *Novel Biological and Biotechnological Applications in Plant Nematode Management*. Springer, Singapore, pp. 3–45. DOI: 10.1007/978-981-99-2893-4_1.

Khan, M.R. (2023b) Plant nematodes, an underestimated constraint in global crop production. Nematode diseases of crops and their sustainable management. In: Khan, M.R. and Quintanilla, M. (eds) *Nematode Diseases of Crops and Their Sustainable Management*. Academic Press, Cambridge, MA, pp. 3–23. DOI: 10.1016/B978-0-323-91226-6.00009-2.

Khan, S.H. and Ahmad, S. (1975) Longidoroidea (Thorne, 1935) n. rank (Nematoda: Dorylaimina) with description of *Xiphinema neoamericanum* n. sp. from India and proposal of a new name for *X. americanum* sensu Carvalho (1956) non Cobb, 1913. *Nematologia Mediterranea* 23–28.

Kis, S., Salamon, P., Kis, V. and Szittya, G. (2017) Molecular characterization of a beet ringspot nepovirus isolated from *Begonia ricinifolia* in Hungary. *Archives of Virology* 162(11), 3559–3562. DOI: 10.1007/s00705-017-3521-z.

Kitajima, E.W. and Costa, A.S. (1969) Association of pepper ringspot virus (Brazilian tobacco rattle virus) and host cell mitochondria. *Journal of General Virology* 4(2), 177–181. DOI: 10.1099/0022-1317-4-2-177.

Koonin, E.V., Senkevich, T.G. and Dolja, V.V. (2006) The ancient virus world and evolution of cells. *Biology Direct* 1, 29. DOI: 10.1186/1745-6150-1-29.

Kubina, J., Hily, J., Mustin, P., Komar, V., Garcia, S. *et al.* (2023) Characterization of grapevine fanleaf virus isolates in chardonnay vines exhibiting severe and mild symptoms in two vineyards. *Viruses* 14(10), 2303. DOI: 10.3390/v14102303.

Lamberti, F. and Bleve-Zacheo, T. (1979) Studies on *Xiphinema americanum* sensu lato with descriptions of fifteen new species (Nematoda, Longidoridae). *Nematologia Mediterranea* 7, 51–106.

Lammers, A., Allison, R. and Ramsdell, D. (1999) Cloning and sequencing of peach rosette mosaic virus. *Virus Research* 65, 57–73. DOI: 10.1079cabicompendium.3922.

Lamprecht, R., Spaltman, M., Stephan, D., Wetzel, T. and Burger, J. (2013) Complete nucleotide sequence of a fan leaf virus and its associated satellite virus. *Viruses* 5, 1815–1823. DOI: 10.3390/v5071815.

Lawaju, B.R., Yan, G.P. and Jonathan, W. (2023) Development of a droplet digital PCR assay for detection and quantification of stubby root nematode, *Paratrichodorus allius* in soil. *Plant Disease* 107, 3344–3353. DOI: 10.1094/PDIS-03-23-0439-SR.

Le Gall, O., Sanfacon, H., Ikegami, M., Iwanami, T., Jones, T. *et al.* (2007) *Cheravirus and sadwavirus*: Two unassigned genera of plant positive-sense single-stranded RNA viruses formerly considered atypical members of the genus *Nepovirus* (family Comoviridae). *Archives of Virology* 152(9), 1767–1774. DOI: 10.1007/s00705-007-1015-0.

Lecoq, H. (2001) Discovery of the first virus, the tobacco mosaic virus: 1892 or 1898? *Académie Des Sciences. Série III, Sciences de La Vie III* 324, 9290933. DOI: 10.1016/S0764-4469(01)01368-3.

Link, A.C., Laporte, L., Valat, C., Ritenthaler, G., Demangeat, E. *et al.* (2004) Invited review of grapevine fanleaf virus still a major threat to the grapevine industry. *Journal of Plant Pathology* 86, 183–195.

Lister, R. (1964) Strawberry latent ringspot: A new nematode-borne virus. *Annals of Applied Biology* 54, 167–176. DOI: 10.1079/cabicompendium.52200.

Loof, P.A.A. (1975) Taxonomy of Trichodoridae. In: Lambreti, F., Taylor, C.E. and Seinhorst, J.W. (eds) *Nematode Vectors of Plant Viruses*. Plenum, London, pp. 103–127. DOI: 10.1007/978-1-4684-0841-6_7.

MacFarlane, S. and Neilson, R. (2009) Testing transmission of tobraviruses by nematodes. *Current Protocols in Microbiology* 16, B.5. DOI: 10.1002/9780471729259.mc16b05s12.

Macho, A.P. and Lozano-Duran, R. (2019) Molecular dialogues between viruses and receptor-like kinases in plants. *Molecular Plant Pathology* 20(9), 1191–1195. DOI: 10.1111/mpp.12812.

Malek, R. (1969) Population fluctuations and observations of the life cycle of *Xiphinema americanum* associated with cottonwood (*Populus deltoides*) in South Dakota. *Proceedings of the Helminthological Society of Washington* 36, 270–274.

Martelli, G. (1965) Preliminary report on purification and serology of a virus associated with Hungarian grapevines showing macroscopic symptoms of fanleaf and yellow mosaic. In: *International Conference on Virus and Vector on Perennial Hosts with Special Reference to Vitis*, University of California, Davis, CA, pp. 402–410.

Martelli, G. (1978) Nematode-borne viruses of grapevine, their epidemiology and control. *Nematologia Mediterranea* 6(1), 1–27.

Martelli, G. and Sarospataki, G. (1969) Grapevine chrome mosaic virus. *Phytopathologia Mediterranea* 8, 1.

Martin, R.R., Tzanetakis, I.E., Barnes, J.E. and Elmhirst, J.F. (2004) First report of strawberry latent ringspot virus in strawberry in the United States and Canada. *Plant Disease* 88(5), 575–575. DOI: 10.1094/PDIS.2004.88.5.575A.

Mayo, M.A. and Jones, A.T. (1999) Nepovirus (comoviridae). In: Granoff, A. and Webster, R.G. (eds) *Encyclopedia of Virology*, 2nd edn. Elsevier, Amsterdam, p. 1007.

Mazyad, A., Khederr, A., El-Attart, A., Amer, W., Mahasen, I. *et al.* (2014) Characterization of strawberry latent ringspot virus (SLRSV) on strawberry in Egypt. *Egyptian Journal of Virology* 219–235.

Merny, G. (1966) Nématodes d'Afrique tropicale: Un nouveau *Paratylenchus* (Criconematidae), deux nouveaux *Longidorus* et observations sur *Longidorus laevicapitatus* williams, 1959 (Dorylaimidae). *Nematologica* 12(3), 385–395. DOI: 10.1163/187529266X00860.

Michaels, T., Hoover, E. and Irish, L. (2022) *The Science of Plants*. University of Minnesota Libraries Publishing, Minneapolis, MN.

Micoletzky, H. (1922) Die freilebenden Erd-nematoden. *Archiv Für Naturgeschichte A* 87, 1–650.

Ouibrahim, L. and Caranta, C. (2013) Exploitation of natural genetic diversity to study plant–virus interactions: What we can learn from *Arabidopsis thaliana*? *Molecular Plant Pathology* 14, 844–854. DOI: 10.111/mpp.12052.

Pennisi, E. (2018) Land plants arose earlier than previously thought – and may have had a bigger impact on the evolution of animals. *Science* aat3642. DOI: 10.1126/science.aat3642.

Pinkerton, J.N., Kraus, J., Martin, R.R. and Schreiner, R.P. (2008) Epidemiology of *Xiphinema americanum* and tomato ringspot virus on red raspberry, *Rubus idaeus*. *Plant Disease* 92(3), 364–371. DOI: 10.1094/PDIS-92-3-0364.

Ramsdell, D.C. and Meyers, R. (1974) Peach rosette mosaic virus, symptomatology and nematodes associated with grapevine degeneration in Michigan. *Phytopathology* 64(9), 1174–1178. DOI: 10.1094/Phyto-64-1174.

Raski, D.J., Goheen, A., Lider, L. and Mereidth, C. (1983) Strategies against grapevine fanleaf virus and its nematode vector. *Plant Disease* 67(3), 335. DOI: 10.1094/PD-67-335.

Robinson, D. (2003) Tobacco rattle virus. *DPVweb.net* (*Database of plant viruses*) DPV No. 398.

Roca, F., Rana, G.L. and Kyriakopoulou, P.E. (1982) *Longidorus fasciatus* Roca *et* Lamberti vector of a serologically distinct strain of artichoke Italian latent virus in Greece. *Nematologia Mediterranea* 10(10), 65–69.

Rupnik-Cigoj, M., Jez-Krebelj, A., Castellarin, S.D., Trošt, K., Sivilotti, P. *et al.* (2018) Grapevine fanleaf virus affects grape (*Vitis vinifera*) berry anthocyanin content via the transcriptional regulation of anthocyanin biosynthetic genes. *Functional Plant Biology* 45(7), 771. DOI: 10.1071/FP18014.

Rush, M. and Gooding, G. (1970) The occurrence of tobacco ringspot virus strains and tomato ringspot virus in hosts indigenous to North Carolina. *Phytopathology* 60, 1756–1760. DOI: 10.1079/cabicompendium.54202.

Schellenberger, P., Sauter, C., Lorber, B., Bron, P., Trapani, S. *et al.* (2017) Structural insights into viral determinants of nematode mediated grapevine fanleaf virus transmission. *PLoS Pathogens* 13(3), e1006268. DOI: 10.1371/journal.ppat.1006268.

Schneider, I.R., Hull, R. and Markham, R. (1972) Multidense satellite of tobacco ringspot virus: A regular series of components of different densities. *Virology* 47(2), 320–330. DOI: 10.1016/0042-6822(72)90267-X.

Service, R.F. (1997) Microbiologists explore life's rich, hidden kingdoms. *Science* 275(5307), 1740–1740. DOI: 10.1126/science.275.5307.1740.

Siddiqi, M.R. (1974) Systematics of the genus *Trichodorus* Cobb, 1913 (Nematoda, Dorylaimida), with descriptions of three new species. *Nematologica* 19, 259–278.

Smith, K.M. (1946) Tomato black-ring: A new virus disease. *Parasitology* 37(3–4), 126–130. DOI: 10.1017/s0031182000013263.

Širca, S., Stare, B.G., Plesko, I.M., Marn, M.V., Urek, G. *et al.* (2007) *Xiphinema rivesi* from Slovania transmit tobacco ringspot virus and tomato ringspot virus to cucumber bait plants. *Plant Disease* 91(6), 770. DOI: 10.1094/PDIS-91-6-0770B.

Smith, K. and Markham, R. (1944) Two new viruses affecting tobacco and other plants. *Phytopathology* 34, 324–329.

Sokora, R., Helder, J., Leendert, P., Desaeger, J. and Mahlein, A. (2023) Integrated nematode management in transition: Constraints, policy, processes, and technologies for the future. *Annual Review of Phytopathology* 61, 209–230. DOI: 10.1146/annurev-phyto-021622-113058.

Taylor, C. and Raski, D. (1964) On the transmission of grape fanleaf by *Xiphinema index*. *Nematologica* 10(3), 489–495.

Taylor, C. and Robertson, W. (1975) Acquisition retention and transmission of viruses by nematodes. In: Lamberti, F., Taylor, CE. and Seinhorst, J.W. (eds) *Nematode Vectors of Plant Viruses*. Plenum, London, pp. 253–276. DOI: 10.1007/978-1-4684-0841-6_17.

Tsuchizaki, T., Hibino, H. and Saito, Y. (1971) Mulberry ringspot virus isolated from mulberry showing ringspot symptom. *Japanese Journal of Phytopathology* 37(4), 266–271. DOI: 10.3186/jjphytopath.37.266.

Uys, D. and Esterhuizen, L. (2024) Pepper ringspot virus in the spotlight. *Chips (Research and Technical)* 38(1), 26–27.

Walker, M., Chisholm, J., Wei, T., Ghoshal, B., Saeed, H. *et al*. (2014) Complete genome sequence of three tomato ringspot virus isolates: Evidence for reassortment and recombination. *Archives of Virology* 160(2), 543–547. DOI: 10.1007/s00705-014-2240-y.

Weischer, B. (1993) Nematode–virus interactions. In: Khan, M.W. (ed.) *Nematode Interactions*. Chapman & Hall, London, pp. 217–231. DOI: 10.1007/978-94-011-1488-2_10.

Wessels, T. (2006) *The Myth of Progress: Towards a Sustainable Future*. University of Vermont Press, Burlington, VT.

Yagita, H. and Komuro, Y. (1972) Transmission of mulberry ringspot virus by *Longidorus martini merny*. *Japanese Journal of Phytopathology* 38(4), 275–283. DOI: 10.3186/jjphytopath.38.275.

Yan, G.P., Plaisance, A., Huang, D., Upadhaya, A., Gudmestad, N.C. *et al*. (2016) First report of the stubby root nematode *Paratrichodorus allius* on potato in North Dakota. *Plant Disease* 100(6), 1247–1247. DOI: 10.1094/PDIS-11-15-1350-PDN.

Zhang, Y., Wang, Y., Xie, Z., Wang, R., Guo, Z. *et al*. (2020) Rapid detection of lily mottle virus and arabis mosaic virus infecting lily (*Lilium* spp.) using reverse transcription loop-mediated isothermal amplification. *The Journal of Plant Pathology* 36(2), 170–178. DOI: 10.5423/PPJ.OA.04.2019.0096.

Zhao, F., Lim, S., Igori, D., Yoo, R.H., Kwon, S.-Y. *et al*. (2016) Development of tobacco ringspot virus-based vectors for foreign gene expression and virus-induced gene silencing in a variety of plants. *Virology* 492, 166–178. DOI: 10.1016/j.virol.2016.02.025.

10 Fungi and Bacteria Vectored by Plant Nematodes

Mujeebur Rahman Khan[1], Frank Kagoda[2] and Ziaul Haque[1]*
[1]Department of Plant Protection, Faculty of Agricultural Sciences, Aligarh Muslim University, Aligarh, India; [2]Crops and Natural Resources Program, NARO-Buginyanya ZARDI, Mbale, Uganda

Abstract

Plant-parasitic nematodes act as carriers or vectors for certain plant-pathogenic bacteria and fungi with an obligatory etiological link between them that leads to new disease conditions in the host plants. The significance of this vector role is that the nematode enables access of the inocula to the susceptible tissue otherwise inaccessible to the pathogen. Yellow ear rot or "tundu" disease of wheat may serve as a classic example to demonstrate this relationship. The disease results from the combined invasion by *Anguina tritici* and *Rathayibacter tritici*. In this disease, the bacterial spores are carried by the nematode to the apical meristem which is tightly ensheathed by the leaf scales and bud scales of the developing wheat plants. There are some other similar diseases such as annual rye grass toxicity, alfalfa wilt, etc., where the inocula are carried by nematodes to the susceptible tissue. The majority of the research that has been done on the vector role of nematodes is confined to a limited number of diseases, and other relationships are not explored. This chapter examines the well-known diseases as well as the lesser studied bacterial and fungal diseases that develop due to the involvement of plant nematodes as a vector for carrying the inocula. The vector relationships between yellow ear rot disease of wheat, cauliflower disease of strawberries and twist disease of cereals are also presented in detail.

Introduction

A plant in nature is a potential host to various microorganisms sharing a common habitat, and a diverse microbial community develops in the rhizosphere which forms different relationships, benefitting or antagonizing each other (Khan, 1995; Topalović and Heuer, 2019). The vector relationship is one of the important ecological requirements of non-motile microorganisms (Parrado and Quintanilla, 2024). Plant nematodes, besides attacking the host directly, also act as vectors for carrying other pathogens to susceptible hosts or tissues (Khan, 2008). It is assumed that in the vector relationship, the nematode feeding on bacteria or fungi initially had a predatory relationship, which later evolved to the vector relationship introducing the bacterial or fungal inocula into the environment by their defecation. Several microbiotrophic free-living nematodes are known to be vectors of bacteria and fungi because they feed on microbial cells that colonize the injured plant tissues (Khan and Pathak, 1993). The damaged or injured plant tissue offers a suitable habitat for microbial growth to which saprozoic or bacterivorous nematodes become attracted (Bird and Riddle, 1984).

*Corresponding author: zia_haq07@yahoo.com

 Nematode Disease Complexes in Agricultural Crops (ed. Mujeebur Rahman Khan)
DOI: 10.1079/9781800625228.0010

During this process of feeding on fungi or bacteria, it is quite possible that these nematodes also guard and transfer the harmful bacteria. Free-living nematodes are likely to be involved as a vector in the disease complexes, although it has been reported that microbiotrophic nematodes consume and eject live bacteria (Wasilewska and Webster, 1975). However, several typical examples are documented where plant nematodes carry the plant-pathogenic bacteria and fungi (Khan and Pathak, 1993). In this chapter, the disease complexes formed due to the vector role of nematodes in transmitting bacterial and fungal pathogens such as *Rathayibacter tritici* (yellow ear rot), *Rhodococcus fascians* (cauliflower disease of strawberry), *Rathayibacter toxicus* (annual ryegrass toxicity), *Clavibacter michiganensis* (alfalfa wilt), *Pseudomonas salomonii* ('Café au lait' bacteriosis) and *Dilophospora alopecuri* (ear twist disease) are discussed in detail.

Anguina–Bacteria

Yellow ear rot

Yellow ear rot disease of wheat is manifested due to a combined infection of the nematode *Anguina tritici* and the bacterium *Rathayibacter* (= *Clavibacter*) *tritici*. The disease is often referred to as 'tundu' in India and also as spike blight or yellow slime. Hutchinson (1917) originally described this disease complex as a bacterial disease from India. Other reports of similar diseases came from Egypt, Australia, Pakistan and China (Khan and Pathak, 1993). In this disease, the vector role of the nematode is essential. The infective juveniles of *A. tritici* carry the bacteria when they migrate to invade the shoot meristem. During this process, the bacterial cell reaches the apical meristem which is inaccessible to the bacteria because of its protection by the compact leaf scales (Khan, 2024). At the initial stage (vegetative stage), typical symptoms of the nematode invasion appear (Fig. 10.1), but at the later stage under low temperature, high humidity and cloudy weather, bacterial infestation intensifies and a bright yellow slime is oozed profusely on the infected ear heads, leaves and stems, which becomes hard and brittle on drying (Nandal *et al.*, 2010). In this complex, grains are not produced, due to the bacterial infection of the floral parts. However, partially diseased ears may produce some wrinkled grains, but in such grains galls are not formed (Paruthi *et al.*, 1987). Sahebani *et al.* (2007) reported the attachment of numerous cells of *R. tritici* on the body and these may cause weakness and death to the nematode larvae. The fresh cockles containing the bacterium cause toxicity to cattle and sheep which feed upon them (Anwar *et al.*, 2001).

Anguina tritici

The *Anguina tritici* was the first nematode found parasitizing plants (Needham, 1743). The nematode incites ear cockle disease in wheat and barley (Haque and Khan, 2021). This nematode was named *Vibrio tritici* by Steinbuch (1799) and later Chitwood (1935) revived the genus *Anguina*, originally proposed by Scopoli (1777). The seed gall nematode has been documented in around 100 countries worldwide, but it has been successfully controlled using clean and certified seeds (Khan, 2024). At present it occurs in isolated pockets in the countries where certified seeds are not used (Haque and Khan, 2021). In northern India, ear cockle disease appears sporadically in pockets especially in tribal areas (Nandal *et al.*, 2010; Khan *et al.*, 2023b). Apart from its primary host, i.e. wheat, other known hosts include spelt, rye, emmer, barley and triticale (Paruthi and Bhatti, 1992). The life cycle of *A. tritici* synchronizes with the development of wheat plants. The seed galls containing nematode juveniles in the anhydrobiotic quiescent state are the source of the nematode, which reaches the field whenever cockles contaminated wheat seeds are sown (Khan and Athar, 1996). The cockles become soft when they absorb soil moisture. The quiescent juveniles also revive and become active. They migrate out of the cockles into the soil within 2 weeks.

The infective juveniles (J_2) ascend to the growing point of the seedlings and remain ectoparasitic until the initiation of floral primordia. The majority of J_2 can be seen on the inner-most layer of the leaf scale (Paruthi *et al.*, 1990) and invade the floral tissues, where flower primordia are formed. This also initiates the development of galls in place of normal grains. At this stage, the nematodes start feeding endoparasitically on the floral initials and become

Fig. 10.1. (A) Wheat plant earheads showing crinkling and twisting of leaves due to ectoparasitic feeding by *Anguina tritici*. (B) Seed galls and healthy grains of wheat. (C) Torn seed gall showing the mass of quiescent larvae of *A. tritici* under a stereomicroscope. [Figure courtesy of M.R. Khan, Aligarh Muslim University, India.]

enclosed in the cavities formed by over-growth of floral initials which eventually are turned into cockles or galls. Three successive moltings take place within 3–5 days of endoparasitic feeding and the juveniles become adults in the ratio of 1:2 male:female (Midha and Swarup, 1972). Reproduction is amphimictic and egg laying starts within 6–12 days of mating. A female may lay up to 2000 eggs. The eggs hatch quickly giving rise to second-stage juveniles (Khan,

2008). This stage coincides with the drying and maturing of wheat plants, which force the juveniles to undergo quiescence for anhydrobiosis. The nematode completes only one generation in a growing season (Nandal *et al.*, 2010).

Rathayibacter tritici

The bacterium is soil-inhabiting, Gram-positive and causes 'tundu' or spike blight or yellow slime disease in wheat (Fig. 10.2). Earlier, the bacterium was named as *Corynebacterium tritici* (Hutchinson, 1917). Later, Carlson and Vidaver (1982) and other researchers proposed that the bacterium should be designated as *Clavibacter tritici* (Davis *et al.*, 1984; Riley and McKay, 1990). Zgurskaya *et al.* (1993) proposed the genus *Rathayibacter* to accommodate three species of *Clavibacter*, one of them was *C. tritici*.

Relationship between the bacteria and nematode

When *A. tritici* and *R. tritici* cohabit on wheat plants, 'tundu' syndrome develops under ideal conditions of 60–85% relative humidity and 18–24°C temperature. The development of bright yellow slime or a gummy appearance on the leaf surface of immature plants, as well as of the abortive ears and leaves of maturing plants, are characteristic symptoms of the disease. Either the sprouting spikes appear narrow and small, or the infected ears do not erupt at all. The grains, partially or fully, are replaced by bacterial deposition. The disease development is influenced by temperature, humidity, seedling age and the source of the galls as well as the inoculum level of the nematode and bacteria (Khan and Pathak, 1993).

The juveniles (J_2) of *A. tritici* carry the bacterial cells from the soil to the plant or from the plant surface to shoot meristem. The nematode is necessary for the manifestation of 'tundu.' The inoculum of *R. tritici* without the aid of the nematode is not able to reach the shoot meristem, hence the disease does not develop in the absence of the nematode. To prove this, Gupta and Swarup (1972) utilized surface-sterilized juveniles of *A. tritici* as inocula, but 'tundu' did not develop. Additionally, the 'tundu' bacterium was isolated from *A. tritici* juveniles that were taken from dry galls. The bacterial cells adhered to the nematode body and were carried to the growth points of wheat plants or from the soil or plant surface.

The shoot meristem is tightly covered by leaf scales and bud scales which make it inaccessible to non-motile bacterial cells (Zgurskaya *et al.*, 1993). The bacterial cells riding over the second-stage juveniles are able to reach the apical meristem because the nematode juveniles also attack the developing flower buds which are very close to the tissue susceptible to the bacteria. Gokte and Swarup (1988) demonstrated that the bacteria can also come in contact with the juvenile nematodes inside the galls and may become infested. In the wheat seed galls, Pathak and Swarup (1984) discovered a 40–55% occurrence of the bacteria. They also found that the bacterium did not originate from the field soil and that *A. tritici* and *R. tritici* have an obligatory link for cohabitation in the gall. As a result,

Fig. 10.2. (A) Symptoms of yellow ear rot or 'tundu' disease caused by *Rathayibacter tritici* in wheat. (B) Gumming disease of ryegrass due to *R. toxicus*. [Figure courtesy of NASI, India and J. Allen.]

the juveniles that contain bacterial cells emerge from the galls. Gokte and Swarup (1988) also discovered many bacterial cells adhering to the body surface of nematode juveniles. They also provided proof that the nematode body did not contain any bacterial spores internally. Later, Riley and McKay (1990) also established that the plant-pathogenic bacteria only adhere to the external cuticle of nematodes.

Annual ryegrass toxicity

Annual ryegrass toxicity (ARGT) is a disease complex formed due to the vector relationship of the nematodes *Anguina agrostis* or *A. funesta* and the bacterium *Rathayibacter toxicus* (= *Corynebacterium rathayi*, *Clavibacter toxicus* and *Corynebacterium toxicus*) in annual ryegrass (*Lolium rigidum*). Since toxicity symptoms develop in cattle and sheep which graze on the infected ryegrass, the disease is called ARGT. The nematode incites the seed galls, in which the bacteria produce toxin, as a result of which the seed galls become toxic to the animals when they consume the galls (Bird and Riddle, 1984). The cells of *R. toxicus* are carried on the external cuticle of the nematode juveniles. Upon contact, the cells of *R. toxicus* adhere to the nematode cuticle surface (Bird and Riddle, 1984). In Oregon, USA it was found that cattle, horses and sheep when grazed on *Festuca nigrescens* infested with *A. agrostis* galls exhibited nervous disorders, muscle trembling and lack of coordination, which led to death in some cases (Shaw and Muth, 1949). Riley and Ophel (1992) reported the death of livestock in Southern Australia due to ARGT.

Anguina agrostis and *Anguina funesta*

Annual ryegrass toxicity was attributed to the two nematode species *A. agrostis* and *A. funesta* along with the bacteria *R. toxicus*. Both nematode species are associated with ARGT independently (Haque and Khan, 2021). Morphologically, *A. agrostis* and *A. funesta* can be distinguished from *A. tritici*, but are similar to each other and are identified by biochemical tools such as isozymes (Riley *et al.*, 1988) and molecular analyses (Powers *et al.*, 2001; Li *et al.*, 2015). Taxonomic details of these two nematodes can be found in the *Handbook of Invasive Plant-Parasitic Nematodes* (Haque and Khan, 2021).

Rathayibacter toxicus

The bacteria *R. toxicus* is known to cause ARGT. Previously the bacterium was classified as *Corynebacterium rathayi* (Bird, 1981) and *Clavibacter rathayi* (Davis, 1986). Later Riley and Ophel (1992) proposed a new species *Clavibacter toxicus*. However, Sasaki *et al.* (1998) reclassified the bacterium as *Rathayibacter toxicus* which belongs to the family Microbacteriaceae. The bacterium is known to infect the floral parts of Poaceae grasses, especially annual ryegrass (*Lolium rigidum*), annual blown grass (*Agrostis avenacea*), annual veldt grass (*Ehrharta longiflora*) and annual beard grass (*Polypogon monspeliensis*). The ARGT bacteria is Gram-positive, obligate aerobic, irregular rod shaped, ~1.1–2.0 µm long, 0.5–0.7 µm diameter with blunt and rounded ends. It is non-spore forming and non-motile bacterium possessing a 0.08–0.2 µm thick capsule around the cell which enables it to survive under hot and arid conditions or in the absence of host plants. Complete genomes of four strains of *R. toxicus* (WAC3373, 70137, DSM 7488 and FH142) have been sequenced with an average genome size of 2.325 MB, 61.5% GC content, 2165 total genes, 2069 protein coding genes, 54 RNA genes (45 tRNA, six rRNA and three other RNAs) and 42 pseudogenes.

Relationship between the bacteria and nematode

Ryegrass infection with *R. toxicus* does not take place in the absence of *A. agrostis* or *A. funesta*. The bacterium infects annual ryegrass and a mass of bacteria covers the glumes and fills the spikelets (Khan and Pathak, 1993). The bacteria-colonized seed galls have a vivid yellow appearance (Fig. 10.2). The gall becomes poisonous to animals because it contains a toxin, a class of glycolipid poisons (Lanigan *et al.*, 1976). According to Stynes *et al.* (1979), the plant portion of the gall is 20–30 times more poisonous than the bacterial part. The J_2 can acquire bacterial cells from the soil or from infected plants during the movement in the soil or on the plant surface. The bacterial cells (spores) adhere to the external cuticle of the nematode body. The spore-encumbered juveniles move between the compact leaves and leaf sheath to approach the shoot meristem where floral primordia are formed. The nematode juveniles invade the young flower buds, and at this point of nematode pathogenesis, the bacterial spores (cells) reach the developing shoot tip in a similar way as seen in 'tundu' disease of wheat (Khan, 2024). Later, the bacterial cells enter into the galls by surface contamination. The seed galls containing the bacteria have toxins fatal to grazing livestock. According to Bird *et al.* (1980), bacteria that colonized seed galls were found inside the galls as well as along the wall of the cells that made up the gall walls. These bacteria were closely packed in a regular array. They believed that the survival of infectious juveniles and bacteria under harsh environmental conditions depended on their close proximity and anhydrous state. The second-stage juveniles undergo quiescence whereas the bacteria form capsules to survive under unfavorable conditions.

In the next season, when cockles are sown along with healthy seeds or already present in the soil, the nematode juveniles migrate out from the bacteria colonized galls, and the bacterial cells stick to the nematode cuticle. The juveniles move to the seedlings, feed on the growing points and form galls in the floral parts. The *R. toxicus* cells grow quickly inside the galls and fill them with the spores. The nematode cuticle and bacterial capsule interact specifically during adhesion, which involves a significant attractive force (Bird, 1981). Riley and McKay (1990) ascertained a high degree of specificity of *R. toxicus* with the nematode. The compatibility of nematode populations and bacterial strains is considered a significant factor in determining the degree of adherence.

Aphelenchoides–Bacteria

Cauliflower disease of strawberry

The term cauliflower in this disease refers to the symptom that resembles the cauliflower crown. The name 'cauliflower' was first used by Ritzema-Bos (1891) to characterize the shortening and thickening of the aerial parts of strawberry plants, as well as the deformity of the inflorescence. In severe situations, blooms become malformed and plants become stunted. The petals stop growing or shrink to a greenish size and the sepals grow significantly. Fruits do not develop because of the deformed stamens and receptacles. In the crowns, axillary buds are constantly being formed (Khan and Pathak, 1993).

It has been established that both the nematode, *Aphelenchoides fragariae* or *A. ritzemabosi*, and the bacterium, *Rhodococcus* (= *Corynebacterium*) *fascians*, are necessary to produce cauliflower disease of strawberry (Crosse and Pitcher, 1952). Since *A. fragariae* was usually present in the affected strawberry plants, it was assumed that this nematode is the primary cause of disease. Later etiological investigations revealed that *A. fragariae* was not exclusively responsible for the disease (Crosse and Pitcher, 1952). Lacey (1942) hinted that the syndrome linked to a bacterium, *R. fascians*. However, her studies were unable to ascertain the role of *A. fragariae* in the cauliflower syndrome. Pitcher (1965) described cauliflower disease as a complex involving the nematode and specific bacterial strains, with both organisms actively contributing to the manifestation of the disease. *R. fascians* alone causes secondary crowns and galls (Sitaramaiah and Pathak, 1993). In Russia, the virulent strains of *R. fascians* along with the nematode incited cauliflower symptoms, whereas the less virulent strains caused alamination of leaves or red plant disease (Drozdovskii *et al.*, 1971).

Aphelenchoides *species*

A. ritzemabosi and *A. fragariae* serve as vectors of *R. fascians*, and are essential for the formation of strawberry cauliflower disease complex (Khan *et al.*, 2023a). The nematode is an obligate parasite of foliar parts of the plant, and feeds ectoparasitically on buds, apical meristem, leaf tips and other actively growing parts, and endoparasitically on mesophyll cells inside the leaves (Westerdahl and Bello, 2023). Strawberry is a major host of *A. fragariae*. During the active phase of parasitism, the nematode larvae can be seen within the folded

Fig. 10.3. Symptoms of crimp disease of strawberry caused by *Aphelenchoides fragariae*. [Figure courtesy of The Food and Environment Research Agency, UK.]

crown and runner buds, causing crimp or spring dwarf in strawberries (Franklin, 1950) (Fig. 10.3). The infected strawberry plants exhibit twisted and puckered leaves, discolored patches with a rough surface, dwarf leaves with crinkled margins, reddening of petioles in some cultivars, short internodes of runners, diminished flower trusses and death of the crown buds. Numerous ferns and angiosperms are among the wide range of hosts of *A. fragariae*. The nematode can successfully move from strawberries to ferns and from ferns to strawberries, and has also been reported from the USA, Japan and several European countries (Westerdahl and Bello, 2023).

Rhodococcus fascians

R. fascians is a Gram-positive, non-sporulating, aerobic, pleiomorphic and non-motile bacterium of the Actinomyces. The bacterium alone causes leafy gall disease in both dicotyledonous and monocotyledonous host plants. Earlier this bacterium was known as *Corynebacterium fascians*, but Goodfellow (1984) re-described it as *R. fascians*. The bacterium is widespread, and commonly persists in healthy strawberries as saprophytes (epiphyte). Only a small number of strains can cause cauliflower symptoms, and for the disease to manifest, the nematode must transport these strains to the meristem of a crown bud (Khan and Pathak, 1993). The virulent strain of *R. fascians* contains the genes located on a plasmid, whereas the non-virulent strains lack a plasmid. The transcription of the genes can be induced on cultures containing certain carbon sources such as glucose, sucrose, arabinose, glycerol, pyruvate, mannitol, mannose or histidine nitrogen sources.

Relationship between the bacteria and nematode

Lacey (1936) was the first to attribute *R. fascians* to cauliflower disease in strawberry. However, the etiological studies of Crosse and Pitcher (1952) proved that the cauliflower symptoms appear when the strawberry plants are infected with *A. ritzemabosi* and specific strains of *R. fascians*, while the strawberry plants that had only been infected by nematodes (*A. ritzemabosi* or *A. fragariae*) display alaminate leaves with a very short petiole and decreased lamina. Further, they demonstrated two ways of formation of the cauliflower-like crowns in strawberry plants. The crown may either be formed directly from a secondary crown bud or by trailing behind alaminate leaves. The symptoms of alaminate leaves are exclusively caused by the nematode, which acts as the main pathogen and continues to produce alaminate leaves even in the absence of *R. fascians*. In the absence of nematode, the bacterium is not able to cause any abnormalities. The nematode and bacterium are required to be associated for the development of cauliflower symptoms (Pitcher and Crosse, 1958).

Ditylenchus–Bacteria

"Café au lait" bacteriosis

Caubel and Samson (1984) reported that the stem nematode *Ditylenchus dipsaci* and *Pseudomonas salomonii* (= *Pseudomonas fluorescens*) are jointly involved in the manifestation of 'Café au lait' bacteriosis in garlic. The nematode pathogenesis promotes the access of *P. salomonii* into the garlic bulbs. However, the diseased plants did not always confirm the association of *D. dipsaci*. Further, the severity of browning of bulbs (an indicator symptom of infected bulbs) was not linearly correlated with 'Café au lait' bacteriosis symptoms in the plants developed from infected bulbs. Jacques *et al.* (2009) observed that treatments with soil fumigants or other nematicides did not affect the bacterial infection in garlic. Gardan *et al.* (2002) characterized *Pseudomonas* garlic and rice strains using a polyphasic taxonomic approach using 16S rRNA gene sequences. The garlic strain was characterized as *P. salomonii* whereas the rice strain was characterized as *P. palleroniana*. The garlic leaves turned yellow and wilted due to infection of *P. salomonii*, and the foliage part collapsed and died. However, in some cases, the bulbs appeared normal but when dried, some papery leaves exhibited discoloration and browning (Sawada *et al.*, 2020) (Fig. 10.4).

Bacterial wilt of alfalfa (*Clavibacter michiganensis* subsp. *insidiosus*)

The stem and bulb nematode *Ditylenchus dipsaci* also transmits another plant-pathogenic bacterium, *Clavibacter michiganensis* subsp. *insidiosus*. (= *Corynebacterium insidiosum*), into the crown buds of alfalfa plants, increasing the incidence of bacterial wilt (Fig. 10.5).

The bacterium is Gram-positive, aerobic, non-sporulating with coryneform morphology and lacks a type III secretion system. The cell wall is a B2γ type. The complete genome of *C. michiganensis* has been sequenced (Gartemann *et al.*, 2008). The bacterium uses hydrolysis products as carbon and energy sources, suggesting that the cells can survive in the soil on the decaying host material (Gartemann *et al.*, 2008). Hawn (1963) discovered that bacterial wilt of alfalfa was often associated with *C. michiganensis* subsp. *insidiosus* as well as *D. dipsaci* in southern Alberta, Canada. The nematode *D. dipsaci* is the main means of bacteria transfer into the host. Hawn (1971) isolated *C. michiganensis* subsp. *insidiosus* from homogenized nematode bodies and extracted from

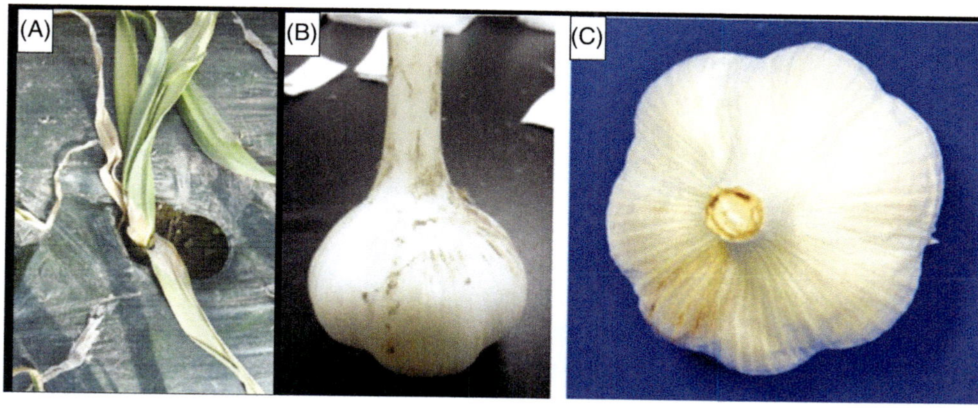

Fig. 10.4. Symptoms of 'Café au lait' bacteriosis of garlic caused by *Pseudomonas salomonii* (= *P. fluorescens*). (A) Yellowing and wilting in the infected plant. (B) Normal appearing bulbs from an infected plant. (C) Plant when dried showing discoloration and browning in some papery leaves. [Figure courtesy of Hiroyuki Sawada.]

the buds of alfalfa plants afflicted with the wilt disease. Since lumen diameters of the adult male and female nematodes were too small to permit the ingestion of mature bacterial cells, it was thought that the nematodes carried *C. michiganensis* subsp. *insidiosus* on the external cuticle, which was proved by culturing of surface-disinfected and non-disinfected nematode larvae on artificial media. No colonization of the bacteria from the surface-sterilized larvae ruled out the internal transmission of the bacterial cells (Khan and Pathak, 1993). Hawn (1971) showed that nematodes exposed to a suspension of the bacteria had a large number of bacterial cells adhering to their external cuticles.

Other Bacteria Vectored by Nematodes

There are few other nematode–bacterial disease complexes where nematodes serve as vector of plant-pathogenic bacteria that are less explored and lack sufficient details (Table 10.1). Riedel and Larsen (1974) observed that bacterial leaf spot disease caused by *Xanthomonas begonia* became highly severe, and the disease progressed rapidly in cohabitance with *A. fragariae* on Rieger begonias. Similarly, *Bacterium rhaponticum*, the causative agent of rhubarb crown rot, was able to invade into the host with the vector role by *D. dipsaci* (Metcalfe, 1940). *Ralstonia* (= *Pseudomonas*) *solanacearum* in potatoes is reported to be spread by *Globodera pallida* (Jensen, 1978). It was assumed that the stylet-bearing nematodes can engulf bacterial cells and introduce them intact into plant cells through fecal matter. However, this mode of transmission is unlikely to occur because the majority of tylenchids have stylet lumen of 0.2–1.0 µm in diameter, which is too narrow to allow passage for most plant-pathogenic bacteria. In another report, *Tylenchus* and *Aphelenchus* species were found to be associated with *Xanthomonas phaseoli* and *Erwinia carotovora*, respectively, on the rubber plants *Taraxacum koksaghys* and *Scorzonera tau* (Kalinenko, 1936). This research claimed that these bacteria might infect rubber plants because they were transported both internally as well as externally by the nematodes.

According to Wasilewska and Webster (1975), female *Pristionchus iheritieri* consumed *A. tumefaciens* and kept them alive for up to 27 hour. Several phytobacteria such as *Pseudomonas phaseolicola*, *P. syringae*, *Erwinia amylovora* and *E. carotovora* are reported to be transported by the free-living nematodes. Poinar and Hansen (1986) reported that the nematode *Cruznema tripartitum* (= *Rhabditis lamdiensis*)

Fig. 10.5. Wilt symptoms in alfalfa caused by *Clavibacter michiganensis* subsp. *insidiosus*. [Figure courtesy of Jim Sheppard, CABI.]

transmitted *Pseudomonas tolasii* which causes blackspot diseases in mushrooms.

Nematodes as Vectors of Plant-Pathogenic Fungi

There are some plant-pathogenic fungi which need nematodes as a vehicle to reach the susceptible tissue. However, the vector role is limited to a few centimeters from the soil or plant surface to the susceptible tissue of the host plant.

Anguina–Dilophospora

The ear twist disease caused by *Dilophospora alopecuri* is an important disease in cereals such as wheat, barley, oats and a number of grasses (Asad *et al.*, 2007). The disease is wide and prevalent throughout Europe, the

Table 10.1. Bacterial diseases vectored by nematodes.

Nematode	Bacteria	Disease and crop	Mode of adherence	Status	Reference
Anguina tritici	*Rathayibacter tritici*	Yellow ear rot or 'tundu' disease of wheat	External cuticle	Confirmed	Hutchinson, 1917
Anguina agrostis, *Anguina funesta*	*Rathayibacter toxicus*	Annual ryegrass toxicity	External	Confirmed	Bird *et al*., 1980
Aphelenchoides fragariae, *Aphelenchoides ritzemabosi*	*Rhodococcus fascians*	Cauliflower disease of strawberry	External	Confirmed	Crosse and Pitcher, 1952
Ditylenchus dipsaci	*Pseudomonas salomonii*	'Café au lait' bacteriosis of garlic	External	Confirmed	Caubel and Samson, 1984
Ditylenchus dipsaci	*Clavibacter michiganensis* subsp. *insidiosus*	Wilt of alfalfa	External	Confirmed	Hawn, 1971
Aphelenchoides fragariae	*Xanthomonas begonia*	Rieger begonias	–	Suspected	Riedel and Larsen, 1974
Ditylenchus dipsaci	*Bacterium rhaponticum*	Rhubarb crown rot	–	Suspected	Metcalfe, 1940
Globodera pallida	*Ralstonia solanacearum*	Potato wilt	–	Suspected	Jensen, 1978

USA, India and Australia. The disease is quite common among the members of the Poaceae family (Paruthi and Bhatti, 1990). The diseased plant first exhibits rolling and spiraling of young leaves. The upper leaves of plants initially develop small, round or oblong bright color spots or blotches. The center of the spots shows black specks containing one or more pycnidia. The pycnidia soon spread and grow larger until the entire blotch is covered. The blotches are prominently noticeable on the underside of the leaf. The diseased leaves wither and turn yellow. The fungus may also infect the leaf sheaths and the young leaves (Asad *et al.*, 2007). Affected plants stay extremely stunted, without ears and die later. The ears may remain attached to the top leaf sheath, if infection occurs after the ear formation. The ears start to twist and grow abnormally. There could be partial or complete ear malformation (Fig. 10.6).

Dilophospora alopecuri

The description of *D. alopecuri* was provided by Atanasoff (1925) and re-described by Walker and Sutton (1974). The pycnidia of the fungus are brown, globose, glabrous, unilocular, very rarely multilocular. The middle ostiole is frequently papillate. Pycnidia are grouped into rows between the veins in more mature leaf lesions and leaf sheaths. Conidiogenous cells can be enteroblastic, phialidic, separate or grouped into short conidiophores. They originate straight from the inner cells of the complete pycnidial wall. Conidiophores are conidiogenous, meaning that their individual cells produce conidia one after the other. Conidia have multiple, simple to dichotomously or irregularly branched appendages. They are hyaline, smooth, cylindrical or tapering towards the base. When conidia reach their maximum size, septa and

Fig. 10.6. Ear twist symptoms on (A) young and (B) mature wheat plants caused by *Dilophospora alopecuri*. The blackish colonization of the fungal spores (pycnidia) can be seen on the ear heads and panicle. [Figure courtesy of M.R. Khan, Aligarh Muslim University, India.]

appendages are formed (Walker and Sutton, 1974).

Relationship between the fungus and nematode

The pathogenicity of *D. alopecuri* on wheat and barley depends on the co-occurrence of *A. tritici*. Fungal conidia are carried by the nematode to the growing tips of the plant (shoot meristem and apical meristem). The conidia attach to the nematode cuticle with the help of appendages. The conidia cannot reach the apical meristem by themselves because this tissue is well protected by tightly arranged leaf scales and leaf sheaths. The nematode juveniles carrying fungal conidia move between the leaf sheaths, and eventually reach to apical meristem and young ears of the plant. The conidial appendages or bristles aid in the attachment to the nematode body. The nematode juveniles are necessary for accessing the conidia to the shoot meristem of the plant. Conidia can be seen on the nematode's body under a microscope. Bird and McKay (1987) investigated the attachment of *D. alopecuri* conidia to the cuticle of *A. agrostis* (or *A. funesta*) and found that their attachment was caused by the adhesin receptor pathway. On fungus-infected ears, the nematodes are unable to produce seed galls; if they do, the fungus eventually colonizes and destroys the galls. The nematodes may also perish if the twist disease develops in a severe form.

Other fungi vectored by nematodes

Some other fungi, such as *Tilletia foetida*, *Ustilago tritici*, *U. nuda*, *Neovossia indica* and *Alternaria triticina*, are suspected to be transmitted by *A. tritici* (Nandal *et al.*, 2010). However, this interaction forms non-essential associations, as these fungi can also cause disease without the aid of the nematode. In such associations, the relationship between both the nematode and fungus has not been thoroughly studied (Table 10.2). Paruthi *et al.* (1990) found that the cockles of wheat infected with *A. triticina* inhibited the development and reproduction of *A. tritici*. Jensen (1967) proved that *Fusarium*

Table 10.2. Fungal diseases vectored by nematodes.

Nematode	Fungus	Disease and crop	Mode of adherence	Status	Reference
Anguina tritici	*Dilophospora alopecuri*	Twist disease of wheat and barley	External cuticle	Confirmed	Bird and McKay, 1987
Anguina tritici	*Tilletia laevis* (= *T. foetida*)	Bunt disease of wheat	External	Suspected	Nandal *et al.*, 2010
Anguina tritici	*Ustilago tritici*, *Ustilago nuda*	Smut disease of wheat	External	Suspected	Nandal *et al.*, 2010
Anguina tritici	*Neovossia indica*	Karnal bunt disease of wheat	External	Suspected	Nandal *et al.*, 2010
Anguina tritici	*Alternaria triticina*	Leaf blight of wheat	External	Suspected	Paruthi *et al.*, 2009

and *Verticillium* spores could withstand transit through a microbiotrophic movement via the alimentary canal of the nematodes.

Management of Diseases Vectored by Nematodes

The management of multipathogenic diseases is difficult because of the lack of information on all the pathogens involved in the disease. The disease complexes emerging through the vector role of nematodes usually cause prominent symptoms of the secondary pathogen. For this reason, bactericide or fungicide is applied which does not give proper control of the disease. Hence, correct diagnosis of all the participating pathogens in the disease is essential for proper management of the disease complexes. Hence, the treatments should target both the nematode and the bacteria/fungi involved in the disease. Further, different management options may be integrated to achieve sustainable management of the disease. Some of the important methods are summarized below.

Physical and cultural control

Physical and cultural control includes soil solarization, seed-cleaning techniques, hot water treatment, use of disinfested planting material, summer ploughing, host resistance, trap cropping, intercropping, etc. (Khan *et al.*, 2012). These methods may help in suppressing and preventing the spread of the pathogens by reducing the nematode population in the soil as well as the in the planting materials. Use of certified or disinfested seeds or planting material is strongly recommended. For example, in the case of 'tundu' disease, both the pathogens (*A. tritici* and *R. tritici*) are seed-borne in nature, and the use of disinfested seeds or certified seeds (free from cockles) will prove extremely effective in growing the disease-free crop even in disease-prone areas. In Western regions such as Europe and the USA, seed gall disease has been eliminated by adopting crop rotation and seed-cleaning techniques (Brown, 1987). There are effective mechanical sieving and seed flotation techniques to separate cockles from healthy seeds (Paruthi *et al.*, 2009; Nandal *et al.*, 2010). Similarly, hot water treatment or use of microwaves are good techniques to disinfest the seeds (Strümpel, 1967; Khan and Hayat, 1999).

Chemical control

Foliar spraying with 0.05% ethyl parathion (Singh and Prasad, 1972) or monocrotophos (1 kg/ha) during the vegetative stage (Khan *et al.*, 1998) partially suppressed the seed gall in wheat varieties. However, Paruthi *et al.* (2009)

achieved successful control by applying carbofuran (1 kg a.i./ha) at 50 days age, and recorded 18.3% incidence of seed gall vs 46.7% in control. Than *et al.* (1998) attained some success against ARGT in grazing cattle by administering tunicamycin conjugated to a carrier protein for inducing toxin immunity. However, chemical treatments are costly and toxic, their application may not prove economic and sustainable against AGRT in the large grazing areas.

Application of aldicarb in September or October effectively controlled *A. fragariae* and *A. ritzemabosi* in strawberry runners grown for the production of planting materials (runners) in southern Germany (Blank, 1985). Dipping fresh strawberry plants in thionazin before freeze preservation virtually eradicated *A. fragariae* infection (Tacconi *et al.*, 1982). The soaking of *A. fragariae*-infested planting materials of strawberry cv. Redgauntlet and Tioga in fenamiphos solution greatly controlled the disease and gave significantly higher yields (Soteros, 1985). Two foliar applications of oxamyl, with a 2-week interval, provided excellent control of *A. fragariae* in two begonia cultivars, and completely checked the symptom development (Hansen *et al.*, 1972).

Chemical control strategies for bacterial diseases vectored by nematodes involve the topical applications of various compounds, such as copper-containing compounds, or an antibiotic (gentamicin, oxytetracycline, streptomycin and oxolinic acid). Application of copper compounds significantly prevented the infection of *R. fascians* in strawberry plants (Cornelis *et al.*, 2001).

Biological control

The biological control of plant diseases is a sustainable method based on ecological principles. For example, high occurrence of *D. alopecuri* in Western Australia is attributed to displacement of *C. toxicus*, leading to significant ARGT decline in the region (Riley, 1994). There are a number of antagonistic microorganisms such as *Aspergillus niger*, *Purpureocillium lilacinum*, *Pochonia chlamydosporia*, *Pasteuria penetrans*, *Trichoderma* spp., etc. (Jatala, 1986; Stirling, 1991; Kerry, 2000; Khan, 2016) and phosphate solubilizing microorganisms such as *Pseudomonas*, *Bacillus*, *Penicillium*, etc. (Stirling, 1991; Khan *et al.*, 2009, 2016a, 2016b), which may suppress phytonematodes and plant-pathogenic fungi and bacteria. These antagonists, if applied singly or in consortium at the time of planting, may help in achieving sustainable management of disease complexes (Khan, 2007; Khan and Anwer, 2011; Sikora and Roberts, 2018; Khan *et al.*, 2019, 2021). Multifaceted beneficial microbes such as *Trichoderma* spp. (Mohiddin *et al.*, 2010; Khan and Mohiddin, 2018) or *Pseudomonas* (Stirling, 1991; Khan and Mohiddin, 2023) may play an important role in achieving sustainable management of disease complexes because of their proven antagonism against plant-pathogenic fungi and bacteria along with the phytonematodes (Sarangi *et al.*, 2018; Mohammed and Khan, 2021). Commercial formulations of *T. virens*, *T. harzianum*, *P. fluorescens*, *P. putida*, etc., are available in the market and may be applied against disease complexes (Khan *et al.*, 2011; Rao *et al.*, 2014; Kumar *et al.*, 2018).

Limited research has been done on the use of biocontrol agents against *A. tritici* infecting wheat crop (Owen *et al.*, 2023). Gokte and Swarup (1988) observed that *Arthrobotrys oligospora* parasitized *A. tritici* larvae. Seed application of *T. harzianum* was found to fully colonize the cockles, inducing 100% juvenile mortality in *A. tritici* (Khan *et al.*, 2023b). The predatory nematode *Paractinolamus elongates* has demonstrated marked predation on *A. tritici* (Jairajpuri and Ahmad, 1992). Bilgrami *et al.* (2005) reported that *A. tritici* larvae were highly preferred as prey of *Mononchoides gaugleri*. Haraguchi and Yoshiga (2020) explored the potential of the fungivore nematode *Aphelenchus avenae* in parasitizing *Ditylenchus destructor* in garlic crop.

Integrated disease management

Various integrated disease management modules have been suggested for the management of disease complexes involving nematodes and other fungal or bacterial microbes (Naik *et al.*, 2015; Kavitha and Thirukumaran, 2019; Dhir *et al.*, 2020; Khan, 2023a, Khan, 2023b). Sharma *et al.*, 2007 described that the combination of neem kernel powder (0.3%), carbofuran (0.5 kg/ha) and phorate (1 kg/ha) greatly reduced the fungus–nematode disease complex. Similarly, microbial consortia along with carbofuran or bavistin and neem seed powder or FYM were reported highly effective in suppressing the complex diseases in several crops (Kumar *et al.*, 2009; Rao *et al.*, 2017; Priti *et al.*, 2018).

Conclusion and Future Prospects

Nematodes play an important role as vector for transmitting some plant-pathogenic fungi and bacteria. In most such cases, the secondary pathogens on their own are incapable of reaching the susceptible host tissue. This is especially true when novel disease syndromes are formed where the host tissue susceptible to the fungus or bacteria is protected by tight leaf scales and/or bud scales, as is the apical shoot meristem in wheat. Although in a few cases the mechanisms of infestation and transmission have been well studied, there are still certain aspects which need to be fully understood. The ear cockle phase in 'tundu' complex is critical for nematodes as well as bacteria. During this phase alone, seed galls are generated. 'Tundu' bacteria are present on the external cuticle of *A. tritici* juveniles emerging from these galls. It is not known fully how the bacteria spread to the juveniles in the following generation. Further, information on the molecular and genetic basis of the compatibility between the nematode population and bacteria isolates that determines the successful attachment is not available. Little is known about the role of physiological and histological changes brought about by nematode feeding in the development of bacterial infection. It seems that the vector relationships is advantageous only to the bacteria or fungi. Nematodes are found to be highly sensitive to the harmful effects of bacteria/fungi. Wheat tillers with *R. tritici* adversely impact the development of nematode galls, indicating that the bacteria cause a negative impact on the nematode, but its exact mode of action is not fully understood. Substantial research and development efforts are required to target management aspects of the disease complexes. Breeding programs should consider the intricate relation of the nematodes and other pathogens.

References

Anwar, S.A., McKenry, M.V., Riaz, A. and Khan, M.S.A. (2001) Evaluation of wheat cultivars for *Anguina tritici* resistance, development and influence of nematode on wheat growth. *International Journal of Nematology* 11, 150–156.

Asad, S., Sultan, A., Iftikhar, S., Munir, A., Ahmad, I. *et al.* (2007) First report of dilophospora leaf spot (twist) disease of wheat in Pakistan. *Pakistan Journal Botany* 39(4), 1387–1389.

Atanasoff, D. (1925) The diplophospora disease of cereals. *Phytopathology* 15, 11–40.

Bilgrami, A., Gaugler, R. and Brey, C. (2005) Prey preference and feeding behaviour of the diplogastrid predator *Mononchoides gaugleri* (Nematoda: Diplogastrida). *Nematology* 7(3), 333–342.

Bird, A.F. and Riddle, L.D. (1984) Effect of attachment of *Corynebacterium rathayi* on movement of *Anguina agrostis* larvae. *International Journal for Parasitology* 14(5), 503–511.

Bird, A.F. and McKay, A.C. (1987) Adhesion of conidia of the fungus *Dilophospora alopecuri* to the cuticle of the nematode *Anguina agrostis*, the vector in annual ryegrass toxicity. *International Journal for Parasitology* 17(7), 1239–1247.

Bird, A.F., Stynes, B.A. and Thomson, W.W. (1980) A comparison of nematode and bacteria-colonized galls induced by *Anguina agrostis* in *Lolium rigidum*. *Phytopathology* 70, 1104–1109.

Bird, A.F. (1981) The *Anguina–Corynebacterium* association. In: Zuckerman, B.M. and Rohde, R.A. (eds) *Plant Parasitic Nematodes Volume 3*. Academic Press, London, pp. 303–323.

Blank, W. (1985) Leaf and stem nematodes in strawberries – a serious problem. *Mitteilungen Des Obstbauversuchsringes Des Alten Landes* 40(6), 229–234.

Brown, R.H. (1987) Control strategies in low value crops. In: Brown, R.H. and Kerry, B.R. (eds) *Principles and Practice of Nematode Control in Crops*. Academic Press, Sydney, pp. 351–388.

Carlson, R.R. and Vidaver, A.K. (1982) Taxonomy of *Corynebacterium* plant pathogens, including a new pathogen of wheat, based on polyacrylamide gel electrophoresis of cellular proteins. *International Journal of Systematic Bacteriology* 32(3), 315–326.

Caubel, G. and Samson, R. (1984) Effect of the stem nematode *Ditylenchus dipsaci* on the development of "Cafe au lait" bacteriosis in garlic (*Allium sativum*) caused by *Pseudomonas fluorescens*. *Agronomie* 4, 311–313.

Chitwood, B.G. (1935) Nomenclatorial notes. *Proceedings of the Helminthological Society of Washington* 2, 51–54.

Cornelis, K., Ritsema, T., Nijsse, J., Holsters, M., Goethals, K. *et al*. (2001) The plant pathogen *Rhodococcus fascians* colonizes the exterior and interior of the aerial parts of plants. *MPMI* 14(5), 599–608.

Crosse, J.E. and Pitcher, R.S. (1952) Studies in the relationship of eelworms and bacteria to certain plant diseases. The etiology of strawberry cauliflower disease. *Annals of Applied Biology* 39, 475–486.

Davis, M.J. (1986) Taxonomy of plant-pathogenic coryneform bacteria. *Annual Review of Phytopathology* 24(1), 115–140.

Davis, M.J., Gillaspie, A.G., Vidaver, A.K. and Harris, R.W. (1984) Clavibacter: A new genus containing some phytopathogenic coryneform bacteria including *Clavibacter xyli* subsp. *xyli* sp. nov.; subsp. nov. and *Clavibacter xyli* subsp. *cynodontis* subsp. nov., pathogens that cause ratoon stunting disease of sugarcane and bermudagrass stunting disease. *International Journal of Systematic Bacteriology* 107–117.

Dhir, B.C., Panigrahi, D., Alim, M.A. and Samant, T.K. (2020) Integrated management of wilt and root knot nematode complex of brinjal in mid-central zone of Odisha. *Annals of Plant Protection Sciences* 28(2), 151.

Drozdovskii, E.M., Blinov, B.A. and Yakovleva, V.A. (1971) Study of *Aphelenchoides–Corynebacterium* complex. *Sbornik Nauchnykh Rabot Nauchno-Issledovatel'skogo Zonal'nogo Instituta Sadovodstva Nechernozemnoi Polosy (Plodovodstvo i Yagodovodstvo Nechernozemnoi Polosy) Moscow, USSR* 3, 393–402.

Franklin, M.T. (1950) Two species of *Aphelenchoides* associated with strawberry bud disease in Britain. *Annals of Applied Biology* 37(1), 1–10.

Gardan, L., Bella, P., Meyer, J.M., Christen, R., Rott, P. *et al*. (2002) *Pseudomonas salomonii* sp. nov., pathogenic on garlic, and *Pseudomonas palleroniana* sp. nov., isolated from rice. *International Journal of Systematic and Evolutionary Microbiology* 52(6), 2065–2074.

Gartemann, K.-H., Abt, B., Bekel, T., Burger, A., Engemann, J. *et al*. (2008) The genome sequence of the tomato-pathogenic actinomycete *Clavibacter michiganensis* subsp. *michiganensis* NCPPB382 reveals a large island involved in pathogenicity. *Journal of Bacteriology* 190(6), 2138–2149.

Gokte, N. and Swarup, G. (1988) On the association of bacteria with larvae and galls of *Anguina tritici*. *Indian Journal of Nematology* 18, 313–318.

Goodfellow, M. (1984) Reclassification of *Corynebacterium fascians* (tilford) Dowson in the genus *Rhodococcus*, as *Rhodococcus fascians* comb. nov. *Systematic and Applied Microbiology* 5(2), 225–229.

Gupta, P. and Swarup, G. (1972) Ear-cockle and yellow ear-rot disease of wheat. II. Nematode bacterial association. Nematodes as vectors of plant pathogens. *Nematologica* 18, 320–324.

Hansen, T., Rasmussen, A.N. and Schadegg, E. (1972) Trials with plant protection materials in fruit growing and nurseries 1970. *Tidsskrift for Planteavl* 76(1), 77–104.

Haque, Z. and Khan, M.R. (2021) *Handbook of Invasive Plant-Parasitic Nematodes*. CAB International, Wallingford, UK.

Haraguchi, S. and Yoshiga, T. (2020) Potential of the fungal feeding nematode *Aphelenchus avenae* to control fungi and the plant parasitic nematode *Ditylenchus destructor* associated with garlic. *Biological Control* 143, 104203.

Hawn, E.J. (1963) Transmission of bacterial wilt of alfalfa by *Ditylenchus dipsaci* (Kühn). *Nematologica* 9(1), 65–68.

Hawn, E.J. (1971) Mode of transmission of *Corynebacterium insidiosum* by *Ditylenchus dipsaci*. *Journal of Nematology* 3(4), 420–421.

Hutchinson, O.M. (1917) A bacterial disease of wheat in the Punjab. *Memoires of the Department of Agriculture in India. Bacteriological Series* 169–175.

Jacques, M.-A., Luçon, N. and Houdault, S. (2009) Clove-transmissibility of *Pseudomonas salomonii*, the causal agent of “Café au lait” disease of garlic. *European Journal of Plant Pathology* 124(4), 695–700.

Jairajpuri, M.S. and Ahmad, W. (1992) Dorylaimida: *Free Living, Predaceous and Plant-Parasitic Nematodes*. E.J. Brill, Leiden, The Netherlands.

Jatala, P. (1986) Biological control of plant-parasitic nematodes. *Annual Review of Phytopathology* 24(1), 453–489.

Jensen, H.J. (1967) Do saprozoic nematodes have a significant role in epidemiology of plant diseases? *Plant Disease Reporter* 51, 98–102.

Jensen, H.J. (1978) Interrelations of nematodes and other organisms in disease complexes. In: *International Potato Centre Report of the Second Planning Conference on the Developments in the Control of Nematode Pests of Potatoes*, International Potato Center, Lima Peru.

Kalinenko, V.O. (1936) The inoculation of phytopathogenic microbes into rubber bearing plants by nematodes. *Phytopathologische Zeitschrift* 9, 407–416.

Kavitha, K. and Thirukumaran, K. (2019) Integrated management of nematode disease complex in tuberose (*Polianthes tuberose* L). *Journal of Krishi Vigyan* 8(1), 170.

Kerry, B.R. (2000) Rhizosphere interactions and the exploitation of microbial agents for the biological control of plant-parasitic nematodes. *Annual Review of Phytopathology* 38, 423–441.

Khan, M.R. (1995) Fungus–nematode interactions. In: Swarup, G., Dasgupta, D.R. and Gill, J.S. (eds) *Nematode Pest Management – An Appraisal of Eco-Friendly Approaches*. Nematological Society of India, New Delhi, pp. 70–78.

Khan, M.R. (2007) Prospects of microbial control of root-knot nematodes infecting vegetable crops. In: Sharma, N. and Singh, H.B. (eds) *Biotechnology: Plant Health Management*. International Book Distributing, Biotechnology, pp. 643–665.

Khan, M.R. (2008) *Plant Nematodes – Methodology, Morphology, Systematics, Biology and Ecology*. Science Publishers, Enfield, NH. DOI: 10.1201/b10766.

Khan, M.R. (2016) Nematode biocontrol agents: Diversity and effectiveness against phytonematodes in sustainable crop protection. *Indian Phytopathology* 69(4s), 453–463.

Khan, M.R. (2023a) Plant nematodes, an underestimated constraint in the global food production. In: Khan, M.R. and Quintanilla, M. (eds) *Nematode Diseases of Crops and Their Sustainable Management*. Academic Press, Cambridge, MA, pp. 3–26.

Khan, M.R. (2023b) Nematode pests of agricultural crops, a global overview. In: Khan, M.R. (ed.) *Novel Biological and Biotechnological Applications in Plant Nematode Management*. Springer, Singapore, pp. 3–45.

Khan, M.R. (2024) Wheat seed-gall nematode: Going to archives or have local significance in the developing countries? *Indian Phytopathology* 77(1), 1–11.

Khan, M.R. and Anwer, A. (2011) Fungal bioinoculants for plant disease management. In: Paul, M., Clinton, M. and Ahmad, I. (eds) *Microbes and Microbial Technology*. Springer, Dordrecht, The Netherlands, pp. 447–488.

Khan, M.R. and Athar, M. (1996) Response of wheat cultivars to different inoculum levels of *Anguina tritici*. *Nematologia Mediterranea* 24, 269–272.

Khan, M.R. and Hayat, F. (1999) Effect of microwave irradiation on the ear cockle disease and yield of wheat. *Nematologia Mediterranea* 27, 225–231.

Khan, M.W. and Pathak, K.N. (1993) Nematodes as vectors of bacterial and fungal plant pathogens. In: Khan, M.W. (ed.) *Nematode Interactions*. Chapman & Hall, London, pp. 42–54.

Khan, M.R. and Mohiddin, F.A. (2018) Trichoderma: Its multifarious utility in crop improvement. In: Prasad, R., Gill, S.S. and Tuteja, N. (eds) *New and Future Developments in Microbial Biotechnology and Bioengineering: Crop Improvement Through Microbial Biotechnology*. Elsevier, Amsterdam, pp. 263–291.

Khan, M.R. and Mohiddin, F.A. (2023) Biocontrol strategies for nematode management, an overview. In: Khan, M.R. (ed.) *Novel Biological and Biotechnological Applications in Plant Nematode Management*. Springer, Singapore, pp. 113–131.

Khan, M.R., Athar, M. and Khan, B. (1998) Effect of foliar application of monocrotophos on the earcockle disease of wheat. *Nematologia Mediterranea* 26(2), 173–177.

Khan, M.R., Altaf, S., Mohiddin, F.A., Khan, U. and Answer, A. (2009) Biological control of plant nematodes with phosphate solubilizing microorganisms. In: Khan, M.S. and Zaidi, A. (eds) *Phosphate Solubilizing Microbes for Crop Improvement*. Nova Science Publishers, Hauppauge, NY, pp. 395–426.

Khan, M.R., Majid, S., Mohidin, F.A. and Khan, N. (2011) A new bioprocess to produce low cost powder formulations of biocontrol bacteria and fungi to control fusarial wilt and root-knot nematode of pulses. *Biological Control* 59(2), 130–140.

Khan, M.R., Ramakrishnan, S., Rajan, K. and Krishnamurthy, V. (2012) Cultural and physical control of nematode infestation in horticultural crops. In: Khan, M.R. and Jairajpuri, M.S. (eds) *Nematode Infestations Part III: Horticultural Crops. National Academy of Sciences*. Allahabad, India, pp. 529–551.

Khan, M.R., Mohiddin, F.A. and Ahamad, F. (2016a) Inoculant rhizobia suppressed root-knot disease, and enhanced plant productivity and nutrient uptake of some field-grown food legumes. *Acta Agriculturae Scandinavica, Section B – Soil & Plant Science* 68(2), 166–174.

Khan, M.R., Mohidin, F.A., Khan, U. and Ahamad, F. (2016b) Native *Pseudomonas* spp. suppressed the root-knot nematode in *in-vitro* and *in-vivo*, and promoted the nodulation and grain yield in the field grown mungbean. *Biological Control* 101, 159–168.

Khan, M.R., Mukhopadhyay, A.N., Pandey, R.N., Thakur, M.P., Singh, D. *et al.* (2019) *Bio-Intensive Approaches: Application and Effectiveness in Management of Plant Nematodes, Insects and Weeds*. Today & Tomorrow's Printers and Publishers, New Delhi.

Khan, M.R., Ahmad, I. and Shah, M.H. (2021) Emerging important nematode problems in field crops and their management. In: Singh, K.P., Jahagirdar, S. and Sarma, B.K. (eds) *Emerging Trends in Plant Pathology*. Springer, Dordrecht, The Netherlands, pp. 33–62.

Khan, M.R., Mohiddin, F.A., Haque, Z. and Sharma, R.K. (2023a) Disease complexes and their sustainable management. In: Khan, M.R. and Quintanilla, M. (eds) *Nematode Diseases of Crops and Their Sustainable Management*. Academic Press, Cambridge, MA, pp. 65–93.

Khan, M.R., Manzoor, S. and Ansari, M.S.A. (2023b) Effectiveness of *Trichoderma* species in controlling the seed-borne infestation of *Anguina tritici* in wheat seed-galls. *Indian Phytopathology* 76(4), 1083–1090.

Kumar, N.U., Ravichandra, N.G. and Nataraja, A. (2018) Ecofriendly management of wilt complex in black pepper (*Piper nigrum* L). *Indian Journal of Nematology* 48(2), 146–155.

Kumar, V., Haseeb, A. and Sharma, A. (2009) Integrated management of *Meloidogyne incognita–Fusarium solani* disease complex of brinjal cv Pusa Kranti. *Annals of Plant Protection Sciences* 17(1), 192–194.

Lacey, M.S. (1936) Studies in bacteriosis. XXII. I. The isolation of a bacterium associated with fasciation of sweet peas, cauliflower of strawberry plants and "leafy gall" of various plants. *Annals of Applied Biology* 23(4), 743–751.

Lacey, M.S. (1942) Studies in bacteriosis. XXV. Studies on a bacterium associated with leafy galls, fasciations and cauliflower disease of various plants. Part IV. The inoculation of strawberry plants with *Bacterium fascians* tilford. *Annals of Applied Biology* 29(1), 11–15.

Lanigan, G.W., Payne, A.L. and Frahn, J.L. (1976) Origin of toxicity in parasitized annual ryegrass (*Lolium rigidum*). *Australian Veterinary Journal* 52(5), 244–246.

Li, W., Zonghe, Y., Nakhla, M.K. and Skantar, A.M. (2015) Real-time PCR for detection and identification of *Anguina funesta*, *A. agrostis*, *A. tritici*, and *A. pacificae*. *Plant Disease* 99, 1584–1589.

Metcalfe, G. (1940) *Bacterium rhaponticum* (Millard) Dowson, a cause of crown-rot disease of rhubarb. *Annals of Applied Biology* 27(4), 502–508.

Midha, S.K. and Swarup, G. (1972) Factors affecting development of earcockle and tundu diseases of wheat. *Indian Journal of Nematology* 2, 97–104.

Mohammed, R.K.A. and Khan, M.R. (2021) Management of root-knot nematode in cucumber through seed treatment with multifarious beneficial microbes in polyhouse under protected cultivation. *Indian Phytopathology* 74(4), 1035–1043.

Mohiddin, F.A., Khan, M.R., Khan, S.M. and Bhat, B.H. (2010) Why *Trichoderma* is considered super hero (super fungus) against the evil parasites? *Plant Pathology Journal* 9(3), 92–102.

Naik, V., Nishitha Sharma, D.D., Thippeswamy, T. and Sivaprasad, V. (2015) Integrated approach for management of root disease complex in mulberry (*Morus* spp). *Indian Journal of Nematology* 45(2), 207–212.

Nandal, S.N., Paruthi, I.J. and Khan, M.R. (2010) Seed gall nematode infestation in wheat. In: Khan, M.R. and Jairajpuri, S. (eds) *Nematode Infestations Part I: Food Crops*. National Academy of Sciences, Allahabad, India, pp. 135–150.

Needham, T. (1743) A letter concerning certain chalky tubulous concretions called maim, with some microscopical observations on the farina of the red lily, and References 271 of worms discovered in smutty corn. *Philosophical Transactions of the Royal Society of London* 42(471), 634–641.

Owen, K., Walia, R.K., Yan, G., Khan, M.R. and Khan, M.R. (2023) Nematode problems in wheat and barley and their sustainable management. In: Khan, M.R. and Quintanilla, M. (eds) *Nematode Diseases of Crops and Their Sustainable Management*. Academic Press, Cambridge, MA, pp. 97–131.
Parrado, L.M. and Quintanilla, M. (2024) Plant-parasitic nematode disease complexes as overlooked challenges to crop production. *Frontiers in Plant Science* 15, 1439951.
Paruthi, I.J. and Bhatti, D.S. (1990) Wheat seed gall nematode *Anguina tritici*: A review. *Current Nematology* 1, 107–136.
Paruthi, I.J. and Bhatti, D.S. (1992) Efficacy of different practices for the control of earcockle in wheat caused by *Anguina tritici*. *Current Nematology* 3, 1–4.
Paruthi, I.J., Singh, M. and Gupta, D.C. (1987) Quantitative and qualitative losses in wheat (*Triticum aestivum*) grains due to earcockle and "tundu". *Seed Research* 15, 83–86.
Paruthi, I.J., Bajaj, H.K. and Bhatti, D.S. (1990) Studies on some factors affecting the emergence of *Anguina tritici* larvae from wheat seed galls and their ascendance on to seedlings. *Indian Journal of Ecology* 17(1), 48–50.
Paruthi, I.J., Nandal, S.N. and Kumar, A. (2009) Wheat seed gall nematode (*Anguina tritici*), its status and management. In: Sharma, G.L. (ed.) *Phytonematode Management in Field Crops*. Oxford Book, Jaipur, India, pp. 1–20.
Pathak, K.N. and Swarup, G. (1984) Incidence of *Corynebacterium michiganense* pv. *Tritici* in the ear-cockle nematode (*Anguina tritici*) galls and pathogenicity. *Indian Phytopathology* 37, 267–270.
Pitcher, R.S. (1965) Interrelationships of nematodes and other pathogens of plants. *Helminthological Abstracts (Series B)* 34, 1–17.
Pitcher, R.S. and Crosse, J.E. (1958) Studies in the relationship of eelworms and bacteria to certain plant diseases. II. Further analysis of the strawberry cauliflower disease complex. *Nematologica* 3(3), 244–256.
Poinar, G.O. and Hansen, E.C. (1986) Associations between nematodes and bacteria. *Helminthological Abstracts (Series B)* 55, 62–91.
Powers, T.O., Szalanski, A.L., Mullin, P.G., Harris, T.S., Bertozzi, T. *et al.* (2001) Identification of seed gall nematodes of agronomic and regulatory concern with PCR-RFLP of ITS1. *Journal of Nematology* 33(4), 191–194.
Priti, K., Rao, M.S., Pavani, K.V., Umamaheswari, R. and Gopalakrishnan, C. (2018) Combination of biopesticide formulations to manage *Meloidogyne incognita* and *Fusarium oxysporum* f. sp. *cepae* disease complex in onion. *Pest Management in Horticultural Ecosystems* 24(2), 163–172.
Rao, M.S., Dwivedi, M.K., Kumar, R.M., Chaya, M.K., Rathnamma, K. *et al.* (2014) Evaluation of bio-efficacy of *Bacillus subtilis* (NBAIMCC-B-01211) against disease complex caused by *Meloidogyne incognita* and *Fusarium oxysporum* f. sp. *vasinfectum* in okra. *Pest Management in Horticultural Ecosystems* 20(2).
Rao, M.S., Umamaheswari, R., Prabu, P., Priti, K., Chaya, M.K. *et al.* (2017) Field performance of *Pseudomonas putida* (IIHR pp-2) for the management of *Meloidogyne incognita* and *Fusarium oxysporum* f. sp. *vasinfectum* disease complex in okra (*Abelmoschus esculentus*). *Vegetos – An International Journal of Plant Research* 30(4), 33.
Riedel, R.M. and Larsen, P.O. (1974) Interrelationship of *Aphelenchoides fragariae* and *Xanthomonas begoniae* on Rieger begonia. *Journal of Nematology* 6(4), 215–216.
Riley, I.T. (1994) *Dilophospora alopecuri* and decline in annual ryegrass toxicity in Western Australia. *Australian Journal of Agricultural Research* 45(4), 841.
Riley, I.T. and McKay, A.C. (1990) Specificity of the adhesion of some plant pathogenic micro-organisms to the cuticle of nematodes in the genus *Anguina* (Nematoda: Anguinidae). *Nematologica* 36(1–4), 90–103.
Riley, I.T. and Ophel, K.M. (1992) *Clavibacter toxicus* sp. nov., the bacterium responsible for annual ryegrass toxicity in Australia. *International Journal of Systematic Bacteriology* 42(1), 64–68.
Riley, I.T., Reardon, T.B. and McKay, A.C. (1988) Electrophoretic resolution of species boundaries in seed-gall nematodes, *Anguina* spp. (Nematodes: Anguina), from some graminaceous hosts in Australia and New Zealand. *Nematologica* 34(4), 401–411.
Ritzema-Bos, J. (1891) Zwei neu Nematodenkrankheiten der Erdbeerpflanze. *Zeitschrift Pflanzenkrankheiten* 1, 1–16.
Sahebani, N., Kheiri, A., Rahimian, H. and Tehrani, A.S. (2007) Effect of rathayibacter tritici on movement of anguina tritici larvae and nematode function as vector of ear rot bacterium. *Journal of Science and Tech Agriculture and Natural Resources* 10(4B), 357–365.

Sarangi, T., Ramakrishnan, S. and Nakkeeran, S. (2018) Effectiveness of *Bacillus* spp. in management of nematode fungal disease complex of tomato. *Annals of Plant Protection Sciences* 26(1), 196.

Sasaki, J., Chijimatsu, M. and Suzuki, K. -i. (1998) Taxonomic significance of 2,4-diaminobutyric acid isomers in the cell wall peptidoglycan of actinomycetes and reclassification of *Clavibacter toxicus* as *Rathayibacter toxicus* comb. nov. *International Journal of Systematic Bacteriology* 48(2), 403–410.

Sawada, H., Horita, H., Nishimura, F. and Mori, M. (2020) *Pseudomonas salomonii*, another causal agent of garlic spring rot in Japan. *Journal of General Plant Pathology* 86(3), 180–192.

Scopoli, J.A. (1777) *Introductio ad historiam naturalem sistens genera lapidum, plantarum, et animalium*. Apud Wolfgangum Gerle, Prague.

Sharma, H.K., Singh, S. and Meher, H.C. (2007) Management of root-knot nematode, *Meloidogyne incognita* and wilt fungus, *Fusarium oxysporum* complex on okra. *Pesticide Research Journal* 19(2), 176–179.

Shaw, H.M. and Muth, O.H. (1949) Some types of forage poisoning in Oregon cattle and sheep. *Journal of the American Medical Association* 114, 315–317.

Sikora, R.A. and Roberts, P.A. (2018) Management practices: An overview of integrated nematode management technologie. In: Sikora, R.A., Coyne, D., Hallmann, J. and Timper, P. (eds) *Plant Parasitic Nematodes in Subtropical and Tropical Agriculture*, 3rd edn. CAB International, Wallingford, UK, pp. 795–838.

Singh, I. and Prasad, S.K. (1972) Effect of some pesticides and growth regulators on *Anguina tritici* infesting wheat. *Indian Journal of Entomology* 34(1), 77–79.

Sitaramaiah, K. and Pathak, K.N. (1993) Nematode bacterial disease interactions. In: Khan, M.W. (ed.) *Nematode Interactions*. Chapman & Hall, London, pp. 232–250.

Soteros, J.J. (1985) Strawberry nematode. *New Zealand Commercial Grower* 40, 44.

Steinbuch, J.G. (1799) Das grasalchen *Vibrio agrostis*. *Naturoforschung Halle* 28, 233–250.

Stirling, G.R. (1991) *Biological Control of Plant Parasitic Nematodes: Progress, Problems and Prospects*. CAB International, Wallingford, UK.

Strümpel, H. (1967) Beobachtungen zur lebensweise von *Aphelenchoides fragariae* in Lorraine-Begonien. *Nematologica* 13(1), 67–72.

Stynes, B.A., Petterson, D.S., Lloyd, J., Payne, A.L. and Lanigan, G.W. (1979) The production of toxin in annual ryegrass, *Lolium rigidum*, infected with a nematode, *Anguina* sp., and *Corynebacterium rathayi*. *Australian Journal of Agricultural Research* 30(1), 201.

Tacconi, R., Zani, A. and Pasini, P. (1982) Control trial against *Aphelenchoides fragaria* on strawberry using chemical and physical means. *Frutticoltura* 44(5), 102–106.

Than, K.A., Cao, Y., Michalewicz, A. and Edgar, J.A. (1998) Development of a vaccine against annual ryegrass toxicity. In: Garland, T. and Barr, A.C. (eds) *Toxic Plants and Other Natural Toxicants*. CAB International, Wallingford, UK, pp. 165–168.

Topalović, O. and Heuer, H. (2019) Plant–nematode interactions assisted by microbes in the rhizosphere. *Current Issues in Molecular Biology* 30, 75–88.

Walker, J. and Sutton, B.C. (1974) *Dilophia* sacc. and *Dlophospora* desm. *Transactions of the British Mycological Society* 62(2), 231–241.

Wasilewska, L. and Webster, J.M. (1975) Free-living nematodes as disease factors of man and his crops. *International Journal of Environmental Studies* 7(3), 201–204.

Westerdahl, B.B. and Bello, O.M. (2023) Leaf and bud nematodes in agricultural crops and their management by biotechnological approaches. In: Khan, M.R. (ed.) *Novel Biological and Biotechnological Applications in Plant Nematode Management*. Springer, Singapore, pp. 359–376.

Zgurskaya, H.I., Evtushenko, L.I., Akimov, V.N. and Kalakoutskii, L.V. (1993) *Rathayibacter* gen. nov., including the species *Rathayibacter rathayi* comb. nov., *Rathayibacter tritici* comb. nov., *Rathayibacter iranicus* comb. nov., and six strains from annual grasses. *International Journal of Systematic Bacteriology* 43(1), 143–149.

11 Disease Complexes Involving Multiple Nematodes

Mahfouz M.M. Abd-Elgawad*
Plant Pathology Department, National Research Centre, Giza, Egypt

Abstract
Plant nematodes are important pests of agricultural crops, and inflict yield losses of economic significance to food crops worldwide. The yield losses to agricultural crops due to nematode attacks range from 7% to 21% depending on the crop species, nematode species, population density and environmental conditions, with an estimated annual value of US$125–200 billion, considering various kinds of commercial and non-commercial crops. Nonetheless, the yield losses are mostly based on individual nematode species on a specific plant species or groups of crops. However, a recent estimate on the crop losses incited by nematodes to 40 life-sustaining crops has shown 13.5% losses as an average, with a value of about US$358 billion annually. Plant nematodes are obligate parasites with mostly polyphagous host specificity, hence different genera and species aggregate in the root zone of a plant species and coinhabit. However, one or two species of plant nematodes attain threshold levels and are able to cause a disease, but other species present in the rhizosphere influence the population development of the entire community and cause cumulative impacts on the host's health. This chapter examines the ecological and etiological interactions among cohabiting plant nematodes, and discusses the management and future prospects for disease complexes involving multiple nematodes

Introduction

Nematodes that parasitize plants are frequently found in polyspecific populations causing 13.5% losses as an average for the 40 life-sustaining crops; about USD 358.2 billion annually (Abd-Elgawad, 2023). The occurrence of plant nematodes in polyspecific communities may be attributed to their wide distribution, persistence without host, polyphagous nature and weak interspecific competition (Manachini *et al.*, 2005). Manachini *et al.* (2005) recorded nematodes belonging to 14 families and 24 genera from a canola field. However, despite weak competition, the nematode community is dynamic, and its populations interact continuously, developing different relationships among the genera, species and even races. In interaction, one population of the nematode influences the other; as a result, reproduction of a population may be enhanced or inhibited (Eisenback, 1993; Khan, 1993). Entomopathogenic nematodes also cohabit with plant-parasitic nematodes in the soil, but they usually do not interact mutually (Neumann and Shields, 2006). The entomopathogenic nematodes may, however, affect the pathogenesis and reproduction of phytonematodes (Khan *et al.*, 2005). Although disease complexes have mostly been addressed in the context that involves plant nematodes and other pathogenic taxa, e.g. fungi (see Chapter 7), bacteria (see Chapter 8) and viruses (see Chapter 9), the complexities of diseases engaging multiple nematode species/races deserve independent attention.

*Corresponding author: mahfouzian2000@yahoo.com

 Nematode Disease Complexes in Agricultural Crops
(ed. Mujeebur Rahman Khan)
DOI: 10.1079/9781800625228.0011

Nematode groups have varied feeding habits, survival mechanisms and ecological requirements. Nematodes characteristics such as dissemination, host suitability, host range and response to other factors determine the occurrence of a species/ races at a location at a given time (Weischer, 1993). In addition, the ability to compete with other organisms, including nematodes themselves, is also important in the interaction. The nature and intensity of the competition determines the occurrence and abundance of an organism at a location. Usually, strong competition occurs among organisms having similar ecological and physiological demands. Nematode species usually coexist if they have similar life cycle strategies or utilize different ecological niches. A species may dominate at a location if it is adapted to the host and the ecological conditions better than other cohabitants. Plant nematodes in cohabitance may interact ecologically as competitors and etiologically through their effects on the host plant. Both these relationships are interrelated, since disease severity or host damage depends on the population of the nematode, and is regulated by the host, environment, rhizospheric community, etc. Hence, nematodes in the root zone of a plant may develop ecological and etiological interactions, which are discussed separately in this chapter.

Ecological Interactions Among Cohabiting Plant Nematodes

The ecological interactions among plant nematodes have been less studied in comparison to monospecific situations, although disease complexes by various nematode species represent the true scenario under actual field conditions (Helder and Heuer, 2022). Within these situations, competition occurs between cohabitant plant-parasitic nematodes for space and nutrition. Basically, such competition among the nematode species is apparent when the multiplication capacity of an individual nematode species becomes superior or inferior to its natural capacity under cohabitance with another nematode species. Earlier reported factors that contribute to such negative or positive interactions may also comprise those resulting from the physical breaking of attacked plant tissues, spatial entry and nematode distribution, feeding sites, shifts in the host physiology, etc. (Eisenback, 1993). Usually, when two nematode species engage to compete for limited resources (food, space, etc.) the species with the stronger competitive ability commands the relationship. The competition among the nematode genera, species and races is mainly governed by persistence, pathogenicity and host suitability, and the interaction may be dependent on both time and relative density. The dynamic status of the soil setting makes it hard to determine and delineate niches. Thus, Norton (1978) pointed out that applying competitive exclusion rules is complex to specifically match plant-parasitic nematode communities. Interactions may also take place among nematode species that occupy different sites on the root in close proximity or on the same rootlet; such interactions are not permanently antagonistic. Hence, the interaction may be useful for either or both nematode species due to physiological or mechanical shifts in the host plant. Thus, plants boosted by nutritional value may become more suitable hosts via decreased natural defenses (Eisenback, 1993).

Further, the nature of nematode biology and parasitism may considerably impact the competition. Based on the nature or mode of their parasitism, plant-parasitic nematodes are categorized into four wide divisions: sedentary endoparasites, sedentary semi-endoparasites, migratory endoparasites and ectoparasites (Fig. 11.1). However, the most harmful plant-parasitic nematode genera are found in the two categories that possess an endoparasitic life stage, i.e. sedentary endoparasites (e.g. *Meloidogyne*, *Globodera* and *Heterodera*) and migratory endoparasites (e.g. *Radopholus* and *Pratylenchus*). Yet, huge differences exists in lifestyle between these two categories. The act of moving of the second-stage (J_2) juveniles in the sedentary class ceases at an adequate feeding site to grow and lay eggs, whereas the migratory endoparasitic forms can migrate into and out of the host plant tissue at any stage during the entire lifespan (Khan, 2008). Both root-knot nematodes and cyst nematodes secrete effector proteins to ease invading the host root, avoiding plant defense responses and inducing roots to form specific feeding cells (Abd-Elgawad, 2022). Generally, both migratory and sedentary endoparasitic nematodes are more harmful than ectoparasites (Abd-Elgawad,

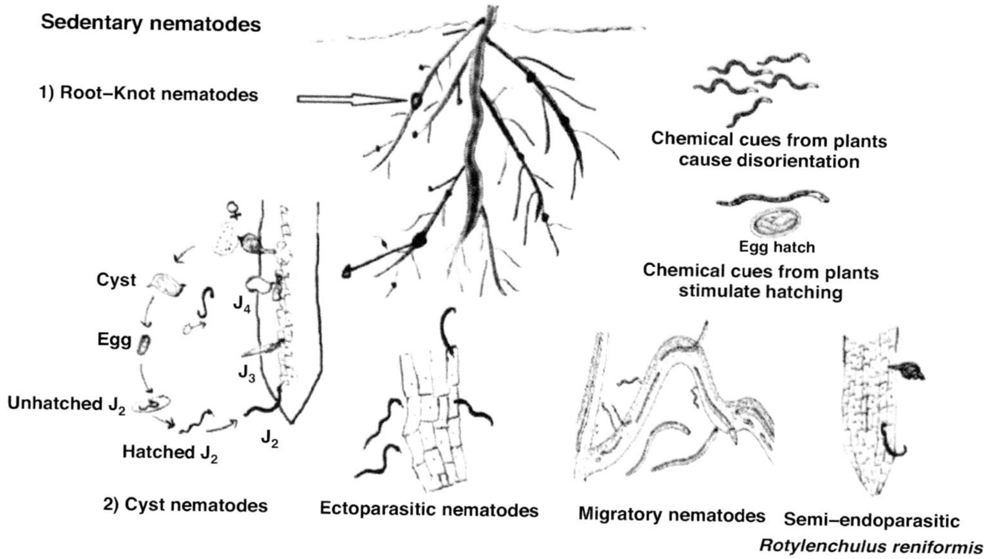

Fig. 11.1. Various divisions of plant-parasitic nematodes manifesting their interactions with host plants where plant metabolites can function as repellents, killers, attractants or inhibitors (Abd-Elgawad, 2022).

2022). However, despite intense diversity in life strategies, the above nematode groups frequently coinhabit. For example, *Tylenchulus semipenetrans* is a very common species in the soil of citrus orchards as perennial crops but may occur together with species of the genera *Tylenchorhynchus*, *Pratylenchus*, *Criconemoides*, *Xiphinema* and *Helicotylenchus*. Similarly, *Meloidogyne* spp. *Rotylenchulus reniformis*, *Pratylenchus*, *Hoplolaimus*, *Helicotylenchus*, etc., are commonly encountered infesting vegetable crops such as tomato, okra, eggplant (Desaeger *et al.*, 2023) and cucurbits (Khan *et al.*, 2023b), with a usual predominance of root-knot nematodes. Likewise, Hammam *et al.* (2023) observed the existence of *Meloidogyne arenaria* as the prevalent nematode species in all soil samples collected from a *Vicia faba* field, but other species of the genera *Helicotylenchus*, *Xiphinema*, *Tylenchorhynchus*, *Criconemoides* and *Pratylenchus* also cohabited in the same field for several years with such a seasonal crop. Also, several species of the same genus, e.g. *M. incognita*, *M. javanica* and *M. arenaria*, can cohabit in a cultivated field parasitizing vegetables or other crops (Khan *et al.*, 1984, 2019; Hammam *et al.*, 2022). The above examples reveal the ecological interactions among the plant nematodes as a natural phenomenon, but these interactions are highly diverse, varying with the feeding and parasitic behaviors of the cohabiting nematode species and races, and are described below and summarized in Table 11.1. Further, to enable understanding of the ecological interactions, the effects as the interacting nematode genera or species are grouped under mutually beneficial, mutually harmful and beneficial to any one of the interacting species within each category.

Interactions among ectoparasites

Ectoparasites may be surface feeders, subsurface feeders or deep feeders, and parasitize epidermal cells, root hairs and parenchyma to a depth of a few cells (Khan and Jairajpuri, 2010a). They have a relatively short (surface feeders) or long stylet (deep feeders), and their attachment to the host is generally for a short duration, lasting for a few seconds. However, for deep feeders such as *Xiphinema* and *Longidorus*, the interaction with the host is complex as it also involves hypertrophy and hyperplasia, and

Table 11.1. Important examples of ecological interactions between plant-parasitic groups of nematodes.

Interacting nematode species	Host plant	Interaction	Reference(s)
Ectoparasitic nematodes			
Tylenchorhynchus martini and *Belonolaimus longicaudatus*	Bermudagrass	No interaction; cultivar-based reactions were found	Johnson, 1970
Belonolaimus longicaudatus and *Scutellonema brachyurum*	Cotton	Stimulatory; both species were stimulated after 90 days	Yang *et al*., 1976
*Paratylenchus nanus*a and *Rotylenchus robustus*	Grass	Antagonistic; *P. nanus* was less pathogenic	Boag and Alphey, 1988
Ectoparasitic and migratory endoparasitic nematodes			
Paratrichodorus minor and *Pratylenchus zeae*	Corn	Stimulatory; cultivars could define the level of the two species	Johnson and Nusbaum, 1968
Tylenchorhynchus agri and *Pratylenchus penetrans*	Creeping bentgrass	No interaction; good host for *T. agri*; poor host for *P. penetrans*	Sikora *et al*., 1972
Pratylenchus zeae and *Quinisulcius acutus*	Sorghum	Antagonistic; after 42 days in greenhouse tests	Cuarezma-Terán and Trevathan, 1985
Ectoparasitic and sedentary endoparasitic nematodes			
Meloidogyne hapla and *Criconemoids xenoplax*a	Grape	Stimulatory; reproduction of *C. xenoplax* was stimulated	Santo and Bolander, 1977
Meloidogyne incognita and *Tylenchorhynchus brassicaea*	Vegetables	Antagonistic or mutually antagonistic; sex ratio increased towards males	Khan *et al*., 1986c, 1986d, 1995
Between migratory endoparasitic nematodes			
Radopholus similis and *Pratylenchus coffeae*	Citrus	Mutually antagonistic; *R. similis* prevails in coarse soil, but *P. coffeae* in fine-textured soil	O'Bannon *et al*., 1976
Scutellonema bradys and *Pratylenchus coffeaea*	Guinea yam	Antagonistic; same observation in greenhouse and field tests	Acosta and Ayala, 1976
Migratory and sedentary endoparasitic nematodes			
Meloidogyne incognita and *Pratylenchus penetrans*	Tomato	Mutually antagonistic; also split-root experiment	Estores and Chen, 1972
*Meloidogyne incognita*a and *Pratylenchus brachyurus*	Soybean	Antagonistic; split-root for translocatable factor inhibited *P. brachyurus*	Herman *et al*., 1988

Continued

Table 11.1. Continued

Interacting nematode species	Host plant	Interaction	Reference(s)
Meloidogyne hapla and *Pratylenchus brachyurus*	Tobacco	No interaction; used resistant cultivars	Eisenback, 1993
Meloidogyne incognita and *Pratylenchus brachyurus*a	Cotton	Stimulatory; *M. incognita* stimulates *P. brachyurus* penetration	Eisenback, 1993
Between sedentary endoparasitic nematodes			
Meloidogyne incognita and *Heterodera cajani*	Cowpea	Neutral; no mutual reaction was noted	Sharma and Sethi, 1978
Meloidogyne incognita and *Rotylenchulus reniformis*	Grape	Mutually antagonistic; *M. incognita* is more competitive over time	Rao and Seshadri, 1981
*Heterodera schachtii*a and *Meloidogyne hapla*	Sugarbeet	Stimulatory; prior addition of with *M. hapla* stimulated *H. schachtii*	Jatala and Jensen, 1983
*Meloidogyne hapla*a and *Heterodera schachtii*	Tomato	Antagonistic; *M. hapla* added 20 days before *H. schachtii*	Griffin, 1985
Between species of the same genus			
*Meloidogyne javanica*a and *Meloidogyne hapla*	Tomato	Antagonistic; *M. javanica* dominated at all temperatures	Kinloch and Allen, 1972
Meloidogyne hapla and *Meloidogyne hapla*	Sugarbeet	No interaction; double inoculations 10 days apart	Jatala and Jensen, 1976a, 1976b
Heterodera schachtii and *Heterodera schachtii*	Sugarbeet	Stimulatory; infections 3–5 times	Jatala and Jensen, 1976a, 1976b
Globodera rostochiensis and *Globodera pallida*	Potato	Hybridization; natural hybrids occurred	Olsson, 1988
Meloidogyne hapla and *Meloidogyne hapla*	Tomato	Inhibitory; diploid isolate more competitive than polyploid	Triantaphyllou, 1991

[a]Dominant species.

the feeding lasts for a relatively longer period, minutes to several hours on terminal galls (Khan and Jairajpuri, 2010b). Hence, the mode of parasitism may greatly influence the competition between species; apparently, the surface feeders may disturb or interrupt the feeding of deep feeders. The cohabitance of different species of ectoparasitic nematodes may lead to beneficial or inhibitory effects on the interacting nematode species as described under the following categories.

Mutually beneficial

In a few instances it has been found that the interaction between ectoparasites may be beneficial to all the cohabiting nematode species, as observed by Kraus-Schmidt and Lewis (1981) who found that interactions between two deep feeders may prove favorable to all the interacting species. They reported that when cotton plants were inoculated concomitantly with *H. columbus* and *S. brachyurum*, their reproductions were mutually promoted significantly. Yang *et al.* (1976) also reported an increase in the reproduction of *H. galeatus* and *B. longicaudatus* on cotton roots.

Beneficial or inhibitory to either nematode species

It is common observance that the interaction among different species of ectoparasites can stimulate the reproduction of any one of the participating nematode species. In a greenhouse trial, Johnson and Nusbaum (1968) observed that *Tylenchorhynchus claytoni* adversely affected the reproduction of *Paratrichodorus minor* on a maize cultivar, whereas *P. minor* proved suppressive to *T. claytoni* on three cultivars of *Zea mays*. The reproduction of *Paratylenchus projectus* significantly suppressed cohabitance with *Helicotylenchus pseudorobustus* and *Criconemella similis* in the soyabean root zone (McGawley and Chapman, 1983). This was apparently due to host preferences, as soyabean was a poor host for *P. projectus*, while a good host for the latter two species.

The interaction among the deep feeders may be governed by the relative pathogenicity and competitive nature of the cohabiting nematode species. Rhoades (1985) observed that reproduction of a relatively less pathogenic nematode species, *Dolichodorus heterocephalus*, was significantly suppressed in the presence of highly pathogenic *B. longicaudatus* on corn, but the reproduction of the latter remained unaffected. In another study, *Paratylenchus neoamblycephalus* suppressed a smaller population of *Criconemella xenoplax* at 22°C, whereas a higher population of *C. xenoplax* adversely affected the lower density of *P. neoamblycephalus* (Braun *et al.*, 1975). Similarly, the higher density of *P. neoamblycephalus* at 26°C proved suppressive to a lower density of *C. xenoplax*. The high density of *C. xenoplax* greatly checked the reproduction of *P. neoamblycephalus* when cohabiting at a low population. The higher populations of both the nematode species were found more competitive with relatively lesser effects of prevailing temperatures. The above examples demonstrate that relative population densities and abiotic factors such as temperature may alter the interaction among ectoparasites.

Mutually suppressive

The interaction among ectoparasites is seldom suppressive to both the interacting nematode species. However, Johnson (1970) reported intense competition among *T. martini* and *Criconemella ornata* on Bermudagrass varieties which led to suppressive effects on the reproduction of both nematode species. The competition among *C. ornata* and *B. longicaudatus* or *T. martini* and *B. longicaudatus* was not so intense, and led to relatively lower suppressive effects on both interacting nematodes.

Interaction between ectoparasites and migratory endoparasites

The interacting nematodes under this category generally feed on the cells at different locations, although closely located. However, the interacting nematode species may reach a face-to-face position while making perforations on the root surface for feeding (ectoparasites) or for penetration (endoparasites). Further, migratory endoparasitic nematodes may damage the root cortex or induce necrosis as well as alter the root physiology which may have direct implications on the parasitism of ectoparasites. Usually, this interaction has generally been found suppressive to either one or all of the interacting nematode species, but seldom mutually beneficial.

Mutually beneficial

A study carried out by Johnson and Nusbaum (1968) reported that reproduction of *P. minor* and *Pratylenchus zeae* infesting corn in cohabitance was promoted. The exact mechanism(s) involved in this mutually positive interaction was not known, but it may be due to perforations made by the ectoparasitic nematode and other morphological modifications induced by the endoparasitic nematode might had helped mutually at some stage.

Beneficial or suppressive effects to other nematode species

Host suitability often proves critical in determining the fate of interaction. Further, a stronger pathogen may suppress a relatively weaker parasite. Red clover is a good host to both *P. penetrans* and *T. martini*; reproduction of the latter was inhibited by 75–90% in cohabitance with *P. penetrans* (Chapman, 1959). In addition, feeding by one species may affect root suitability to other species, or the ectoparasitic feeding may affect the root penetration ability of the endoparasitic nematode. The penetration of *P. penetrans* in the roots of tobacco cv. WS117 was inhibited by 25–90% in the presence of *T. claytoni* (Miller and Mcintyre, 1975).

Johnson and Nusbaum (1968) observed that *P. minor* or *H. dihystera* adversely affected the population buildup of *P. brachyurus* on most soyabean varieties, but *P. brachyurus* checked the reproduction of *H. dihystera* on two varieties. The populations of *T. agri* and *P. penetrans* on red clover were not affected during the first 3 months, but during the next 2 months, the population of *T. agri* was adversely affected (Amosu and Taylor, 1975).

No interactive effects

In a few instances, nematode species sharing a common host do not develop any relationship and act independently. Sikora *et al.* (1979) observed that *T. agri* and *P. penetrans* infecting creeping bentgrass were not affected in cohabitance. Although, for the reproduction or population buildup of the two nematode species, bentgrass is an excellent host for *T. agri* and a poor host for *P. penetrans*.

Interaction between ectoparasites and sedentary endoparasites

These two groups of nematodes cohabit frequently on the common root zone without affecting each other, apparently because of their different feeding sites. However, in some situations they interact and develop suppressive or beneficial relationships. The feeding of ectoparasite nematodes on the perforation sites made by the juveniles of sedentary endoparasitic nematodes may affect the parasitism of the latter. Further, the roots damaged by ectoparasites may directly reduce the healthy root mass available for the endoparasites. Hence, the interaction may be positive or negative to either nematode species.

Beneficial to either nematode species

The population of *Hoplolaimus columbus* in cohabitance with *M. incognita* on cotton was increased considerably (Kraus-Schmidt and Lewis, 1981). In other studies, the presence of *M. hapla* led to promotion in the reproduction of *Tylenchorhynchus vulgaris* on bajra (Vaishnav and Sethi, 1978) and *C. xenoplax* on Concord grape (Santo and Bolander, 1977). Guy and Lewis (1987a) recorded a density-dependent interaction between *H. columbus* and *M. incognita*. Guy and Lewis (1987b) observed roots infected with *M. incognita* to be more attractive for invasion by *H. columbus* than non-infected soyabean roots. Sometimes competition may exert serious challenges to even strong pathogens such as root-knot nematodes, e.g. *H. galeatus* became a dominant species by replacing *M. incognita*. Its population increased gradually from an extremely low level in seasons in a cotton field (Bird *et al.*, 1974).

Suppressive to either nematode species

Sedentary endoparasites can decrease the adverse effect of the parasitism of ectoparasites, modifying the host substrate and making the roots less suitable for the invasion and feeding. *M. hapla* outclassed *Xiphinema americanum* in 4 years in an alfalfa field. The former nematode species may have decreased the feeding sites otherwise available to the lesser or may have altered the host physiology, rendering the root unsuitable for the feeding of *X. americanum* (Norton, 1978). Similarly, the population of *Scutellonema brachyurum* was decreased on cotton while in cohabitance with *M. incognita* (Kraus-Schmidt and Lewis, 1981).

In a tomato field, root-knot nematode suppressed the population of *T. brassicae* (Khan and Haq, 1979) and *T. agri* in a red clover field (Amosu and Taylor, 1975). Santo and Bolander (1977) observed a significant reduction in the reproduction of *C. xenoplax* when coexisting with *M. hapla* on Concord grape. The cohabitance of *P. minor* decreased the *M. javanica* population in a tomato field (Van Gundy and Kirkpatrick, 1975). Ishibashi and Choi (1991) recorded fewer galls incited by *M. incognita* on cucumbers in cohabitance with *Aphelenchus avenae* over root-knot nematode alone.

Sikora *et al.* (1979) observed that reproduction of *M. naasi* was suppressed in the presence of *P. minor* on creeping bentgrass. Similarly, the reproduction of *M. naasi* was also suppressed in the presence of *T. agri*. This kind of interaction has been recorded with a number of nematode combinations on different crops. The cohabitance of *M. incognita* with *T. vulgaris* on pear millet (Vaishnav and Sethi, 1978), *T. brassicae* on vegetables (Khan and Khan, 1986), *T. brassicae* on cauliflower (Khan *et al.*, 1987) and *H. galeatus* or *B. longicaudatus* on cotton (Yang *et al.*, 1976) resulted in a notable decrease in the reproduction of the root-knot nematode population.

Mutually inhibiting effects

In a few cases inhibitory effects were recorded on all the interacting nematode species. Misra and Das (1979) found that eggplant (*Solanum melongena*) was inoculated concomitantly with *M. incognita* and *C. ornata*; the reproduction of both nematode species was reduced evidenced by a significant decrease in the final population of the nematodes compared to the respective initial populations. Similarly, Van Gundy and Kirkpatrick (1975) recorded significant suppression in the reproduction of *M. javanica* and *Hemicycliophora arenaria* on tomato.

Interaction among migratory endoparasites

Intensive competition may take place among the migratory endoparasites as they feed on the same kind of cell and share common feeding strategies. This kind of interaction is quite common in occurrence in crop fields. Two or more species of *Pratylenchus* are often encountered in a field. These species may develop usually negative relationships. Host suitability or edaphic factors may serve as important determinant factors for the interaction among migratory endoparasitic nematodes and determine which species may be suppressed most. In a mixed population experiment, Ferris *et al.* (1967) found that the sex ratio in *P. penetrans* was altered towards females when in cohabitance with *P. alleni* on soyabean, but reproduction of the latter was inhibited. O'Bannon *et al.* (1976) observed that populations of both *P. coffeae* and *Radopholus similis* were decreased in cohabitance on citrus. However, fine-textured soil favored the population buildup of *P. coffeae*, whereas *R. similis* was stimulated in coarse-textured soil.

Interaction between migratory endoparasites and sedentary endoparasites

The normal feeding strategies of migratory endoparasites may lead to disruption of plant tissue which may cause hindrance or disturb feeding of sedentary endoparasites on specialized nurse cells. Further, the morphological modifications made by sedentary endoparasites may affect the penetration of the migratory species. There may be varied effects of the interactions.

Mutually suppressive

P. brachyurus and *M. incognita* were found suppressive to each other on tobacco cv. NC2326, susceptible to *M. incognita* (Johnson and Nusbaum, 1970). Chapman and Turner (1972, 1975) examined the impact of host suitability and nematode population on the reproduction of root-lesion nematodes. They found that *M. incognita* at low and high populations caused 37% and 57% lower deposition of eggs by *P. penetrans*, respectively. However, in another study, high populations of *P. penetrans* decreased the reproduction of *Globodera tabacum* on tobacco, whereas at low populations of *P. penetrans*, the reproduction of *G. tabacum* was also low (Miller and Wihrheim, 1968; Miller, 1970b). Griffin (1980) found that prior inoculation of *Ditylenchus dipsaci* increased the galling and reproduction of *M. hapla* on alfalfa. There are a number of other interactions where sedentary and migratory endoparasites were mutually affected while coexisting. Kaplan and Timmer (1982) reported that the populations of *P. coffeae* and *T. semipenetrans* in cohabitance were reduced on citrus. Similarly, on black pepper the reproduction of *R. similis* and *M. incognita* was decreased (Sheela and Venkitesan, 1981). Herman *et al.* (1988) observed that populations of both *P. brachyurus* and *M. incognita* infecting soyabean were mutually decreased.

Beneficial or suppressive to either nematode species

It has been found that relative host suitability may provide an edge to one species over the other species and vice versa in cohabitance. For example, tomato is a good host for *M. incognita*, better than *P. brachyurus*, on which pre-inoculation of the root-knot nematode reduced the root penetration of the latter nematode species. In a similar study on cotton, which is a better host for *P. brachyurus* than *M. incognita*, pre-inoculation by the root-knot nematode slightly promoted root penetration of *P. brachyurus* (Gay and Bird, 1973). On tobacco, the population of *P. brachyurus* was increased in cohabitance with *M. incognita* (Johnson and Nusbaum, 1970). Similarly, the population of *P. brachyurus* was increased on creeping bentgrass in co-occurrence with *M. naasi* (Sikora *et al.*, 1972). In addition, the differential pace of penetration of larvae in the host roots may play an important role in the interaction of migratory and sedentary endoparasitic nematodes. It was found that migratory endoparasitic nematodes often penetrate the roots faster, leading to inhibition in the penetration of sedentary nematodes (Turner and Chapman 1971, 1972; Gay and Bird, 1973).

Chitamber and Raski (1984) reported significant suppression in the population of *P. vulnus* in co-occurrence of *M. incognita* when it was pre-inoculated (30 days) 125 days after the start of the experiment. In concomitant inoculations the reproduction of *P. vulnus* was inhibited in 250 days. This shows that the suppressive effect of one of the interacting species may also depend on the time and sequence of inoculations. Sedentary endoparasitic nematodes are more advanced parasites than migratory endoparasites, and induce several morphological and physiological modifications in the host, which may influence the suitability of the host to the later. For example, *Meloidogyne* spp. induce translocatable factors which can cause discernible adverse effects on the reproduction of migratory endoparasites. However, the reproduction and population of *Meloidogyne* spp. may also be decreased due to the disturbance to the feeding sites caused by the migratory endoparasites (Estores and Chen, 1970, 1972). Egunjobi *et al.*, 1986 observed that *M. javanica* suppressed the penetration of *P. sefaensis* in corn roots. Other studies have also shown suppressive effects on one of the cohabitants in the interaction. Reproduction of *M. incognita* was severely suppressed on cotton co-infected by *P. brachyurus* (Gay and Bird, 1973). Similarly, reproduction of another sedentary endoparasite, *H. glycines*, was reduced in the presence of *P. scribneri* on soyabean (Lawn and Noel, 1990). *M. incognita* caused inhibitory effects on *P. brachyurus* on susceptible tobacco cv. Hicks. But on tobacco cvs NC95 and NC2512 (resistant), the population of *P. brachyurus* was enhanced in cohabitance with *M. incognita*. However, suppression in *M. hapla* reproduction due to *P. brachyurus* was recorded only on cvs Hicks and NC95. Similarly, *P. brachyurus* was suppressed by *M. hapla* on only cvs Hicks and NC2326.

When shifts in population levels and impacts on plant growth of each *M. incognita*, *B. longicaudatus* and *P. brachyurus* on soybean and corn alone and in mixed populations were compared, no interaction was recorded on corn in 1987 season (Dickson and McSorley, 1990). However, the population of *B. longicaudatus* was significantly higher in *M. incognita*-infested corn plots in 1988 midseason. Moreover, *B. longicaudatus* populations were decreased significantly *P. brachyurus*-infested corn plots at harvest. Interestingly, this study reported that midseason of *M. incognita* population densities were enhanced ($P \leq 0.05$) as combined with *H. glycines*, although this trend reversed at harvest in the soybean field. The yield of soybean was decreased in *H. glycines*-infested plots ($P \leq 0.001$). Yet, *M. incognita* or *B. longicaudatus* had no impact on soybean growth parameters either singly or in combination (Dickson and McSorley, 1990).

No interaction

Differences in the host tissue preference and feeding strategies may minimize the chances of interaction among migratory and sedentary nematodes. The pathogenesis of *D. dipsaci* and *H. schachtii* progressed normally without affecting each other in cohabitance (Griffin,

1987). Both the nematode species occupied different sites for feeding inside the host.

Interaction between semi-endoparasites and sedentary endoparasites

Semi-endoparasite and sedentary endoparasite nematodes have highly specialized, complex and long-lasting host–parasite relationships. Sedentary endoparasites enjoy the ensured protection from the outer biotic and abiotic factors as they becomes fully internal from the second stage onwards. The feedings sites of these two kinds of parasites lie quite close to each other. Sedentary endoparasites feed on nurse cells in phloem (Khan and Khan, 1997), whereas semi-endoparasites feed on nurse cells in the pericycle, endodermis or cortex (Taha and Sultan, 1977). Hence, complex interactions may occur. In this kind of interaction, usually one or both of the interacting nematode species are adversely affected.

Beneficial or suppressive to either nematode species

Reniform nematodes frequently occur in cohabitance with root-knot nematodes on vegetables, but their occurrence is commonly overlooked due to massive galling caused by the latter (Desaeger *et al.*, 2023). This co-occurrence has been observed to be usually suppressive to reniform nematodes. However, Pathak *et al.* (1985) stated that interactions between *Meloidogyne* and *Rotylenchulus* can be inhibitory to either interacting species. These interactions may also depend on time of invasion and population densities as well as host suitability (Khan *et al.*, 1986a, 1986b, 1987). Singh, 1976 observed that the population of *R. reniformis* was decreased in the presence of *M. incognita* on soyabean, but the latter was not affected by the former. A similar impact of cohabitance of the two above species was also recorded on black gram (Mishra and Gaur, 1981).

Low populations of *M. incognita* did not affect the reproduction of the reniform nematode, *R. reniformis*, and vice versa on sweet potato (Thomas and Clark, 1983a, 1983b). The reproduction and population of root-knot nematode were also not altered when a high level was used; however, in such situations, the population of *R. reniformis* was decreased (Thomas and Clark, 1980, 1981). It has been found that *R. reniformis* reduced the reproduction of *M. incognita* on tomato (Winoto and Lim, 1972; Khier and Osman, 1977).

Mutually suppressive effects

The interaction between semi-endoparasites and sedentary endoparasitic nematodes is generally found mutually inhibitory because they feed very close to each other (Norton, 1978). Both the nematodes also induce significant physiological modifications, which may lead to negative impacts on both inside the host tissues (Khan *et al.*, 1985). The populations of reniform and root-knot nematodes were initially inhibited mutually on tomato, but later become more competitive over time (Taha and Kassab, 1979, 1980). The relatively short life cycle of *R. reniformis* allows it to dominate over *M. javanica* up to around 2 months, thereafter the root-knot nematode dominated over the former. Both these nematodes have also been found mutually inhibitory on grapevines, but *M. incognita* was less affected due to its stronger competitiveness (Rao and Seshadri, 1981).

Interactions between citrus nematodes and reniform nematodes on grape were also found suppressive to each other, as revealed from their final populations in the concomitantly inoculated set which were significantly less than monospecific treatments (Taha and Sultan, 1977). Likewise, in a three-species test, *Nacobbus aberrans* was found to be suppressive to both *H. schachtii* and *M. hapla*, especially the latter (Inserra *et al.*, 1984), but reproduction of *N. aberrans* was not affected, though it was less aggressive among three species.

Interaction among sedentary endoparasites

Sedentary endoparasite nematodes are highly evolved parasites and establish a complex relationship with the susceptible host-like vegetables that lasts for the entire lifespan

(Khan and Khan, 1994a, b). Interactions that occur among various sedentary nematode species from the same or different genera are presented below.

Beneficial or suppressive to either nematode species

Jatala and Jensen (1976a, 1976b, 1983) reported that pre-inoculation of *H. schachtii* reduced reproduction of *M. hapla* on sugarbeet, but pre-inoculation with *M. hapla* stimulated cyst nematode reproduction. Griffin (1985) also observed that *H. schachtii* inhibited the reproduction of *M. hapla* on tomato when temperatures were raised.

Interactions among species of a genus

Intensive competition occurs between different species from a genus, due to aggregation at the same site because of same feeding site and identical feeding preferences, strategies and other similar capabilities. The cohabitants from the same genus, however, show diverse interactions.

Mutually beneficial

Sometimes the competition among different species may prove beneficial to both the species, probably due to interspecific hybridization (Eisenback, 1993), as observed with *Globodera* spp. Where *G. rostochiensis* and *G. pallida* produced fertile hybrid progenesis (Miller, 1983). The interspecific fertilization among the field populations of *Globodera* species resulted in the siblings based on host specificity and juvenile styles (Kort and Jaspers, 1973) and spicule morphology (Olsson, 1988). These morphological divergent hybrids revealed the cross-breeding between *G. rostochiensis* and *G. pallida* under natural conditions. Greenhouse tests between different *Heterodera* species have also yielded morphologically divergent hybrids (Miller, 1983). The crosses between *H. schachtii* and *H. glycines* (Potter and Fox, 1965; Miller, 1988) and *H. glycines* and *H. cruciferae* (Miller, 1989) produced fertile hybrids. Interestingly, the hybrids obtained from *H. cruciferae* and *H. schachtii* crosses were intermediate in their morphology, and reproduced on both sugarbeet and soyabean, whereas the parents reproduced on either host (Miller, 1988). Similarly, the hybrids obtained from *H. glycines* and *H. cruciferae* crosses reproduced on soyabean, cabbage and sugarbeet, but *H. glycines* and *H. cruciferae* reproduced only on soyabean and cabbage, respectively (Miller, 1989).

Beneficial or inhibitory effects to either nematode species

Meloidogyne spp. such as *M. incognita, M. javanica* and *M. arenaria* are commonly found in tomato, eggplant and okra fields in the subtropics and tropics, and concomitantly infect these plants (Khan *et al.*, 1984). Environmental factors such as temperature, rain fall, host species, etc., may contribute to their adoptability and relative dominance (Khan and Khan, 1989). For example, at high temperatures, *M. incognita* dominated over *M. javanica* and *M. hapla*, whereas at low temperatures, *M. hapla* dominated over *M. incognita* and *M. javanica* (Khan and Khan, 1991). Sometimes, it has been found that factors other than temperature may also contribute to relative abundance. It was observed that *M. javanica* dominated over *M. hapla* at 20°C, although *M. hapla* is a temperate species (Minz and Strich-Harari, 1959). When in cohabitance, *M. hapla* adversely affected the population of *M. incognita* on tobacco cvs NC95 and NC2512 (nematode-resistant cultivars) at 25°C. When *M. incognita* and *M. hapla* were inoculated concomitantly, the root penetration of the former species was faster than the latter, but feeding sites were destroyed by the hypersensitive response of the cultivar leading to lesser galling (Johnson and Nusbaum, 1970). Similarly, the galling and egg mass production of *M. arenaria* decreased when *M. incognita* was pre-inoculated on a soyabean cultivar susceptible to *M. arenaria* (Ibrahim, 1987).

Intense completion among *Globodera* spp. may occur because they share the same ecological niche (Parrot *et al.*, 1976). To avoid direct completion, *G. rostochiensis* and *G. pallida* co-occur over different time periods, and one species dominates over the other cohabitant at one time (Kort and Bakker, 1980). The reproduction of *Heterodera oryzicola* was reduced in the presence of *M. graminicola* on rice in sequential and concomitant inoculations (Rao and Prasad, 1981). Sharma and Sethi (1978) observed pre-inoculation of *M. incognita* to be injurious to *H. cajani* on cowpea but pre-inoculation of *H. cajani* was suppressive to *M. incognita*. Jatala and Jensen (1983) reported that pre-inoculation with *M. hapla* promoted the reproduction of *H. schachtii* on sugarbeet. *Heterodera* species have been found suppressive to the pathogenesis of *Meloidogyne* species. Kaul and Sethi (1982) observed that the inoculation of *H. zeae* 3 days prior to *M. incognita* led to suppression of reproduction of the latter. Inoculation with *H. schachtii* also inhibited the infection and reproduction of *M. hapla* on tomato (Griffin and Waite, 1982). Jatala and Jensen (1976a, 1976b) also observed similar interaction between the above nematodes on sugarbeet. The relationship between the *Meloidogyne* and *Heterodera/Globodera* species appears to depend on density and time. In one study, it was found that low populations of *M. incognita* did not affect the reproduction of *H. glycines* on soyabean, but high populations caused suppressive and stimulatory effects in the early and late season, respectively (Ross, 1964).

Mutually inhibitory

Marshall (1989) observed that reproduction and population buildup of *G. pallida* were suppressed in the presence of *G. rostochiensis*, although a low population of *G. pallida* was found to be more competitive and persisted in the soil. However, Parrot *et al.* (1976) found *G. pallida* to be more competitive than *G. rostochiensis*. The eggs of *G. rostochiensis* rapidly hatch and their survival is low at colder temperatures in England. In the Netherlands, the juveniles of *G. pallida* expressed greater competitiveness than *G. rostochiensis* (Seinhorst and Oostrom, 1989).

No interaction

Sometimes the sedentary endoparasites in cohabitance act independently without affecting each other. Cyst-forming nematode *H. cajani* and root-knot nematode *M. incognita* on cowpea did not develop any interaction and acted independently (Sharma and Sethi, 1978). In another study, *H. schachtii* and *M. hapla* on sugarbeet did not affect the reproduction and pathogenesis of each other (Jatala and Jensen, 1976a, b, 1976b).

Interracial competition

Mutually beneficial

Interactions among the individuals of a species have generally been found to be density dependent. At a normal population, neither too high (overcrowding) or too low (under population), the individuals of a species flourish well, and the reproduction and population dynamics follow a positive and linear relationship over time. The races of the same species hybridize with each other (Price *et al.*, 1978). Possessing the same feeding strategies, races compete with each other while searching the feeding sites inside the host tissue (Price *et al.*, 1976). The multiple cross matings may further complicate the competition (Triantaphyllou and Esbenshade, 1990), as the inbreeding yields isolates with varied capabilities with regard to morphology and parasitism. Miller (1970a) observed isolates to be morphologically distinct, and capable of successful inbreeding among the 11 isolates of the nematodes collected from a site. Likewise, intercrossing of two races of *H. avenae* yielded viable progenies with parasitic capabilities resembling either parent (Andersen, 1965). The races of *D. dipsaci* also intercrossed and produced fertile hybrids (Eriksson, 1965, 1974).

Mutually inhibitory

Overcrowding is one of the commonly occurring scenarios in agricultural fields, and leads to intensive intraspecific or interracial competition for feeding sites, and generally proves to be mutually suppressive. The competitiveness of the races within a species varies considerably and leads to mutually suppressive effects. Khan and Haider (1991) observed that *M. javanica* and races of *M. incognita* were usually mutually inhibitory, although race 2 of *M. incognita* was more competitive than its other races. The interaction of these races with *M. javanica* was found mutually negative.

Intraspecific competition

Overcrowding of nematodes in or around roots leads to a decrease in the population. The decrease results due to intense competition between individual of different species or within a species for space and food, and only healthy ones are able to survive while the rest die due to injury or starvation, e.g. the *G. rostochiensis* male to female ratio increases with an increase in the population density, and tends to reduce the reproductive rate. It has been found that in a field heavily infested with *G. rostochiensis*, only 1 in 40 or 1 in 80 larvae could reach maturity while the remaining die due to intense intraspecific completion for space and/or root invasion. At a moderate population of 1 in about 5–8 larvae, *G. rostochiensis* attains maturity. Intense competition occurs due to mass invasion of the roots and reduction in tuber production, as a result nutrient supply becomes limited and nematodes starve. In addition, scramble competition among *Globodera* larvae occurs that has been held responsible for the decreases in egg production per cyst in high density populations. In *Meloidogyne* species, high population brings about sex reversal towards maleness (Fassuliotis, 1970). Low populations also lower the reproductive rates in both parthenogenetic and amphimictic species (Jaenike and Selander, 1979).

Etiological Interactions of Cohabitant Nematodes

Nematode-induced diseases are usually reported to be caused by a single plant-parasitic nematode species (Abd-Elgawad *et al.*, 2010), although under natural condition quite often groups of nematode species are involved, inciting the symptoms and impacting the plant growth and yield parameters (Norton, 1978; Abd-Elgawad *et al.*, 2016; El-Nagdi *et al.*, 2019). For this reason, researchers have also evaluated the effect of polyspecific nematode communities including interactions among the species to understand the natural disease scenarios (Duncan and Ferris, 1983; McSorley and Dickson, 1989; Dickson and McSorley, 1990). When two or more pathogens (nematode species) attack a plant, there are three possibilities with regard to the etiological interaction that leads to antagonistic, synergistic or additive interaction. In antagonistic interaction, crop damage caused by the two nematode species in cohabitance is less than the sum of the damage incited by the two species singly (negatively interaction). In synergistic interaction, the damage by the two pathogens in cohabitance is greater than the additive individual effects (positive interaction). The third possibility is no interaction, and the combined impact is equal to the total amount of the damage caused by each species individually, as expressed in the following equations:

$$\text{Nemotode} - \text{A} \times \text{Nemotode} - \text{B} > \text{A} + \text{B} \longrightarrow \text{Synergistic/positive interaction}$$

$$\text{Nemotode} - \text{A} \times \text{Nemotode} - \text{B} < \text{A} + \text{B} \longrightarrow \text{Antagonistic/negative interaction}$$

$$\text{Nemotode} - \text{A} \times \text{Nemotode} - \text{B} = \text{A} + \text{B} \longrightarrow \text{Additive/neutral interaction}$$

However, statistically, as the impacts of two or more determinants (e.g. two or more plant-parasitic nematode species) on plant yield or growth parameters are additive, the hypothesis can be explained in the following equation:

$$Y = b_1x_1 + b_2x_2 + (or-)b_nx_n + e$$

where Y is a measure of the dependent variable (e.g. plant yield or growth parameter), $b_1, b_2, \ldots, b_n$ are constants, $x_1, x_2, \ldots, x_n$ are the independent variables (e.g. numbers of each plant-parasitic nematode genus/species/race in a nematode community, temperature, etc.) and e is the error term.

However, as two of these independent variables interact, a further term (x_1x_2) is incorporated to get the following equation for interaction:

$$Y = b_1x_1 + b_2x_2 + b_3(x_1x_2) + (or-)b_nx_n + e$$

These interactions occur where the impact of such variables and their interactive effects on crop yields/growth parameters are quantified and in tests where variables are set to provide several treatment combinations. In these analyses, the estimated variance of *Y* is frequently divided into portions (e.g. various growth parameters), which offer unbiased measures of the variation resulting from the different sources, e.g. ecological factors, biocontrol agents or experimental treatments (Duncan and Ferris, 1983; McSorley and Dickson, 1989; El-Nagdi *et al.*, 2019; Giese *et al.*, 2022). In these multiple biological systems, interactions are always dictated by several factors. The interactions of mixed plant-parasitic nematode populations that affect the expressions of disease complexes are summarized in Table 11.2. Their major etiological interactions are described below.

Antagonistic (negative) interactions

Disease complexes involving multiple nematode species inflict less damage in situations involving robust competition among the involved nematode genera/species/races. Interestingly, in such interactions the lesser pathogenic nematode species usually render the feeding sites less suitable for the feeding of the stronger nematode pathogenic species (Duncan and Ferris, 1981, 1982). This leads to a lower population buildup and lesser crop damage due to negative or antagonistic interactions among the nematode species. Physiological shifts within the plant also appear to be an additional factor in such interaction. In addition, either of the two different pathogenic species can prevail, relying on their associated factors such as inoculum level and edaphic factors (Eisenback, 1983). Tomato growth was more inhibited by *M. incognita* singly than with *P. penetrans* (Estores and Chen, 1970).

Synergistic (positive) interactions

In synergistic (positive) interactions, nematodes may cause a shift in the host physiology or modify the host cells that make it easier for the other partner to invade the host (Kimenju *et al.*, 2022; Khan *et al.*, 2023a). Interruption in host defense mechanisms may also be implicated. In most synergistic interactions with two or more plant-parasitic nematode species, a species of root-knot nematode has figured as one of the coinhabiting nematodes (Table 11.2). Griffin and Waite (1982) reported that the mixed population of *M. hapla* and *H. schachtii* interacted synergistically, and caused very high reductions in plant growth of tomato. Similarly, *M. hapla*, *H. schachtii* and *N. aberrans*, when inoculated jointly, induced extremely high inhibition in the plant growth of sugarbeet (Inserra *et al.*, 1984). The combined inoculation with *P. penetrans*, *M. naasi* and *Tylenchorhynchus agri* was found to be suppressive to the growth of bentgrass, significantly greater than their individual effects (Sikora *et al.*, 1972). In another, three-species interaction involving *M. naasi*, *T. agri* and *P. minor*, the highest decrease in plant growth occurred when *M. naasi* was inoculated 3 weeks before the other two nematode species (Sikora *et al.*, 1979).

Griffin (1980) reported that mixed populations of *M. hapla* and *D. dipsaci* greatly suppressed the plant growth of alfalfa plants susceptible to both species. The host reaction of alfalfa cv. Ranger to *D. dipsaci* was not influenced but the reaction of alfalfa cv. Vemal 298 resistant to *M. hapla* was decreased in cohabitance. In a study on tobacco cv. NC 95, Eisenback (1983) observed that resistance to *M. incognita* race one was partially decreased under pre-inoculation with *M. hapla* or *M. arenaria*. In contrast, pre-infection with *M. javanica* or *M. incognita* race four did not influence the resistance to *M. incognita* race 1. McGawley and Winchell (1987) observed that *M. javanica* alone caused moderate galling on soyabean roots while *M. incognita* alone did not incite galls. However, the two species in cohabitance interacted synergistically causing extensive galling on the roots of soybean.

Management of Disease Complexes Involving Multiple Nematodes

Recent advances in research on molecular mechanisms of nematode–nematode interactions can be exploited for novel management strategies such as interfering RNA methods, RNA sequencing in developing biocontrol agents and targeted genome editing (Haque and Khan, 2023; Khan *et al.*, 2023b; Abd-Elgawad, 2024). Polymerase chain reaction-based technologies

Table 11.2. Important examples of etiological antagonistic interactions between plant-parasitic nematodes.

Interacting nematode species	Host plant species	Effects	Reference
Synergistic interactions			
Meloidogyne incognita and *Belonolaimus longicaudatus* and *Pythium aphanidermatum*	Chrysanthemum	All pathogens jointly caused decreased plant growth greater than any pathogen alone or with fungus	Johnson and Littrell, 1970
Meloidogyne incognita and *Rotylenchulus reniformis* and *Rhizoctonia solani*	Cowpea	Three species jointly caused more disease than two or one species	Khan and Husain, 1989
Meloidogyne incognita and *Hoplolaimus indicus* and *Macrophomina phaseoli*	Jute	All pathogens jointly incited the most severe symptoms	Haque and Mukhopadhyaya, 1979
Meloidogyne incognita and *Meloidogyne javanica*	Soybean	Greater galling with combined inoculation than with either alone	McGawley and Winchell, 1987
Meloidogyne javanica and *Meloidogyne incognita*	Tobacco	Significant decrease in plant growth of tobacco due to combined inoculations over single inoculations	Paez *et al*., 1976
Meloidogyne hapla and *Heterodera schachtii*	Tomato	Combined inoculation reduced root growth by 65%, 64 and 61% over controls	Griffin and Waite, 1982
Radopholus similis and *Meloidogyne incognita* and *Fusarium* spp.	Pepper	All three pathogens most severely suppressed the plant growth in comparison to other combinations	Mustika, 1984
Pratylenchus penetrans and *Pratylenchus coffeae* and *Fusarium oxysporum*	Tomato	Three pathogens jointly reduced the plant growth with more severe wilt symptoms	Hirano and Kawamura, 1972
Antagonistic interactions			
Meloidogyne incognita and *Pratylenchus penetrans*	Tomato	Both species together suppressed plant growth less than that by *M. incognita*	Estores and Chen, 1970
Meloidogyne incognita and *Rotylenchulus reniformis*	Black gram	Plant growth reduced in combined inoculations greater than in single species	Mishra and Gaur, 1981
Meloidogyne incognita and *Rotylenchulus reniformis*	Grape	Plant growth reduced in combined inoculations greater than in single species	Rao and Seshadri, 1981

may further assist in identifying and quantifying the beneficial and harmful microbiota at the rhizosphere level for better nematode management. Further, the technique may prove of great help in detecting mixed populations of plant nematodes, as has been achieved for *Meloidogyne fallax*, *M. minor* and/or *M. chitwoodi* in a sample (Anderson and Gleason, 2023). Molecular identification might help in precisely distinguishing beneficial microorganisms colonizing the root zone of plants infested with multiple nematode species. A consortium of suitable microorganisms for nematode management may be developed to deal with mixed population situations (Khan, 2016).

Change in host reaction to mixed populations is a major issue facing breeding programs in recent times. The resistance-breaking isolates of serious plant nematodes have been documented and tactics/strategies addressing such issues as well as for managing mixed multiple nematodes populations in key crops likes potato, pepper and tomato have already been reported (Abd-Elgawad, 2020a, b, c, d). Host-specific resistance is a strong tool of management that should be exploited in light of recent advancements in host resistance genes, plant-parasitic nematode effectors and quantitative trait loci to enhance their effective application especially in key crops (Abd-Elgawad, 2022). However, Simon *et al.* (2022) recommended cultural practices such as planting date, tillage, crop rotation, using organic amendments, weed control, applying cover crops and sanitation for their control. They also suggested seeking proper nematode management products and resistant/tolerant genotypes for better plant-parasitic nematode management.

Application of biocontrol agents at the time of planting is gaining popularity in achieving sustainable nematode management (Stirling, 1991; Khan, 2007; Khan and Anwer, 2011; Sikora and Roberts, 2018; Abd-Elgawad and Askary, 2020; Khan *et al.*, 2021). Nematode antagonists, such as *Pochonia chlamydosporia* (Stirling, 1991), *Purpureocillium lilacinum* (Jatala, 1986), *Aspergilus niger* (Khan and Anwer, 2008) and *Pasteuria penetrans* (Kerry, 2000), and phosphate solubilizing microorganisms, such as *Pseudomonas*, *Bacillus*, *Penicillium* and *Aspergillus* species/strains (Khan *et al.*, 2009, 2016a, 2016b; Sikora and Roberts, 2018), may greatly contribute in protecting the plants at the early stage of growth, if applied at pre- or planting stage. In recent decades, *Trichoderma* spp. have emerged as a hero in the management plant diseases caused by fungi (Mohiddin *et al.*, 2010; Khan and Mohiddin, 2018) and nematodes (Khan, 2016), and commercial formulations of *T. virens*, *T. harzianum* and *T. hamatum* are available (Khan *et al.*, 2011), which have been found highly effective against fungus–nematode disease complexes (Khan *et al.*, 2023a), besides the monopathogenic diseases (Sikora and Roberts, 2018; Mohammed and Khan, 2021). Sikora *et al.* (2022) have advocated for integrated pest management to be the most effective tool for multiples nematode management. Further, integrated nematode management may prove more effect and offer maximum reliability of nematode control both in mono-and multispecies situations.

Conclusion and Future Prospects

Mono- and multispecies populations in cohabitance continuously interact giving a variety of outputs. Mixed populations of nematodes may mutually or differentially influence the reproductive ability of either or all species in cohabitance or the etiology of the disease complex may be modified. With an enormous complexity of interactions in mixed nematode populations, only general conclusions may be derived regarding ecological and etiological aspects. The interactions between plant-parasitic nematode species are generally weak, may be neutral, but can become inhibitory or stimulatory to one or all the cohabiting species. The species with similar feeding habits manifest intense competition; greater competition occurs among endoparasites compared to ectoparasitic nematodes. Further, stronger competition can be seen among sedentary endoparasites than migratory endoparasites. However, interactions between the nematode species have generally been seen to be dependent on density, time and several biotic/abiotic factors such as host suitability, soil properties, associated biota, etc. Modes of actions among nematode species in mixed populations may involve mechanical destruction, competition for feeding sites, physiological modifications or shift in the host suitability. The interactive effects of the nematodes in cohabitance are generally additive or antagonistic, but sometimes synergistic interaction, in which mostly sedentary nematodes are involved, may occur. However, despite significant improvement in the detection techniques, it still remains unresolved to correctly understand the varied effects of different combinations of nematode species on plant growth as well as the reproduction and population growth of the involved species in cohabitance. It is presumed that pathogenicity and population densities, host–parasite relationships and other factors regulate these interactions and outputs. In view of the involvement of nematode species with diverse feeding strategies, management of the resulting disease complex becomes challenging to scientists and growers. Integrated strategies involving field sanitation strategies, solarization and use of multifaceted microbial antagonists considering various ecological and etiological aspects will provide economic and sustainable management of nematode–nematode disease complexes. Such approaches will also certainly and substantially help in improving the efficiency of the existing nematode management modules in monospecific-dominated situations.

Acknowledgments

Mahfouz M.M. Abd-Elgawad acknowledges support through In-House project No. 13050112, US–Egypt Project cycle 17 (No. 172), the Subject Data funded in part by the NAS and USAID. However, any opinions, findings, conclusions or recommendations expressed in this chapter do not reflect the views of the USAID, NAS or any other agencies.

References

Abd-Elgawad, M.M.M. (2024) Nematode spatial distribution in the service of biological pest control. *Egyptian Journal of Biological Pest Control* 34, 3. DOI: 10.1186/s41938-024-00768-6.

Abd-Elgawad, M.M.M. (2020b) Biological control agents in the integrated nematode management of potato in Egypt. *Egyptian Journal of Biological Pest Control* 30(1), 121. DOI: 10.1186/s41938-020-00325-x.

Abd-Elgawad, M.M.M. (2020c) Biological control agents in the integrated nematode management of pepper in Egypt. *Egyptian Journal of Biological Pest Control* 30(1), 70. DOI: 10.1186/s41938-020-00273-6.

Abd-Elgawad, M.M.M. (2020d) Optimizing biological control agents for controlling nematodes of tomato in Egypt. *Egyptian Journal of Biological Pest Control* 30(1), 58. DOI: 10.1186/s41938-020-00252-x.

Abd-Elgawad, M.M.M. (2022) Understanding molecular plant–nematode interactions to develop alternative approaches for nematode control. *Plants* 11(16), 2141. DOI: 10.3390/plants11162141.

Abd-Elgawad, M.M.M. (2023) Spatial distribution of nematodes to improve their sampling and management decision. *Pakistan Journal of Nematology* 41(2), 144–152. DOI: 10.17582/journal.pjn/2023/41.2.144.152.

Abd-Elgawad, M.M.M. and Askary, T.H. (2020) Factors affecting success of biological agents used in controlling the plant-parasitic nematodes. *Egyptian Journal of Biological Pest Control* 30(1), 17. DOI: 10.1186/s41938-020-00215-2.

Abd-Elgawad, M.M., El-Mougy, N.S., El-Gamal, N.G., Abdel-Kader, M.M. and Mohamed, M.M. (2010) Protective treatments against soilborne pathogens in citrus orchards. *Journal of Plant Protection Research* 50(4), 512–519. DOI: 10.2478/v10045-010-0079-0.

Abd-Elgawad, M.M.M., Koura, F.F.H., Montasser, S.A. and Hammam, M.M.A. (2016) Distribution and losses of *Tylenchulus semipenetrans* in citrus orchards on reclaimed land in Egypt. *Nematology* 18(10), 1141–1150. DOI: 10.1163/15685411-00003020.

Abd-Elgawad, M.M.M. (2020a) Plant-parasitic nematodes and their biocontrol agents: Current status and future vistas. In: Ansari, R.A., Rizvi, R. and Mahmood, I. (eds) *Management of Phytonematodes: Recent Advances and Future Challenges*. Springer, Singapore, pp. 171–204.

Acosta, N. and Ayala, A. (1976) Effects of *Pratylenchus coffeae* and *Scutellonema bradys* alone and in combination on Guinea Yam (*Dicorea rotunda*). *Journal of Nematology* 8, 315–317.

Amosu, J.O. and Taylor, D.P. (1975) Interaction of *Meloidogyne hapla*, *Pratylenchus penetrans* and *Tylenchorhynchus agri* on Kenland clover, *Trifolium pratense*. *Indian Journal of Nematology* 4, 124–131.

Andersen, S. (1965) Heredity of race 1 or race 2 in *Heterodera avenae*. *Nematologica* 11(1), 121–124. DOI: 10.1163/187529265X00555.

Anderson, S.D. and Gleason, C.A. (2023) A molecular beacon real-time polymerase chain reaction assay for the identification of *M. chitwoodi, M. fallax*, and *M. minor*. *Frontiers in Plant Science* 14, 1096239. DOI: 10.3389/fpls.2023.1096239.

Bird, G.W., Rich, J.R. and Glover, S.U. (1974) Increased endomycorrhizae of cotton roots in soil treated with nematicides. *Phytopathology* 64(1), 48. DOI: 10.1094/Phyto-64-48.

Boag, B. and Alphey, T.J.W. (1988) Can interspecific competition be utilized to control plant-parasitic nematodes. *Nematologica* 34, 257–258.

Braun, A.L., Mojtahedi, H. and Lownsbery, B.F. (1975) Separate and combined effects of *Paratylenchus neoamblycephalus* and *Criconemoides xenoplax* on "Myrobalan" plum. *Phytopathology* 65(3), 328. DOI: 10.1094/Phyto-65-328.

Chapman, R.A. (1959) Development of *Pratylenchus penetrans* and *Tylenchorynchus martini* on red clover and alfalfa. *Phytopathology* 49, 357–359.

Chapman, R.A. and Turner, D.R. (1972) Effect of entrant *Meloidogyne incognita* on reproduction on concomitant *Pratylenchus penetrans* in red clover. *Journal of Nematology* 4, 221.

Chapman, R.A. and Turner, D.R. (1975) Effect of *Meloidogyne incognita* on reproduction of *Pratylenchus penetrans* in red clover and alfalfa. *Journal of Nematology* 7(1), 6–9.

Chitamber, N. and Raski, D.J. (1984) Reactions of grape rootstocks to *Pratylenchus vulnus* and *Meloidogyne* spp. *Journal of Nematology* 16, 166–170.

Cuarezma-Terán, J.A. and Trevathan, L.E. (1985) Effects of *Pratylenchus zeae* and *Quinisulcius acutus* alone and in combination on sorghum. *Journal of Nematology* 17(2), 169–174.

Desaeger, J., Coburn, J., Freeman, J. and Brym, Z. (2023) Plant-parasitic nematodes associated with *Cannabis sativa* in Florida. *Journal of Nematology* 55(1), 2023–0018. DOI: 10.2478/jofnem-2023-0018.

Dickson, D.W. and McSorley, R. (1990) Interaction of three plant-parasitic nematodes on corn and soybean. *Journal of Nematology* 22(4S), 783–791.

Duncan, L. and Ferris, H. (1981) Preliminary considerations of a model of multiple nematode species-plant growth relationships. *Journal of Nematology* 13, 435.

Duncan, L.W. and Ferris, H. (1983) Validation of a model for prediction of host damage by two nematode species. *Journal of Nematology* 15(2), 227–234.

Duncan, L. and Ferris, H. (1982) Interactions between phytophagous nematodes. In: Freckman, D. (ed.) *Nematodes in Soil Ecosystems*. University of Texas Press, Austin, TX, pp. 29–51.

Egunjobi, O.A., Akonde, P.T. and Caveness, F.E. (1986) Interaction between *Pratylenchus sefaensis, Meloidogyne javanica* and *Rotylenchulus reniformis* in sole and mixed crops of maize and cowpea. *Revue de Nematologie* 9, 61–70.

Eisenback, J.D. (1983) Loss of resistance in tobacco cultivar "NC95" by infection of *Meloidogyne arenaria* or *M. hapla*. *Journal of Nematology* 15, 478.

Eisenback, J.D. (1993) Interactions between nematodes in cohabitance. In: Kahn, M.W. (ed.) *Nematode Interactions*. Springer, Dordrecht, The Netherlands, pp. 134–174. DOI: 10.1007/978-94-011-1488-2_7.

El-Nagdi, W.M.A., Youssef, M.M.A., El-Khair, H.A. and Abd-Elgawad, M.M.M. (2019) Effect of certain organic amendments and *Trichoderma* species on the root-knot nematode, *Meloidogyne incognita*, infecting pea (*Pisum sativum* L.) plants. *Egyptian Journal of Biological Pest Control* 29(1), 75. DOI: 10.1186/s41938-019-0182-0.

Eriksson, K.B. (1965) Crossing experiments with races of *Ditylenchus dipsaci* on callus tissue cultures. *Nematologica* 11(2), 244–248. DOI: 10.1163/187529265X00131.

Eriksson, K.B. (1974) Intraspecific variation in *Ditylenchus dipsaci*. 1. Compatibility tests with races. *Nematologica* 20(2), 147–162. DOI: 10.1163/187529274X00122.

Estores, R.A. and Chen, T.A. (1970) Interaction of *Pratylenchus penetrans* and *Meloidogyne incognita acrita* as cohabitants on tomatoes. *Phytopathology* 60, 1291.

Estores, R.A. and Chen, T.A. (1972) Interactions of *Pratylenchus penetrans* and *Meloidogyne incognita* as coinhabitants in tomato. *Journal of Nematology* 4(3), 170–174.

Fassuliotis, G. (1970) Resistance of *Cucumis* spp. to the root-knot nematode, *Meloidogyne incognita acrita*. *Journal of Nematology* 2(2), 174–178.

Ferris, V.R., Ferris, J.M. and Bernard, R.L. (1967) Relative competition on two species of *Pratylenchus* in soybeans. *Nematologica* 13, 143.

Gay, C.M. and Bird, G.W. (1973) Influence of concomitant *Pratylenchus brachyurus* and *Meloidogyne* spp. on root penetration and population dynamics. *Journal of Nematology* 5(3), 212–217.

Giese, G., Saldaña Zepeda, D.P., Beacham, J. and Velasco Cruz, C. (2022) Modeling nematode population dynamics using a multivariate poisson model with spike and slab variable selection. *Journal of Applied Statistics* 49(12), 3195–3214. DOI: 10.1080/02664763.2021.1935800.

Griffin, G.D. (1980) Interrelationship of *Meloidogyne hapla* and *Ditylenchus dipsaci* on resistant and susceptible alfalfa. *Journal of Nematology* 12, 287–293.

Griffin, G.D. (1985) Interrelationship of *Heterodera schachtii* and *Meloidogyne hapla* on tomato. *Journal of Nematology* 17, 385–388.

Griffin, G.D. (1987) Interaction of *Ditylenchus dipsaci* and *Meloidogyne hapla* on resistant and susceptible plant species. *Journal of Nematology* 19, 441–446.

Griffin, G.O. and Waite, W.W. (1982) Pathological interaction of a combination of *Heterodera schachtii* and *Meloidogyne hapla* on tomato. *Journal of Nematology* 14, 182–187.

Guy, D.W. and Lewis, S.A. (1987a) Interaction between *Meloidogyne incognita* and *Hoplolaimus* on Davis soybean. *Journal of Nematology* 19, 346–351.

Guy, D.W. and Lewis, S.A. (1987b) Selective migration and root penetration by *Meloidogyne incognita* and *Hoplolaimus columbus* on soybean roots *in vitro*. *Journal of Nematology* 19, 390–392.

Hammam, M.M.A., Abd-El-Khair, H., El-Nagdi, W.M.A. and Abd-Elgawad, M.M.M. (2022) Can agricultural practices in strawberry fields induce plant–nematode interaction towards *Meloidogyne*-suppressive soils? *Life* 12(10), 1572. DOI: 10.3390/life12101572.

Hammam, M.M.A., Mohamed, M.M.M. and Abd-Elgawad, M.M.M. (2023) The damage and gain threshold of *Meloidogyne arenaria* on faba bean favors the use of a safe biological control over chemical nematicides. *Pakistan Journal of Nematology* 41(1), 8–17. DOI: 10.17582/journal.pjn/2023/41.1.8.17.

Haque, M.S. and Mukhopadhyaya, M.C. (1979) Pathogenicity of *Macrophomina phaseoli* on jute in the presence of *Meloidogyne incognita* and *Hoplolaimus indicus*. *Journal of Nematology* 11, 318–321.

Haque, Z. and Khan, M.R. (2023) Optimization of different application methods of multi-facial bacterial and fungal antagonists against sheath blight pathogen of rice, *Rhizoctonia solani* AG1-IA. *Journal of Phytopathology* 171(1), 23–35. DOI: 10.1111/jph.13151.

Helder, J. and Heuer, H. (2022) Let's be inclusive – the time of looking at individual plant parasitic nematodes is over, and new technologies allow for it. In: Sikora, R.A., Desaeger, J. and Molendijk, L.P.G. (eds) *Integrated Nematode Management: State-of-the-Art and Visions for the Future*. CAB International, Wallingford, UK, pp. 403–407.

Herman, M., Hussey, R.S. and Boerma, H.R. (1988) Interactions between *Meloidogyne incognita* and *Pratylenchus brachyurus* on soybean. *Journal of Nematology* 20(1), 79–84.

Hirano, K. and Kawamura, T. (1972) Incidence of complex diseases caused by *Pratylenchus penetrans* or *P. coffeae* and fusarium wilt fungus in tomato seedlings. *Technical Bulletin, Faculty of Horticulture, Chiba University* 20, 37–43.

Ibrahim, I.K.A. (1987) Interaction between *Meloidogyne arenaria* and *M. incognita* on tobacco. *Nematologia Mediterranea* 15, 287–291.

Inserra, R.V., Griffin, G.D., Vovlas, N., Anderson, J.L. and Kerr, E.D. (eds) (1984) Relationships between *Heterodera schachtii*, *Meloidogyne hapla*, and *Nacobbus aberrans* on sugarbeet. *Journal of Nematology* 16, 135–140.

Ishibashi, N. and Choi, D.R. (1991) Biological control of soil pests by mixed application of entomopathogenic and fungivorous nematodes. *Journal of Nematology* 23(2), 175–181.

Jaenike, J. and Selander, R.K. (1979) Evolution and ecology of parthenogenesis in earthworms. *American Zoologist* 19(3), 729–737. DOI: 10.1093/icb/19.3.729.

Jatala, P. (1986) Biological control of plant-parasitic nematodes. *Annual Review of Phytopathology* 24(1), 453–489. DOI: 10.1146/annurev.phyto.24.1.453.

Jatala, P. and Jensen, H.I. (1976a) Self-interactions of *Meloidogyne hapla* and of *Heterodera schachtii* on beta vulgaris. *Journal of Nematology* 8, 43–48.

Jatala, P. and Jensen, H.J. (1976b) Parasitism of *Beta vulgaris* by *Meloidogyne hapla* and *Heterodera schachtii* alone and in combination. *Journal of Nematology* 8, 200–205.

Jatala, P. and Jensen, H.J. (1983) Influence of *Meloidogyne hapla* Chitwood, 1949 on development and establishment of *Heterodera schachtii* Schmidt, 1871 on *Beta vulgaris* L. *Journal of Nematology* 15, 564–566.

Johnson, A.W. (1970) Pathogenicity and interaction of three nematode species on six bermudagrasses. *Journal of Nematology* 2, 36–41.

Johnson, A.W. and Nusbaum, C.J. (1968) The activity of *Tylenchorhynchus claytoni*, *Trichodorus christiei*, *Pratylendrus brachyurus*, *P. zeae*, and *Helicotylenchus dihystera* in single and multiple inoculations on corn and soybean. *Nematologica* 14, 9.

Johnson, A.W. and Littrell, R.H. (1970) Pathogenicity of *Pythium aphanidermatum* to chrysanthemum in combined inoculations with *Belonolaimus longicaudatus* or *Meloidogyne incognita*. *Journal of Nematology* 2(3), 255–259.

Johnson, A.W. and Nusbaum, C.J. (1970) Interactions between *Meloidogyne incognita*, *M. hapla*, and *Pratylenchus brachyurus* in tobacco. *Journal of Nematology* 2(4), 334.

Kaplan, D.T. and Timmer, L.W. (1982) Effects of *Pratylenchus coffeae–Tylenchulus semipenetrans* interactions on nematode population dynamics in citrus. *Journal of Nematology* 14(3), 368–373.

Kaul, R.K. and Sethi, C.L. (1982) Effect of simultaneous inoculations and prior establishment of *Heterodera zeae* and *Meloidogyne incognita*, and *Tylenchorhynchus vulgaris* singly and in combination on penetration of *H. zeae* and *M. incognita* into maize roots. *Indian Journal of Nematology* 11, 100.

Kerry, B.R. (2000) Rhizosphere interactions and the exploitation of microbial agents for the biological control of plant-parasitic nematodes. *Annual Review of Phytopathology* 38, 423–441. DOI: 10.1146/annurev.phyto.38.1.423.

Khan, M.W. (ed.) (1993) *Nematode Interactions*. Springer, Dordrecht, The Netherlands. DOI: 10.1007/978-94-011-1488-2.

Khan, M.R. (2007) Prospects of microbial control of root-knot nematodes infecting vegetable crops. In: Sharma, N. and Singh, H.B. (eds) *Biotechnology: Plant Health Management*. International Book Distributing, Co., Lucknow, India, pp. 643–665.

Khan, M.R. (2008) *Plant Nematodes – Methodology, Morphology, Systematics, Biology and Ecology*. Science Publishers, Enfield, NH. DOI: 10.1201/b10766.

Khan, M.R. (2016) Nematode biocontrol agents: Diversity and effectiveness against phytonematodes in sustainable crop protection. *Indian Phytopathology* 69(4s), 453–463.

Khan, M.R. and Anwer, M.A. (2008) DNA and some laboratory test of nematode suppressing efficient soil isolates of *Aspergillus niger*. *Indian Phytopathology* 61(2), 59–73.

Khan, M.R. and Anwer, A. (2011) Fungal bioinoculants for plant disease management. In: Paul, M., Clinton, M. and Ahmad, I. (eds) *Microbes and Microbial Technology*. Springer, Dordrecht, The Netherlands, pp. 447–488. DOI: 10.1007/978-1-4419-7931-5_17.

Khan, M.W. and Haider, S.R. (1991) Interaction of *Meloidogyne javanica* with different races of *Meloidogyne incognita*. *Journal of Nematology* 23(3), 298.

Khan, M.W. and Haq, S. (1979) Interaction between *Meloidogyne incognita* and *Tylenchorynchus brassicae* on tomato. *Libyan Journal of Agriculture* 8, 181–186.

Khan, M.R. and Jairajpuri, S. (2010a) *Nematode Infestations Part I: Food Crops*. National Academy of Sciences, Allahabad, India.

Khan, M.R. and Jairajpuri, S. (2010b) *Nematode Infestations Part II: Industrial Crops*. National Academy of Sciences, Allahabad, India.

Khan, M.R. and Khan, M.W. (1986) Antagonistic behaviour of root-knot, reniform, and stunt nematodes in mixed infection. *International Nematology Network Newsletter* 3, 3.

Khan, M.R. and Khan, M.W. (1989) Interaction of simulated acid rain and *Meloidogyne incognita* (race 1) on tomato. 28th Annual Meeting, Society of Nematologists, Davis, U.S.A. *Journal of Nematology* 21, 569.

Khan, M.R. and Khan, M.W. (1991) Impact of air pollutants emanating from a thermal power plant on tomato. *Journal of Indian Botanical Society* 70, 373–378.

Khan, M.R. and Khan, M.W. (1994a) Effects of simulated acid rain and root-knot nematode on tomato. *Plant Pathology* 43, 41–49.

Khan, M.R. and Khan, M.W. (1994b) Single and interactive effects of root-knot nematode and coal-smoke on okra. *New Phytologist* 126, 337–342.

Khan, M.R. and Khan, M.W. (1997) Effects of the root-knot nematode, *Meloidogyne incognita*, on the sensitivity of tomato to sulfur dioxide and ozone. *Environmental and Experimental Botany* 38, 117–130.

Khan, M.R. and Mohiddin, F.A. (2018) Trichoderma: Its multifarious utility in crop improvement. In: Prasad, R., Gill, S.S. and Tuteja, N. (eds) *New and Future Developments in Microbial Biotechnology and Bioengineering: Crop Improvement Through Microbial Biotechnology*. Elsevier, Amsterdam, pp. 263–291. DOI: 10.1016/B978-0-444-63987-5.00013-X.

Khan, M.W., Khan, M.R. and Khan, A.A. (1984) Identity of root-knot nematodes on certain vegetables of Aligarh district in Northern India. *International Nematological Network Newsletter* 1, 6–7.

Khan, M.R., Khan, M.W. and Khan, A.M. (1985) Cohabitation of *Meloidogyne incognita* and *Rotylenchulus reniformis* in tomato roots and effect on multiplication and plant growth. *Nematologia Mediterranea* 13, 153–159.

Khan, M.R., Khan, M.W. and Khan, A.A. (1986a) Effect of soil salinity on hatching, mortality, penetration and development of root-knot nematode, *Meloidogyne javanica*. In: *Proceedings of the National Conference on Plant Parasitic Nematodes*, New Delhi, p. 13.

Khan, M.R., Khan, M.W. and Khan, A.A. (1986b) Effect of mercury, a heavy metal, on hatching, mortality and penetration of *Meloidogyne incognita* on pepper. In: *Proceedings of the National Conference on Plant Parasitic Nematodes*, New Delhi, pp. 12–13.

Khan, M.R., Khan, A.M. and Khan, M.W. (1986c) Interaction between *Meloidogyne incognita*, *Rotylenchulus reniformis* and *Tylenchorynchus brassicae* on tomato. *Revue de Nematologie* 9, 245–250.

Khan, M.R., Khan, M.W. and Khan, A.M. (1986d) Interactions of *Meloidogyne incognita*, *Rotylenchulus reniformis* and *Tylenchorynchus brassicae* as cohabitants on eggplant. *Nematologia Mediterranea* 14, 201–206.

Khan, M.R., Khan, M.W. and Khan, A.M. (1987) Competitive interaction between *Meloidogyne incognita*, *Rotylenchulus reniformis* and *Tylenchorynchus brassicae* on cauliflower. *Pakistan Journal of Nematology* 5, 73–83.

Khan, M.R., Khan, M.W. and Khan, A.A. (1995) Response of some cabbage varieties to *Tylenchorhynchus brassicae*. *Annals of Applied Biology* 70–71. DOI: 10.1111/j.1744-7348.1996.tb07914.x.

Khan, M.R., Mohiddin, F.A., Khan, S.M. and Khan, B. (2005) Effect of seed treatment with certain biopesticides on root-knot of chickpea. *Nematologia Mediterranea* 32, 107–112.

Khan, M.R., Altaf, S., Mohiddin, F.A., Khan, U. and Anwer, A. (2009) Biological control of plant nematodes with phosphate solubilizing microorganisms. In: Khan, M.S. and Zaidi, A. (eds) *Phosphate Solubilizing Microbes for Crop Improvement*. Nova Science Publishers, Hauppauge, NY, pp. 395–426.

Khan, M.R., Majid, S., Mohidin, F.A. and Khan, N. (2011) A new bioprocess to produce low cost powder formulations of biocontrol bacteria and fungi to control fusarial wilt and root-knot nematode of pulses. *Biological Control* 59(2), 130–140. DOI: 10.1016/j.biocontrol.2011.04.007.

Khan, M.R., Mohiddin, F.A. and Ahamad, F. (2016a) Inoculant rhizobia suppressed root-knot disease, and enhanced plant productivity and nutrient uptake of some field-grown food legumes. *Acta Agriculturae Scandinavica, Section B – Soil & Plant Science* 68(2), 166–174. DOI: 10.1080/09064710.2017.1374448.

Khan, M.R., Mohidin, F.A., Khan, U. and Ahamad, F. (2016b) Native *Pseudomonas* spp. suppressed the root-knot nematode in *in-vitro* and *in-vivo*, and promoted the nodulation and grain yield in the field grown mungbean. *Biological Control* 101, 159–168. DOI: 10.1016/j.biocontrol.2016.06.012.

Khan, M.R., Ansari, R.A., Rizvi, T.F. and Rafeeq, K.A.M. (2019) Nematode problems in vegetable crops and its biomanagement. In: Khan, M.R., Mukhopadhyay, A.N., Pandey, R.N., Thakur, M.P., Singh, D. et al. (eds) *Bio-Intensive Approaches: Application and Effectiveness in the Management of Plant Nematodes, Insects and Weeds*. Today & Tomorrow's Printers and Publishers, New Delhi, pp. 67–99.

Khan, M.R., Haque, Z., Ahamad, F. and Zaidi, B. (2021) Biomanagement of rice root-knot nematode *Meloidogyne graminicola* using five indigenous microbial isolates under pot and field trials. *Journal of Applied Microbiology* 130(2), 424–438. DOI: 10.1111/jam.14788.

Khan, M.R., Mohiddin, F.A., Haque, Z. and Sharma, R.K. (2023a) Disease complexes and their sustainable management. In: Khan, M.R. and Quintanilla, M. (eds) *Nematode Diseases of Crops and Their Sustainable Management*. Academic Press, Cambridge, MA, pp. 65–93. DOI: 10.1016/B978-0-323-91226-6.00016-X.

Khan, M.R., Haque, Z. and Sharma, R.K. (2023b) Novel biotechnological interventions in plant nematode management technologies. In: Khan, M.R. (ed.) *Novel Biological and Biotechnological Applications in Plant Nematode Management*. Springer, Singapore, pp. 167–186.

Khan, T.A. and Husain, S.I. (1989) Relative resistance of six cowpea cultivars as affected by the concomitance of two nematodes and a fungus. *Nematologia Mediterranea* 17, 39–41.

Khier, A.M. and Osman, A.A. (1977) Interaction of *Meloidogyne incognita* and *Rotylenchulus reniformis*. *Nematalogia Mediterranea* 5, 113–116.

Kimenju, J.W., Wendot, P.K. and Thuo, A.K. (2022) Cumulative damage impact of plant parasitic nematodes in smallholder maize cropping systems in East Africa. In: Sikora, R.A., Desaeger, J. and Molendijk,

L.P.G. (eds) *Integrated Nematode Management: State-of-the-Art and Visions for the Future*. CAB International, Wallingford, UK, pp. 48–54. DOI: 10.1079/9781789247541.0007.

Kinloch, R.A. and Allen, M.W. (1972) Interaction of *Meloidogyne hapla* and *M. javanica* infecting tomato. *Journal of Nematology* 4(1), 7–16.

Kort, J. and Jaspers, C.P. (1973) Shift of pathotypes of *Heterodera rostochiensis* under susceptible potato cultivars. *Nematologica* 19(4), 538–545. DOI: 10.1163/187529273X00556.

Kort, J. and Bakker, J. (1980) The occurrence of mixtures of potato cyst-nematode pathotypes or species. *Nematologica* 26(2), 272–274. DOI: 10.1163/187529280X00152.

Kraus-Schmidt, H. and Lewis, S.A. (1981) Dynamics of concomitant populations of *Hoplolaimus columbus*, *Scutellonema brachyurum*, and *Meloidogyne incognita* on cotton. *Journal of Nematology* 13(1), 41–48.

Lawn, D.A. and Noel, G.R. (1990) Effects of temperature on competition between *Heterodera glycines* and *Pratylenchus scribneri* on soybean. *Nematropica* 20, 57–69.

Manachini, B., Landi, S. and Tomasini, V. (2005) Biodiversity of nematofauna of oilseed rape (*Brassica napus* L.). *Communications in Agricultural and Applied Biological Sciences* 70(4), 927–935.

Marshall, J.W. (1989) Changes in relative abundance of two potato cyst nematode species *Globodera rostochiensis* and *G. pallida* in one generation. *Annals of Applied Biology* 115(1), 79–87. DOI: 10.1111/j.1744-7348.1989.tb06814.x.

McGawley, E.C. and Chapman, R.A. (1983) Reproduction of *Criconemoides simile*, *Helicotylenchus pseudorobustus*, and *Paratylenchus projectus* on soybean. *Journal of Nematology* 15(1), 87–91.

McGawley, E.C. and Winchell, K.L. (1987) Greenhouse reproduction of single and combined *Meloidogyne incognita* and *M. javanica* populations on soybean. *Journal of Nematology* 19, 542.

McSorley, R. and Dickson, D.W. (1989) Effects and dynamics of a nematode community on soybean. *Journal of Nematology* 21(4), 490–499.

Miller, L.I. (1970a) Intraspecific variation in *Heterodera glycines* Ichinohe, summaries. In: *X International Symposium, Society of European Nematologists*, Pescara, p. 11.

Miller, L.I. (1988) Morphological comparisons of cyst cone structures of one isolate each of *Heterodera schachtii*, *H. glycines* and one of their isolates. *Journal of Nematology* 20, 648–649.

Miller, L.I. (1989) Morphological comparisons of second-stage juveniles of one isolate each of *Heterodera glycines*, *H. cruciferae* and one of their hybrids. *Journal of Nematology* 21, 574.

Miller, P.M. (1970b) Rate of increase of a low population of *Heterodera tabacum* reduced by *Pratylenchus penetrans* in the soil. *Plant Disease Reporter* 54, 25–26.

Miller, P.M. and Wihrheim, S.E. (1968) Mutual antagonism between *Heterodera tabacum* and some other parasitic nematodes. *Plant Disease Reporter* 52, 57–58.

Miller, P.M. and Mcintyre, J.L. (1975) *Tylenchorhynchus claytoni* feeding on tobacco root inhibits entry of *Pratylenchus penetrans*. *Journal of Nematology* 17, 327.

Miller, L.I. (1983) Diversity of selected taxa of *Globodera* and *Heterodera* and their interspecific and intergeneric hybrids. In: Stone, A.R., Platt, A.M. and Khalil, L.F. (eds) *Concepts in Nematode Systematics*. Academic Press, New York, pp. 207–220.

Minz, G. and Strich-Harari, D. (1959) Inoculation experiments with a mixture of *Meloidogyne* spp. on tomato roots. *Ktavim Records Agricultural Experiment Station* 9, 275–279.

Mishra, S.D. and Gaur, H.S. (1981) Effect of individual and concomitant inoculation with *Meloidogyne incognita* and *Rotylenchulus reniformis* on the growth of black gram (*Vigna mungo*). *Indian Journal of Nematology* 11, 25–28.

Misra, C. and Das, S.N. (1979) Interaction of some plant parasitic nematodes on the root-knot development in brinjal. *Indian Journal of Nematology* 7, 46–53.

Mohammed, R.K.A. and Khan, M.R. (2021) Management of root-knot nematode in cucumber through seed treatment with multifarious beneficial microbes in polyhouse under protected cultivation. *Indian Phytopathology* 74(4), 1035–1043. DOI: 10.1007/s42360-021-00422-3.

Mohiddin, F.A., Khan, M.R., Khan, S.M. and Bhat, B.H. (2010) Why *Trichoderma* is considered super hero (super fungus) against the evil parasites? *Plant Pathology Journal* 9(3), 92–102. DOI: 10.3923/ppj.2010.92.102.

Mustika, I. (1984) Effects of nematodes and fungi on the growth of pepper and yellow disease. *Pemberitaan Lembaga Penelitian Tanaman Industri, Bogor, Indonesia* 8, 28–37.

Neumann, G. and Shields, E.J. (2006) Interspecific interactions among three entomopathogenic nematodes, *Steinernema carpocapsae* Weiser, *S. feltiae* Filipjev, and *Heterorhabditis bacteriophora* Poinar,

with different foraging strategies for hosts in multipiece sand columns. *Environmental Entomology* 35(6), 1578–1583. DOI: 10.1603/0046-225X(2006)35[1578:IIATEN]2.0.CO;2.
Norton, D.C. (1978) *Ecology of Plant-Parasitic Nematodes*. Wiley, New York.
O'Bannon, J.H., Radewald, J.D., Tomerlin, A.T. and Inserra, R.N. (1976) Comparative influence of *Radopholus similis* and *Pratylenchus coffeae* on citrus. *Journal of Nematology* 8(1), 58–63.
Olsson, E. (1988) Hybridization experiments between *Globodera rostochiensis* and *G. pallida*. *Nematologica* 34, 285.
Paez, N., Arcia, M.A. and Meredith, J.A. (1976) Individual and combined effects of *Meloidogyne incognita* and *M. javanica* on four tobacco (*Nicotiana tabacum* L.) cultivars in Venezuela. *Nematropica* 6, 68–76.
Parrot, D.M., Berry, M.M. and Farrell, K.M. (1976) Competition between *Globodera rostochiensis* Ro 1 and *G. pallida* Pa3. *Rothamsted Experimental Station Report* 197–198.
Pathak, K.N., Nath, R.P. and Haider, M.G. (1985) Effect of initial inoculum level of *Meloidogyne incognita* and *Rotylenchulus reniformis* on pigeonpea and their interrelationship. *Indian Journal of Nematology* 15, 177–179.
Potter, J.W. and Fox, J.A. (1965) Hybridization of *Heterodera schachtii* and *H. glycines*. *Phytopathology* 55, 800–801.
Price, M., Riggs, R.D. and Caveness, C.E. (1976) Races of soybean-cyst nematode compete. *Arkansas Farm Research* 25, 16.
Price, M., Riggs, R.D. and Caveness, CE. (1978) Hybridization of races of *Heterodera glycines*. *Journal of Nematology* 10, 114–118.
Rao, K.T. and Seshadri, A.R. (1981) Studies on interaction between *Meloidogyne incognita* and *Rotylenchulus reniformis* on grapevine seedlings. *Indian Journal of Nematology* 11, 101–102.
Rao, Y.S. and Prasad, J.S. (1981) Interaction of *Heterodera oryzicola* and *Meloidogyne graminicola* in artificial inoculation in rice. *Indian Journal of Nematology* 11, 104.
Rhoades, H.L. (1985) Effects of separate and concomitant population of *Belonolaimus longicaudatus* and *Dolichodorus heterocephalus* on *Zea mays*. *Nematropica* 15, 171–174.
Ross, J.P. (1964) Interaction of *Heterodera glycines* and *Meloidogyne incognita* on soybeans. *Phytopathology* 54, 304–307.
Santo, G.S. and Bolander, W.J. (1977) Separate and concomitant effects of *Macroposthonia xenoplax* and *Meloidogyne hapla* on concord grapes. *Journal of Nematology* 9, 282–283.
Seinhorst, J.W. and Oostrom, A. (1989) Reproduction of a mixture of pathotypes Ro 1 of *Globodera rostochiensis* and Pa 3 of *G. pallida* and its offspring on cvs. Cardinal and Irene. *Nematologica* 35, 469–474.
Sharma, N.K. and Sethi, C.L. (1978) Interaction between *Meloidogyne incognita* and *Heterodera cajani* on cowpea. *Indian Journal of Nematology* 6, 1–12.
Sheela, M.S. and Venkitesan, T.S. (1981) Interrelationships of infectivity between the burrowing and root-knot nematode in black pepper, *Piper nigrum*. *Indian Journal of Nematology* 11, 105.
Sikora, R.A., Taylor, D.P., Malek, R.B. and Edwards, D.I. (1972) Interaction of *Meloidogyne naasi*, *Pratylenchus penetrans*, and *Tylenchorhynchus agri* on creeping bentgrass. *Journal of Nematology* 4, 162–165.
Sikora, R.A., Malek, R.B., Taylor, D.P. and Edwards, D.I. (1979) Reduction of *Meloidogyne naasi* infection of creeping bentgrass by *Tylenchorhynchus agri* and *Paratrichodorus minor*. *Nematologica* 25(2), 179–183. DOI: 10.1163/187529279X00208.
Sikora, R.A. and Roberts, P.A. (2018) Management practices: An overview of integrated nematode management technologies. In: Sikora, R.A., Coyne, D., Hallmann, J. and Timper, P. (eds) *Plant Parasitic Nematodes in Subtropical and Tropical Agriculture*. CAB International, Wallingford, UK, pp. 795–838.
Sikora, R.A., Desaeger, J. and Molendijk, L.P.G. (eds) (2022) *Integrated Nematode Management: State-of-the-Art and Visions for the Future*. CAB International, Wallingford, UK. DOI: 10.1079/9781789247541.0000.
Simon, A.C.M., Lopez-Nicora, H.D. and Niblack, TL. (2022) Impact of plant parasitic nematodes on maize in mid-western USA: An unrecognized or ignored threat to production. In: Sikora, R.A., Desaeger, J. and Molendijk, L.P.G. (eds) *Integrated Nematode Management: State-of-the-Art and Visions for the Future*. CAB International, Wallingford, UK, pp. 34–40.
Singh, N.D. (1976) Interaction of *Meloidogyne incognita* and *Rotylenchulus reniformis* on soybean. *Nematropica* 6, 76–81.
Stirling, G.R. (1991) *Biological Control of Plant Parasitic Nematodes: Progress, Problems and Prospects*. CAB International, Wallingford, UK.

Taha, A.H.Y. and Sultan, S.A. (1977) Populations of *Rotylenchulus reniformis* and *Tylenchulus semipenetrans* on grape seedlings as influenced by coincident infestations. *Nematologia Mediterranea* 5, 253–257.

Taha, A.H.Y. and Kassab, A.S. (1979) The histopathological reactions of *Vigna sinensis* to separate and concomitant parasitism by *Meloidogyne javanica* and *Rotylenchulus reniformis*. *Journal of Nematology* 11(2), 117–123.

Taha, A.H.Y. and Kassab, A.S. (1980) Interrelations between *Meloidogyne incognita*, *Rotylenchulus reniformis*, and *Rhizobium* sp. on *Vigna sinensis*. *Journal of Nematology* 12, 57–62.

Thomas, R.J. and Clark, C.A. (1980) Interactions between *Meloidogyne incognita* and *Rotylenchulus reniformis* on sweet potato. *Journal of Nematology* 12, 239.

Thomas, R.J. and Clark, C.A. (1981) *Meloidogyne incognita* and *Rotylenchulus reniformis* interactions in a sweet potato field. *Phytopathology* 71, 908.

Thomas, R.J. and Clark, C.A. (1983a) Effects of concomitant development on reproduction of *Meloidogyne incognita* and *Rotylenchulus reniformis*. *Journal of Nemaology* 15, 215–221.

Thomas, R.J. and Clark, C.A. (1983b) Population dynamics of *Meloidogyne incognita* and *Rotylenchulus reniformis* alone and in combination, and their effects on sweet potato. *Journal of Nematology* 15, 204–211.

Triantaphyllou, A.C. (1991) Further studies on the role of polyploidy in the evolution of *Meloidogyne*. *Journal of Nematology* 23(2), 249–253.

Triantaphyllou, A.C. and Esbenshade, P.R. (1990) Demonstration of multiple mating in *Heterodera glycines* with biochemical markers. *Journal of Nematology* 22(4), 452–456.

Turner, D.R. and Chapman, R.A. (1971) Infection of alfalfa and red clover by concomitant populations of *Meloidogyne incognita* and *Pratylenchus penetrans*. *Journal of Nematology* 3, 332.

Turner, D.R. and Chapman, R.A. (1972) Infection of seedlings of alfalfa and red clover by concomitant populations of *Meloidogyne incognita* and *Pratylenchus penetrans*. *Journal of Nematology* 4(4), 280–286.

Vaishnav, M.U. and Sethi, C.L. (1978) Pathogenicity of *Meloidogyne incognita* and *Tylenchorhynchus vulgaris* on bajra and their interrelationship. *Indian Journal of Nematology* 8, 1–8.

Van Gundy, S.D. and Kirkpatrick, J.D. (1975) Nematode–nematode interactions on tomato. *Journal of Nematology* 7, 330.

Weischer, B. (1993) Nematode–virus interactions. In: Khan, M.W. (ed.) *Nematode Interactions*. Chapman & Hall, London, pp. 217–231. DOI: 10.1007/978-94-011-1488-2_10.

Winoto, S. and Lim, T.K. (1972) Interaction of *Meloidogyne incognita* and *Rotylenchulus reniformis* on tomato. *Malaysian Agricultural Research* 1, 6–13.

Yang, H., Powell, N.T. and Barker, K.R. (1976) Interactions of concomitant species of nematodes and *Fusarium oxysporum* f. sp. *vasinfectum* on cotton. *Journal of Nematology* 8(1), 74–80.

12 Disease Complexes Involving Nematodes and Insects

Carla S. Pimentel[1]*, Mujeebur Rahman Khan[2] and Natsumi Kanzaki[3]
[1]*Forest Research Centre (CEF), School of Agriculture, University of Lisbon, Lisbon, Portugal;* [2]*Department of Plant Protection, Faculty of Agricultural Sciences, Aligarh Muslim University, Aligarh, India;* [3]*Kansai Research Centre, Forestry and Forest Products Research Institute, Kyoto, Japan*

Abstract

Tritrophic associations between nematodes, insects and plants mainly involve a small subset of lineages within the Aphelenchida and Tylenchida and fly, beetle and wasp insect vectors. Despite their relative rarity among plant dweller nematodes, a few high-profile communities have attracted researchers' attention over the past century, such as nematodes vectored to woody tissues by wood beetles, nematodes transmitted by Agaonidae pollinating wasps to fig fruits and the association of *Fergusobia* nematodes/*Fergusonina* flies with the galls in Myrtaceae trees. These tripartite interactions encompass a wide ecological spectrum, from entomoparasitism to mutualism, commensalism and plant parasitism, and have been the focus of research due to their phylogenetic, evolutionary and co-evolutionary implications. However, two nematode–insect–plant complexes have emerged as significant global pest-related issues: pine wilt, affecting coniferous trees, and red ring, affecting palms. Furthermore, although not parasitic to plants, there is the interesting case of the gill-knot nematodes, infecting fungus gnats and economically important varieties of mushrooms. This chapter explores the complexities of these interactions, including their potential ecological and economic impacts, and management strategies.

Introduction

Associations between insects and nematodes are known from the fossil record dating back more than 100 million years, with insect-parasitic nematodes appearing in the early Cretaceous. Evidence of tritrophic interactions between nematodes, insects and plants, precursors of plant parasitism, appeared a few million years later, in the mid-Cretaceous (Poinar, 2015). Giblin-Davis *et al.* (2013) made a conservative estimate of around 40,000–500,000 nematode species currently associated with insects worldwide, with the independent evolution of interactions that range from purely phoretic to entomoparasitic. However, these associations do not occur with the same frequency in all insect lineages, being rare or even absent in Lepidoptera (butterflies and moths), Orthoptera (grasshoppers) and Hemiptera (true bugs), while being common in Diptera (flies), Coleoptera (beetles) and Hymenoptera (mainly bees and wasps) (Giblin-Davis *et al.*, 2013). On the other hand, most nematodes associated with insects belong to the class Secernentea (Poinar, 1975), and tritrophic nematode–insect–plant associations only occur in a very small subset of lineages within two of its orders: stylet-bearing

*Corresponding author: carlapimentel@isa.ulisboa.pt

 Nematode Disease Complexes in Agricultural Crops (ed. Mujeebur Rahman Khan)
DOI: 10.1079/9781800625228.0012

Aphelenchida and Tylenchida (Giblin-Davis, 1993; Giblin-Davis *et al.*, 2003). Consequently, plant disease complexes integrating insects and nematodes generally involve flies, beetles or wasps associated with Aphelenchida or Tylenchida nematodes (Giblin-Davis *et al.*, 2003; Khan, 2008).

The order Aphelenchida comprises the larger number of the currently known nematode species involved in tritrophic interactions with pathogenic or potentially pathogenic effects on plants. The occurrence of these interactions is, however, restricted to only two families: Parasitaphelenchidae and Aphelenchoididae, in the superfamily Aphelenchoidoidea (Hunt, 1993; Giblin-Davis *et al.*, 2003; Haque and Khan, 2021). All aphelenchoid insect-vectored plant parasites probably evolved from mycophagous ancestors that were transmitted by phoresy to the breeding sites of their insect associates on plants. The subsequent evolution of plant parasitism was probably facilitated by pre-adaptations to phoretic transport, including the synchronization of the nematode's development with that of the vector and the inclusion in its life cycle of resistant stages adapted to surviving transport for some time, such as dauer juveniles or adult females (Giblin-Davis *et al.*, 2003).

The Parasitaphelenchidae includes the two nematode–insect–plant complexes with the greatest impact worldwide: the pinewood nematode, *Bursaphelenchus xylophilus*, and the red ring nematode, *B. cocophilus* (Khan 2020, 2023a). PWN is transmitted by beetles, the pine sawyer, *Monochamus* spp. (Cerambycidae), whereas RRN is transmitted by the South American palm weevil, *Rhynchophorus palmarum* (Curculionidae) (Giblin-Davis *et al.*, 2003). They arise from a large community of nematodes that inhabit woody plants at some stage in their life cycle. The vector beetles are themselves the primary or secondary pests of the nematode host plant species (Pimentel *et al.*, 2023). These includes mycophages, bacteriophages or predators (Tylenchida and Rhabditida) which are transported by phoresis to new habitats (susceptible trees and shrubs), but also entomoparasites that subsist directly on the beetles (mainly Tylenchida: Neotylenchoidea) (Massey, 1974; Giblin-Davis *et al.*, 2013). Currently, although it is generally believed that these nematodes still primarily follow a mycophagous lifestyle, several species of *Bursaphelenchus* associated with tree declines in the field exhibit slight to moderate pathogenicity when inoculated on to seedlings (Kanzaki *et al.*, 2011; Dayi and Akbulut, 2012; Kanzaki and Giblin-Davis, 2018). In addition, several species of the genera *Aphelenchoides* and *Laimaphelenchus* (Aphelenchoididae) have been described worldwide, associated with bark beetles and declining trees, although their relationship with tree diseases is unclear (Maleita *et al.*, 2018; Gu *et al.*, 2022; Pimentel *et al.*, 2023). Overall, the emergence of *B. xylophilus* and *B. cocophilus* as high-impact mortality agents, and the description of several other potential plant pathogens originating from this wide community of aphelenchid insect-vectored tree dwellers, highlights the potential for the selection for virulence within this group of nematodes.

As happens with the phloem of declining trees, the short-lived syconium of *Ficus* spp. (Moraceae) is home to several species of nematodes, which find here a habitat with reliable moisture content and abundant food items, including yeasts, fungi, bacteria and plant and pollen cells (Davies *et al.*, 2015; Van Goor *et al.*, 2023). This specious genus of trees, widely distributed in warmer climates, is pollinated by species-specific fig wasps (Chalcidoidea: Agaonidae), and these obligatory mutualisms have become an important model system for studying co-speciation due to the ubiquity of such associations (Cruaud *et al.*, 2012). Additionally, figs also support a large assemblage of non-pollinating chalcidoid wasps, which are plant parasitic (gall inducers) or parasitoids (Compton and Hawkins, 1992; Kerdelhué and Rasplus, 1996). Fig nematodes belong to the Tylenchida, Aphelenchida or Rhabditida, groups known to form associations with insects, occupying various niches within the syconium (Giblin-Davis, 1993; Giblin-Davis *et al.*, 2003; Davies *et al.*, 2010; Van Goor *et al.*, 2023). Similarly, the fig wasp species tend to form associations with several nematode species (Giblin-Davis *et al.*, 2013). It is probable that the abundance of wasps using the fig syconium had a significant role in the evolution of tritrophic associations, but the diversity of these associations is far from being fully understood. As an emblematic example of Aphelenchoidoidea insect-vectored plant parasites, the large radiation of species-specific Agaonidae-vectored *Schistonchus* spp. and closely related *Ficophagus* spp. and *Martininema* spp. are the only known nematode plant parasites dependent on an obligatory mutualism and are able to affect the plant directly while feeding on the florets (Davies *et al.*, 2015). This association indirectly reduces the lifespan and dispersal ability of pollinator wasps, hence impacting seed production (Shi *et al.*, 2019). Additionally, the fig syconium often harbors different species of apparently obligate plant-parasitic *Bursaphelenchus* spp., exhibiting convergent morphologies adapted to this habitat. However, the insect–vector relations associated with these nematodes are still largely unknown (Khan, 2012, 2023b; Kanzaki *et al.*, 2014, 2022).

Despite the huge importance of the tylenchids as a source of crop and tree pests, the evolution of nematode insect-vectored-plant parasitism is a rare event within this order, occurring only in one of its four suborders, the Hexatylina, which are primarily entomoparasites, most with a free-living mycetophagous phase (Siddiqi, 1986, 2000; Quist *et al.*, 2015). The most prevalent example is always the radiation of *Fergusobia* (Neotylenchidae) associated with *Fergusonina* flies (Diptera: Fergusoninidae). This association results in gall formation on flower, bud and leaf of Myrtaceae trees due to mutualism because the survival of the vector fly depends on the galls induced by the nematode, while the nematode uses the insects for its ride to new host plants as well as a source of nutrition during the transport (Giblin-Davis *et al.*, 2001). A lesser-known example of a Tylenchida/insect association is the nematode *Iotonchium* (Iotonchoidea: Iotonchiidae). While not parasites of plants, they infect fungus gnats (Diptera) and mushrooms, including economically important varieties such as the oyster mushroom, *Pleurotus* (Agaricomycetes), and for this reason are included in this chapter (Kanzaki *et al.*, 2021).

This chapter will explore know associations between nematodes and insects that are confirmed or potentially pathogenic to plants (or mushrooms, in the case of *Iotonchium*). It will cover important aspects for understanding their pathogenesis and potential impact, including the life cycle, biological relationships, etiology and symptoms, as well as environmental factors. Furthermore, the overall ecological and economic importance along with management strategies will also be discussed.

Pinewood Nematode (*Bursaphelenchus Xylophilus*)

Pinewood nematode (PWN) is a most important forest pest worldwide, inflicting death to millions of pine trees annually (Khan, 2020). The nematode incites wilt to coniferous trees, especially pines, for which the disease is commonly known as pine wilt disease (PWD) (Zheng and Khan, 2023). PWN was first described in Japan in 1905 (Mamiya, 1988), and it continues to devastate pine forests, having spread to other areas, including East Asia (China, Korea and Taiwan) and Southern Europe (Portugal and Spain) (Mota *et al.*, 1999; Abelleira *et al.*, 2011; Pimentel *et al.*, 2023). It is considered to be native to North America, where it is not viewed as a pest but rather as an innocuous saprophytic organism (Atkins *et al.*, 2021). However, in other global forests, PWD stands out as a major forest invasive disease, alongside other significant global forest diseases such as Dutch elm disease, white pine blister rust and chestnut blight. Notably, these diseases are invasive in Europe and North America, caused by Asian fungi, whereas PWD originated in North America and has become invasive in Eurasia. Due to its considerable economic and environmental impacts, numerous studies on PWD have been published over the past decades (Kishi, 1995; Mota and Vieira, 2008; Zhao *et al.*, 2008; Futai, 2013; Kanzaki and Giblin-Davis, 2016). Therefore, this chapter aims to summarize just the fundamental aspects of the disease.

Life cycle of *Bursaphelenchus xylophilus*

The life cycle of *B. xylophilus* is shown in Fig. 12.1. The dauer larvae (J_4) of PWN are carried to pine hosts by *Monochamus* spp., in their elytra and in the tracheal system (Zheng and Khan, 2023). At emergence from the diseased tree, the reproductive system of the beetle is not fully developed, requiring to feed on the bark of twigs of pine trees to achieve sexual maturity, a process

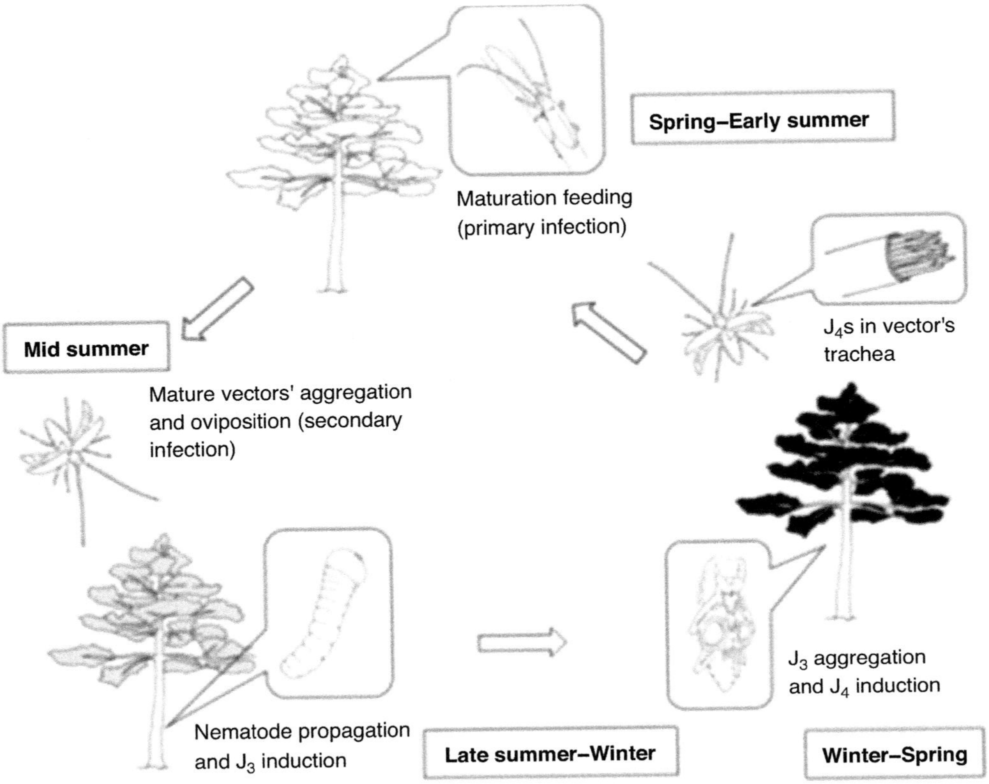

Fig. 12.1. Life cycle of the pinewood nematode (*Bursaphelenchus xylophilus*) and its insect vector the pine sawyer beetle (*Monochamus* spp.). [Figure courtesy of Natsumi Kanzaki.]

known as maturation feeding. The dauers (J_4) present in the beetle at this time infiltrate into the host tree through feeding wounds, lured by the pine volatiles (Zhang *et al.*, 2020). This process is commonly termed 'primary infection.' The earliest flying beetles in the spring carry the most PWNs (Pimentel *et al.*, 2014; Firmino *et al.*, 2017; Togashi *et al.*, 2023). This results in most nematodes being transmitted into the host tree during spring/earlier summer, allowing PWN populations to build up on the host during the hottest season, which, in mediterranean climates, coincides with the driest period. Thus, the wilting symptoms typically occur during the summer, a time favorable for PWN development (Pimentel and Ayres, 2018; Pimentel *et al.*, 2023).

After being inoculated into the host, the dauer larvae molt into adult males and females, which feed and propagate on plant tissues, gradually increasing their population. Symptoms of decline, such as discoloration of the host canopy (Fig. 12.2A) and cessation of resin exudation from external wounds, is observable 1–2 months after infection. As summer transitions into autumn, symptoms worsen, culminating in the browning of all leaves and eventual death. Dying or declining trees emit various volatiles, such as α-pinene, which attracts mature adults of *Monochamus* spp., leading them to aggregate for mating and oviposition on the susceptible host (Pajares *et al.*, 2004). In some cases, nematode dauers retained in the vector may enter the dying tree through the beetles' oviposition marks, resulting in secondary infection (Naves *et al.*, 2007). As their populations increase and the host environment becomes depleted, PWNs develop into a pre-dauer stage for overwintering (Khan, 2020). *Monochamus* larvae feed on the inner bark of the pine trees and bore into the

Fig. 12.2. Symptoms of pine wilt caused by *Bursaphelenchus xylophilus* in a pine forest area south of Lisbon in Portugal: (A) wilting maritime pines *Pinus pinaster* and (B) exit hole of pine sawyer beetle *Monochamus galloprovincialis*. [Figure courtesy of Carla Pimental.]

sapwood for overwintering and pupation in late spring. Meanwhile, the predauers aggregate around *Monochamus* pupal chambers and develop into dauers (J_4) in response to C_{16} and C_{18} fatty acid ethyl esters emitted at a late stage from the pupae (Zhao *et al.*, 2013). The J_4 enter the tracheal system of the adult beetle before they emerge from the pupae/hosts, eventually exiting the tree along with them, leaving behind the characteristic round emergence hole in the stem of the dead trees (Fig. 12.2B).

The life cycle can exhibit variations as adaptations to environmental conditions. For instance, in temperate zones, primary infections may occur later in the summer due to the slower development of vectors compared to warmer regions. Consequently, some infected trees may not display external symptoms such as discoloration and cessation of resin exudation in the year of infection, allowing them to survive for more than a year. These asymptomatic carriers, often referred to as 'latent carriers,' are frequently overlooked during efforts to eradicate infected hosts (Futai, 2003).

Detection and identification of PWN

PWN can be isolated from samples of host tissues using adaptations to the Baermann funnel method and can be identified morphologically based on four key characters observable in adults (Nickle *et al.*, 1981): a typical aphelenchoidid stylet and median bulb (Fig. 12.3A), discernible vulval flap (Fig. 12.3D), a broad and rounded tail of the female, sometimes with a very short mucron (Fig. 12.3E and F), and a male spicule with a somewhat trapezoidal shape (Fig. 12.3B and C). The male bursal flap is an important diagnostic feature of the genus (Fig. 12.3), but confirming its presence solely through microscopic observation can be challenging.

The dauers (J_4) of PWN can also be easily isolated from the vector *Monochamus* species, whereas the pre-dauers (J_3) can be isolated from the dying tree trunk as they are commonly found in dead wood. However, the absence of adults renders morphological identification impossible. Under this situation, the J_3 or J_4 should be cultured to obtain adults for identification (Pimentel *et al.*, 2022). Thus, molecular methods can be crucial for rapid and accurate identification of the pathogen. Polymerase chain reaction–restriction fragment length polymorphism (PCR-RFLP) and sequencing of the internal transcribed spacer (ITS) region have been widely used in the past (Burgermeister *et al.*, 2009), but currently, a loop-mediated isothermal amplification (LAMP) kit is commercially available (Kikuchi *et al.*, 2009), providing a more accessible and rapid detection method.

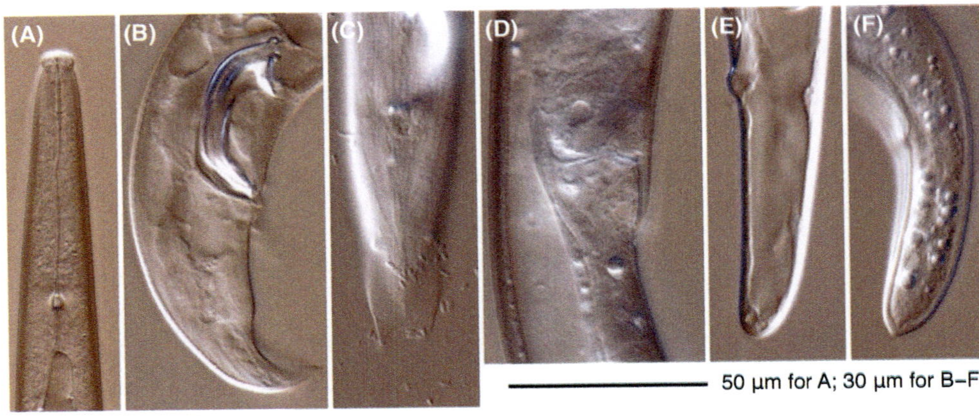

Fig. 12.3. Key diagnostic characters of the pinewood nematode. (A) Anterior region. (B) Right lateral view of a male tail. (C) Ventral view of a male tail with a concave terminus on the bursal flap, although this shape can vary among individuals, often presenting a bluntly pointed terminus. (D) Right lateral view of the vulval region. (E, F) Left lateral view of female tails showing variation between individuals. [Figure courtesy of Natsumi Kanzaki.]

Eco- and Pathophysiology of PWN

PWN is one of only two known lethal plant-pathogenic *Bursaphelenchus* species, alongside RRN, *B. cocophilus* (discussed in the next section), that infects stem tissues, causing a sort of sudden death syndrome in their respective hosts, pine trees and palms. This contrasts with the more well-known plant-parasitic tylenchids, such as *Meloidogyne*, *Heterodera* and *Globodera*, etc., which primarily affect root tissues, leading to plant death due to malnutrition. The mechanisms for the sudden death of trees are less well known, but the rapid onset of symptoms suggests a hypersensitive reaction of susceptible host pines to the infection by PWN. While the host defense responses are triggered, they fail to effectively eliminate the nematode, resulting in massive destruction of the conducting tissues of the host tree. Numerous researchers have conducted extensive molecular and chemical ecological analyses to understand the mechanisms of virulence in PWN and the onset of PWD (Kikuchi *et al.*, 2011; Zhao *et al.*, 2013, 2016; Cardoso *et al.*, 2024), but it still remains incompletely understood, necessitating concerted studies, particularly genetic and histochemical analyses employing the most up-to-date analytical methodologies (Cardoso *et al.*, 2024). The development of PWD relies on a multipartite system involving insect vectors and fungi, which serve as alternative nutritional substrates for PWN. Therefore, understanding the eco-physiology of the nematode is crucial when addressing PWD, alongside its direct effects on plant tissues.

PWN consistently adapts to local pine sawyers, hence it can be vectored by any species within the genus *Monochamus*. In its native continent of the North Americas, PWN was found to be abundantly associated with *M. carolinensis* and *M. titillator* in the southeastern USA (Pimentel *et al.*, 2014). After invading East Asia, PWN has adapted to local beetle species *M. alternatus* (Mamiya and Enda, 1972), and in Europe to *M. galloprovincialis* (Firmino *et al.*, 2017). PWN is attracted to the developing vector and aggregate around pupae inside the tree trunk. The development of the dispersal adopted stage of PWN (J_4) occurs in close proximity to the pupae, which is induced by a wide range of *Monochamus* species (Zhao *et al.*, 2013, 2014). This contrasts with other closely related nematodes in the *xylophilus* group, which exhibit species-specific induction. For example, development of J_4 of *B. conicaudatus* is specifically induced by its native vector, *Psacothea hilaris* (Kanzaki and Futai, 2001). This broad tolerance of vector species observed in PWN likely has facilitated the global expansion of PWD, contributing to its status as a global invasive species.

PWN belongs to a genus primarily comprised of mycophagous nematodes, and to a significant extent, it remains primarily mycophagous, as it is often cultured on *Botrytis cinerea* mats *in vivo*. Evidence suggests that the fungi colonizing the drying or recently dead pines serves as an important resource for the completion of the PWN's life cycle, although further research is needed to fully understand these interactions (Pimentel *et al.*, 2021, 2023; Ning *et al.*, 2023). Moreover, extensive research has been conducted on the chemical ecological interactions between PWN, its vector beetle, food fungus and other microbes, which may shape the development of PWD (e.g. Alves *et al.*, 2016; Cai *et al.*, 2021).

Management and control of PWD

PWN subsists in concealed habitats, under the bark of dying and dead wood, and is transported within the tracheal system of pine sawyer beetles. This poses challenges for conventional control methods such as chemical pesticides, e.g. methyl parathion, fenthion, carbaryl and thiacloprid, making them ineffective against the nematode. Consequently, targeting the vector becomes the primary approach for controlling PWD. However, given the vast forested regions impacted by PWD, with diverse geographical terrains with slopes, cliffs and valleys, ground spraying also presents significant challenges. Aerial spraying of chemical pesticides can be more effective in such cases, albeit with a potentially higher environmental burden and restrictions in some countries. Conversely, in flatland forests or garden areas, ground spraying of low-impact pesticides is feasible. Pesticides such as fenthion, thiacloprid, carboryl, etc., may prove effective in significantly suppressing the vector population. The nematode can also be targeted directly through trunk injection of commercial nematicides such as milbemectin, morantel tartrate, avermectin, etc. This method minimizes the environmental impact as the chemicals are enclosed within the host tree, although it requires labor-intensive procedures and relatively high costs, limiting its use to ornamental trees.

Biological and cultural methods are also viable. Commercial applications of entomopathogenic fungi such as *Beauveria bassiana* may prove effective against the sawyer beetle. However, the potential of parasites and parasitoids against the beetles has not been explored and needs to be duly researched. To prevent the spread of PWD, the declining and recently dead trees harboring the PWN and *Monochamus* larvae must be cut and removed from the site, followed by burial, burning or fumigation; however, this procedure is cumbersome, involving labor and other resources.

As a cultural control strategy, tolerant variants of susceptible pine species have been developed for replanting in pine forests and significant garden areas. However, further improvement in tolerance levels is necessary. Since pine trees are found in various environments, including natural forests, plantations, coastal windbreak forests and gardens, management strategies should be tailored to each environment, considering costs and environmental impacts.

Red Ring Nematode (*Bursaphelenchus cocophilus*)

Red ring nematode (RRN) causes red ring disease (RRD) in various palm species of the families Arecaceae and Cycadaceae. RRD leads to wilt and death of affected palms, often accompanied by the formation of a red band in the outer circumference of the trunk which appears as a 'red ring' on the cross-section of the dead stems. This symptom likely arises from the trees' resistance reaction to the disease and has become both the name and a diagnostic feature of RRD. Both the RRN and its insect vector, *R. palmarum*, are native to Central and South America, with local palm species exhibiting varying degrees of resistance or tolerance to RRD, which primarily affects exotic palms grown in commercial plantations in the area (Kanzaki *et al.*, 2023a) (Table 12.1).

Extensive research has been conducted to understand various aspects of RRD since its discovery at the beginning of the 20th century, owing to its economic significance. Extensive reviews of the effects of the pathogen, vector, hosts, their interactions and management

Table 12.1. Native distribution and susceptibility of major palm species to *Bursaphelenchus cocophilus*.

Palm species	Susceptible (reference)	Native range
Gru-gru palm (*Acrocomia aculeata*)	No	American tropics
Macaúba-barriguda tree (*Acrocomia intumescens*)	No	South American tropics
Arecanut palm (*Areca catechu*)	Yes (CABI, 2019)	Africa to South-East Asia
Cohune palm (*Attalea cohune*)	No	American tropics
Peach palm (*Bactris gasipaes*)	No	American tropics
Bactris sp.	No	American tropics
Coconut palm (*Cocos nucifera*)	Yes (Gunn *et al*. (2011)	South-East Asia
Cocos sp.	Yes	Asia-Pacific tropics
African oil palm (*Elaeis guineensis*)	Yes (Bakewell-Stone, 2023)	West African tropics
Açai (*Euterpe precatoria*)	No	American tropics
Patawa (*Oenocarpus bataua*)	No	South American tropics
Ita palm (*Mauritia flexuosa*)	No	South American tropics
Caraná (*Mauritia carana*)	No	South American tropics
Cocorite (*Attalea maripa*)	No	South American tropics
Bacaba (*Oenocarpus distichus*)	No	South American tropics
Canary Island date palm or pineapple palm (*Phoenix canariensis*)	Yes (Saro *et al*., 2019)	Canary Islands (NW coastal African tropics)
Date palm (*Phoenix dactylifera*)	Yes (Gros-Balthazard and Flowers, 2021)	Middle East to African region
Caribbean royal palm or cabbage palm (*Roystonea oleracea*)	No	American tropics
Cuban or Florida royal palm (*Roystonea regia*)	No	North and Central American tropics
Sabal palm (*Sabal palmetto*)	No	American tropics
Sabal sp.	No	American tropics

Table adapted from Kanzaki *et al*. (2023a).

methodologies can be found elsewhere (e.g. Giblin-Davis, 1993; Kanzaki and Giblin-Davis, 2016; Griffith *et al.*, 2018; Khan, 2020). Therefore, this chapter offer a concise account on the current understanding on RRD, its management strategies and future research prospects.

Life cycle of *Bursaphelenchus cocophilus*

RRN is an obligate nematode that requires a vector for spread from diseased tree to another tree. The dauers (J_3) of RRN are transmitted by black palm weevil (*Rhynchophorus* spp.), sugarcane weevil (*Metamasius* spp.) and possibly by other weevils. But *R. palmarum* is a major vector of RRN. The dauer larvae enter the weevil's body cavity (haemocoel) as well as its tracheal and reproductive systems when the weevil feeds on RRN-infected wood (Gerber and Giblin-Davis, 1990; Griffith *et al.*, 2018). RRN can potentially burden its vector, with the number of phoretic J_3 reaching up to 5000–13,000 individuals per beetle (Blair, 1969; Giblin-Davis, 1993). Furthermore, evidence suggests that the nematode body size increases within the vector, hinting at a potential parasitic relationship, although entomoparasitism has never been experimentally confirmed (Gerber *et al.*, 1989). The weevils are attracted to the volatiles from the healthy palms and enter or oviposit on the soft tissues of the leaf base and leaf bud. Furthermore, the identification of a *R. palmarum* aggregation pheromone, rhynchophorol (2-methyl-(5*E*)-hepten-4-ol), suggests aggregation behavior in selected hosts (Giblin-Davis *et al.*, 1996; Oehlschlager *et al.*, 2002). After entry of the weevil in the host, the nematode dauers leave the weevil if present in the tracheal system. The dauer may also be introduced along with the eggs laid by the weevil if present in the reproductive system. According to Giblin-Davis (1993) the main transmission may occur through the ovipositor, as 13,000 J_3 per weevil have been recorded in the region of the ovipositor. It enters the host, then feeds and disperses within the plant tissue, eventually developing into adults through J_4 propagative juveniles. Symptoms of RRD usually appear in the infected tree 1–2 months after infection, becoming more extensive as the nematodes spread and multiply (Goberdhan, 1964). Concurrently, the weevil eggs hatch and larvae develop by feeding on palm tissues. As the palm starts to deteriorate, it attracts even more weevils, which gather, enter the weakened palm and lay eggs there, leading to secondary infection (Giblin-Davis *et al.*, 1996). Populations of RRN develop to dauers (J_3) in a depleted host, which then enter the weevil larvae. The nematode dauers may survive in the decomposed coconut tissue for up to 8 months, and upon having the opportunity they enter in to the proliferated tracheal sac of the weevil before pupation (Griffith, 1987). The dauers may stay in the tracheal sac or may move to the site of the ovipositor through loops of the gut tract or haemocoel (Griffith, 1987). These weevil larvae mature and form cocoons for pupation, eventually emerging as adults, carrying 5000–13,000 J_3 that individually transmit RRN to new healthy or declining host palms (Gerber *et al.*, 1989). The life cycle of RRN is outlined in Fig. 12.4. The weevils infested with the nematode are found to be significantly smaller in size than non-infested weevils, apparently due to minimum fat bodies (Griffith, 1987), and 40–47% of the vector weevil population was found infested with RRN (Hagley, 1962; Weissling *et al.*, 1994).

Detection and identification of RRN

RRN can be identified on the basis of morphological characters (Hunt, 1993). RRN has a typical aphelenchoidid anterior region, somewhat mitten-shaped male spicule and very slender body with an *a* value (body length/diameter ratio) of around 100. In addition, a relatively long male tail ($c' > 3$) (Fig. 12.5A), spicule with clear dorsal and ventral limbs (Fig. 12.5A, B), and oval-shaped bursal flap (Fig. 12.5C) are the important diagnostic characters of *B. cocophilus*. This slender body is also a characteristic of the dispersal juveniles.

Molecularly, the nematode can be identified through ribosomal DNA sequencing (Ye *et al.*, 2009). A commercial kit has not been developed, but specific LAMP primers for the species have been established (Ide *et al.*, 2017).

Management and control of RRD

As in the case of PWN, RRNs remain consistently sheltered by either the host plant or the vector weevil. Therefore, targeting and managing the nematode directly is challenging, hence the insect vector becomes the primary target for control measures. Considering that numerous palm plantations play a role in food production, it is advisable to limit chemical controls such as aerial and ground spraying. Consequently, a widely adopted approach involves the use of pheromone traps combined with rhynchophorol, partially fermented palm or sugarcane tissues (used as oviposition substrates), alongside chemical insecticides such as methionyl, for trapping and killing the vector (Giblin-Davis *et al.*, 1996; Oehlschlager *et al.*, 2002). The prompt removal of dead or declining infected palms is also a proven method for controlling and preventing the spread of RRD (Warwick and Bezerra, 1992). However, before uprooting the tree, a 2-cm block from the middle of the trunk should be cut and examined to confirm the infestation of RRN (Khan, 2020) Additionally, while not yet widely implemented, exploring biological control options such as entomopathogenic fungi and parasites/parasitoids of weevils is worthwhile (Kanzaki *et al.*, 2023a, 2023b). Furthermore, in garden trees not involved in

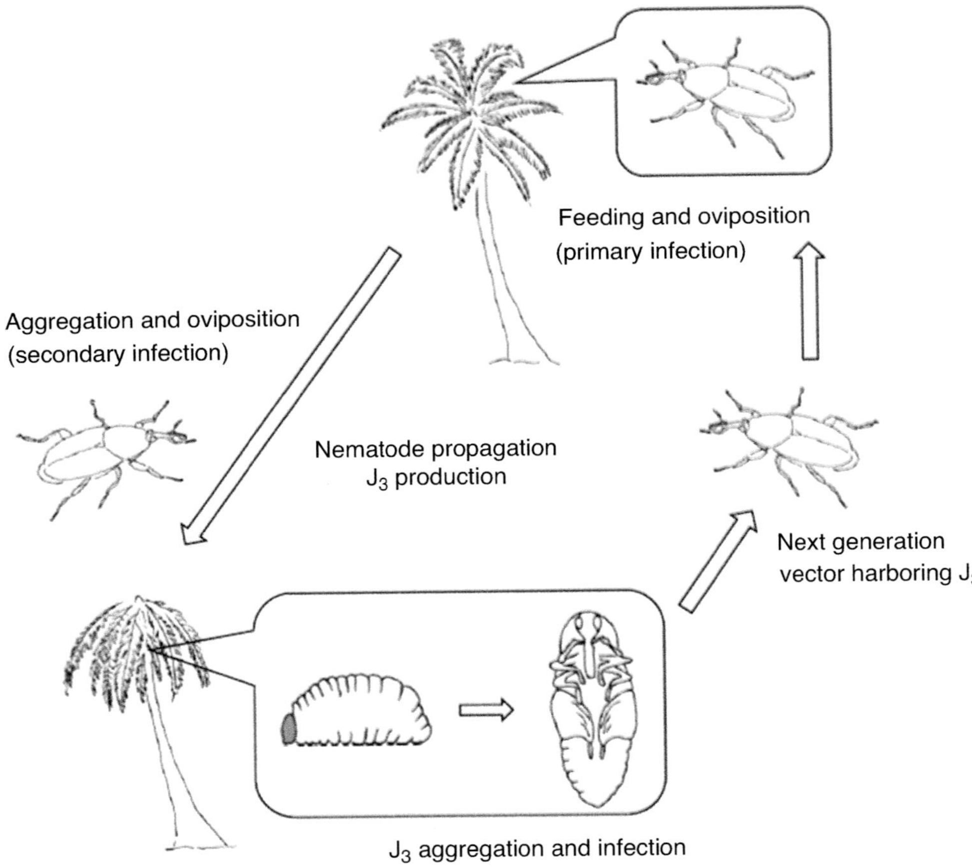

Fig. 12.4. Life cycle of the red ring nematode on palms. [Figure courtesy of Natsumi Kanzaki.]

food production, ground spraying of commonly used insecticides and/or trunk injection of nematicides could be considered.

Unlike PWN, which remains primarily a mycophagous organism that can be readily cultured on fungi, RRN has lost its ancestral capacity to feed on fungi and has become an obligate plant parasite. Currently, it can only be cultured on living tissue (cells) of palms. This limitation hampers comprehensive laboratory experiments, including physiological studies, assessments of resistance to chemicals, investigations into vector interactions (including potential vector range), production of dauer juveniles and examination of pathogenicity to plant tissue, all of which require stable and uncontaminated laboratory cultures. Therefore, the next crucial step in researching and managing RRD would involve establishing an accessible and dependable method for culturing the nematode.

The expansion of the distribution range of *Rhynchophorus ferrugineus*, another palm weevil that has emerged as a major global pest (Atakan *et al.*, 2012; Peri *et al.*, 2013; Rochat *et al.*, 2017), alongside the example of PWN, which can adapt to be vectored by congeneric *Monochamus* species (Mamiya and Enda, 1972; Linit, 1988; Pimentel *et al.*, 2014; Maehara *et al.*, 2020), raises concerns regarding the potential adaptation of RRN to other vectors within the genus *Rhynchophorus*. Such adaptation could elevate its risk of becoming a significant global pest. Additionally, ecological studies of RRN, including investigations into its natural distribution range and host/vector associations in the field, remain insufficiently explored. These essential

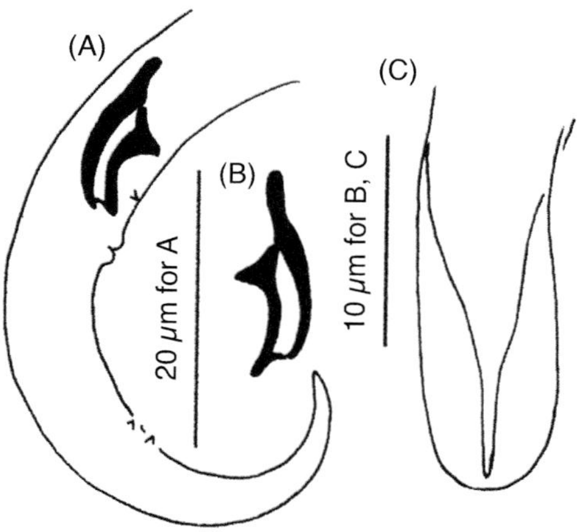

Fig. 12.5. key diagnostic characteristics in the male tail of *Bursaphelenchus cocophilus*. (A) Right lateral view. (B) Spicule in left lateral view. (C) Bursal flap. Illustrations were redrawn after the description of *Rhadinaphelenchus cocophilus* by Brathwaite and Siddiqi (1975). [Figure courtesy of Natsumi Kanzaki.]

biological inquiries may offer further insights into the ecological risks and potential control strategies for managing RRD.

Other Aphelenchids Potentially Pathogenic to Trees and Shrubs

In view of highly plant-parasitic nature of *B. xylophilus* and *B. cocophilus*, other *Bursaphelenchus* species have also been considered as potential plant pathogens, and serious efforts have been made worldwide to characterize the community of insect-vectored tree-dweller nematodes and their potential relationship with forest decline (Braasch, 2001; Gu *et al.*, 2006; d'Errico *et al.*, 2015; Kanzaki and Giblin-Davis, 2018). Pine forests were particularly targeted, not only because of the status of PWN as one of the main quarantine tree pests worldwide, but also because of the wide distribution and economic importance of pines, including the silviculture *Pinus* species, namely *P. radiata*, *P. taeda* and *P. patula* (Richardson and Rundel, 1998; Keeley, 2012). Extensive surveys on the nematodes associated with these *Pinus* spp. revealed numerous *Bursaphelenchus* spp., both in areas affected by PWN, such as Portugal (Penas *et al.*, 2004), as well as in areas not yet invaded by this pest (Akbulut *et al.*, 2008; Đođ *et al.*, 2015; Prospero *et al.*, 2015). Special attention has been paid to the *B. xylophilus* sister species *B. mucronatus* which is also morphologically similar to *B. fraudulentus*. *B. mucronatus* is also phoretic on pine sawyers *Monochamus* spp. but has not been proven to wither trees in the field, being apparently outcompeted by invading *B. xylophilus* (Tomminen, 1990; Jikumaru and Togashi, 2003; Vincent *et al.*, 2008). It has been found to be somewhat pathogenic to seedlings of some host species in greenhouse inoculation trials (Bakke *et al.*, 1991; Tomminen, 1993; Braasch, 2000; Akbulut *et al.*, 2007; Dayi and Akbulut, 2012). Furthermore, isolates of *B. mucronatus* were found to be able to hybridize with *B. xylophilus*, resulting in some pathogenic hybrids (Liu *et al.*, 2019; Tomalak and Filipiak, 2021).

Inoculation assays in seedlings, with other *Bursaphelenchus* spp. isolated from coniferous and/or bark and wood borer insects, have generally shown pathogenicity among most of them, indicating that virulence on these organisms could be regarded as a pre-adaptive character (Skarmoutsos and Michalopoulos-Skarmoutsos, 2000; Kanzaki *et al.*, 2011; Dayi and Akbulut, 2012). The *B. sexdentati* group has stood out, because these nematodes very commonly occur in Europe, and have been found associated with declining conifers and bark beetles. This group has also been found in the Middle East, Asia and Africa, often showing low to moderate virulence (Skarmoutsos and Michalopoulos-Skarmoutsos, 2000; Lange *et al.*, 2008; Slonim *et al.*, 2018). However, the pathogenicity of these organisms has not being proven in a forest habitat (Öztürk *et al.*, 2019), hence *Bursaphelenchus* spp., apart from *B. xylophilus*, are not generally considered to be pests of conifers. *Aphelenchoides* and *Laimaphelenchus* species (Aphelenchoididae) have also been frequently found associated with declining pine trees around the world, including in production forestry (Zhao *et al.*, 2007; Afshar *et al.*, 2021; Gu *et al.*, 2022; Pimentel *et al.*, 2023). However, no inoculation experiments have been carried out with these nematodes, which are considered to be mainly fungal feeders. Thus, a clear relationship between them and tree pathology has not been established.

Fewer studies are available on broadleaf trees, but apparently *Bursaphelenchus*, *Aphelenchoides* and *Laimaphelenchus* are also important components of the insect-vectored wood-inhabiting nematode community associated with broad leaf trees (Lazarova *et al.*, 2004; Taşdemir *et al.*, 2020). Notorious examples include *B. eremus*, transmitted by *Scolytus intricatus* (bark beetle), which has been isolated from declining oak species in Europe (Carletti *et al.*, 2007); however, *in vitro* experiments have not supported involvement of this nematode in the decline (Carletti *et al.*, 2011). Another example is *B. luxuriosae*, which was found in the haemocoel and tracheal system of *Acalolepta luxuriosa* (longhorn beetle). This beetle feeds and oviposits on leaves and shoots of *Araliacea* spp., supposedly transmitting the nematode to its host during these activities (Kanzaki and Futai, 2003). *B. luxuriosae* has been shown to be somewhat pathogenic on pine seedlings (Kanzaki *et al.*, 2011), and there is currently a need for further *in vitro* and *in vivo* experiments to test the pathogenicity of these interesting groups of nematodes on broadleaf trees. Nematodes of the genus *Laimaphelenchus*, vectored by beetles, have also often been found associated with declining and dying trees of *Quercus suber*, but their relationship with the disease is unclear (Maleita *et al.*, 2018).

In short, several aphelenchids transmitted to trees by insects have shown the potential to be pathogenic in inoculation experiments in seedlings, but were not found to be virulent in mature trees and forests. Further, they are vectored by bark and wood borer beetles, themselves generally considered to be primary or secondary tree pests. These pests form phoretic associations with mites, nematodes and/or fungi, which use the insects as a means of transport to colonize tree hosts (Hofstetter *et al.*, 2015). Thus, the potential for pathogenesis of these aphelenchids is not only intrinsic, but also dependent on interactions with vectors and other hitchhiking organisms. Thus, monitoring and control of wood aphelenchids should also consider their insect vectors and community interactions. Environmental factors are also determinant on the setting of tree disease, with water stress and high temperatures not only favoring pathogenesis by aphelenchids (Akbulut *et al.*, 2007), but may also promote fungal infection and bark beetles attack on the bee (Croisé *et al.*, 2001; Erbilgin *et al.*, 2021). Projections of future climate change predict that some areas may see increases in temperature and frequency of severe droughts (Cook *et al.*, 2018), hence elevating the risk of the emergence of novel disease complexes involving insect-vectored tree killing aphelenchoids. In addition, globalization, with the intensification of the movement of goods and people, further promotes the risk of the emergence of invasive species originating from these communities (Humble and Allen, 2006; Giblin-Davis *et al.*, 2013). This situation warrants much greater attention of the state agencies on monitoring forest nematodes and their insect vectors, to guarantee the biosecurity of forests worldwide. This may include vigilance on entry ports and monitoring forests through remote sensing and regular sampling of declining trees (Poland and Rassati, 2019). Similarly, controlling the populations of these nematodes will involve managing their insect vectors. Prompt removal of freshly killed and wilting trees should be considered as a standard measure for preventing the spread as well as control of the tree dwelling nematodes and their vectors.

Fig Nematodes and Fig Wasps

Ficus is the most diversified genus of woody plants in the world, containing over 750 species, predominantly pantropical, including trees, shrubs, climbers, epiphytes and strangler hemi-epiphytes. The genus not only has a high diversity of vegetative structures, but also of reproductive structures and strategies, with approximately the same number of species falling into two systems: monoecious (hermaphrodite) and gynodioecious (hermaphrodite and female) (Berg, 1989; Harrison, 2005). The *Ficus* syconium is the stem of an inverted, enlarged and fleshy inflorescence, which encloses small flowers inside, with a tiny opening (ostiole) at the apex, the entrance for the species-specific pollinating wasps Agaonidae (Anstett *et al.*, 1997). Figs are generally a nutritious and unprotected fruit, easy to handle and consume, and can be very abundant throughout the year, especially those of monoecious species that must maintain a constant population of pollinating wasps. Consequently, they can support a wide variety of frugivorous seed-dispersing animals and are considered keystone trees in many tropical forests (Shanahan *et al.*, 2001). They have also been an important component of the human diet since ancient times (Kislev *et al.*, 2006) and although *F. carica* is by far the most cultivated, many other species have been traditionally used by local populations in tropical and subtropical regions (Kunwar and Bussmann, 2006; Mon *et al.*, 2020). Thus, pests and pathogens affecting the production of edible figs have the potential to exert strong ecological and economic impacts in tropical and subtropical areas.

Life cycle of fig wasps and associated nematodes

The syconia of monoecious figs contains both male and female flowers. The maturation of the flowers can be categorized into five stages (Galil and Eisikowitch, 1968):

A. Pre-female: small syconium with the ostiole still closed.
B. Female: the female flowers reach maturity, the pollinating female wasps enter the syconium, pollinating flowers with long styles and ovipositing in those with short styles, dying shortly afterwards
C. Inter-floral: *Ficus* embryos develop in pollinated flowers, at the same time as wasp larvae develop inside galled flowers
D. Male: wasps complete their development before the seeds ripen, the wingless males emerge first, immediately entering the galls of the female wasps and fertilizing them, then they aggregate in the upper part of the syconium where the male flowers are located, they begin to tunnel, piercing the flowers and spreading pollen in the process, eventually opening an exit to the outside of the syconium, dying shortly afterwards; the fertilized female wasps use this tunnel to exit the syconium, collecting pollen in the process.
E. Post-flowering: the figs grow and ripen to be dispersed by foraging animals.

In gynodioecious *Ficus* species, different plants produce different types of syconia: female plants only produce syconia that exclusively bear female flowers with long styles, which allows for pollination by wasps and seed production, but not oviposition and galling; other plants produce syconia that bear both male and female flowers, but the latter all have short styles and can all be oviposited and galled by wasps, producing no seeds. Thus, gynodioecious *Ficus* species are functionally dioecious, with female plants producing seeds, and male plants producing pollens and wasps. The best-known example of this reproductive strategy is *F. carica*, which produces the world's most consumed edible figs. The edible figs are the mature syconia of the female trees, while the male trees produce an inedible syconium, commonly known as caprifig.

Around 30 species of *Schistonchus* and closely related *Ficophagus* spp. and *Martininema* spp. were described infesting figs in Europe, Asia, Australia, Africa, and North and Central America (Davies *et al.*, 2015; Kanzaki *et al.*, 2023b). The nematode species were found in all types of syconia, females, caprifigs and monoecious (Davies *et al.*, 2015; Van Goor *et al.*, 2023). In general, female pollinating wasps are the most reliable vectors for nematode transmission in figs, since the pollinating and non-pollinating male wasps are generally wingless and never disperse from figs, and the females of non-pollinating fig wasps oviposit by piercing the syconia and not by entering it (Krishnan *et al.*, 2010). Pollinating females disperse the nematodes from the figs in phase D, introducing them into fresh syconia in phase B, with the vector and the nematode developing at the expense of the plant tissues, mainly the flowers, during phase C (Reddy and Rao, 1986; Vovlas *et al.*, 1992, 1998; Giblin-Davis *et al.*, 1995).

The diversity among the vector, nematode and plant hosts species leads to extensively diverse mechanisms operative in this tripartite interaction, which is far from being fully understood (Van Goor *et al.*, 2023). A few of these tripartite systems have been studied on monoecious *Ficus* plants. On Asian *F. racemosa*, the dauers (J_3) of *S. racemosa* are carried in the abdominal folds of the female

Ceratosolen fusciceps (pollinating wasp). The parents of these nematodes reach maturity and mate and lay eggs in the syconium of the fig at stage C (Reddy and Rao, 1986; Krishnan *et al.*, 2010), whereas in African *F. thonningii*, juveniles, females and males of the nematode *S. africanus* n. sp. were found both in the abdomen of the female wasp *Elisabethiella stuckenbergi*, as well as in female and male florets of the fig. Nematode densities were far higher on the florets (150–2000 nematodes/g fresh weight of flower tissue) compared to the vector, having only a nematode to be transmitted (Vovlas *et al.*, 1998). On the New World *Ficus laevigata*, the nematode *Ficophagus laevigatus* is transported as mated entomogenous females in the haemocoel of *Pegoscapus* sp. The nematodes complete the life cycle on male flowers causing epidermal cell hypertrophy of the anthers, while the wasps develop on female flowers (Giblin-Davis *et al.*, 1995, 2023). The most well-known gynodioecious *Ficus* and economically most important is *F. carica*, on which the first ever species of *Schistonchus* as *S. caprifici* was described in the south of Europe. This nematode species is vectored by the pollinator wasp *Blastophaga psenes*. The activity of *S. caprifici* leads to necrosis and cavitation in the cortical parenchyma of female flowers on both caprifigs and commercial female figs (Khan, 2020). However, a lower population is attained by the nematode in the edible figs which is a reproductive dead end 12,000 versus 550 nematodes/g fresh floret tissue, respectively). The wasp develops in the inflorescences of caprifigs, whereas *S. caprifici* reproduces in the haemocoel of the wasp larvae. All stages of the nematode (e. ggs, juveniles and adults) persist in the pupae and adults (winged females) of the vector, suggesting an entomoparasitic phase (Vovlas *et al.*, 1992).

Potential ecological and economic impact of fig nematodes

Although limited, research has demonstrated the potential for high densities of *Schistonchus* to affect the lifespan of pollinating wasps, as well as their reproductive success, dispersal capacity and risk of predation, impairing pollination services and reducing seed production in figs (Gupta and Borges, 2019; Shi *et al.*, 2019). Further, the quality of pollination services has the potential to affect the quality of the figs produced, as observed in at least some varieties of *F. carica* (Gaaliche *et al.*, 2011). However, fig nematodes are not generally considered as crop pests and have attracted research mainly as a model to study the evolution of species interactions, with a large proportion of recent studies focusing on species descriptions and not so much on their ecological role in fig systems (Van Goor *et al.*, 2023). Still, the widespread distribution, abundance and diversity of *Schistonchus* spp., *Ficophagus* spp. and *Martininema* spp. (Davies *et al.*, 2015) and the recent discovery of other plant parasites that are part of the syconia community, such as *Bursaphelenchus* spp. (Kanzaki *et al.*, 2014; Kanzaki *et al.*, 2022), make this system worthy of further investigation for better understanding the evolution, ecology and etiology of the tripartite interaction involving nematodes, vector insects and *Ficus* species, which are likely to be future major plant pests.

Fergusobia Nematodes and *Fergusonina* Flies

A large number of hosts, mostly eucalypts from the genus *Angophora*, *Corymbia* and *Eucalyptus* (over 90 species), susceptible to *Fergusonina* fly and/or *Fergusobia* nematode have been registered in Australia. However, other Myrtaceae members including *Leptospermum*, *Melaleuca*, *Metrosideros* and *Syzygium* can also be affected, not only in Australasia but also in South and South-East Asia (Nelson *et al.*, 2014; Khan, 2020). In general, each *Fergusonina* fly/*Fergusobia* nematode pair is host-specific, causing a characteristic type of gall on a specific plant tissue; very rarely does a pair affect different tissues of its Myrtaceae host (Davies *et al.*, 2016). The different types of galls are classified according to their location: leaf galls (on the petiole or on the lamina), shoot galls (axillary or terminal) and flower bud galls (on the stigma, stamen or ovary). In addition, each of these types can be either unilocular or multilocular (Taylor, 2004; Nelson *et al.*, 2014).

Life cycle of *Fergusonina* flies and associated *Fergusobia* nematodes

Fergusobia nematodes complete the life cycle on two hosts (dicyclic). In one generation the life cycle starts in the fly and ends in the plant's gall, and another generation starts in the gall and ends in the fly. The *Fergusonina* fly carries dauer nematodes (J_4) in its haemocoel and introduces them into plant tissue through small wounds to lay its eggs. The dauers begin to feed immediately and introduce esophageal gland secretions. The secretions contain plant growth hormones which accelerate cell division into the host tissues inducing (hyperplasia) that leads to gall formation. The fly larvae enter the gall, generally one larva/locule inside the gall chamber. The fly larvae feed on the cells lining the wall (Giblin-Davis *et al.*, 2004; Davies *et al.*, 2016). The nematode dauers also feed in the locules and give rise to parthenogenetic females, which produce eggs. The eggs hatch giving rise to larvae that develop into amphimytic males or females, which mate inside the locules (Fisher and Nickle, 1968). The galls mature during these events, generally by the end of winter or early spring, before the advent of the summer when high temperatures prevail. The hot climate negatively affects nematode development (Goolsby *et al.*, 2000; Nelson *et al.*, 2014). The fertilized female nematodes enter the third instars and lay eggs in the haemocoel before the instar pupate. The nematode overwinters as eggs which hatch and molt to become adult (J_4) in the spring, synchronized with the termination of insect pupation. The adult fly containing nematode dauers bores its way out of the galls to continue the life cycle (Siddiqi, 2000; Giblin-Davis *et al.*, 2001).

Ecological and economic impacts

The complex ancient evolutionary relationship of *Fergusonina* and *Fergusobia* has attracted much research on the phylogeny, evolution and co-evolution of this dynamic tritrophic mutualistic interaction (Taylor, 2004; Ye *et al.*, 2007; Davies *et al.*, 2010). Generally, galls may function as a metabolic sink, which contains high densities of nematodes and insect larvae which potentially affect seed production and tree vigor (Dorchin *et al.*, 2006). *Fergusobia*/*Fergusonina*-induced galls have been of some concern on commercial plantations of eucalyptus in tropical regions (Safitri *et al.*, 2019). However, the extent of suppression of plant growth and yield (wood, oil, etc.) due to the disease complex incited by *Fergusobia* nematodes and *Fergusonina* flies is not well known. Limited research on the negative effects of these tritrophic interaction including the management aspects has been done (Goolsby *et al.*, 2000; Giblin-Davis *et al.*, 2001).

Gill-Knot Nematodes, *Iotonchium* Spp.

The genus *Iotonchium* currently comprises ten species (Poinar, 1991; Siddiqi, 2000; Tsuda and Futai, 2005), with the most well known being *Iotonchium ungulatum* vectored by the fungus gnat *Allodiopsis domestica* (Diptera: Mycetophilidae). The fungus gnats are small, dark and short-lived flying insects from seven families in the superfamily Sciaroidae, which attack fungi. Infection by the gnat/*Iotonchium* association results in the development of galls on the lamellae (gills) beneath the caps of mushrooms (Fig. 12.6). Fungus hosts are typically the common wood-rotting basidiomycetes *Pleurotus* spp., found on dead logs or stumps of various broad-leaved trees, such as *Quercus* spp. Several of these species are edible mushrooms, particularly the oyster mushroom *Pleurotus ostreatus*, and can be cultivated using logs or wood chips/flakes. The infected mushrooms remain edible, but their commercial value is diminished due to their altered appearance, which may deter consumers and they may also decay more rapidly than healthy mushrooms. Consequently, this relatively understudied insect–nematode disease complex poses economic concerns for mushroom cultivation. The nematode is referred to here as gill-knot nematode (GKN), responsible for causing mushroom gill-knot disease (GKD).

Besides its potential and economic impact, GKN and GKD have several biologically interesting

Fig. 12.6. Oyster mushroom *Pleurotus ostreatus* with galls on its lamellae (gills) as indicated by arrows (white spots) caused by the gnat/*Iotonchium* association. [Figure courtesy of Natsumi Kanzaki.]

characteristics worthy of research. First, *Pleurotus* spp. are well known carnivorous fungi (Barron and Thorn, 1987). Their hyphae have drop-like structures known as toxocysts on their surface, which contain toxins such as ostreatin (Kwok *et al.*, 1992) and 3-octanone (Lee *et al.*, 2023). Upon contact, a wide range of nematodes, including GKN, are quickly paralyzed. Subsequently, the fungus extends its hyphae to enter the paralyzed nematode's body by either penetrating the cuticle or exploiting natural openings, ultimately digesting the contents to obtain nutrients, particularly nitrogen. The toxocysts are continuously produced on the filamentous hyphae tissue, but several laboratory experiments have suggested that unfavorable conditions (e.g. poor nutrient availability) and mechanical damage enhance fungal predation of nematodes (Truong *et al.*, 2007). Although toxins are not produced by the fruiting bodies (mushrooms), nematode feeding injures on these tissues may lead to their regeneration through the formation of hyphae, which subsequently secrete toxins. The galls induced by GKN serve as its feeding structure, shielding them from the toxic hyphae resulting from feeding injury. This bears some resemblance to root galls induced by *Meloidogyne* spp., where root tissue is modified without causing hypersensitive cell death to become nematode feeding sites. Other *Iotonchium* spp. that feed on the fruiting bodies of non-carnivorous mushroom species do not induce galls or other feeding structures (Poinar, 1991; Tsuda and Futai, 1999, 2005). Thus, GKN gall induction on *Pleurotus* mushrooms can be regarded as the result of their evolutionarily established histochemical and physiological interactions.

GKN is also an interesting animal model to study life history strategies due to its unusual reproductive strategies, alternating between gonochoristic and parthenogenetic reproduction, a rarity among nematodes. A comparable phase-dependent alternation of reproductive form is observed in various groups of insects, such as gall wasps parasitizing *Quercus* trees (Ide and Abe, 2021). In this system, the parthenogenetic generation produces male and female progenies through gonochoristic reproduction, followed by the subsequent generation reverting to parthenogenesis. This strategy is, however, rare among nematodes.

Life cycle of *Iotonchium ungulatum* and fungal gnats

GKN has both insect-parasitic and mycophagous phases. In the entomoparasitic form, it uses its vector gnat to obtain nutrients and for transportation. During its mycophagous phase, it feeds and reproduces on the fruiting body or mushrooms (Tsuda *et al.*, 1996; Tsuda and Futai, 2000). The host mushroom can be found throughout the year, but it is mostly observed in the field from late autumn to winter.

The female *A. domestica*, once parasitized by the entomoparasitic female GKN, becomes sterilize as the nematode absorbs nutrients, leading to the atrophy of her reproductive system. Subsequently, the entomoparasitic female nematode lays eggs within the body cavity of the host gnat, where these eggs develop into entomoparasitic juveniles. Despite being sterilize, parasitized female gnats are still attracted to the mushroom and exhibit mating and oviposition behaviors. These infected females deposit entomoparasitic juvenile nematodes on to the mushroom gill, rather than laying gnat eggs. Subsequently, GKN penetrates the mushroom tissues and mature into large parthenogenetic mycophagous females, which induce gall formation. Within the gall, nematode females lay eggs, which then develop into gonochoristic free-living or infective adults through mycophagous juvenile stages. Simultaneously, healthy female gnats also visit the same mushroom and lay eggs, which hatch and the emerging gnat larvae feed on the mushroom tissues. The mushrooms hosting both nematodes and gnat larvae subsequently begin to rot and eventually collapse and may also fall on the soil. The gnat larvae emerge from the decaying mushroom and burrow into the soil for pupation. Concurrently, the nematodes also burrow into the soil, tracking the gnat larvae and parasitize their pupae. Parasitized pupae emerge as adults, yet gnat males often succumb to parasitism during development, presumably due to malnutrition. Because GKN is vectored to host mushrooms by female fungus gnats via their oviposition, nematodes parasitizing male gnats are in a reproductive dead end. The life cycle of GKN is outlined in Fig. 12.7.

Detection and identification

Over the past decades, there has been a significant revision and consolidation of the identification and classification of the genus Iotonchium. Tsuda and Futai (2005) addressed the generic status of Iotonchium and Fungiotonchium,

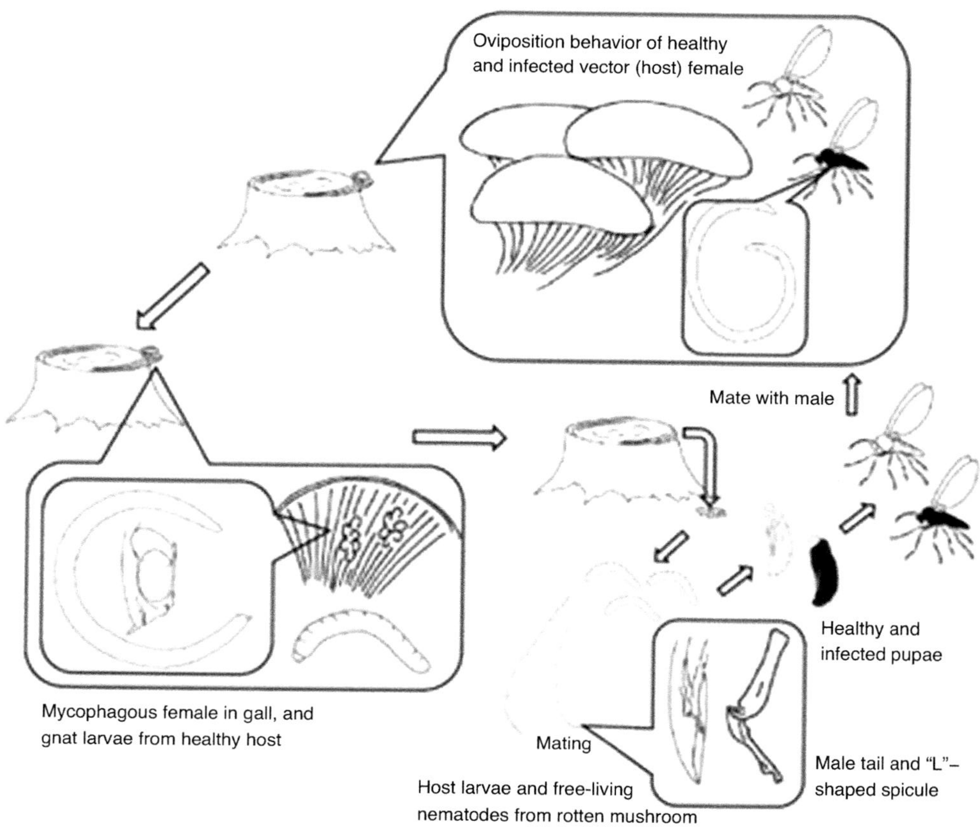

Fig. 12.7. Life cycle of gill-knot nematode *Iotonchium ungulatum* and its insect vector the fungus gnat *Allodiopsis domestica*. [Figure courtesy of Natsumi Kanzaki.]

initially established by Siddiqi (1986), and amalgamated these two genera into *Iotonchium*. This decision was prompted by the observation of several species displaying intermediate characteristics between the two genera. Recently, a phylogenetic analysis of four *Iotonchium* species confirmed the existence of these clades, thus supporting the generic revision by Tsuda and Futai (2005) (Kanzaki *et al.*, 2021).

The occurrence of both a fungus gnat and a mushroom parasitic phase is a defining feature of the genus *Iotonchium*. A distinct 'L'-shaped male spicule, lacking a gubernaculum, is its most notable morphological trait (see male phase in Fig. 12.7). The morphology of an entomoparasitic female of GKN was first described by Tsuda and Futai (2000). Subsequently, the nematode was taxonomically classified as *I. ungulatum* by Aihara (2001), based on the morphological characteristics of mycophagous and infective females, as well as males. These females are, so far, exclusively found within the body cavity of *A. domestica*, exhibiting the typical form of entomoparasitic females in the superfamily Tylenchoidea. They possess a large (approximately 5 mm in length) ventrally coiled or spiral body and are oviparous (Siddiqi, 2000; Tsuda and Futai, 2000). Identifying the entomoparasitic female GKN without first identifying the host gnat, which is also challenging, poses significant difficulties. Therefore, to accurately identify GKNs, it is preferable to locate either the mycophagous female from the mushroom gall or the free-living males.

The potential for molecular identification of GKN lies in the ribosomal DNA, which includes approximately 1.7 kb of near full-length 18S, about 1 kb of ITS, roughly 1.6 kb of the 5′ half of 28S and 660 bp. These sequences have been confirmed to be nearly identical among four geographically distant populations of GKNs (Kanzaki *et al.*, 2021). Therefore, molecular sequencing has the potential to be the simplest and most widely applicable method for identifying GKD.

Management of GKD

In GKD, both the nematode and vector gnat are considered as pests of mushroom; the nematode induces galls, while gnat larvae feed on the mushroom. Since GKN can only be transmitted by the gnats (Tsuda *et al.*, 1996), management of the nematode mainly involves the control of vector gnats. However, because oyster mushrooms are consumed as food, chemical control of the disease has toxicological limitations. Moreover, biological control using other insects or nematode-parasitizing fungi may affect mushroom production, hence GKN management is challenging. Consequently, the primary approach is to produce mushrooms in a controlled environment, where access of gnats to their host is physically prevented. This is often achieved in buildings with screened doors allied with chemical or physical gnat-trapping methods. Alternatively, in outdoor production, substrates for mushroom cultivation (e.g. logs or wood chips) should be covered with plastic mesh to prevent gnat infestation. These control measures have been successfully applied to oyster mushroom production in Japan, the only country where major damages by GKD have been reported (Tanaka *et al.*, 2002).

Due to the limitations in culturing the nematode and vector gnat because of the complexity of their life cycles, adequate physiological, histochemical and genetics (genomic and transcriptomic) data are lacking. However, the host mushrooms are relatively easy to culture, and it is technically feasible to maintain this tritrophic relationship in the laboratory or outdoor conditions. Therefore, establishing a culture system for the nematode and vector followed by further analyses on the above aspects will help in fully understanding this complex tritrophic interaction.

Conclusion and Future Prospects

Nematode plant diseases vectored by insects are of lesser occurrence compared to other plant diseases caused by nematodes and distributed in specific geographical areas. However, such diseases have high economic significance, e.g. PWD and RRD. To this list, the GKN and fig gall nematode can also be added, which affect mushroom and fig crops, respectively. A few tritrophic interactions involving nematodes, insects and plants have been extensively researched, which include nematodes vectored to woody tissues of trees and shrubs by wood beetles, nematodes transmitted to the syconia of fig trees by the vector wasps, and gall-inducing *Fergusobia* nematodes associated with *Fergusonina* flies and *Myrtaceae* trees. Detailed studies on these disease complexes have disentangled complex issues of the phylogenetic, evolutionary and co-evolutionary interactions, encompassing entomoparasitism, mutualism and commensalism. Plant parasitism has frequently evolved within these communities, and thus there is always the possibility of the emergence of the next high-impact pest. However, comprehensively understanding the emergence of virulence in insect-transmitted nematodes necessitates a thorough understanding of the intricate community structure and species interactions. The fact that these nematodes are dispersed by flying insects, coupled with the potential for vector and host shifts, increases the risk of making global invasive pests, akin to PWN. Globalization, with the intensified movement of goods and people, further amplifies the risk of invasive species emerging from these communities. Therefore, while it may seem uncommon for economically significant plant pests to arise from insect-borne nematodes, it remains imperative to maintain vigilance through ongoing research and surveillance efforts.

References

Abelleira, A., Picoaga, A., Mansilla, J.P. and Aguin, O. (2011) Detection of *Bursaphelenchus xylophilus*, causal agent of pine wilt disease on *Pinus pinaster* in north-western Spain. *Plant Disease* 95(6), 776.

Afshar, F.J., Rashidifard, M., Abolafia, J., Zouhar, M., Fourie, H. *et al.* (2021) *Laimaphelenchus africanus* n. sp. (Tylenchomorpha: Aphelenchoididae) from South Africa, a morphological and molecular phylogenetic study, with an update to the diagnostics of the genus. *Journal of Nematology* 53, 1–14.

Aihara, T. (2001) *Lotonchium ungulatum* n. sp.(Nematoda: Iotonchiidae) from the oyster mushroom in Japan. *Japanese Journal of Nematology* 31(1–2), 1–11.

Akbulut, S., Yüksel, B., Serin, M., Baysal, I. and Erdem, M. (2007) Pathogenicity of *Bursaphelenchus mucronatus* in pine seedlings under greenhouse conditions. *Turkish Journal of Agriculture and Forestry* 31, 169–173.

Akbulut, S., Vieira, P., Ryss, A., Valadas, V., Keten, A. *et al.* (2008) *Bursaphelenchus* Fuchs, 1937 (Nematoda: Parasitaphelenchidae) species associated with *Pinus* species in northern Turkey. *Helminthologia* 45(2), 89–95.

Alves, M., Pereira, A., Matos, P., Henriques, J., Vicente, C. *et al.* (2016) Bacterial community associated to the pine wilt disease insect vectors *Monochamus galloprovincialis* and *Monochamus alternatus*. *Scientific Reports* 6, 23908.

Anstett, M.-C., Hossaert-McKey, M. and Kjellberg, F. (1997) Figs and fig pollinators: Evolutionary conflicts in a coevoled mutualism. *Trends in Ecology & Evolution* 12(3), 94–99.

Atakan, E., Yüksel, O. and Soroker, V. (2012) Current status of the red palm weevil in canary island date palms in Adana. *Türkiye Entomoloji Bülteni* 2, 11–22.

Atkins, D.H., Davis, T.S. and Stewart, J.E. (2021) Probability of occurrence and phenology of pine wilt disease transmission by insect vectors in the Rocky Mountains. *Ecological Solutions and Evidence* 2(1), e12044.

Bakewell-Stone (2023) Elaeis guineensis (African oil palm). *CABI Compendium* (Online). Available at: https://doi.org/10.1079/cabicompendium.20295 (accessed March 2024).

Bakke, A., Anderson, R.V. and Kvamme, T. (1991) Pathogenicity of the nematodes *Bursaphelenchus xylophilus* and *B. mucronatus* to *Pinus sylvestris* seedlings: A greenhouse test. *Scandinavian Journal of Forest Research* 6, 407–412.

Barron, G.L. and Thorn, R.G. (1987) Destruction of nematodes by species of *Pleurotus*. *Canadian Journal of Botany* 65, 774–779.

Berg, C.C. (1989) Classification and distribution of *Ficus*. *Experientia* 45(7), 605–611.

Blair, G.P. (1969) Studies of red ring disease of coconut palm. In: *Conference Abstract in Symposium on Tropical Nematology*, Mayaguez, Puerto Rico, pp. 89–106.

Braasch, H. (2000) Influence of temperature and water supply on mortality of 3-year-old pines inoculated with *Bursaphelenchus xylophilus* and *B. mucronatus*. *Nachrichtenblatt Des Deutschen Pflanzenschutzdienstes* 2, 244–249.

Braasch, H. (2001) *Bursaphelenchus* species in conifers in Europe: Distribution and morphological relationships. *EPPO Bulletin* 31(2), 127–142.
Brathwaite, C.W.D. and Siddiqi, M.R. (1975) *Rhadinaphelenchus cocophilus*. In: Gooch, P.S., Siddiqi, M.R. and Franklin, M.T. (eds) *CIH Descriptions of Plant-Parasitic Nematodes*. Commonwealth Agricultural Bureaux, Farnham Royal, UK, p. 72.
Burgermeister, W., Braasch, H., Metge, K., Gu, J., Schröder, T. *et al.* (2009) ITS-RFLP analysis, an efficient tool for differentiation of *Bursaphelenchus* species. *Nematology* 11(5), 649–668.
CABI (2019) *Areca catechu* (betelnut palm). *CABI Compendium* (Online). Available at: https://doi.org/10.1079/cabicompendium.6981 (accessed 30 March 2024).
Cai, S., Jia, J., He, C., Zeng, L., Fang, Y. *et al.* (2021) Multi-omics of pine wood nematode pathogenicity associated with culturable associated microbiota through an artificial assembly approach. *Frontiers in Plant Science* 12, 798539.
Cardoso, J.M.S., Manadas, B., Abrantes, I., Robertson, L., Arcos, S.C. *et al.* (2024) Pine wilt disease: What do we know from proteomics. *BMC Plant Biology* 24, 98.
Carletti, B., Ambrogioni, L., Irdani, T., Brandstetter, M., Puleri, F. *et al.* (2007) Morphometrics and molecular identification of some Italian populations of *Bursaphelenchus eremus* Ruhm (Goodey) associated with *Quercus* spp. *Redia* 90, 3–21.
Carletti, B., Cotroneo, A. and Roversi, P.F. (2011) Inoculation experiments of *Bursaphelenchus eremus* Ruhm (Goodey) (Nematoda: Parasitaphelenchidae) on *Quercus robur* L. *Redia* 94, 159–162.
Compton, S.G. and Hawkins, B.A. (1992) Determinants of species richness in southern African fig wasp assemblages. *Oecologia* 91(1), 68–74.
Cook, B.I., Mankin, J.S. and Anchukaitis, K.J. (2018) Climate change and drought: From past to future. *Current Climate Change Reports* 4(2), 164–179.
Croisé, L., Lieutier, F., Cochard, H. and Dreyer, E. (2001) Effects of drought stress and high density stem inoculations with *Leptographium wingfieldii* on hydraulic properties of young scots pine trees. *Tree Physiology* 21(7), 427–436.
Cruaud, A., Rønsted, N., Chantarasuwan, B., Chou, L.S., Clement, W.L. *et al.* (2012) An extreme case of plant–insect codiversification: Figs and fig-pollinating wasps. *Systematic Biology* 61(6), 1029–1047.
Davies, K.A., Ye, W., Giblin-Davis, R.M., Taylor, G.S., Scheffer, S. *et al.* (2010) The nematode genus *Fergusobia* (Nematoda: Neotylenchidae): molecular phylogeny, descriptions of clades and associated galls, host plants and *Fergusonina* fly larvae. *Zootaxa* 2633(1), 1–66.
Davies, K.A., Ye, W., Kanzaki, N., Bartholomaeus, F., Zeng, Y. *et al.* (2015) A review of the taxonomy, phylogeny, distribution and co-evolution of *Schistonchus* cobb, 1927 with proposal of *Ficophagus* n. gen. and *Martininema* n. gen. (Nematoda: Aphelenchoididae). *Nemotology* 17(7), 761–829.
Davies, K.A., Ye, W., Taylor, G.S., Scheffer, S. and Giblin-Davis, R.M. (2016) Galling problems – the *Fergusobia* nematode/*Fergusonina* fly mutualism on myrtaceous hosts. *Nematology* 18(6), 629–649.
Dayi, M. and Akbulut, S. (2012) Pathogenicity testing of four *Bursaphelenchus* species on conifer seedlings under greenhouse conditions. *Forest Pathology* 42(3), 213–219.
d'Errico, G., Carletti, B., Schröder, T., Mota, M., Vieira, P. *et al.* (2015) An update on the occurrence of nematodes belonging to the genus *Bursaphelenchus* in the mediterranean area. *Forestry: An International Journal of Forest Research* 88(5), 509–520.
Đođ, N., Cota, E. and Pernek, M. (2015) Wood nematode species spectrum in the mediterranean pine forests of Croatia. *Periodicum Biologorum* 117(4), 505–512.
Dorchin, N., Cramer, M.D. and Hoffmann, J.H. (2006) Photosynthesis and sink activity of wasp-induced galls in *Acacia pycnantha*. *Ecology* 87(7), 1781–1791.
Erbilgin, N., Zanganeh, L., Klutsch, J.G., Chen, S.H., Zhao, S. *et al.* (2021) Combined drought and bark beetle attacks deplete non-structural carbohydrates and promote death of mature pine trees. *Plant, Cell & Environment* 44(12), 3636–3651.
Firmino, P.N., Calvão, T., Ayres, M.P. and Pimentel, C.S. (2017) *Monochamus galloprovincialis* and *Bursaphelenchus xylophilus* life history in an area severely affected by pine wilt disease: Implications for forest management. *Forest Ecology and Management* 389, 105–115.
Fisher, J.M. and Nickle, W.R. (1968) On the classification and life history of *Fegusobia curriei* (Sphaerulariidae: Nematoda). *Proceedings of the Helminthological Society of Washington* 35, 40–46.
Futai, K. (2003) Role of asymptomatic carrier trees in epidemic spread of pine wilt disease. *Journal of Forest Research* 8(4), 253–260.
Futai, K. (2013) Pine wood nematode, *Bursaphelenchus xylophilus*. *Annual Review of Phytopathology* 51, 61–83.

Gaaliche, B., Trad, M. and Mars, M. (2011) Effect of pollination intensity, frequency and pollen source on fig (*Ficus carica* L.) productivity and fruit quality. *Scientia Horticulturae* 130(4), 737–742.

Galil, J. and Eisikowitch, D. (1968) On the pollination ecology of *Ficus sycomorus* in East Africa. *Ecology* 49(2), 259–269.

Gerber, K. and Giblin-Davis, R.M. (1990) Association of the red ring nematode, *Rhadinaphelenchus cocophilus*, and other nematode species with *Rhynchophorus palmarum* (Coleoptera: Curculionidae). *Journal of Nematology* 22, 143–149.

Gerber, K., Giblin-Davis, R.M., Griffith, R., Escobar-Goyes, J. and Cartaya, A.D. (1989) Morphometric comparisons of geographic and host isolates of the red ring nematode, *Rhadinaphelenchus cocophilus*. *Nematropica* 19, 151–159.

Giblin-Davis, R.M. (1993) Interactions of nematodes with insects. In: Khan, M.W. (ed.) *Nematode Interactions*. Chapman & Hall, London, pp. 302–344.

Giblin-Davis, R.M., Center, B.J., Nadel, H., Frank, J.H. and Ramírez, B.W. (1995) Nematodes associated with fig wasps, *Pegoscapus* spp. (Agaonidae), and syconia of native Floridian figs (*Ficus* spp.). *Journal of Nematology* 27(1), 1–14.

Giblin-Davis, R.M., Oehlschlager, A.C., Perez, A., Gries, G., Gries, R. *et al.* (1996) Chemical and behavioral ecology of palm weevils (Curculionidae: Rhynchophorinae). *The Florida Entomologist* 77(2), 153.

Giblin-Davis, R.M., Makinson, J., Center, B.J., Davies, K.A., Purcell, M. *et al.* (2001) *Fergusobia/fergusonina*-induced shoot bud gall development on *Melaleuca quinquenervia*. *Journal of Nematology* 33(4), 239–247.

Giblin-Davis, R.M., Davies, K.A., Morris, K. and Thomas, W.K. (2003) Evolution of parasitism in insect-transmitted plant nematodes. *Journal of Nematology* 35(2), 133–141.

Giblin-Davis, R.M., Center, B.J., Davies, K.A., Purcell, M.F., Scheffer, S.J. *et al.* (2004) Histological comparisons of *Fergusobia/fergusonina*-induced galls on different myrtaceous hosts. *Journal of Nematology* 36(3), 249–262.

Giblin-Davis, R.M., Kanzaki, N. and Davies, K.A. (2013) Nematodes that ride insects: Unforeseen consequences of arriving species. *Florida Entomologist* 96(3), 770–780.

Giblin-Davis, R.M., Kanzaki, N. and Williams, D.S. (2023) Ultrastructural study of anther parasitism of *Ficus laevigata* by *Ficophagus laevigatus* (Aphelenchoididae). *Journal of Nematology* 55, 20230010.

Goberdhan, L.C. (1964) Observations on coconut palms artificially infested by the nematode *Rhadinaphelenchus cocophilus* (Cobb, 1919) Goodey, 1960. *Journal of Helminthology* 38(1–2), 25–30.

Goolsby, J.A., Makinson, J. and Purcell, M. (2000) Seasonal phenology of the gall-making fly *Fergusonina* sp. (Diptera: Fergusoninidae) and its implications for biological control of *Melaleuca quinquenervia*. *Australian Journal of Entomology* 39(4), 336–343.

Griffith, R. (1987) Red ring disease of coconut palm. *Plant Disease* 71(2), 193–196.

Griffith, R., Giblin-Davis, R.M., Koshy, P.K., Sosamma, V.K. and Kanzaki, N. (2018) Nematode parasites of coconut and other palms. In: Sikora, R.A., Coyne, D., Hallmann, J. and Timper, P. (eds) *Plant Parasitic Nematodes in Subtropical and Tropical Agriculture*, 3rd edn. CAB International, Wallingford, UK, pp. 504–535.

Gros-Balthazard, M. and Flowers, J.M. (2021) A brief history of the origin of domesticated date palms. In: Al-Khayri, J.M., Jain, S.M. and Johnson, D.V. (eds) *The Date Palm Genome Volume I. Phylogeny, Biodiversity and Mapping*. Springer, Dordrecht, The Netherlands, pp. 55–74.

Gu, J., Braasch, H., Burgermeister, W. and Zhang, J. (2006) Records of *Bursaphelenchus* spp. intercepted in imported packaging wood at Ningbo, China. *Forest Pathology* 36(5), 323–333.

Gu, J., Maria, M., Fang, Y., Liu, L., Bian, Y. *et al.* (2022) Description of *Laimaphelenchus sinensis* n. sp. (Nematoda: Aphelenchoididae) from declining Chinese pine, *Pinus tabuliformis* in Beijing, China. *Journal of Nematology* 52, 1–9.

Gunn, B.F., Baudouin, L. and Olsen, K.M. (2011) Independent origins of cultivated coconut (*Cocos nucifera* L.) in the old world tropics. *PLoS ONE* 6(6), e21143.

Gupta, S. and Borges, R.M. (2019) Density-dependent fitness effects stabilize parasitic hitchhiking within a mutualism. *Functional Ecology* 33(12), 2304–2315.

Hagley, E.A.C. (1962) The palm weevil, *Rhynchophorus palmarum* L., a probable vector of red ring disease of coconuts. *Nature* 193(4814), 499–499.

Haque, Z. and Khan, M.R. (2021) *Handbook of Invasive Plant-Parasitic Nematodes*. CAB International, Wallingford, UK.

Harrison, R.D. (2005) Figs and the diversity of tropical rainforests. *Bioscience* 55(12), 1053–1064.

Hofstetter, R.W., Dinkins-Bookwalter, J., Davis, T.S. and Klepzig, K.D. (2015) Symbiotic associations of bark beetles. In: Vega, F.E. and Hofstetter, R.W. (eds) *Bark Beetles: Biology and Ecology of Native and Invasive Species*. Academic Press, New York, pp. 209–245.

Humble, L.M. and Allen, E.A. (2006) Forest biosecurity: Alien invasive species and vectored organisms. *Canadian Journal of Plant Pathology* 28, S256–S269.

Hunt, D.J. (1993) *Aphelenchida, Longidoridae, and Trichodoridae: Their Systematics and Bionomics*. CAB International, Wallingford, UK.

Ide, T. and Abe, Y. (2021) The heterogonic life cycles of oak gall wasps need to be closed: A lesson from two species of *Dryophanta* (Hymenoptera: Cynipidae: Cynipini). *Annals of the Entomological Society of America* 114(4), 489–500.

Ide, T., Kanzaki, N., Parra Giraldo, P.P. and Giblin-Davis, R.M. (2017) Loop-mediated isothermal amplification (LAMP) for detection of the red ring nematode, *Bursaphelenchus cocophilus*. *Nematology* 19(5), 559–565.

Jikumaru, S. and Togashi, K. (2003) Boarding abilities of *Bursaphelenchus mucronatus* and *B. xylophilus* (Nematoda: Aphelenchoididae) on *Monochamus alternatus* (Coleoptera: Cerambycidae). *Nematology* 5(6), 843–849.

Kanzaki, N. and Futai, K. (2001) Life history of *Bursaphelenchus conicaudatus* (Nematoda: Aphelenchoididae) in relation to the yellow-spotted longicorn beetle, *Psacothea hilaris* (Coleoptera: Cerambycidae). *Nematology* 3(5), 473–479.

Kanzaki, N. and Futai, K. (2003) Description and phylogeny of *Bursaphelenchus luxuriosae* n.sp. (Nemotoda: Aphelenchoididae) isolated from *Acalolepta luxuriosa* (Coleoptera: Cerambycidae). *Nematology* (5), 556–572.

Kanzaki, N. and Giblin-Davis, R.M. (2018) Diversity and plant pathogenicity of *Bursaphelenchus* and related nematodes in relation to their vector bionomics. *Current Forestry Reports* 4(2), 85–100.

Kanzaki, N., Aikawa, T., Maehara, N. and Ichihara, Y. (2011) An inoculation experiment of Japanese *Bursaphelenchus* nematodes on Japanese black and red pine *Pinus thunbergii* and *P. densiflora*. *Journal of Forest Research* 16, 325–330.

Kanzaki, N., Tanaka, R., Giblin-Davis, R.M. and Davies, K.A. (2014) New plant-parasitic nematode from the mostly mycophagous genus *Bursaphelenchus* discovered inside figs in Japan. *PLoS ONE* 9(6), e99241.

Kanzaki, N. and Giblin-Davis, R.M. (2016) Pine wilt and red ring, lethal plant diseases caused by insect-mediated *Bursaphelenchus* nematodes. In: Brown, J.K. (ed.) *Vector-Mediated Transmission of Plant Pathogens*. APS Press, Saint Paul, MN, pp. 87–107.

Kanzaki, N., Tsuda, K., Akiba, M., Ekino, T., Sawahata, T. *et al.* (2021) Phylogenetic status of *Iotonchium* Cobb, 1920 and a note on the distribution of *I. ungulatum* Aihara, 2001. Nematology 23(10), 1205–1209.

Kanzaki, N., Kruger, M.S., Greeff, J.M. and Giblin-Davis, R.M. (2022) *Bursaphelenchus suri* n. sp.: A second *Bursaphelenchus syconial* parasite of figs supports adaptive radiation among section sycomorus figs. PLoS ONE 17(4), e0265339.

Kanzaki, N., Banu, G. and Anes, K.M. (2023a) Nematode problems in palms and their sustainable management. In: Khan, M.R. and Quintanilla, M. (eds) *Nematode Diseases of Crops and Their Sustainable Management*. Academic Press, Cambridge, MA, pp. 247–455.

Kanzaki, N., Tanaka, R. and Giblin-Davis, R.M. (2023b) *Schistonchus pumilae* n. sp. isolated from syconia of a creeper fig, *Ficus pumila* L. from the Japanese warm temperate zone. Nematology 25(5), 509–529.

Keeley, J.E. (2012) Ecology and evolution of pine life histories. *Annals of Forest Science* 69(4), 445–453.

Kerdelhué, C. and Rasplus, J.-Y. (1996) Non-pollinating afrotropical fig wasps affect the fig-pollinator mutualism in *Ficus* within the subgenus *sycomorus*. *Oikos* 75(1), 3–14.

Khan, M.R. (2008) *Plant Nematodes – Methodology, Morphology, Systematics, Biology and Ecology*. Science Publishers, Enfield, NH. DOI: 10.1201/b10766.

Khan, M.R. (2012) Nematodes, an emerging threat to global forests: Assessment and management. *Plant Pathology Journal* 11(4), 99–113.

Khan, M.R. (2020) Nematode infestation, a potential threat to Indian forests. *Indian Phytopathology* 73(3), 397–414.

Khan, M.R. (2023a) Nematode pests of agricultural crops, a global overview. In: Khan, M.R. (ed.) *Novel Biological and Biotechnological Applications in Plant Nematode Management*. Springer, Singapore, pp. 3–45.

Khan, M.R. (2023b) Plant nematodes, an underestimated constraint in the global food production. In: Khan, M.R. and Quintanilla, M. (eds) *Nematode Diseases of Crops and Their Sustainable Management*. Academic Press, Cambridge, MA, pp. 3–26.

Kikuchi, T., Aikawa, T., Oeda, Y., Karim, N. and Kanzaki, N. (2009) A rapid and precise diagnostic method for detecting the pinewood nematode *Bursaphelenchus xylophilus* by loop-mediated isothermal amplification (LAMP). *Phytopathology* 99(12), 1365–1369.

Kikuchi, T., Cotton, J.A., Dalzell, J.J., Hasegawa, K., Kanzaki, N. *et al.* (2011) Genomic insights into the origin of parasitism in the emerging plant pathogen *Bursaphelenchus xylophilus*. *PLoS Pathogens* 7(9), e1002219.

Kishi, Y. (1995) *The Pine Wood Nematode and the Japanese Pine Sawyer*. Thomas Company, Tokyo.

Kislev, M.E., Hartmann, A. and Bar-Yosef, O. (2006) Early domesticated fig in the Jordan Valley. *Science* 312(5778), 1372–1374.

Krishnan, A., Muralidharan, S., Sharma, L. and Borges, R.M. (2010) A hitchhiker's guide to A crowded syconium: How do fig nematodes find the right ride? *Functional Ecology* 24(4), 741–749.

Kunwar, R.M. and Bussmann, R.W. (2006) *Ficus* (fig) species in Nepal: A review of diversity and indigenous uses. *Lyonia* 11, 85–97.

Kwok, O.C.H., Plattner, R., Weisleder, D. and Wicklow, D.T. (1992) A nematicidal toxin from *Pleurotus ostreatus* NRRL 3526. *Journal of Chemical Ecology* 18(2), 127–136.

Lange, C., Burgermeister, W., Metge, K. and Braasch, H. (2008) Molecular characterization of isolates of the *Bursaphelenchus sexdentati* group using ribosomal DNA sequences and ITS-RFLP. In: Mota, M.M. and Vieira, P. (eds) *Pine Wilt Disease: A Worldwide Threat to Forest Ecosystems*. Springer, Dordrecht, The Netherlands, pp. 165–173.

Lazarova, S.S., Goede, R.G.M., Peneva, V.K. and Bongers, T. (2004) Spatial patterns of variation in the composition and structure of nematode communities in relation to different microhabitats: A case study of *Quercus dalechampii* Ten. *Forest. Soil Biology and Biochemistry* 36(4), 701–712.

Lee, C.H., Lee, Y.-Y., Chang, Y.-C., Pon, W.-L., Lee, S.-P. *et al.* (2023) A carnivorous mushroom paralyzes and kills nematodes via a volatile ketone. *Science Advances* 9(3), eade4809.

Linit, M.J. (1988) Nematode–vector relationships in the pine wilt disease system. *Journal of Nematology* 20, 227–235.

Liu, K.C., Ben, A., Han, Z., Guo, Y. and Cao, D. (2019) Interspecific hybridization between *Bursaphelenchus xylophilus* and *Bursaphelenchus mucronatus*. *Journal of Forestry Research* 30(2), 699–707.

Maehara, N., Kanzaki, N., Aikawa, T. and Nakamura, K. (2020) Potential vector switching in the evolution of *Bursaphelenchus xylophilus* group nematodes (Nematoda: Aphelenchoididae). *Ecology and Evolution* 10(24), 14320–14329.

Maleita, C.M.N., Costa, S.R. and Abrantes, I. (2018) *Laimaphelenchus suberensis* sp. nov. associated with *Quercus suber* in Portugal. *European Journal of Plant Pathology* 150(3), 747–758.

Mamiya, Y. (1988) History of pine wilt disease in Japan. *Journal of Nematology* 20(2), 219–226.

Mamiya, Y. and Enda, N. (1972) Transmission of *Bursaphelenchus lignicolus* (Nematoda: Aphelenchoididae) by *Monochamus alternatus* (Coleoptera: Cerambycidae). *Nematologica* 18(2), 159–162.

Massey, C.L. (1974) *Biology and Taxonomy of Nematode Parasites and Associates of Bark Beetles in the United States*. US Department of Agriculture, Forest Service,Washington, DC.

Mon, A.M., Shi, Y., Yang, X., Hein, P.P., Oo, T.N. *et al.* (2020) The uses of fig (*Ficus*) by five ethnic minority communities in Southern Shan state Myanmar. *Journal of Ethnobiology and Ethnomedicine* 16(1), 55.

Mota, M.M., Braasch, H., Bravo, M.A., Penas, A.C., Burgermeister, W. *et al.* (1999) First report of *Bursaphelenchus xylophilus* in Portugal and in Europe. *Nematology* 1(7), 727–734.

Mota, M.M. and Vieira, P. (eds) (2008) *Pine Wilt Disease: A Worldwide Threat to Forest Ecosystems*. Springer, Dordrecht, The Netherlands.

Naves, P., Camacho, S., Sousa, E. and Quartau, J. (2007) Transmission of the pine wood nematode *Bursaphelenchus xylophilus* through oviposition activity of *Monochamus galloprovincialis* (Cleoptera: Cerambycidae). *Entomologica Fennica* 18(4), 193–198.

Nelson, L.A., Davies, K.A., Scheffer, S.J., Taylor, G.S., Purcell, M.F. *et al.* (2014) The fly-nematode mutualism on *Myrtaceae* host plants: An emerging example of tritrophic coevolution (Diptera: Fergusoninidae; Nematoda: Neotylenchidae). *Biological Journal of the Linnean Society* 111(4), 699–718.

Nickle, W.R., Golden, A.M., Mamiya, Y. and Wergin, W.P. (1981) On the taxonomy and morphology of the pine wood nematode, *Bursaphelenchus xylophilus* (Steiner & Buhrer 1934) Nickle 1970. *Journal of Nematology* 13(3), 385–392.

Ning, J., Gu, X., Zhou, J., Zhang, H., Sun, J. *et al.* (2023) Palmitoleic acid as a coordinating molecule between the invasive pinewood nematode and its newly associated fungi. *The ISME Journal* 17(11), 1862–1871.

Oehlschlager, A.C., Chinchilla, C., Castillo, G. and Gonzalez, L. (2002) Control of red ring disease by mass trapping of *Rhynchophorus palmarum* (Coleoptera: Curculionidae). *Florida Entomologist* 85(3), 507–513.

Öztürk, N., Akbulut, S. and Baysal, İ. (2019) Determination of pathogenicity of *Bursaphelenchus* species on different pine species under natural conditions in Düzce. *Phytoparasitica* 47(1), 89–97.

Pajares, J.A., Ibeas, F., Díez, J.J. and Gallego, D. (2004) Attractive responses by *Monochamus galloprovincialis* (Col., Cerambycidae) to host and bark beetle semiochemicals. *Journal of Applied Entomology* 128(9–10), 633–638.

Penas, A.C., Correia, P., Bravo, M.A., Mota, M. and Tenreiro, R. (2004) Species of *Bursaphelenchus fuchs*, 1937 (Nematoda: Parasitaphelenchidae) associated with maritime pine in Portugal. *Nematology* 6(3), 437–453.

Peri, E., Colazza, S., Guarino, S., Lo Bue, P., Suma, P. *et al.* (2013) The red palm weevil in Sicily: The introduction and spread of an invasive alien species. In: *Proceedings of the AFPP Palm Pest Mediterranean Conference*, Nice, France.

Pimentel, C.S. and Ayres, M.P. (2018) Latitudinal patterns in temperature-dependent growth rates of a forest pathogen. *Journal of Thermal Biology* 72, 39–43.

Pimentel, C., Khan, M.R., Zheng, Y. and Quintanilla, M. (2023) Nematode problems in forests and their sustainable management. In: Khan, M.R. and Quintanilla, M. (eds) *Nematode Diseases of Crops and Their Sustainable Management*. Academic Press, Cambridge, MA, pp. 457–493.

Pimentel, C.S., Ayres, M.P., Vallery, E., Young, C. and Streett, D.A. (2014) Geographical variation in seasonality and life history of pine sawyer beetles *Monochamus* spp: Its relationship with phoresy by the pinewood nematode *Bursaphelenchus xylophilus*. *Agricultural and Forest Entomology* 16(2), 196–206.

Pimentel, C.S., Firmino, P.N. and Ayres, M.P. (2021) Interactions between pinewood nematodes and the fungal community of pine trees. *Fungal Ecology* 51, 101046.

Pimentel, C.S., Firmino, P.N. and Ayres, M.P. (2022) Obtaining and maintaining cultures of pinewood nematodes *Bursaphelenchus xylophilus* from wild dauers. *Methods in Molecular Biology* 25, 3–11.

Poinar, G.O., Jr (1975) *Entomogenous Nematodes: A Manual and Host List of Insect–Nematode Associations*. E.J. Brill, Leiden, The Netherlands.

Poinar, G.O., Jr (1991) The mycetophagous and entomophagous stages of *Iotonchium californicum* n. sp. (Lotonchiidae: Tylenchida). *Revue de Nématologie* 14, 565–580.

Poinar, G.O., Jr (2015) The geological record of parasitic nematode evolution. *Advances in Parasitology* 90, 53–92.

Poland, T.M. and Rassati, D. (2019) Improved biosecurity surveillance of non-native forest insects: A review of current methods. *Journal of Pest Science* 92(1), 37–49.

Prospero, S., Polomski, J. and Rigling, D. (2015) Occurrence and its diversity of wood-associated *Bursaphelenchus* nematodes in Scots pine forests in Switzerland. *Plant Pathology* 64(5), 1190–1197.

Quist, C.W., Smant, G. and Helder, J. (2015) Evolution of plant parasitism in the phylum Nematoda. *Annual Review of Phytopathology* 53, 289–310.

Reddy, Y.N. and Rao, P.N. (1986) *Schistonchus racemosa* sp. n., a nematode parasite of wasp (*Ceratosolen* sp.) associated with the fig, *Ficus racemosa* L. *Indian Journal of Nematology* 16, 135–137.

Richardson, D.M. and Rundel, P.W. (1998) Ecology and biogeography of *Pinus*: An introduction. In: Richardson, D.M. (ed.) *Ecology and Biogeography of Pinus*. Cambridge University Press, Cambridge, UK, pp. 3–46.

Rochat, D., Dembilio, O., Jaques, J.A., Suma, P., Pergola, A. *et al.* (2017) *Rhynchophorus ferrugineus*: Taxonomy, distribution, biology, and life cycle. In: Soroker, V. and Colazza, S. (eds) *Handbook of Major Palm Pests*. Wiley, New York, pp. 69–104.

Safitri, B., Hidayat, P. and Buchori, D. (2019) Resistance of some new clones of eucalyptus to insects causes gall in Mount Mutis, NTT. *Jurnal Hama Dan Penyakit Tumbuhan Tropika* 19(2), 109.

Saro, I., García-Verdugo, C., González-Pérez, M.A., Naranjo, A., Santana, A. *et al.* (2019) Genetic structure of the Canarian palm tree (*Phoenix canariensis*) at the island scale: Does the "island within islands" concept apply to species with high colonisation ability? *Plant Biology* 21(1), 101–109.

Shanahan, M., So, S., Compton, S.G. and Corlett, R. (2001) Fig-eating by vertebrate frugivores: A global review. *Biological Reviews of the Cambridge Philosophical Society* 76(4), 529–572.

Shi, R.R., Miao, B.G., Segar, S.T., Zeng, Y., Wang, B. *et al.* (2019) Are nematodes costly to fig tree–fig wasp mutualists? *Entomologia Experimentalis et Applicata* 167(12), 1000–1011.

Siddiqi, M.R. (1986) *Tylenchida: Parasites of Plants and Insects*, 1st edn. Commonwealth Institute of Parasitology, Farnham Royal, UK.

Siddiqi, M.R. (2000) *Tylenchida: Parasites of Plants and Insects*, 2nd edn. CAB International, Wallingford, UK.

Skarmoutsos, G. and Michalopoulos-Skarmoutsos, H. (2000) Pathogenicity of *Bursaphelenchus sexdentati*, *Bursaphelenchus leoni* and *Bursaphelenchus hellenicus* on European pine seedlings. *Forest Pathology* 30, 149–156.

Slonim, O., Bucki, P., Mendel, Z., Protasov, A., Golan, O. *et al.* (2018) First report on *Bursaphelenchus sexdentati* (Nematoda: Aphelenchoididae) in Israel. *Forest Pathology* 48, e12431.

Tanaka, H., Tsuda, K., Yamamoto, K., Isowa, I. and Shibao, M. (2002) Occurrence and control of the gill-knot disease of the oyster mushroom, *Pleurotus ostreatus* (Jacq.: Fr.) Kummer, in Osaka prefecture. *Proceedings of Kansai Plant Protection* 44, 47–48.

Taşdemir, S., Akbulut, S., Kanzaki, N. and Öztürk, N. (2020) Preliminary survey of nematodes associated with broadleaved trees in Izmit forest management directorate, Turkey. *Forest Pathology* 50(6), e12642.

Taylor, G.S. (2004) Revision of *Fergusonina malloch* gall flies (Diptera: Fergusoninidae) from *Melaleuca* (myrtaceae). *Invertebrate Systematics* 18(3), 251.

Togashi, K., Appleby, J.E., Oloumi-Sadeghi, H. and Malek, R.B. (2023) Frequency distribution of the initial number of *Bursaphelenchus xylophilus* carried by adult *Monochamus carolinensis* in Illinois, with reference to that by *M. alternatus* in Japan. *Nematology* 25(3), 259–274.

Tomalak, M. and Filipiak, A. (2021) Effects of inter-specific crossbreeding between the invasive pine wood nematode, *Bursaphelenchus xylophilus* and native *B. mucronatus* on morphology and reproduction of the hybrid offspring. *Forest Pathology* 51(2), e12676.

Tomminen, J. (1990) Presence of *Bursaphelenchus mucronatus* (Nematoda: Aphelenchoididae) fourth dispersal stages in selected conifer beetles in Finland. *Silva Fennica* 24(3), 273–278.

Tomminen, J. (1993) Pathogenicity studies with *Bursaphelenchus mucronatus* in Scots pine in Finland. *European Journal of Forest Pathology* 23(4), 236–243.

Truong, B.-N., Suzuki, A., Truong, B.-N., Okazaki, K., Fukiharu, T. *et al.* (2007) Characterization of the nematocidal toxocyst in *Pleurotus* subgen. *coremiopleurotus*. *Mycoscience* 48(4), 222–230.

Tsuda, K. and Futai, K. (1999) *Iotonchium cateniforme* n. sp.(Tylenchida: Iotonchiidae) from fruiting bodies of *Cortinarius* spp. and its life cycle. *Nematological Research (Japanese Journal of Nematology)* 29(1), 24–31.

Tsuda, K. and Futai, K. (2000) The insect-parasitic stage and life cycle of *Iotonchium ungulatum* (Tylenchida: Iotonchiidae), the causal agent of gill-knot disease of the oyster mushroom. *Nematological Research (Japanese Journal of Nematology)* 30(1–2), 1–7.

Tsuda, K. and Futai, K. (2005) Description of two new species of *Iotonchium* Cobb, 1920 (Tylenchida: Iotonchiidae) from Japan. *Nematology* 7, 789–801.

Tsuda, K., Futai, K. and Kosaka, H. (1996) The tripartite relationship in gill-knot disease of the oyster mushroom, *Pleurotus ostreatus* (Jacq.: Fr.) Kummer. *Canadian Journal of Zoology* 74(8), 1402–1408.

Van Goor, J., Kanzaki, N. and Woodruff, G. (2023) How to be a fig nematode. *Acta Oecologica* 119, 103916.

Vincent, B., Altemayer, V., Roux-Morabito, G., Naves, P., Sousa, E. *et al.* (2008) Competitive interaction between *Bursaphelenchus xylophilus* and the closely related species *Bursaphelenchus mucronatus*. *Nematology* 10(2), 219–230.

Vovlas, N., Inserra, R.N. and Greco, N. (1992) *Schistonchus caprifici* parasitizing caprifig (*Ficus carica sylvestris*) florets and the relationship with its fig wasp (*Blastophaga psenes*) vector. *Nematologica* 38, 215–226.

Vovlas, N., Troccoli, A., Noort, S. and Berg, E. (1998) *Schistonchus africanus* n. sp. (Aphelenchida: Aphelenchoididae) associated with *Ficus thonningii* (Moraceae) and its pollinator wasp *Elisabethiella stuckenbergi* (Chalcidoidea: Agaonidae). *Journal of Nematology* 30, 404.

Warwick, D.R.N. and Bezerra, A.P.T. (1992) Possible root transmission of the red ring nematode (*Rhadinaphelenchus cocophilus*) to coconut palms. *Plant Disease* 76(8), 809–811.

Weissling, T.J., Giblin-Davis, R.M., Gries, G., Gries, R., Perez, A.L. *et al.* (1994) Aggregation pheromone of palmetto weevil, *Rhynchophorus cruentatus* (F.) (Coleoptera: Curculionidae). *Journal of Chemical Ecology* 20(3), 505–515.

Ye, W., Giblin-Davis, R.M., Davies, K.A., Purcell, M.F., Scheffer, S.J. *et al.* (2007) Molecular phylogenetics and the evolution of host plant associations in the nematode genus *Fergusobia* (Tylenchida: Fergusobiinae). *Molecular Phylogenetics and Evolution* 45(1), 123–141.

Ye, W., Davies, K., Giblin-Davis, R. and Thomas, K. (2009) *Ficotylus congestae* gen. n., sp. n. (Anguinata), from *Ficus congesta* (moraceae) sycones in Australia. *Nematology* 11, 1–75.

Zhang, W., Li, Y., Pan, L., Wang, X., Feng, Y. *et al.* (2020) Pine chemical volatiles promote dauer recovery of a pine parasitic nematode, *Bursaphelenchus xylophilus*. *Parasitology* 147(1), 50–57.

Zhao, L., Zhang, S., Wei, W., Hao, H., Zhang, B. *et al.* (2013) Chemical signals synchronize the life cycles of a plant-parasitic nematode and its vector beetle. *Current Biology* 23(20), 2038–2043.

Zhao, L., Mota, M., Vieira, P., Butcher, R.A. and Sun, J. (2014) Interspecific communication between pinewood nematode, its insect vector, and associated microbes. *Trends in Parasitology* 30(6), 299–308.

Zhao, L., Zhang, X., Wei, Y., Zhou, J., Zhang, W. *et al.* (2016) Ascarosides coordinate the dispersal of a plant-parasitic nematode with the metamorphosis of its vector beetle. *Nature Communications* 7, 12341.

Zhao, Z.Q., Davies, K.A., Riley, I.T. and Nobbs, J.M. (2007) *Laimaphelenchus heidelbergi* sp. nov. (Nematoda: Aphelenchina) from Victoria, Australia, and emendment of the diagnosis of the genus. *Transactions of the Royal Society of South Australia* 131(2), 182–191.

Zhao, B.G., Futai, K., Sutherland, J.R. and Takeuchi, Y. (eds) (2008) *Pine Wilt Disease*. Springer, Dordrecht, The Netherlands.

Zheng, Y. and Khan, M.R. (2023) Pine wood nematode in coniferous forests and their management by novel biological and biotechnological interventions. In: Khan, M.R. (ed.) *Novel Biological and Biotechnological Applications in Plant Nematode Management*. Springer, Singapore, pp. 489–514.

13 Management of Disease Complexes

B. B. Westerdahl[1]* and Deliang Peng[2]
[1]Department of Entomology and Nematology, University of California, Davis, CA, USA; [2]Institute of Plant Protection, Chinese Academy of Agricultural Sciences, Beijing, China

Abstract
The chapter explores the core principles of managing plant nematodes and disease complexes, exemplifying practical applications of management techniques for disease control and providing perspectives for enhancing future management strategies. It also delves into the fundamental principles of plant nematodes and disease management, showcasing practical examples of how management techniques have been employed to control disease complexes and offering insights for the future advancement of management strategies.

Introduction

The initial phase of effective management of disease complexes involves the identification of plant-parasitic nematodes and an understanding of their potential interactions with disease-causing agents such as fungi, bacteria and viruses. Recognition typically begins when growers observe symptoms or indicators of crop distress. If a grower suspects the presence of nematodes or a disease issue, referencing a list of common symptoms and signs associated with nematode and disease problems often helps in discerning whether these manifestations may be attributed to other factors such as suboptimal irrigation or fertilization practices. Once the pest complex present is identified, the next step involves evaluating various components of a comprehensive management program including prevention, physical control, biological control, cultural practices and use of commercially available products (Powell, 1971; Bergeson, 1972; Hirano, 1975; Pitcher, 1978).

Symptoms and Signs of Disease Complexes

Some indicators suggestive of nematode infestation include reduced yield, premature maturity, delayed maturity, stunting, chlorosis, mid-day wilting, leaf drop, small fruit, yellowing, curling and twisting of leaves and stems, localized patches of plants with stunted growth, failure to respond to treatments targeting other issues, and lack of thrift (Khan, 2023a; Westerdahl *et al.*, 2023). Due to the disruptive impact of nematode feeding and movement on plant vascular tissue, plant roots may struggle to absorb sufficient water during hot periods, resulting in wilting. As the water stress diminishes by evening, plants regain vigor and appear healthy (Thorne, 1961; Westerdahl, 2011; Khan, 2023b). The following are some common symptoms of plant diseases (Agrios, 2005):

- Leaf spot (dead, discolored or injured areas with distinct margins on leaves or fruit).

*Corresponding author: bbwesterdahl@ucdavis.edu

 Nematode Disease Complexes in Agricultural Crops (ed. Mujeebur Rahman Khan)
DOI: 10.1079/9781800625228.0013

- Blight (rapid yellowing, browning, collapse and death of plant parts or the entire plant).
- Chlorosis (yellowing of normally green leaves and stems).
- Necrosis (browning or blackening of plant areas indicating cell death).
- Wilting (drooping of leaves, shoots or the plant due to water stress).
- Distortion (abnormal traits or twisting of leaves, stems or shoots).
- Mosaic (uneven patterns of yellow, light green or dark green on leaves).
- Canker (dead area on stems or branches distinguished by color).
- Rot (breakdown and decay of plant tissue, often in roots and fruit).
- Dieback (death of leaf tips, shoots or stems).
- Witches' broom (abnormal proliferation of shoots in a bushy appearance).
- Galling (swelling or abnormal growth on plant tissues).
- Stunting (abnormally small plant parts or a whole plant failure to grow to full size).

Upon suspicion of a problem, it is crucial to assess the available local resources that can support the development and execution of a management program, as these resources can vary significantly based on the geographical context. The WWW (World Wide Web) has developed into an excellent source of information for pest management-related information. Various entities such as federal and state environmental protection agencies, universities and agricultural product companies have established online platforms offering valuable insights into pest management practices. The resources available may differ from one country to another. For instance, in the USA, state and federal environmental protection agencies host databases on integrated pest management accessible to the public. Moreover, entities such as the US Department of Agriculture (USDA), county agricultural commissioners, university departments and cooperative extension programs provide expert advisers and researchers to aid in pest management efforts and operate diagnostic laboratories for identifying pest issues.

Those familiar with crop damage patterns would observe that many symptoms and signs observed are not exclusive indicators of nematode presence or plant disease and may stem from other underlying factors such as inadequate nutrition, irrigation practices or alternative pathogens. This underscores that no definitive symptoms or signs exist to solely diagnose a disease complex issue. However, witnessing these symptoms and signs in the field should prompt suspicion of a potential disease complex, subsequently warranting further diagnosis through sampling.

Sampling for Disease Complexes

The initial phase of effective management involves the identification of problematic nematodes or diseases existing within a system. Furnishing soil or plant samples to a diagnostic laboratory stands as a critical first step in this process. Once the identified organisms are determined, a crucial next step is to revisit the foundational biological aspects of nematodes and diseases, which is vital in formulating a robust management plan. In subsequent sections we will delve into the pertinent biological fundamentals essential for an effective management strategy.

Different methods are recommended for collecting the nematode-infested soil and plant samples to be submitted to a diagnostic laboratory (Haque and Khan, 2021). Due to the characteristic spotty, patchy or non-random distribution of nematodes, obtaining an accurate representation of the nematode population typically involves collecting multiple smaller subsamples randomly and amalgamating them into a composite sample as opposed to taking a single sample from a singular location. The standard protocol for soil sampling includes the following steps. Excavate within the root zone where moisture is prevalent. Place the soil along with small roots into a plastic bag. Soil samples from various locations within a field can be combined. Aim to collect approximately 1 kg of soil and roots. Healthy areas can also be sampled for comparison as well and placed in a separate bag. Seal the bags securely and store them in a cool environment (avoid freezing). Properly label the bags with essential information (e.g. name, address, sample location, date, crop history, current crop, next planned crop). Notify the designated laboratory of the sample that is being sent to them for analysis (Wallace, 1978; McSorley, 1987).

Determining the appropriate size of the sampling area is a crucial consideration influenced by several factors; samples obtained from a 2-ha area may yield reliable data for providing recommendations to growers. Sometimes, fewer samples from larger areas, such as one sample from each 8-ha block, considering economic constraints may fulfil the sample requirement. However, prior to initiating the sampling process, it is essential to visually assess the site and identify any discernible variations such as differences in soil texture, soil moisture levels, drainage patterns, presence of healthy and unhealthy plant patches or distinctive cropping histories that may necessitate the collection of separate samples. Each delineated area within the field is referred to as a stratum. For effective nematode population mapping and tailored management strategies, each stratum should be represented by a distinct sample (McSorley, 1987).

To ensure accurate and timely plant disease diagnosis, it is crucial to submit high-quality plant or soil samples following standard guidelines: whole plants with intact roots should be submitted, unless the issue is clearly related to leaves, stems or fruits. Leave the soil attached to the roots to maintain plant viability during transit. Enclose the roots and soil in a plastic bag near the base of the plants. Seedlings should be enclosed in a separate bag. Keep samples cool and away from direct sunlight. Clearly label sample bags on the outside of the bag with a permanent marker. Differentiate between affected and unaffected plants by packaging and labeling them separately. For plants up to 1m in height, submit a minimum of three whole plants with soil or 20 seedlings. For plants over 1m in height, separate the tops and bottoms before submission. Submit at least ten whole plants or 20 seedlings (Wallace, 1978; Grogan, 1981).

Diagnostic Procedures

As with sampling, different analysis techniques are utilized for nematodes than for plant diseases (Brown and Kerry, 1987). For nematodes, the choice of extraction technique depends on the nematode species of interest and the sample material, such as soil or plant tissue (Khan, 2008). Laboratory testing accuracy is enhanced by knowing the history of crops grown in the area, allowing for appropriate selection of extraction methods that target specific nematode genera. Different extraction methods yield varying proportions of nematode genera present in the sample, underscoring the importance of understanding the chosen laboratory's extraction protocol for accurate result interpretation. For example, looking at Fig. 13.1, consider a sample containing lesion, ring and dagger nematodes that is extracted by four different techniques: elutriator followed by sugar centrifugation, Baermann funnel lined with tissue, Baermann funnel lined with cheesecloth and roots from the sample placed on a funnel in a mist chamber (Fig. 13.1A). The relative shape and size of a nematode genus (Fig. 13.1B) combined with activity level influence the effectiveness of an extraction method. A sieving procedure is typically utilized to begin the process of extracting nematodes from soil (Fig. 13.1C). The soil is mixed with a large volume of water and allowed to settle briefly for heavy particles to separate. The mixture is then poured through a series of sieves with mesh sizes tailored to retain larger debris on the upper sieves and the target nematodes on the blower sieves, based on their sizes and characteristics. The suspension collected on the lower sieves are backwashed into containers for further processing. The resulting solution, if clear enough, can be directly observed under a microscope or if not, subjected to additional purification steps such as Baermann funnel (Fig. 13.1D) or elutriation followed by sugar centrifugation or flotation for improving sample clarity (Fig. 13.1E). A misting chamber is utilized to extract nematodes from roots (Fig. 13.1F). The shape, size and activity level of a nematode species affects the relative number of nematodes that will be extracted by each method. Ectoparasitic nematodes residing outside plant roots in the soil will not be detected from root samples placed in a mist chamber, but endoparasitic root-lesion nematodes are well assayed. Each of the other three methods recovered lesion, ring and dagger nematodes but in varying proportions. The relatively larger pore size of the cheesecloth-lined funnel recovers more of the relatively long, thin dagger

	(A) Extraction method			
Nematode	Elutriator-sugar centrifugation (number/l soil)	Baermann funnel (number/l soil)	Cheese cloth lined funnel (number/l soil)	Roots on mist (number/g)
Lesion	2750	7200	280	1.6
Ring	1000	20	150	0
Dagger	150	260	350	0

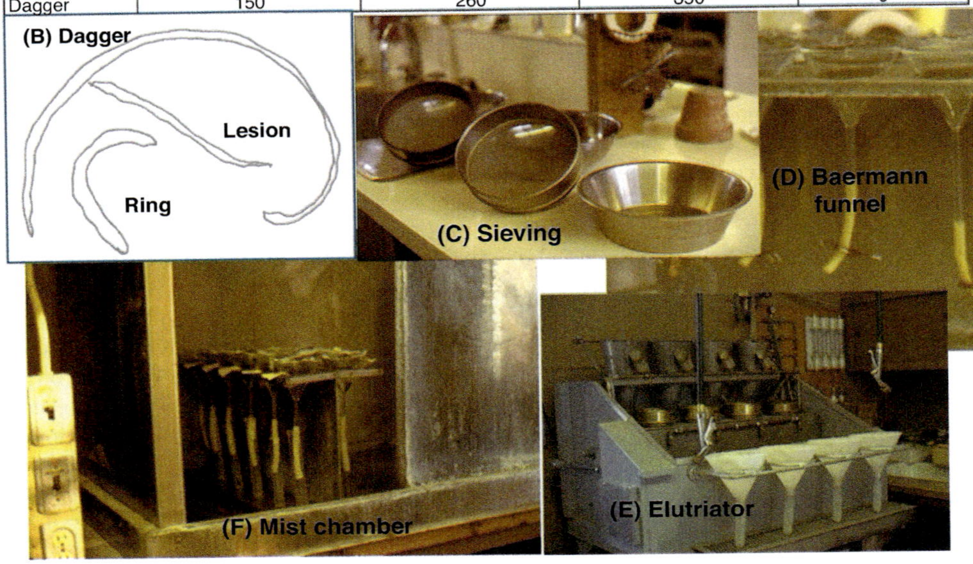

Fig. 13.1. Methods for extracting nematodes from soil and roots. (A) Relative proportions of lesion, ring and dagger nematodes extracted by four different methods. (B) Relative shapes and sizes of lesion, ring and dagger nematodes. (C) Sieving. (D) Baermann funnel. (E) Elutriator. (F) Mist chamber.

nematodes than do the other two methods. Stout, sluggish ring nematodes are best recovered by flotation in a sugar solution rather than the funnel linings.

Biology of Nematodes as Related to Management

Once a diagnosis is obtained, the next step is to become familiar with the biology of the organisms involved as a prerequisite to developing a management program. This is especially true for utilizing non-chemical management techniques. Understanding the phylogenetic relationships of plant-parasitic nematodes has become crucial for devising effective nematode management strategies. A characteristic of all plant-parasitic nematodes is that they utilize a stylet during their interactions with host plants. This stylet is used to puncture the host plant cell wall for nutrient extraction and, in some cases, to deliver secretory molecules into host cells to establish a permanent feeding site (Khan, 2007). The life cycle of nematodes consists of an egg, four juvenile stages and an adult. Molting occurs between each of the juvenile stages and the adult stage (Thorne, 1961; Ferris and Ferris, 1998).

Three basic life history patterns are found among plant parasites: migratory ectoparasitic, migratory endoparasitic and sedentary endoparasitic. All life cycle stages of migratory ectoparasites are present in the soil, while eggs, juveniles and adults of migratory endoparasites can be found either in roots or the soil. In sedentary endoparasites, the second-stage juvenile is typically the migratory and infective stage that emerges from the egg, resides in the soil and penetrates the root to establish a feeding site. Subsequent juvenile stages and adults of sedentary endoparasites are generally non-motile and remain fixed at a single feeding location. Eggs may be located within roots or in the soil. Categorizing nematodes based on their life history habits can offer valuable insights when exploring management strategies. For instance, when considering the application of chemicals or soil amendments, such products are more likely to target nematodes in the soil (migratory ectoparasites) rather than those residing within roots, unless the product has systemic effects. Despite spending some time in the soil during their life cycle, a substantial portion of endoparasites is typically found within roots (Thorne, 1961; Ferris and Ferris, 1998; Westerdahl, 2011).

Nematodes damage roots in several ways. Mechanical injury occurs by penetration and movement through plant tissues leading to cell death or stimulating cell growth. Cellular changes occur by induction of necrosis (death of cells) and alterations in cell growth. Feeding in vascular tissues disrupts the flow of water, nutrients and food, impacting the plant's overall health. Infection is facilitated by the creation of entry points for other microorganisms such as bacteria and fungi. Nematodes can introduce pathogens either on their bodies or within their bodies, or they can aid the entry of pathogens that are already present on the plant cell surface. Transmission of pathogens includes viruses. Stress from nematode feeding leads to heightened vulnerability of plants to environmental stress factors. Nematodes are adept at surviving as obligate parasites, as they have evolved over millions of years to stress their plant hosts rather than kill them outright (Thorne, 1961).

Plant-parasitic nematodes play a crucial role in the development of diseases in plants. They can act independently as pathogens or engage in synergistic relationships with other plant pathogens. In addition to causing damage to plants at own, the nematodes also help other pathogens in invading or infecting the host (Khan *et al.*, 2023b). Some nematodes from Longidoridae and Trichodoridae act as vectors for nepoviruses and tobraviruses, respectively. Interactions between nematodes and various wilt-inducing and root-rot fungi are commonplace in agricultural settings, underscoring their significance in influencing disease dynamics. Within Tylenchida, the following nematode genera with fungal/bacterial pathogens form important disease complexes: *Meloidogyne* (*Alternaria, Fusarium, Rhizoctonia, Phytophthora, Phomopsis, Sclerotium*), *Pratylenchus* (*Pythium, Rhizoctonia, Verticillium*), Trichodoridae (*Rhizoctonia*), Heteroderidae (*Rhizoctonia*), *Rotylenchulus* (*Fusarium, Rhizoctonia*) and *Belonolaimus* (*Fusarium*). Also, in Tylenchida, the following bacteria diseases complexes are found: *Anguina* (*Clavibacter*), *Ditylenchus* (*Pseudomonas, Clavibacter*), *Helicotylenchus* (*Pseudomonas*), *Meloidogyne* (*Pseudomonas, Curtobacter, Agrobacterium, Clavibacter*), *Pratylenchus* (*Pseudomonas, Agrobacterium*), *Rotylenchulus* (*Agrobacterium*), *Meloidogyne* (*Ralstonia*) and *Criconemoides* (*Pseudomonas*). Within the Aphelenchida, *Aphelenchoides* with *Rhodococcus* and *Clavibacter* form important bacteria disease complexes (Sikora and Carter, 1987; Khan, 1993; Sankaralingam and McGawley, 1994; Starr *et al.*, 1996). Plant-parasitic nematodes can also serve as vectors for plant-pathogenic fungi and bacteria (Sitaramaiah and Pathak, 1993).

Methods of Management of Disease Complexes

Management of plant diseases is not an easy task because plant do not speak, rather plant protectionists have to access and diagnose the plant ailments with the help of the symptoms or other diagnostic measures (Khan and Haque, 2023). The management of disease complexes becomes further challenging because of their multipathogenic nature and their intricate interrelationships (Khan and Sharma, 2020). Rarely has a single method been found effective against a disease complex, rather a set of two or more methods starting from prevention of pathogen spread and suppressing the pre-plant population in the field may yield effective control of the disease complexes.

Prevention

Some key techniques to minimize the spread of nematode pests include implementing quarantine measures that can help restrict the movement of potentially infested plants, plant materials or soil to prevent the introduction and spread of pests in new areas, using certified planting materials that are inspected and tested to ensure their freeness from known pests and diseases.

The limited awareness of nematodes and pathogens among the public can lead to the inadvertent introduction of pests into new environments, particularly through the importation of ornamental plants. Growers who have not encountered significant pest issues may lack awareness of the potential threats posed by pests such as nematodes and pathogens. This lack of awareness can result in the unintentional movement of pests through various means such as soil, irrigation water, equipment and planting stock. In agricultural settings such as farms, nurseries and government agencies, there is often a frequent turnover of personnel. This personnel turnover can contribute to a lack of consistent knowledge and understanding of pest management practices.

Effective internal and external sanitation programs play a critical role in preventing the entry of destructive pests and pathogens into protected areas and confining them to limited hectares. A practical illustration of this concept can be seen by considering a nursery with common row and plant spacing practices. For instance, if a nursery employs a typical row spacing of 1m with plants placed 15 cm apart, and approximately half of these plants reach harvestable size, an area of 0.5 ha containing infested nursery stock could result in the infestation of over 40 ha of farmland if the farmland is planted with 300 trees/ha (McNamara, 1995).

Washing soil off equipment before moving it to a new, uninfested field can help prevent the inadvertent spread of nematodes and other pests, reducing the likelihood of infestations in previously unaffected areas. The potential for nematodes to be transmitted through irrigation water poses a significant risk to agricultural systems. Research conducted in Washington state has shown that multiple nematode species in numbers of several million could be introduced into the fields through irrigation water (Faulkner and Bolander, 1966). One practical solution involves the use of settling ponds in conjunction with irrigation practices. By diverting irrigation water into settling ponds before application, growers can allow nematodes and other sediment to settle out, a process that typically only requires a few minutes. Subsequently, water can be drawn off from the top of the settling ponds for irrigation purposes.

Physical methods

Physical methods play a crucial role in nematode management, often complementing cultural and chemical control strategies. These methods offer effective and environmentally friendly approaches to combat nematode infestations (Khan *et al.*, 2012). Some key physical methods utilized in the management of nematodes and other plant pathogens which form disease complexes include heat treatment of planting stock, steam sterilization of soil, soil solarization and root pruning of infected plant material before planting (Maas, 1987).

Heat treatment

Heat treatment of planting stock involves subjecting planting materials to elevated temperatures to eliminate nematodes and other pathogens. This method has demonstrated effectiveness in controlling nematodes and pathogens in various crops, and has been particularly valuable in commercial settings, targeting specific species infesting different plant types. Hot water treatments of planting stock have been a longstanding management practice, dating back to at least the early 1900s. Chemicals may be added to a hot water treatment bath to enhance effectiveness against nematodes, fungi and bacteria. Examples include the use of hot water treatments for managing stem and bulb nematodes in daffodils and garlic, root-knot nematodes in grape rootstocks, and burrowing nematodes in banana corms (Bridge, 1975; Saleh *et al.*, 1988).

The precise temperature and duration are important detriments for successful treatment with hot water. It is critical to determine the optimal combination of temperature and exposure duration that effectively eradicates target organisms while minimizing damage to the plant material. The susceptibility of each pest species and sensitivity of the crop variety to heat must be assessed on

Table 13.1. Hot water treatment in nematode management.

Nematode species	Temperature (°C)	Exposure time (min)	Crop	Reference
Meloidogyne spp.	45	30	Tomatoes	Yun *et al.*, 2021
Heterodera spp.	47	20	Wheat	Ibrahim *et al.*, 2019
Ditylenchus dipsaci	48	30	Onions	Bernard *et al.*, 2023
Globodera spp.	50	20	Potatoes	Perry *et al.*, 2018
Meloidogyne incognita	52	15	Potatoes	McSorley *et al.*, 2020
Pratylenchus spp.	55	10	Bananas	Hallmann and Meressa, 2018
Rotylenchulus reniformis	57	5	Pineapple	Bridge and Starr, 2007
Meloidogyne javanica	58	10	Cucumbers	Thies *et al.*, 2019
Helicotylenchus dihystera	60	5	Sugarcane	Stirling, 2014
Hoplolaimus columbus	62	5	Soybeans	Schmitt and Inserra, 2013
Xiphinema index	65	5	Grapevine	Tzortzakakis *et al.*, 2016

a small scale prior to implementing large-scale treatment programs to ensure efficacy and prevent thermal injury to plants. Using inappropriate temperatures or exposing the plant material for either too long or too short a period can compromise the treatment's effectiveness. Selected hot water treatments against nematode, fungi and bacteria are presented in Table 13.1.

Steam

Utilizing steam to heat the soil to a temperature of 66–71°C for a dwell time of 20 min has proven highly effective in eradicating weed seeds and nutsedge tubers in the soil, along with combating plant-parasitic nematodes, fungi and bacteria. Recent evaluations of steam applicators in crops including carrot, lettuce, spinach and strawberry have shown promising results against pathogens such as *Meloidogyne* spp., *Radopholus* spp., *Rotylenchulus reniformis*, *Sclerotinia minor*, *Phytophthora* spp., *Pythium ultimum*, *Verticillium dahliae* and suppression of *Fusarium* spp. Steam applicators are designed to move continuously to prevent damage to beneficial soil microorganisms. Rapidly heating soil to 66–71°C within 60–90 s typically sustains the required high temperature for the 20 min dwell time above 70°C. Prior to planting, steam is applied in narrow bands at depths of 8–13 cm (Kim *et al.*, 2021).

Soils and planting mixes for potting can be effectively steam-treated in enclosed containers, such as steam chambers, to eliminate pests and pathogens. Treatment of raised beds can be accomplished through methods such as buried perforated tubing or pipes. Another approach involves blowing steam under a plastic sheet secured at the edges to treat the soil (Khan *et al.*, 2023c). Perforated tubing is typically buried 25–80 cm deep, influencing the area that can be effectively treated from a single tube. Steam treatments for larger areas usually require 4–8 hour to complete, aiming to achieve temperatures of 60–70°C for a minimum of 30 min for effective control (Fennimore *et al.*, 2014; Kim *et al.*, 2021; Guerra *et al.*, 2022).

Soil solarization

Soil solarization involves heating the soil through solar energy by covering the ground with clear plastic sheets. This method raises soil temperatures to levels that can effectively kill nematodes, fungi and bacteria present in the top layers of the soil. The process involves using plastic mulches to generate lethal temperatures in the soil and is currently being employed in various countries, especially under tropical, subtropical and mediterranean climates to combat root knot and soil-borne diseases (Katan, 1981; Stapleton and DeVay, 1986; Stapleton *et al.*, 1987; Chellemi *et al.*, 1993; Gamliel and Stapleton, 1993; Noling, 1994).

Soil solarization typically requires 4–6 weeks of coverage, preferably during the hottest time of the year when maximum solar energy can be harnessed for heating the soil. Soil solarization has demonstrated effectiveness in controlling weeds, nematode and fungi due to the elevated temperatures reached during the process, which can significantly reduce their populations in the soil. A deep plough should be done before laying polythene to expose the material residing deeper in the soil profile (Khan *et al.*, 2012). Soil solarization has been proven effective in controlling root-knot nematode, *Verticillium* wilt and weeds in agricultural crops (Overman and Jones, 1986; Stapleton and DeVay, 1986; Saleh *et al.*, 1988).

Remote sensing

Advances in remote sensing technologies, such as infrared and digital thermography, have enabled the detection of areas within fields where plant-parasitic nematodes are causing damage. By integrating this technology with precision farming equipment, it is possible to improve control efficacy by precisely applying nematicides in high nematode density areas (Nutter *et al.*, 2002).

Root pruning

Root pruning is a technique that involves removal of nematode-infested roots from plant material, such as bulbs, before planting. Removing infested roots prior to planting, especially in bulb crops where nematodes such as lesion nematodes may predominantly infest the roots, has shown some promise in reducing nematode populations in roots and subsequently improving bulb growth. This practice helps prevent the introduction of nematodes into the soil and reduces the risk of nematode spread to healthy plants during transplantation (Westerdahl *et al.*, 1998).

Biological control

Basic concepts

Biological control in the context of nematode and pathogen management encompasses a range of methods, including the use of predators, parasites, toxins produced by microorganisms, killed microbial agents, etc. The biocontrol of pests and disease management can be explained as a reduction in the pest population or disease incidence accomplished through the action of naturally occurring parasites or predators by manipulating the environment or by introduction of antagonists (Stirling, 1991). Resource competition, interference and occupation of space contribute to the success of biological control. Living organisms compete for resources, preventing the disease agent from accessing essential nutrients. Some organisms inhibit the growth and development of pathogens through the production of inhibitory compounds.

The observation of population cycling dynamics in parasite–host relationships, where parasites selectively target a portion of the host population, has been a subject of study in ecological research. Over evolutionary time scales, parasites and hosts engage in co-evolutionary interactions that shape their relationship. Well-adapted parasites develop strategies to exploit specific host populations while minimizing the risk of host extinction, maintaining a stable and sustainable coexistence. Host-specific parasites face the risk of host extinction if they excessively parasitize their hosts. This can limit the effectiveness of using a single parasite for biological control. The challenges associated with using host-specific organisms have led to scepticism about the practicality of natural biological control in certain contexts. While relying solely on a single organism may not be feasible, the potential for developing combinations of organisms or integrating biological control agents with cultural practices offers promising avenues for enhancing management outcomes. Various soil-dwelling predators, including mites, tardigrades, turbellarians, enchytraeids, insects and predatory nematodes, have demonstrated potential in providing some level of control over plant-parasitic nematodes in both natural and agricultural environments (Kerry, 1990; Tedford *et al.*, 1993). However, a number of fungi and some bacteria constitute most potential biocontrol agents of plant nematodes (Stirling, 1991; Khan *et al.*, 2023a).

Fungal parasites

A wide range of biocontrol fungi have been found to infect nematodes, which are grouped into nematophagous fungi and nematode-parasitic fungi (Khan, 2016). Nematophagous fungi employ diverse mechanisms to trap the nematodes followed by their parasitism. Nematophagous fungi trap the nematode by different mechanisms, such as adhesive networks (*Arthrobotrys robustas*) or knobs (*Monacrosporium dreshleri*), non-constricting (*Dactylaria candida*) or constricting rings (*M. bembicoides*) or lateral branches (*M. eudermatous*). Each group employs distinct structures and mechanisms to adhere to and attack nematodes, reflecting the diversity of strategies evolved by these fungi to capture and feed on nematodes. The nematode trapping fungi wait for the nematodes to come into contact with the adhesive networks, knob, rings or conidia. Once a nematode interacts with these structures, they may become stuck or wedged, leading to subsequent hyphal penetration and utilization of the nematode as a nutrient source. Studies have shown that nematodes tend to put their heads through constricting rings, potentially leading to entrapment or adherence to adhesive substances. The formation of more rings by fungi in the presence of nematodes suggests a responsive mechanism to increase trapping efficiency. The exact reasons why nematodes engage with these structures, such as putting their heads through rings, are not fully understood but likely involve a combination of innate behavior and environmental cues. Once nematodes are trapped, fungal hyphae penetrate the nematode's body, leading to nutrient extraction and utilization by the fungus (Kerry *et al.*, 1982).

Despite fascinating mechanisms involved in the trapping of nematodes, these fungi have been observed to be poor competitors in the natural environment. As a result, only a few have been used in nematode management at a commercial scale (Khan, 2007). However, the nematode-parasitic fungi have been found to be good competitors in the soil environment. Some of those fungi form motile spores whereas others produce conidia (Money, 2016).

Zoosporic fungi represent a diverse group of fungi characterized by their motile spores, known as zoospores, which have the ability to actively swim through aqueous environments. These fungi have evolved unique strategies to target nematodes, harnessing their swimming capability to seek out potential hosts, attach to their cuticles and subsequently develop hyphae that enter into the nematode body through cuticles or natural openings (Stirling, 1991). Zoosporic fungi exhibit a broad spectrum of host specificity, parasitizing various life stages of nematodes. Some fungi target vermiform adults (Singh *et al.*, 1996) or larval nematodes (Giuma *et al.*, 1973), while others specialize in parasitizing nematode eggs or cysts (Khan and Jairajpuri, 2012). The nematode-parasitic fungi (non-zoosporic) produce non-motile spores, conidia, chlamydospores or sclerotia (Money, 2016). These fungi are highly competitive and show great diversity in distribution. These fungi are specialized parasites of eggs, juveniles, adults, egg masses and cysts (Khan and Jairajpuri, 2012). The important nematode-parasitic fungi include *Purpureocillium lilacinum* (Jatala, 1986). The fungus *Pochonia chlamydosporia*, which was earlier known as *Verticillium chlamydosporium* (Zare *et al.*, 2001), is a promising opportunistic fungus as it readily parasitizes eggs and unhatched larvae of nematodes (Lopez-Llorca and Duncan, 1988; Kerry, 2000). The fungus contends with the nematodes through enzymatic disruption of the eggshell and larval cuticle, and physiological disturbances brought about by biosynthesis of diffusible toxic metabolites (Morgan-Jones *et al.*, 1981; Meyer *et al.*, 1990; Stirling, 1991). Prior to infection, *P. chlamydosporia* in response to nematode eggs forms a branched mycelial network in close contact with the smooth eggshell. The invasion of the eggshell may take place with the help of specialized perforation organs or by lateral branches of the mycelium (Lopez-Llorca and Duncan, 1988; Stirling, 1991). The hyphae readily proliferate within eggs as well as unhatched juveniles. The eggs and larvae infected with *P.*

chlamydosporia generally fail to hatch and invade roots, respectively.

Tribe (1980) considered *P. chlamydosporia* ideal for soil application to control nematode diseases. Nematode eggs in soil provide excellent nutrition to *P. chlamydosporia* and the fungus seems to be quite specific to them. However, some researchers report that *P. chlamydosporia* does not depend on nematode eggs, as the fungus has also been isolated from snail eggs (Barron and Onions, 1966). It also colonizes plant roots and degrades cellulose and chitin. Kerry (2000) has raised doubts about *P. chlamydosporia* being a good bioagent because of the numerous shortcomings the fungus possesses. The biocontrol activity of the fungus has demonstrated satisfactory control of root-knot nematode disease in vegetables under glasshouse conditions. *P. chlamydosporia* appears to contribute towards the natural balance of nematode populations at least in some parts of the world, especially in Europe (Stirling, 1991). The fungus requires a unique ecological niche consisting of specific mycoflora in addition to favorable environments (Morgan-Jones and Rodriguez-Kabana, 1987). This makes *P. chlamydosporia* less suitable for field application in many climates, including India. Apart from this, only a few isolates of the fungus are, in fact, parasitic to eggs and adults of plant-parasitic nematodes.

Kerry *et al.* (1984) used oat seeds to rear *P. chlamydosporia* for field application. Soil application of the colonized oat kernels at 0.5% and 1.0% (w/w, soil:seed) considerably reduced the population of root-knot and cyst nematodes (Rodriguez-Kabana *et al.*, 1984). Khan *et al.* (2001) reared *P. chlamydosporia* on various substrates. Leaf litter, sawdust–soil–molasses and bagasse–soil–molasses revealed 10^{6-7} CFU/g material. Successful encapsulation of liquid suspension of spores and hyphae of *P. chlamydosporia* was done with sodium alginate containing 10% (w/v) Kaolin or wheat bran. On soil application, the fungus proliferated in soil from only those granules which contained wheat bran as energy source. In another study, Kerry (1988) estimated approximately 9×10^4 and 4×10^4 CFU fungus/g soil after 1 and 12 weeks of application of granules, respectively. The chlamydospores introduced through sand/bran, powdered grain or aqueous suspension without any food base established in soil successfully. The fungus has demonstrated satisfactory control of root-knot nematode disease in vegetables under glasshouse condition. A significant decrease in gall formation (17–23%) and egg mass production (11–18%) by *M. incognita* in chickpea and pigeon pea was recorded due to seed treatment with *P. chlamydosporia* (Khan *et al.*, 2011).

Bacteria

Three bacteria, *Pasteuria penetrans*, *Bacillus thuringiensis* and *Streptomyces avermitilis*, have all been extensively studied in relation to nematodes. *P. penetrans* is classified as a true parasite of nematodes (Sayre and Wergin, 1977). It infects nematodes by attaching to and penetrating the cuticles, subsequently colonizing the nematode's body, especially the female reproductive organs, ultimately leading to the nematode's death. *B. thuringiensis* is known for producing insecticidal proteins known as delta endotoxins that target specific insect pests by causing gut paralysis and ultimately leading to their death (Stirling, 1991; Becker and Schwinn, 1993; Kerry, 1995).

P. penetrans acts as a direct parasite of nematodes, whereas *B. thuringiensis* and *S. avermitilis* utilize toxins or metabolites to affect nematodes indirectly by causing toxicity or physiological disruptions. *P. penetrans*, an actinomycete endospore-forming bacterium, was first described by Thorne in 1940. Spores of *P. penetrans* adhere to the cuticle of a nematode, establishing initial contact for infection. Subsequently, a penetration tube is formed, facilitating the entry of the bacterium into the nematode host. Once inside the nematode, *P. penetrans* undergoes reproduction, utilizing the nematode's cuticle as a protective environment. This intracellular replication leads to the formation of a considerable number of spores within the nematode host. While the nematode may initially survive and continue to feed post-infection, the reproductive capacity of the nematode is significantly impaired (Khan *et al.*, 2021). The presence and proliferation of *P. penetrans* within the nematode host disrupts its normal physiology and reproduction, leading to a decrease in nematode population dynamics. By reducing the reproductive output of infected nematodes and potentially causing mortality over time, *P. penetrans* contributes to the regulation of nematode numbers in various ecosystems. Studies have shown that a single nematode infected with *P. penetrans* can harbor a vast number of spores, with estimates reaching up to 2 million spores within an individual nematode (Kerry, 1995; Stirling, 1991). The bacterium possesses all the characteristics for qualifying as a highly effective biocontrol agent against nematodes, but its strict obligatory nature has become a serious constraint in its mass production. As a result, very few or no commercial formulations of *P. penetrans* have been developed (Kutsuwa, 2019).

While *B. thuringiensis* primarily targets insects, some nematode species have been found to be susceptible to certain Cry toxins produced by this bacterium (Khan and Kounsar, 2000). This has led to interest in exploring the potential of *B. thuringiensis* as a biocontrol agent for nematodes.

S. avermitilis is known for producing avermectins, a class of compounds with anthelmintic properties that target parasitic worms, including nematodes. Avermectins produced by *S. avermitilis* have been widely used in veterinary and agricultural applications to control parasitic nematodes in livestock and crops. These compounds disrupt nematode nervous systems, leading to paralysis and death (Stirling, 1991; Kerry, 1995). Avermectin (MK-936) is a potent macrocyclic lactone compound derived from the actinomycete bacterium *S. avermitilis* through fermentation processes. Avermectin blocks the neurotransmitter γ-aminobutyric acid (GABA), and disruption of GABA function leads to paralysis and death in susceptible insects. Avermectin has demonstrated activity against plant-parasitic nematodes. Its practical use in plant nematode control is limited due to the molecule's relatively large size, which hinders its movement through the soil (Stirling, 1991; Kerry, 1995).

Suppressive soils

Suppressive soils owe their ability to suppress the growth, multiplication and pathogenesis of soil-borne pathogens to the action of collective competition and antagonistic activity of the microbial community (microbiome) present in the soil (Schlatter *et al.*, 2015). Suppressive soils offer an ideal example of natural microbiome-based plant defense where rhizodeposition plant roots stimulate, enrich and support soil microorganisms as the first line of defense against soil-borne pathogens (Weller, 1988). In suppressive soils, due to the microbial makeup and their activity, a pathogen is not able to establish, and if it does establish, is only able to cause disease of a lower severity which further declines in successive cropping (Weller, 1988). The concept of suppressive soils suits nematode management well because nematode populations gradually build up, offering the opportunity for suppression to act. Research has indicated that soils exhibiting suppression against sugarbeet cyst nematodes often harbor one or more species of nematode-parasitic fungi. It was discovered that *Hirsutella rhossiliensis*, a species parasitic on juvenile stages of sugarbeet cyst nematode, was prevalent in sugarbeet fields across California, including a high occurrence rate in San Joaquin County where it was found in 80% of sampled fields (Olatinwo *et al.*, 2006). Additionally, other nematode-parasitic fungi such as the ring trapping fungus *Arthrobotrys dactyloides*, as well as species parasitic on cysts and eggs of sugarbeet cyst nematode such as *Hyalorbilia oviparasitica*, *Dactylella oviparasitica* and *Brachypyoris oviparasitica*, are commonly present in fields with a history of sugarbeet cultivation (Weller, 1988; Olatinwo *et al.*, 2006). It has been shown that nematode-parasitic fungi potentially contribute to the observed declines in nematode populations within suppressive soils (Kerry, 1988). It has been proposed that the populations of parasitic fungi act in a self-regulating manner, maintaining a level below that which would completely eliminate the nematode population. This self-regulation ensures that a sustainable food source for the fungi, in the form of nematodes, remains available (Pires *et al.*, 2023). The interplay between nematode-parasitic fungi and nematode populations in these suppressive soils suggests a complex and finely balanced ecological dynamic that contributes to the mitigation of nematode-related issues in agricultural systems. At harvest in chemical control field trials, it is not uncommon to observe higher nematode populations in chemically treated plots compared to untreated controls. This difference can be misinterpreted as the untreated controls having naturally suppressive soils. However, the increased nematode populations in chemically treated plots may be attributed to healthier root systems that can support larger nematode populations compared to those in untreated plots (Gair *et al.*, 1969; Jaffee *et al.*, 1992; Tedford *et al.*, 1993; Kerry, 1995; Chen *et al.*, 2021). Since suppressive soils contain a microbiome of a diverse microbial community generally inhibitory to soil-borne pathogens and growth promoting to plants, this approach would work effectively against disease complexes involving nematodes and soil-borne pathogens and bacteria. Soils in tropical and subtropical regions are often enriched with species of *Bacillus*, *Pseudomonas*, *Aspergillus*, *Penicillium*, *Purpureocillium*, *Trichoderma*, etc., which may suppress plant-pathogenic nematodes and fungi generally involved in wilt and root-rot disease complexes (Khan and Mohiddin, 2023). Further, suppressive soils can also be developed by introducing a consortia of non-mutually competitive biocontrol agents and beneficial rhizobacteria coupled with improving the organic contents of the soil to enable establishment, multiplication and activity of the introduced microbes (Wu *et al.*, 2023). Such approaches may result in long-term sustainable management of nematode disease complexes. While using such strategies, it should be kept in mind that suppressive soils do not eliminate the pathogens but rather maintain their low population in order to obtain nutrition from such pathogens.

Soil amendments

Soil amendments can act through different mechanisms to help suppress nematodes and enhance overall soil health. This practice may also promote the activity of suppressive soils as well as biocontrol agents. Soil amendments include the addition of organic materials and wastes, especially oil cakes, crop wastes, animal wastes, etc. Adding organic matter to soil offers several advantages, including improved soil structure, enhanced water retention and nutrient provision. These enhancements reduce plant stress, potentially mitigating the impact of stress induced by plant-parasitic nematodes. Amendments that enhance soil structure and water-holding capacity can create conditions less favorable for nematodes (Barker and Koenning, 1998). Some amendments can release compounds that have nematicidal properties, either directly or after breakdown in the soil (Akhtar and Malik, 2000). Soil amendment waste materials, such as coffee grounds, newsprint, crab shells and quinoa bran, can adversely affect nematode populations. Substances containing chitin, such as crab and shrimp shells, are believed to stimulate the growth of nematophagous fungi (Singh *et al.*, 2021). These fungi utilize chitin present in nematode eggshells as a food source. Additionally, decomposing plant-based products such as marigolds, brassicas and sesame release chemicals with nematicidal properties. For instance, brassicas have been found to release compounds similar to those in chemical nematicides such as metam-sodium-containing products (Culbreath *et al.*, 1985; Rodriguez-Kabana, 1986; Spiegel *et al.*, 1987; Westerdahl *et al.*, 1992; Kaplan and Noe, 1993; Stapleton *et al.*, 1998).

The mode of action of many natural soil amendments remains unclear. Due to the variability in agricultural soils, the effectiveness of these amendments may vary across different farming scenarios. In developing treatment programs with soil amendments, it is crucial to include untreated replicated areas for comparison. By leaving some sections untreated, growers can accurately assess the success of the treatment program and adjust strategies as needed based on observed outcomes (Muller and Gooch, 1982; Rodriguez-Kabana *et al.*, 1987).

Commercially available products

Formulations of different species/strains of *Trichoderma*, *Pseudomonas*, *Penicillium*, *Bacillus*, *Purpureocillium lilacinum*, *Pochonia chlamydosporia* and *Burkholderia* spp. have shown varying degrees of effectiveness against nematodes (Westerdahl, 2024). Seed treatment with biological nematicides, either *Trichoderma*, *P. lilacinum*, *Pasteuria nishizawae*-Pn1 or abamectin, along with an insecticide and multiple fungicides provides a high degree of protection to corn, cotton, soybeans and other crops against nematodes as well as the disease complexes (Gaspar *et al.*, 2014). Currently a number of commercial formulations based on microbes, microbial toxins and plant products are available in the market: products based on a toxin produced by a *Myrothecium verrucaria* strain AARC-0255 and an extract of the soapbark tree *Saponaria*, *Quillaja* abamectin (from *S. avermitilis*) and azadirachtin (an extract from the neem tree) are available to growers.

Cultural control

A range of cultural practices can be employed for management of nematodes and diseases complexes. These practices encompass a diverse set of strategies such as crop rotation, host resistance, fallowing, using cover crops, carefully determining dates for planting and harvesting, inducing flooding, utilizing trap crops, promptly removing plants showing symptoms, incorporating soil amendments (see above), ensuring proper plant nutrition and watering, maintaining proper plant spacing and good air circulation, pruning infected plant parts to reduce disease spread, disposing of infected plant material, regularly inspecting plants for early signs of disease, and monitoring environmental conditions that may promote disease development. The major cultural practices are described below.

Crop rotation

Though crop rotation may initially seem like a straightforward approach to nematode management, the process of implementing a crop rotation plan reveals that several critical aspects need to be carefully considered. Key factors to take into account include the types of nematodes and pathogens with accurate identity existing in the field, the host range of these species, potential rotation crops, anticipated rates of population growth and decline, availability of resistant crop varieties, timing of planting and harvesting, damage thresholds, and the impact of weeds. Once the pathogen species in the field is identified and its population is determined, one can refer to host lists or databases to determine the availability of non-host crops that could be economically viable or seek out suitable resistant crop varieties. Additionally, consulting these resources may reveal whether local weed species serve as hosts for the pathogen found in the field (Brown, 1987; Roberts, 1993; Ferris, 2024).

Crop rotation is commonly employed for annual crops or long-term perennials such as alfalfa that can be cultivated profitably for several years before being rotated to a different crop. Similarly, producers of enduring perennials such as grapes, stone fruits and nuts should also consider long-range planning (McKenry *et al.*, 1994; Westerdahl *et al.*, 1998). Crop rotation programs rely on the principle that nematodes and pathogens will gradually perish due to starvation in non-host environments. The duration for nematodes to starve varies across different nematode species, with limited research conducted on this aspect for many species. The rate of decline to non-damaging levels varies among nematode types. Root-knot nematodes may reach non-damaging levels in 1–2 years, cyst nematodes require 3–8 years, stem and bulb nematodes take about 4 years, and lesion nematodes may need 4 years or more (Seinhorst, 1965; Duncan and Ferris, 1983; Abawi and Barker, 1984; Ferris *et al.*, 1986; McSorley and Duncan, 1995; Flint, 1998).

Mustard, onion, garlic, sesame, sunhemp, daincha, marigold, radish and pigeon pea are non-host crops for root-knot nematodes and can be used for crop rotation. Depending upon the preference and needs of the farmer or time available between two susceptible crops, a particular crop could be fitted into the cropping sequence. Rice or green gram after banana reduced the population of *Radopholus similis*, *Pratylenchus coffeae* and *Helicotylenchus multicinctus* (Rajendran *et al.*, 1979). Inclusion of resistant tomato variety Hissar lalith, garlic, potato, onion, cauliflower, resistant tomato, fenugreek, wheat, mustard and fallow in suitable cropping sequences helped in reducing the incidence and population of root-knot nematode (Jain, 2003).

Crop rotation with crops such as wheat, strawberry, cabbage, cauliflower, maize and beans reduces the cyst nematode population in potato. Mohandas *et al.* (1999) achieved a marked reduction in the root-knot nematode population under field conditions as equivalent to leaving the land fallow by using resistant tuber crops. Among various tropical tuber crops evaluated, sweet potato appears to be highly suitable as this has a short duration of 90–110 days, has no season and could be grown in most of the places with relatively good economic yield. Fernandez *et al.* (1992) also reported a reduction in the *M. incognita* population in tomato fields and a consequential increase in fruit yield by including sesame and sweet potato as rotation crops. Sheela *et al.* (2002) reported 21% reduction in the root-knot population and 21% increase in okra pod yield by using sweet potato cv. Sree Bhadra as a preceding crop. Vetrivelkalai and Subramanian (2006) reported onion–fallow–cowpea as an ideal cropping sequence against plant-parasitic nematodes in garden land conditions. Dwivedi *et al.* (2010) has reported reduced root-knot nematode populations in kharif okra with garlic (50%) and mustard (47%) as crops preceding rabi.

However, options for crop rotation may often become limited against disease complexes because of the wide host range of certain secondary pathogens, especially the root-rot-causing fungi. Fungi such as *R. solani*, *M. phaseolina*, *Sclerotinia sclerotiorum*, *Pythium aphanidermatum* have a wide host range among the vegetables and pulses. However, most cereals (except rice) and millets exhibit considerable resistance to the above pathogens, hence they can be used as a crop for a rotation against disease complexes in the above crops.

Rouging

The technique of removing and managing infested plants in localized areas of a field can be a practical approach when nematode and disease populations are not widespread. For example, early identification of distinctive symptoms in host plants affected by nematodes can enable growers to take targeted action to prevent the spread of infestation to surrounding vegetation. For instance, in the case of daffodils infested with the stem and bulb nematode (*Ditylenchus dipsaci*), characteristic symptoms such as leaf markings known as 'spikkles' with raised bumps and yellowing can serve as early indicators of nematode presence. Similarly, on alfalfa, the same nematode may induce symptoms such as the shortening of internodes, contrasting with the healthy appearance of unaffected alfalfa.

By promptly recognizing these distinct symptoms and being aware of the nematode species causing the symptoms, growers can strategically identify and remove infested plants within a field. This targeted removal of infested vegetation can help contain and limit the spread of nematodes to adjacent plants, reducing the potential for widespread infestation and mitigating detrimental impacts on crop productivity (Ferris, 2024).

Host resistance

Host resistance has been a valuable tool in effectively managing plant-parasitic nematodes in agricultural cropping systems. Various crop cultivars have been developed with resistance to specific nematode species, such as root-knot nematodes, root-lesion nematodes, cyst-forming nematodes, etc., have been developed (Roberts, 1992). Utilizing resistant plant varieties has proven to be an effective strategy in managing specific nematode species, such as soybean cyst nematode (*Heterodera glycines*) in soybeans and sugarbeet cyst nematode (*Heterodera schachtii*) in sugar beets (Roberts and May, 1986; Anand *et al.*, 1995; Roberts, 2002). The presence of sugarbeet cyst nematodes has been documented in over 40 countries. Over the years, breeding programs have been actively engaged in developing sugarbeet varieties resistant to cyst nematodes to mitigate the economic impact of nematode infestations on sugarbeet production. It is important to note that while nematode-resistant commercial varieties have shown enhanced performance and yield benefits in the presence of high nematode populations, their performance may vary under different nematode pressure levels. Factors such as nematode population density and environmental conditions can influence the effectiveness of nematode-resistant varieties in achieving yield improvements (Kaloshian *et al.*, 1996; Ogallo *et al.*, 1999). A single variety may offer resistance to only one particular nematode genus while remaining vulnerable to others. For example, Nemaguard rootstock, known for its resistance to root-knot nematodes, may paradoxically be more susceptible to damage caused by other nematode species, such as ring nematodes, in comparison to alternative rootstocks such as Lovell (McKenry *et al.*, 1994).

An extensive breeding program to develop grape rootstocks has resulted in a number of releases with resistance to multiple species of nematodes and to *Phylloxera* (Hewitt *et al.*, 1958). Breeding programs are actively working towards developing additional nematode-resistant (or

-tolerant) varieties for various crops such as carrots, beans, tomatoes, wheat and grapes, as well as fruit and nut trees (Gu and Ramming, 2005; Ferris *et al.*, 2012; Foundation Plant Services, 2024). However, there is a scarcity of crop cultivars that have resistance against multiple pathogens such as root-rot-causing fungi which form disease complexes of high economic significance. Multi-nematode-resistant cultivars have also been developed. For example, tomatoes are typically resistant to three out of the four major *Meloidogyne* species that are commonly found in tomato-growing regions, with susceptibility remaining to *M. hapla* and possibly other species. In addition to nematodes, VFN tomatoes have resistance to *Verticillium* and *Fusarium*. However, the situation with cultivars resistant to wilt or root-rot-causing fungi is different. The response of cultivars resistant to the above fungi is not unknown for nematodes or the breeding programs generally do not include nematodes. As a result, the wilt- or root-rot-resistant cultivars become susceptible to the fungus in nematode-infested fields.

Grafting of annual crops

Grafting refers to the intentional fusion of a scion and a rootstock, sourced from different yet compatible plants within close taxonomic proximity, with the aim of creating a composite plant. The scion, which constitutes the upper section, is typically chosen for its favorable characteristics, such as increased yields, larger fruit sizes or superior flavor profile.

The primary objectives of grafting with regard to crop production is to confer resistance to soil-borne pathogens, as well as boosting yields and enhancing resilience to adverse environmental conditions. Enhanced resistance to soil-borne diseases stands out as a key motivation for adopting grafting in vegetables. Rootstocks are chosen based on their ability to withstand infection including those caused by *Verticillium*, *Phytophthora*, *Fusarium* and nematodes (Sakata *et al.*, 2008; Thies *et al.*, 2010).

Grafting watermelons has emerged as a valuable technique that offers multiple benefits including disease resistance, improved tolerance to adverse environmental conditions and increased yields. Bottle gourd rootstocks are known for their resistance to fusarium wilt and ability to withstand chilling conditions (Louws *et al.*, 2010). While the root growth of bottle gourd rootstocks may not be as robust as that of interspecific hybrids, they exhibit early maturity. Grafting on to bottle gourd typically does not adversely affect flowering or fruit quality. Studies illustrate that grafting can effectively address a plethora of soil-borne diseases caused by fungi, bacteria and nematodes. Grafting has proved beneficial in managing various other soil-borne diseases such as *Phytophthora* root and crown rot as well as *Verticillium* wilt (Keinath and Hassell, 2014).

Cover crops

Cover crops can improve soil fertility by adding organic matter when they decompose, enhancing soil structure and promoting microbial activity. Certain cover crops can scavenge and store nitrogen, reducing leaching and making it available for subsequent crops. Cover crops help reduce erosion and can compete with weeds for resources such as light, water and nutrients, thereby reducing weed pressure. Understanding the interactions between cover crops, nematodes and plant pathogens is a critical aspect of cultural management in agricultural systems. Currently, our knowledge of cover crops is primarily focused on predicting which cover crops are less likely to worsen existing issues rather than actively reducing populations beyond what occurs in fallow conditions. Certain cover crops, such as mustard, oil radish and buckwheat, which are known hosts of sugarbeet cyst nematodes, have shown effectiveness in reducing nematode populations and altering their life cycle dynamics (Caswell-Chen and Goodell, 1992; Gardner and Caswell-Chen, 1993). Notably, breeders in Germany have identified specific varieties of these host cover crops that stimulate egg hatching and may even facilitate nematode penetration of roots (Gardner and Caswell-Chen, 1993). This selective breeding has led to significant reductions in nematode populations, with reported decreases of up to 50%. Additionally, the shift in nematode development from females to males or the prevention of regular development indicates a disruption in the nematode life cycle, potentially inhibiting their reproductive capability.

Fallowing

Fallowing, or leaving the soil without a host crop, is a traditional method used to manage nematodes by depriving them of a host plant, ultimately leading to their decline or death due to starvation. This approach is based on the principle that nematodes cannot survive indefinitely without a suitable host (Duncan, 1986). For certain economically important nematodes such as root-knot, sugarbeet cyst, stem and bulb, and *Xiphinema index*, there is existing knowledge regarding the duration nematodes can persist without a host before their populations decline. However, for many other nematode species, this critical information is still lacking, making it challenging to effectively implement fallowing as a nematode management strategy (Johnson *et al.*, 1992).

One common limitation in utilizing fallowing effectively is the oversight of weed hosts that can serve as alternative hosts for nematodes during the fallow period (Khan *et al.*, 2012). Weeds can sustain nematode populations, undermining the efficacy of fallowing in reducing nematode numbers. Additionally, the persistence of root remnants in the soil can continue to serve as hosts for nematodes even in the absence of above-ground plant growth. The effectiveness of dry vs wet fallowing in managing nematodes remains a topic of debate. The efficacy of either dry or wet fallowing can be influenced by various factors such as the specific nematode species present, soil type, climate conditions and other environmental variables (Minton, 1986; Goodell and Ferris, 1989; Smiley *et al.*, 1994).

Flooding

When considering the use of flooding as a potential method to control nematodes, it is important to recognize key factors that influence its effectiveness. Soil-dwelling nematodes, being aquatic organisms that reside in the thin film of moisture surrounding soil particles, possess adaptations that can limit the efficacy of flooding as a nematode management strategy. Additionally, some nematodes have the capability of anaerobic respiration, which allows them to survive under waterlogged conditions (Sikora *et al.*, 2005). However, it is essential to note that while continuous flooding may not be the most effective strategy for nematode management in certain cases, flooding can still play a role in nematode control under specific conditions. For instance, in situations where flooding is implemented for purposes such as salt leaching or waterflow management, incidental reductions in nematode populations can also be realized. This highlights the potential for integrating nematode management objectives with other agricultural practices that involve flooding. Moreover, in different nematode species and geographical locations, short-term flooding has been utilized as a strategy to stimulate nematode egg hatching. Subsequent fallowing following this inundation can lead to a more rapid decline in nematode populations. This approach leverages flooding to disrupt nematode life cycles and create conditions that are less hospitable for their survival, ultimately aiding in population reduction (Sikora *et al.*, 2005).

Advisory service

In view of the hidden nature of damage and lesser awareness of growers to nematodes, an advisory service may play a major role in helping farmers to realize the potential of damage to their crops and to take appropriate measures for nematode management. An advisory service is meant to identify a nematode problem and suggest appropriate control methods to the growers that seek advice from the service. In the countries or states where nematode advisory services exist, the grower either sends soil or root samples from their fields or invites personnel to visit the field, so that the associated problems can be identified and managed. An advisory service serves two purposes: diagnostic and predictive.

The diagnostic service includes assay of the nematodes for the identification of plant disease problems for future crops in the case of annuals or appropriate therapy for perennials, whereas the predictive service, with the help of diagnostic data, estimates the likelihood of nematode damage to future crops and recommends action for avoiding or minimizing yield losses.

The advisory service provides sampling kits to the growers to collect and dispatch soil or plant samples to the advisory stations for assay. The kit contains instructions to the grower for collecting the soil and plant samples. Information on the size of the plots, crops before the last year and the crop to be grown are also required from the grower. The advisory service assays the sample and identifies the population of each nematode species present in the sample and suggests appropriate action to minimize yield losses. The advisory service may be offered to the growers free of cost generally by state agencies or non-governmental organizations, or may charge some fee, especially by the private agencies. Nematode advisory services do not exist in most of the countries in Asia and Africa where nematode damage to crops is quite high. An advisory service may also prove effective particularly against disease complexes because such diseases are often misjudged even by experienced and progressive farmers.

Date of planting

Temperature plays a major role in the rate of development of nematodes and plant pathogens. Each nematode species has a minimum temperature threshold below which they are unable to effectively penetrate the roots and cause damage to the plants. By aligning planting schedules with optimal soil temperatures that deter nematode penetration, growers can potentially minimize nematode-induced losses (Roberts, 1987; Johnson and Motsinger, 1990; Jeffers and Roberts, 1993). For instance, studies have shown that adjusting planting dates by a few weeks to coincide with temperatures unfavorable for nematode activity can lead to significant reductions in nematode damage, as illustrated by a graph depicting soil temperature variations throughout the year relative to nematode infection thresholds (Fig. 13.2) (Van Gundy *et al.*, 1974; Roberts *et al.*, 1981; Roberts, 1987).

Date of harvest

Adjusting the timing of harvest can be a valuable strategy in managing nematode damage in certain situations. By strategically planning the harvest timing based on predictions of

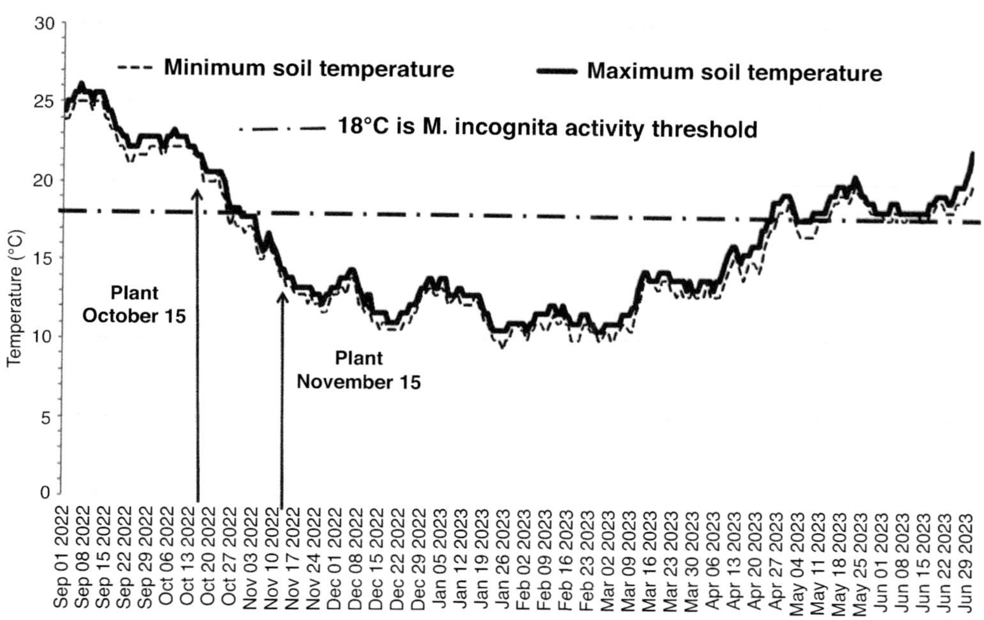

Fig. 13.2. The relationship between soil temperatures and nematode activity threshold is evident in the case of *Meloidogyne incognita*, which cannot penetrate roots when soil temperatures fall below 18°C. Shifting the planting schedule from mid-October to mid-December enables crop cultivation without incurring nematode damage. Daily maximum and minimum soil temperature data for Orange County, California was obtained from UC IPM Online (2024c) and graphed following a method described by Roberts (1987).

nematode population dynamics and economic considerations, growers can potentially minimize the impact of nematodes on crop yield and quality as well as the new crop. Effective utilization of this technique requires the ability to forecast when nematode populations will reach economically significant levels, prompting decisions for an early harvest to mitigate damage (Pinkerton *et al.*, 1991; Carlson *et al.*, 1992). Sufficient information is available for root-knot nematode, for sugarbeet cyst nematode and several fungal diseases to implement degree-day models. By tracking nematode degree-days and monitoring their population dynamics in relation to crop growth stages, growers can make informed decisions regarding harvest timing to minimize nematode damage. Further research and development in nematode degree-day modeling could expand the applicability of this approach to a wider range of nematode species and cropping systems, offering growers additional tools for sustainable nematode management (Ferris and Van Gundy, 1979; Abawi and Barker, 1984; Noling and Ferris, 1987; Schneider and Ferris, 1987; Pinkerton *et al.*, 1991; UC IPM Online, 2024a).

Trap cropping

Trap cropping can be an effective strategy in nematode management due to the specific mechanism through which it targets and controls nematode populations. When an infective juvenile nematode penetrates a host root and begins feeding, it undergoes physiological changes, becoming swollen or rotund, ultimately becoming trapped within the root tissues. By destroying the host plant before nematodes can lay eggs, the trapped nematodes within the roots perish, thereby reducing the nematode population in the soil. One key aspect to note is that trap cropping is generally not effective for managing migratory ectoparasitic or endoparasitic nematodes. Hence, this technique can be employed for reducing populations of sedentary endoparasitic nematodes such as root-knot (*Meloidogyne* sp.) or cyst nematodes (*Heterodera* sp.). This method involves planting a host crop that attracts the nematodes of interest, allowing it to grow for a period, and then terminating it before the development of egg-bearing adults. It is crucial to ensure that the roots of the trap crop are thoroughly destroyed to prevent them from serving as a continued food source for nematode development. The success of trap cropping hinges on the precise timing of crop termination. Nematodes develop more rapidly in warmer soil compared to cooler soil conditions. Therefore, in regions with warmer climates, the termination of the trap crop needs to occur earlier to effectively control nematode populations, compared to regions with cooler climates (Noling and Ferris, 1987; Gardner and Caswell-Chen, 1997; Scholte, 2000; Westerdahl, 2020, 2021; UC IPM Online, 2024b).

Biofumigation

Research efforts focusing on biofumigation as a sustainable approach for managing nematodes, weeds and fungi as well as the disease complexes have provided valuable insights into utilizing *Brassica* species, such as broccoli, for nematode control through the release of glucosinolates and subsequent production of isothiocyanates in the soil during degradation. *Brassica* plants contain glucosinolates, which are secondary metabolites known for their biofumigant properties. When *Brassica* residues are tilled into the soil, glucosinolates break down and release isothiocyanates, which possess nematicidal properties. Intriguingly, metam-sodium, a conventional nematicide, also employs isothiocyanate compounds as its active ingredient for nematode control (Kirkegaard *et al.*, 1998).

In chemical soil fumigation practices, plastic film tarping is commonly used to reduce the volatilization rate of fumigants applied to the soil. Similarly, when incorporating biofumigation techniques in nematode management, utilizing plastic film coverings over the soil can greatly enhance the effectiveness of the biofumigant. This tarping technique helps to contain and concentrate the released isothiocyanates within the soil profile, prolonging their exposure to nematodes and enhancing the overall efficacy of biofumigation for nematode and disease control (Mojtahedi *et al.*, 1991; McFadden *et al.*, 1992; Gamliel and Stapleton, 1993; Špak *et al.*, 1993; Angus et al., 1994; Stapleton et al., 1998; Ploeg and Stapleton, 2001; Ploeg, 2007).

Anaerobic soil disinfestation

The anaerobic soil disinfestation (ASD) method offers a promising approach for managing nematodes, pathogens and weeds. The process involves the application of powdery rice bran to initiate the ASD process. Subsequently, the bran is incorporated into the soil, followed by the application of a clear tarp and auxiliary irrigation system. During the initial 1–2 days, the entire system is irrigated to a depth of 1.5–2 m, ensuring thorough saturation of the targeted soil profile. Daily or alternate-day pulsing of the irrigation system thereafter is essential to maintain soil moisture at levels exceeding field capacity. The final stages encompass the removal of the tarp and irrigation infrastructure, preparing the soil for plant growth suppressive to pathogens (Shennan *et al.*, 2018).

Optimal conditions for ASD application revolve around the availability of ample solar radiation, ideally during the peak of summer or alternatively towards the late summer and early fall seasons. The biological and chemical transformations induced in ASD-treated soils are integral to its efficacy. Through the combined actions of elevated soil temperatures facilitated by the tarp and the saturation of moisture, soil bacteria metabolize the carbon-rich substrate, leading to the generation of organic acids and a reduction in pH levels. These processes result in the release of volatile compounds that collectively modulate the soil microbial community towards a more conducive environment for almond tree cultivation (Shennan *et al.*, 2018).

Chemical control

Chemical disease control primarily involves the application of fungicides, bactericides and nematicides to combat the biotic agents causing the diseases. Pesticides play a significant role in managing certain disease scenarios where their use is indispensable and effective.

Fumigants

The fumigant carbon disulfide was initially trialed in Europe for controlling *Phylloxera* infestations on grapevines. With its wide-ranging effectiveness, researchers discovered its efficacy against nematodes and certain plant-pathogenic fungi. Professionals in nematology who have employed this compound often highlight its flammability, recounting alarming incidents where fields and machinery were consumed by fire if the applicator inadvertently struck a rock during its use. Following World War I, a substantial quantity of chloropicrin remained in the Hawaiian Islands. The pineapple industry there actively backed extensive research endeavors focused on nematode and fungal control. Through these efforts, it was discovered that chloropicrin exhibited both nematicidal and fungicidal properties (Johnson and Feldmesser, 1987; Hague and Gowen, 1987).

The successful utilization of initial fumigants, carbon disulfide and chloropicrin, catalyzed the emergence of various other chemical pesticides in the 1940s. These included methyl bromide, ethylene dibromide (EDB), 1,2-dibromo-3-chlororopane (DBCP), D-D mixture, 1,3-dichloropropene (1,3-D), etc. During the 1960s and 1970s, the use of EDB, D-D mixture, DBCP and later methyl bromide was discontinued due to concerns regarding the potential for environmental contamination or carcinogenic effects (Heald, 1987; Thomason, 1987; Ferguson and Padula, 1994).

Metam-sodium, upon application in the soil, releases methyl isothiocyanate which is a potent broad-spectrum biocide. The chemical is effective against nematodes as well as other invertebrate pests. Among the fumigants, metam-sodium exhibits the least volatility and typically moves around 10 cm within the soil. To optimize its efficacy, as well as that of organophosphates and carbamates, these chemicals must be moved through the soil using methods such as incorporating them into irrigation water or employing mechanical tillage (Heald, 1987). Due to the dependence of the effectiveness of fumigants nematicides on several factors, especially soil type, moisture content and temperature, as well as their ecotoxic effects, use of these chemicals has become restricted to specialized situations.

Liquid and granular products

During the 1950s and 1960s, several organophosphates and carbamates were developed as nematicides (Hough and Thomason, 1975; Davis *et al.*, 1993). Liquid products are often applied by spraying them on to the soil surface either before or after planting. For post-plant applications, many growers utilize sprayers commonly used for herbicide application to treat both sides of tree or vine rows effectively. The granular nematicides are generally used by directly applying in the field. Following the application of liquid products, it is essential to incorporate the product into the soil to enhance its efficacy (Khan *et al.*, 2014). This incorporation can be achieved through irrigation or the use of mechanical incorporation equipment. Mechanical incorporators come in various types, each offering different depths of product integration into the soil (Johnson, 1985). Further new molecules, such as fluopyram, fluensulfone, etc., should be tested in disease complex situations as these products are reported to be nema- and fungi-toxic (Walters *et al.*, 2009; Becker *et al.*, 2019; Haque and Khan, 2023).

The choice of incorporation method depends on factors such as the specific product used, soil type, crop or turfgrass variety and desired depth of distribution. Both irrigation and mechanical incorporation play crucial roles in ensuring proper dispersion of products within the soil, thereby maximizing their effectiveness in controlling populations and protecting plant health. Growers commonly possess granular applicators designed for the precise application of fertilizers and insecticides. These tools can often be adjusted or calibrated to facilitate the accurate application of granular pesticide products as well. Similar to spray applications, when granules are used and applied to the soil surface, it is crucial to promptly incorporate them into the soil for optimal effectiveness. This incorporation can be achieved through irrigation or mechanical devices to ensure proper dispersion and penetration of the granules throughout the root zone where nematodes typically reside.

Liquid products are generally used through furrow and drip irrigation. Drip irrigation minimizes water loss through evaporation and runoff, and can be programmed to deliver water and nutrients according to specific schedules and plant requirements (McKenry *et al.*, 1994).

Conclusion and Future Prospects

In many scenarios, granular and liquid formulations and a few fumigants are still mainstays of management programs against nematodes and disease complexes. However, as reviewed in this chapter, sustainable alternatives in the areas of improved diagnosis, cultural, physical and biological methods are seeing increased usage and integration into management programs. Notable among these molecular diagnostics, improved understanding of biology and population dynamics, online databases and computer-assisted modeling, newly developed resistant varieties, biofumigation and recent nematicides may prove effective if they are exploited in an interactive way. No product or method alone may offer effective management of disease complexes because of their multipathogenic nature. In this situation, a holistic and target-based selection and combination of different management strategies should be made to control the disease complexes (Raski and Allen, 1948; Nyczepir, 1989; Ferris *et al.*, 1993; Westerdahl and Frate, 2007).

References

Abawi, G.S. and Barker, K.R. (1984) Effects of cultivar, soil temperature, and population levels of *Meloidogyne incognita* on root necrosis and fusarium wilt of tomatoes. *Phytopathology* 74(4), 433. DOI: 10.1094/Phyto-74-433.

Agrios, G.N. (2005) *Plant Pathology*, 5th edn. Elsevier Academic Press, Burlington, MA.

Akhtar, M. and Malik, A. (2000) Roles of organic soil amendments and soil organisms in the biological control of plant-parasitic nematodes: A review. *Bioresource Technology* 74(1), 35–47. DOI: 10.1016/S0960-8524(99)00154-6.

Anand, S.C., Koenning, S.R. and Sharma, S.B. (1995) Performance of blends of soybean cyst nematode resistant and susceptible cultivars. *Crop Science* 35(2), 524–528. DOI: 10.2135/cropsci1995.0011183X003500020040x.

Angus, J.F., Gardner, P.A., Kirkegaard, J.A. and Desmarchelier, J.M. (1994) Biofumigation: Isothiocyanates released from *Brassica* roots inhibit growth of the take-all fungus. *Plant and Soil* 162(1), 107–112. DOI: 10.1007/BF01416095.

Barker, K.R. and Koenning, S.R. (1998) Developing sustainable systems for nematode management. *Annual Review of Phytopathology* 36(1), 165–205. DOI: 10.1146/annurev.phyto.36.1.165.

Barron, G.L. and Onions, A.H.S. (1966) *Verticillium chlamydosporium* and its relationships to *Diheterospora*, *Stemphyliopsis*, and *Paecilomyces*. *Canadian Journal of Botany* 44(7), 861–869. DOI: 10.1139/b66-101.

Becker, J.O. and Schwinn, F.J. (1993) Control of soil-borne pathogens with living bacteria and fungi: Status and outlook. *Pesticide Science* 37(4), 355–363. DOI: 10.1002/ps.2780370408.

Becker, J.O., Ploeg, A. and Nuñez, J.J. (2019) Multi-year field evaluation of fluorinated nematicides against *Meloidogyne incognita* in carrots. *Plant Disease* 103(9), 2392–2396. DOI: 10.1094/PDIS-03-19-0489-RE.

Bergeson, G.B. (1972) Concepts of nematode–fungus associations in plant disease complexes: A review. *Experimental Parasitology* 32(2), 301–314. DOI: 10.1016/0014-4894(72)90037-9.

Bernard, G.C., Egnin, M., Mortley, D. and Bonsi, C. (2023) Nematode problems in bulb crops and sustainable management. In: Khan, M.R. and Quintanilla, M. (eds) *Nematode Diseases of Crops and Their Sustainable Management*. Academic Press, Cambridge, MA, pp. 297–309.

Bridge, J. (1975) Hot water treatment to control plant parasitic nematodes of tropical crops. *Mededelingen van de Faculteit Landbouwwetenschappen Rijksuniversiteit Gent* 40, 249–259.

Bridge, J. and Starr, J.L. (2007) *Plant Nematodes of Agricultural Importance: A Color Handbook*. CRC Press, Boca Raton, FL.

Brown, R.H. (1987) Control strategies in low-value crops. In: Brown, R.H. and Kerry, B.R. (eds) *Principles and Practice of Nematode Control in Crops*. Academic Press, Sydney, pp. 351–388.

Brown, R.H. and Kerry, B.R. (eds) (1987) *Principles and Practice of Nematode Control in Crops*. Academic Press, Sydney.

Carlson, H.L., Westerdahl, B.B. and Ferris, H. (1992) Reduction of nematode tuber blemish by early harvest and variety selection. *American Potato Journal* 69, 574.

Caswell-Chen, E.P. and Goodell, P.B. (1992) Potential cover crops for reduction of *Meloidogyne incognita* in California. *Journal of Nematology* 24, 585.

Chellemi, D.O., Olson, S.M., Scott, J.W., Mitchell, D.J. and McSorley, R. (1993) Reduction of phytoparasitic nematodes on tomato by soil solarization and genotype. *Journal of Nematology* 25, 800–805.

Chen, Y.-Y., Koike, S.T., Logan, G.D., Drozd, C., De Oliveira Silva, J. *et al*. (2021) Detection of nematophagous fungi from *Heterodera schachtii* females using a baiting experiment with soils cropped to *Brassica* species from California's central coast. *PhytoFrontiers* 1(1), 4–12. DOI: 10.1094/PHYTOFR-07-20-0009-R.

Culbreath, A.K., Rodriguez-Kabana, R. and Morgan-Jones, G. (1985) The use of hemicellulosic waste matter for reduction of the phytotoxic effect of chitin and control of root-knot nematodes. *Nematropica* 15, 49–75.

Davis, R.F., Johnson, A.W. and Wauchope, R.D. (1993) Accelerated degradation of fenamiphos and its metabolites in soil previously treated with fenamiphos. *Journal of Nematology* 25, 679–685.

Duncan, LW. (1986) Effects of bare fallow on plant-parasitic nematodes in the Sahelian zone of Senegal. *Revue de Nématologie* 9, 75–81.

Duncan, L.W. and Ferris, H. (1983) Validation of a model for prediction of host damage by two nematode species. *Journal of Nematology* 15, 27–234.

Dwivedi, K., Srivastava, J.P. and Verma, K.B. (2010) Effect of cropping sequence on population fluctuation of root knot nematode, *Meloidogyne incognita* in vegetable based cropping system. *Indian Journal of Nematology* 40, 252–254.

Faulkner, L.R. and Bolander, W.J. (1966) Occurrence of large nematode populations in irrigation canals of south central Washington. *Nematologica* 12(4), 591–600. DOI: 10.1163/187529266X00446.

Fennimore, S.A., Martin, F.N., Miller, T.C., Broome, J.C., Dorn, N. *et al.* (2014) Evaluation of a mobile steam applicator for soil disinfestation in California strawberry. *HortScience* 49(12), 1542–1549. DOI: 10.21273/HORTSCI.49.12.1542.

Ferguson, W. and Padula, A. (1994) Economic effects of banning methyl bromide for soil fumigation. Agricultural Economic Report 677. USDA Economic Research Service, Herndon, VA.

Fernandez, E., Perez, A., Lorenzo, E. and Vincent, E. (1992) Effectivity of use of sesame (*Sesamus indicum*) as rotation crop against *Meloidogyne incognita*. *Revista de Protection Vegetal* 7, 39–42.

Ferris, H. (2024) Search options for Nemabase - a database of the host status of plants to nematodes. Available at: http://nemaplex.ucdavis.edu/Nemabase2010/Nemabase%20Search%20Menu.htm (accessed 30 March 2024).

Ferris, H. and Van Gundy, S.D. (1979) *Meloidogyne* ecology and host interrelationships. In: Lamberti, F. and Taylor, C.E. (eds) *Root-Knot Nematodes* (Meloidogyne *spp.) Systematics, Biology and Control*. Academic Press, New York, pp. 205–230.

Ferris, H., Ball, D.A., Beem, L.W. and Gudmundson, L.A. (1986) Using nematode count data in crop management decisions. *California Agriculture* 40, 12–14.

Ferris, H., Carlson, H.L., Viglierchio, D.R., Westerdahl, B.B., Wu, F.W. *et al.* (1993) Host status of selected crops to *Meloidogyne chitwoodi*. *Journal of Nematology* 25, 849–857.

Ferris, H., Zheng, L. and Walker, M.A. (2012) Resistance of grape rootstocks to plant-parasitic nematodes. *Journal of Nematology* 44(4), 377–386.

Ferris, J.M. and Ferris, V.R. (1998) Biology of plant-parasitic nematodes. In: Barker, K.R., Pederson, G.A. and Windham, G.L. (eds) *Plant and Nematode Interactions*. ASA, CSSA and SSSA, Madison, WI, pp. 21–35.

Flint, M.L. (1998) *Integrated Pest Management for Tomatoes*. University of California Statewide Integrated Pest Management Project Division of Agriculture and Natural Resources Publication Number 3274, Oakland, CA.

Foundation Plant Services (2024) University of California, Davis College of Agricultural and Environmental Sciences. Available at: https://fps.ucdavis.edu/fgrmain.cfm (accessed 30 March 2024).

Gair, R., Mathias, P.L. and Harvey, P.N. (1969) Studies of cereal nematode populations and cereal yields under continuous or intensive culture. *Annals of Applied Biology* 63(3), 503–512. DOI: 10.1111/j.1744-7348.1969.tb02846.x.

Gamliel, A. and Stapleton, J.J. (1993) Characterization of antifungal volatile compounds evolved from solarized soil amended with cabbage residues. *Phytopathology* 83(7), 899–905. DOI: 10.1094/Phyto-83-899.

Gardner, J. and Caswell-Chen, E.P. (1993) Penetration, development, and reproduction of *Heterodera schachtii* on *Fagopyrum esculentum*, *Phacelia tanacetifolia*, *Raphanus sativus*, *Sinapis alba*, and *Brassica oleracea*. *Journal of Nematology* 25(4), 695–702.

Gardner, J. and Caswell-Chen, E. (1997) Influence of cyst maturation on apparent population increases of *Heterodera schachtii* on root remnants. *Fundamental and Applied Nematology* 20, 269–276.

Gaspar, A.P., Marburger, D.A., Mourtzinis, S. and Conley, S.P. (2014) Soybean seed yield response to multiple seed treatment components across diverse environments. *Agronomy Journal* 106(6), 1955–1962. DOI: 10.2134/agronj14.0277.

Giuma, A.Y., Hackett, A.M. and Cooke, R.C. (1973) Thermostable nematotoxins produced by germinating conidia of some endozoic fungi. *Transactions of the British Mycological Society* 60(1), 49–56. DOI: 10.1016/S0007-1536(73)80058-0.

Goodell, P.B. and Ferris, H. (1989) Influence of environmental factors on the hatch and survival of meloidogyne incognita. *Journal of Nematology* 21, 328–334.

Grogan, R.G. (1981) The science and art of plant-disease diagnosis. *Annual Review of Phytopathology* 19(1), 333–351. DOI: 10.1146/annurev.py.19.090181.002001.

Gu, S. and Ramming, D. (2005) Viticultural performance of Thompson Seedless grapevines on new USDA-ARS rootstocks for raisin production in the San Joaquin Valley. *American Journal of Enology and Viticulture* 56, 312A.

Guerra, N., Fennimore, S.A., Siemens, M.C. and Goodhue, R.E. (2022) Band steaming for weed and disease control in leafy greens and carrots. *HortScience* 57(11), 1453–1459. DOI: 10.21273/HORTSCI16728-22.

Hague, N.G.M. and Gowen, S.R. (1987) Chemical control of nematodes. In: Brown, R.H. and Kerry, B.R. (eds) *Principles and Practice of Nematode Control in Crops*. Academic Press, Sydney, pp. 131–173.

Hallmann, J. and Meressa, B.H. (2018) Nematode parasites of bananas and plantains. In: Sikora, R.A., Coyne, D., Hallmann, J. and Timper, P. (eds) *Plant Parasitic Nematodes in Subtropical and Tropical Agriculture*, 3rd edn. CAB International, Wallingford, UK, pp. 387–426. DOI: 10.1079/9781786391247.0346.

Haque, Z. and Khan, M.R. (2021) *Handbook of Invasive Plant-Parasitic Nematodes*. CAB International, Wallingford, UK.

Haque, Z. and Khan, M.R. (2023) Nematicidal effect of fluopyram and fluensulfone against rice root-knot nematode, *Meloidogyne graminicola*. *Russian Journal of Nematology* 31, 39–49. DOI: 10.24412/0869-6918-2023-1-39-49.

Heald, C.M. (1987) Classical nematode management practices. In: Veech, J.A. and Dickson, D.W. (eds) *Vistas on Nematology*. Society of Nematologists, Hyattsville, MD, pp. 100–104.

Hewitt, W.B., Raski, D.J. and Goheen, A.C. (1958) Nematode vector of soil-borne fanleaf virus of grapevine. *Phytopathology* 48, 586–595.

Hirano, K. (1975) Interrelationships between plant-parasitic nematodes and other plant pathogenic organisms. *Review of Plant Protection Research* 8, 55–68.

Hough, A. and Thomason, I.J. (1975) Effects of aldicarb on the behavior of *Heterodera schachtii* and *Meloidogyne javanica*. *Journal of Nematology* 7(3), 222–229.

Ibrahim, I.K.A., Mukhtar, T. and Khan, H.U. (2019) Integrated management of cyst nematodes in wheat: Efficacy of soil solarization and hot water treatment. *International Journal of Agriculture and Biology* 21(1), 85–90.

Jaffee, B., Phillips, R., Muldoon, A. and Mangel, M. (1992) Density-dependent host–pathogen dynamics in soil microcosms. *Ecology* 73, 495–506.

Jain, R.K. (2003) Influence of different cropping sequences on population and incidence of root-knot nematode in. In: *Proceedings of National Symposium on Biodiversity and Management of Nematodes in Cropping Systems for Sustainable Agriculture*, Jaipur, India, November 11-13, pp. 84–87.

Jatala, P. (1986) Biological control of plant-parasitic nematodes. *Annual Review of Phytopathology* 24, 453–489. DOI: 10.1146/annurev.py.24.090186.002321.

Jeffers, D.P. and Roberts, P.A. (1993) Effect of planting date and host genotype on the root-knot nematode–*Fusarium* wilt disease complex of cotton. *Phytopathology* 83(6), 645–654. DOI: 10.1094/Phyto-83-645.

Johnson, A.W. and Motsinger, R.E. (1990) Effects of planting date, small grain crop destruction, fallow, and soil temperature on the management of *Meloidogyne incognita*. *Journal of Nematology* 22, 348–355.

Johnson, A.W. (1985) The role of nematicides in nematode management. In: Sasser, J.N. and Carter, C.C. (eds) *An Advanced Treatise on* Meloidogyne. *Volume I. Biology and Control*. North Carolina State University Graphics, Raleigh, NC, pp. 249–267.

Johnson, A.W. and Feldmesser, J. (1987) Nematicides: A historical review. In: Veech, J.J. and Dickson, D.W. (eds) *Vistas on Nematology*. Society of Nematologists, Hyattsville, MD, pp. 448–454.

Johnson, A.W., Golden, A.M., Auld, D.L. and Sumner, D.R. (1992) Effects of rapeseed and vetch as green manure crops and fallow on nematodes and soil-borne pathogens. *Journal of Nematology* 24, 117–126.

Kaloshian, I., Williamson, V.M., Miyao, G., Lawn, D.A. and Westerdahl, B.B. (1996) "Resistance-breaking" nematodes identified in California tomatoes. *California Agriculture* 50(6), 18–19. DOI: 10.3733/ca.v050n06p18.

Kaplan, M. and Noe, J.P. (1993) Effects of chicken-excrement amendments on *Meloidogyne arenaria*. *Journal of Nematology* 25, 71–77.

Katan, J. (1981) Solar heating (solarization) of soil for control of soilborne pests. *Annual Review of Phytopathology* 19, 211–236.

Keinath, A.P. and Hassell, R.L. (2014) Control of *Fusarium* wilt of watermelon by grafting onto bottlegourd or interspecific hybrid squash despite colonization of rootstocks by *Fusarium*. *Plant Disease* 98(2), 255–266. DOI: 10.1094/PDIS-01-13-0100-RE.

Kerry, B. (1988) Fungal parasites of cyst nematodes. *Agriculture, Ecosystems & Environment* 24(1–3), 293–305. DOI: 10.1016/0167-8809(88)90073-4.

Kerry, B.R. (1990) An assessment of progress toward microbial control of plant-parasitic nematodes. *Journal of Nematology* 22, 621–631.

Kerry, B.R. (1995) Microbial agents for the biological control of plant parasitic nematodes. In: *Proceedings International Congress of Tropical Nematology*, Rio Quente, Brazil, pp. 244–250.

Kerry, B.R. (2000) Rhizosphere interactions and the exploitation of microbial agents for the biological control of plant-parasitic nematodes. *Annual Review of Phytopathology* 38(1), 423–441. DOI: 10.1146/annurev.phyto.38.1.423.

Kerry, B.R., Crump, D.H. and Mullen, L.A. (1982) Natural control of the cereal cyst nematode, *Heterodera avenae* Woll., by soil fungi at three sites. *Crop Protection* 1(1), 99–109. DOI: 10.1016/0261-2194(82)90061-8.

Kerry, B.R., Simon, A. and Rovira, A.D. (1984) Observations on the introduction of *Verticillium chlamydosporium* and other parasitic fungi into soil for control of the cereal cyst-nematode *Heterodera avenue*. *Annals of Applied Biology* 105(3), 509–516. DOI: 10.1111/j.1744-7348.1984.tb03077.x.

Khan, M.R. (2008) *Plant Nematodes – Methodology, Morphology, Systematics, Biology and Ecology*. Science Publishers, Enfield, NH. DOI: 10.1201/b10766.

Khan, M.R. (2016) Nematode biocontrol agents: Diversity and effectiveness against phytonematodes in sustainable crop protection. *Indian Phytopathology* 69(4s), 453–463.

Khan, M.R. and Kounsar, K. (2000) Effects of certain bacteria and fungi on the growth of mungbean and reproduction of *Meloidogyne incognita*. *Nematologia Mediterranea* 8, 221–226.

Khan, M.R. and Jairajpuri, S. (2012) *Nematode Infestations Part III: Horticultural Crops*. National Academy of Sciences, Allahabad, India.

Khan, M.R. and Sharma, R.K. (2020) Fusarium–nematode wilt disease complexes, etiology and mechanism of development. *Indian Phytopathology* 73(4), 615–628. DOI: 10.1007/s42360-020-00240-z.

Khan, M.R. (2023a) Nematode pests of agricultural crops, a global overview. In: Khan, M.R. (ed.) *Novel Biological and Biotechnological Applications in Plant Nematode Management*. Springer, Singapore, pp. 3–45. DOI: 10.1007/978-981-99-2893-4_1.

Khan, M.W. (1993) Mechanism of interactions between nematodes and other plant pathogens. In: Khan, M.W. (ed.) *Nematode Interactions*. Chapman & Hall, London, pp. 55–78. DOI: 10.1007/978-94-011-1488-2_4.

Khan, M.R., Khan, N. and Khan, S.M. (2001) Evaluation of agricultural materials as substrate for mass culture of fungal biocontrol agents of fusarial wilt and root-knot nematode diseases. *Annals of Applied Biology* 22, 50–51.

Khan, M.R. (2007) Prospects of microbial control of root-knot nematodes infecting vegetable crops. In: Sharma, N. and Singh, H.B. (eds) *Biotechnology: Plant Health Management*. International Book Distributing, Co., Lucknow, India, pp. 643–665.

Khan, M.R., Majid, S., Mohiddin, F.A. and Khan, N. (2011) A new bioprocess to produce low-cost powder formulations of biocontrol bacteria and fungi. *Biological Control* 59, 130–140. DOI: 10.1016/j.biocontrol.2011.04.007.

Khan, M.R., Sargobhan Ramakrishnan, R. and Krishnamurthy, V. (2012) Cultural and physical control of nematode infestation in horticultural crops. In: Khan, M.R. and Jairajpuri, M.S. (eds) *Nematode Infestations Part III: Horticultural Crops*. National Academy of Sciences, Allahabad, India, pp. 529–551.

Khan, M.R., Haque, Z. and Kausar, N. (2014) Management of the root-knot nematode *Meloidogyne graminicola* infesting rice in the nursery and crop field by integrating seed priming and soil application treatments of pesticides. *Crop Protection* 63, 15–25. DOI: 10.1016/j.cropro.2014.04.024.

Khan, M.R., Ahamad, I. and Shah, M.H. (2021) Emerging important nematode problems in field crops and their management. In: Singh, K.P., Jahagirdar, S. and Sarma, B.K. (eds) *Emerging Trends in Plant Pathology*. Springer, Dordrecht, The Netherlands, pp. 33–62. DOI: 10.1007/978-981-15-6275-4_3.

Khan, M.R. and Haque, Z. (2023) Methods of assay and detection of nematodes in plant and soil. In: Khan, M.R. and Quintanilla, M. (eds) *Nematode Diseases of Crops and Their Sustainable Management*. Academic Press, Cambridge, MA, pp. 27–54. DOI: 10.1016/B978-0-323-91226-6.00022-5.

Khan, M.R. and Mohiddin, F.A. (2023) Biocontrol strategies for nematode management, an overview. In: Khan, M.R. (ed.) *Novel Biological and Biotechnological Applications in Plant Nematode Management*. Springer, Singapore, pp. 113–132. DOI: 10.1007/978-981-99-2893-4_5.

Khan, M.R., Haque, Z. and Sharma, R.K. (2023a) Novel biotechnological interventions in plant nematode management technologies. In: Khan, M.R. (ed.) *Novel Biological and Biotechnological Applications in Plant Nematode Management*. Springer, Singapore, pp. 167–186. DOI: 10.1007/978-981-99-2893-4_7.

Khan, M.R. (2023b) Plant nematodes, an underestimated constraint in the global food production. In: Khan, M.R. and Quintanilla, M. (eds) *Nematode Diseases of Crops and Their Sustainable Management*. Academic Press, Cambridge, MA, pp. 3–26. DOI: 10.1016/B978-0-323-91226-6.00009-2.

Khan, M.R., Mohiddin, F.A., Haque, Z. and Sharma, R.K. (2023b) Disease complexes and their sustainable management. In: Khan, M.R. and Quintanilla, M. (eds) *Nematode Diseases of Crops and Their Sustainable Management*. Academic Press, Cambridge, MA, pp. 65–93. DOI: 10.1016/B978-0-323-91226-6.00016-X.

Khan, M.R., Rizvi, T.F. and Ansari, M.S.A. (2023c) Nematode problems in vegetables and ornamentals under protected cultivation and their sustainable management. In: Khan, M.R. and Quintanilla, M. (eds) *Nematode Diseases of Crops and Their Sustainable Management*. Academic Press, Cambridge, MA, pp. 685–706. DOI: 10.1016/B978-0-323-91226-6.00002-X.

Kim, D.S., Kim, S. and Fennimore, S.A. (2021) Evaluation of broadcast steam application with mustard seed meal in fruiting strawberry. *HortScience* 56(4), 500–505. DOI: 10.21273/HORTSCI15669-20.

Kirkegaard, J.A., Sarwar, M., Matthiessen, J.N., Thomas, G. and Monteiro, A.A. (1998) Assessing the biofumigation potential of crucifers. *Acta Horticulturae* 459, 105–111. DOI: 10.17660/ActaHortic.1998.459.10.

Kutsuwa, K. (2019) *Characterizing the Significance of* Pratylenchus penetrans *on Soybean* (Glycine max (L.) Merr). The University of Wisconsin-Madison, Madison, WI.

Lopez-Llorca, L.V. and Duncan, G.H. (1988) A study of fungal endoparasitism of the cereal cyst nematode (*Heterodera avenae*) by scanning electron microscopy. *Canadian Journal of Microbiology* 34(5), 613–619. DOI: 10.1139/m88-101.

Louws, F.J., Rivard, C.L. and Kubota, C. (2010) Grafting fruiting vegetables to manage soilborne pathogens, foliar pathogens, arthropods and weeds. *Scientia Horticulturae* 127, 127–146.

Maas, P. (1987) Physical methods and quarantine. In: Brown, R.H. and Kerry, B.R. (eds) *Principles and Practice of Nematode Control in Crops*. Academic Press, Sydney, pp. 265–293.

McFadden, W., Potter, J. and Brandle, J.E. (1992) Use of *Brassica* spp. as green manure for biological control of *Pratylenchus penetrans*. *Phytopathology* 82, 246.

McKenry, M., Buzo, T., Kretsch, J., Kaku, S., Otomo, E. *et al*. (1994) Soil fumigants provide multiple benefits: Alternatives give mixed results. *California Agriculture* 48(3), 22–28. DOI: 10.3733/ca.v048n03p22.

McNamara, D.G. (1995) The role of the European and Mediterranean plant protection organization in nematode management. In: *Proceedings International Congress of Tropical Nematology*, Rio Quente, Brazil, pp. 122–127.

McSorley, R. (1987) Extraction of nematodes and sampling methods. In: Brown, R.H. and Kerry, B.R. (eds) *Principles and Practice of Nematode Control in Crops*. Academic Press, Sydney, pp. 13–41.

McSorley, R. and Duncan, L.W. (1995) Economic thresholds and nematode management. In: Andrews, J.H. and Tommerup, I.C. (eds) *Advances in Plant Pathology. Volume 11*. Academic Press, New York, pp. 147–171.

McSorley, R., Dickson, D.W. and Kaplan, D.T. (2020) Hot water treatment for management of meloidogyne incognita in potatoes. *Journal of Nematology* 52(1), 123–130.

Meyer, S.L., Huettel, R.N. and Sayre, R.M. (1990) Isolation of fungi from *Heterodera glycines* and *in vitro* bioassays for their antagonism to eggs. *Journal of Nematology* 22(4), 532.

Minton, N.A. (1986) Impact of conservation tillage on nematode populations. *Journal of Nematology* 18(2), 135–140.

Mohandas, C., Ramakrishnan, S. and Ashadevi, A. (1999) Use of resistant tuber crops for the management of root knot nematode. In: Balagopalan, C. (ed.) *Tropical Tuber Crops in Food Security and Nutrition*. Oxford and IBH Publishing Co., New Delhi, pp. 442–445.

Mojtahedi, H., Santo, G.S., Hang, A.N. and Wilson, J.H. (1991) Suppression of root-knot nematode populations with selected rapeseed cultivars as green manure. *Journal of Nematology* 23(2), 170–174.

Money, N.P. (2016) Spore production, discharge, and dispersal. In: Watkinson, C.S., Boddy, L. and Money, N.P. (eds) *The Fungi*, 3rd edn. Academic Press, New York, pp. 67–97. DOI: 10.1016/B978-0-12-382034-1.00003-7.

Morgan-Jones, G., Godoy, G. and Rodriguez-Kabana, R. (1981) *Verticillium chlamydosporium*, fungal parasite of *Meloidogyne arenaria* females. *Nematropica* 11, 115–119.

Morgan-Jones, G. and Rodriguez-Kabana, R. (1987) Fungal biocontrol for the management of nematodes. In: Veech, J.A. and Dickson, D.W. (eds) *Vistas on Nematology*. Society of Nematologists, Hyattsville, MD, pp. 94–99.

Muller, R. and Gooch, P.S. (1982) Organic amendments in nematode control. An examination of the literature. *Nematropica* 12, 319–326.

Noling, J.W. (1994) Use of hot water, broad spectrum fumigants, and soil solarization for nematode control. *Journal of Nematology* 25, 559.

Noling, J.W. and Ferris, H. (1987) Nematode-degree days, a density-time model for relating epidemiology and crop losses in perennials. *Journal of Nematology* 19(1), 108–118.

Nutter, F.J., Tylka, G.L., Guan, J., Morreira, A.J. and Marett, C.C. (2002) Use of remote sensing to detect soybean cyst nematode-induced plant stress. *Journal of Nematology* 34, 222–231.

Nyczepir, A.P. (1989) Peach tree short life: A nematode associated disease. *Nematology Circular 163*. Florida Department of Agriculture and Consumer Affairs. Division of Plant Industry.

Ogallo, J.L., Goodell, P.B., Eckert, J.W. and Roberts, P.A. (1999) Management of root-knot nematodes with resistant cotton cv. NemX. *Crop Science* 39(2), 418–421. DOI: 10.2135/cropsci1999.0011183X0039000200020x.

Olatinwo, R., Becker, J.O. and Borneman, J. (2006) Suppression of *Heterodera schachtii* populations by *Dactylella oviparasitica* in four soils. *Journal of Nematology* 38(3), 345.

Overman, A.J. and Jones, J.P. (1986) Soil solarization, reaction, and fumigation effects on double-cropped tomato under full-bed mulch. *Proceedings Florida State Horticultural Society* 99, 315–318.

Perry, R.N., Moens, M. and Jones, J.T. (eds) (2018) *Cyst Nematodes*. CAB International, Wallingford, UK. DOI: 10.1079/9781786390837.0000.

Pinkerton, J.N., Santo, G.S. and Mojtahedi, H. (1991) Population dynamics of *Meloidogyne chitwoodi* on russet burbank potatoes in relation to degree-day accumulation. *Journal of Nematology* 23(3), 283–290.

Pires, D., Orlando, V., Collett, R.L., Moreira, D., Costa, S.R. *et al*. (2023) Linking nematode communities and soil health under climate change. *Sustainability* 15(15), 11747.

Pitcher, R.S. (1978) Interactions of nematodes with other pathogens. In: Southy, J.F. (ed.) *Plant Nematology*. Her Majesty's Stationery Office, London, UK, pp. 63–77.

Ploeg, A.T. and Stapleton, J.J. (2001) The effects of temperature, time, and amendment of soil with broccoli residues on the infestation of melon by two root-knot nematode species. *Plant Protection Quarterly* 11, 1–3.

Ploeg, A. (2007) Biofumigation to manage plant-parasitic nematodes. In: Caincio, A. and Mukerji, .K. (eds) *Integrated Management of Plant Pests and Diseases*. Springer, Dordrecht, The Netherlands, pp. 239–249.

Powell, N.T. (1971) Interactions between nematodes and fungi in disease complexes. *Annual Review of Phytopathology* 9(1), 253–274. DOI: 10.1146/annurev.py.09.090171.001345.

Rajendran, G., Naganathan, T.G. and Sivagami, V. (1979) Studies on banana nematodes. *Indian Journal of Nematology* 9, 54.

Raski, D. and Allen, M. (1948) Sugar-beet nematode: Identification and recommendations for control of the pest in California fields. *California Agriculture* 2, 8–16.

Roberts, P.A. (1987) The influence of planting date of carrot on *Meloidogyne incognita* reproduction and injury to roots. *Nematologica* 33(3), 335–342. DOI: 10.1163/187529287X00470.

Roberts, P.A. (1992) Current status of the availability, development, and use of host plant resistance to nematodes. *Journal of Nematology* 24(2), 213–227.

Roberts, P.A. (1993) The future of nematology: integration of new and improved management strategies. Journal of Nematology 25(3), 283–294.

Roberts, P.A. and May, D.D. (1986) *Meloidogyne incognita* resistance characteristics in tomato genotypes developed for processing. *Journal of Nematology* 18(3), 353–359.

Roberts, P.A., Van Gundy, S.D. and McKinney, H.E. (1981) Effects of soil temperature and planting date of wheat on *Meloidogyne incognita* reproduction, soil populations, and grain yield. *Journal of Nematology* 13(3), 338–345.

Roberts, P.A. (2002) Concepts and consequences of resistance. In: Starr, J.L., Cook, R. and Bridge, J. (eds) *Plant Resistance to Parasitic Nematodes*. CAB International, Wallingford, UK, pp. 23–41.

Rodriguez-Kabana, R. (1986) Organic and inorganic nitrogen amendments to soil as nematode suppressants. *Journal of Nematology* 18(2), 129–135.

Rodriguez-Kabana, R., Morgan-Jones, G., Godoy, G. and Gintis, B.O. (1984) Effectiveness of species of *Gliocladium*, *Paecilomyces*, and *Verticillium* for control of *Meloidogyne arenaria* in field soil. *Nematropica* 155–170.

Rodriguez-Kabana, R., Morgan-Jones, G. and Chet, I. (1987) Biological control of nematodes: Soil amendments and microbial antagonists. *Plant and Soil* 100(1–3), 237–247. DOI: 10.1007/BF02370944.

Sakata, Y., Ohara, T. and Sugiyama, M. (2008) The history of melon and cucumber grafting in Japan. *Acta Horticulturae* 767, 217–228. DOI: 10.17660/ActaHortic.2008.767.22.

Saleh, H., Abu-Gharbieh, W.I. and Al-Banna, L. (1988) Effect of solarization combined with solar-heated water on *Meloidogyne javanica*. *Nematologica* 34, 290–291.

Sankaralingam, A. and McGawley, E.C. (1994) Interrelationships of *Rotylenchulus reniformis* with *Rhizoctonia solani* on cotton. *Journal of Nematology* 26, 475–485.

Sayre, R.M. and Wergin, W.P. (1977) Bacterial parasite of a plant nematode: Morphology and ultrastructure. *Journal of Bacteriology* 129(2), 1091–1101. DOI: 10.1128/jb.129.2.1091-1101.1977.

Schlatter, D.C., Bakker, M.G., Bradeen, J.M. and Kinkel, L.L. (2015) Plant community richness and microbial interactions structure bacterial communities in soil. *Ecology* 96(1), 134–142. DOI: 10.1890/13-1648.1.

Schmitt, D.P. and Inserra, R.N. (2013) Managing nematodes with hot water treatments. *Plant Disease* 97(5), 563–569.

Schneider, S.M. and Ferris, H. (1987) Stage-specific population development and fecundity of *Paratrichodorus minor*. *Journal of Nematology* 19, 395–403.

Scholte, K. (2000) Effect of potato used as a trap crop on potato cyst nematodes and other soil pathogens and on the growth of a subsequent main potato crop. *Annals of Applied Biology* 136(3), 229–238. DOI: 10.1111/j.1744-7348.2000.tb00029.x.

Seinhorst, J.W. (1965) The relation between nematode density and damage to plants. *Nematologica* 11(1), 137–154. DOI: 10.1163/187529265X00582.

Sheela, M.S., Nisha, M.S. and Mohandas, C. (2002) Impact of trap cropping and crop rotation for the management of root-knot nematode in vegetables. In: *Proceedings of International Conference on Vegetables*, Bangalore, India, pp. 344–346.

Shennan, C., Muramoto, J., Koike, S., Baird, G., Fennimore, S. *et al*. (2018) Anaerobic soil disinfestation is an alternative to soil fumigation for control of some soilborne pathogens in strawberry production. *Plant Pathology* 67(1), 51–66. DOI: 10.1111/ppa.12721.

Sikora, A.A., Bridge, J. and Starr, J.L. (2005) Management practices: An overview of integrated nematode management technologies. In: Luc, M., Sikora, R.A. and Bridge, J. (eds) *Plant Parasitic Nematodes in Subtropical and Tropical Agriculture*, 2nd edn. CAB International, Wallingford, UK, pp. 793–825.

Sikora, R.A. and Carter, W.W. (1987) Nematode interactions with fungal and bacterial plant pathogens – facts or fantasy. In: Veech, J.A. and Dickson, D.W. (eds) *Vistas on Nematology*. Society of Nematologists, Hyattsville, MD, pp. 307–320.

Singh, K.P., Stephen, R.A. and Vaish, S.S. (1996) Pathogenicity and development of *Catenaria anguillulae* on some nematodes. *Mycological Research* 100, 1204–1206.

Singh, R., Upadhyay, S.K., Singh, M., Sharma, I., Sharma, P. *et al*. (2021) Chitin, chitinases and chitin derivatives in biopharmaceutical, agricultural and environmental perspective. *Biointerface Research in Applied Chemistry* 11(3), 9985–10005.

Sitaramaiah, K. and Pathak, K.N. (1993) Nematode bacterial disease interactions. In: Khan, M.W. (ed.) *Nematode Interactions*. Chapman & Hall, London, pp. 232–250.

Smiley, R.W., Ingham, R.E., Uddin, W. and Cook, G.H. (1994) Crop sequences for managing cereal cyst nematode and fungal pathogens of winter wheat. *Plant Disease* 78(12), 1142–1149. DOI: 10.1094/PD-78-1142.

Spiegel, Y., Chet, I. and Cohn, E. (1987) Use of chitin for controlling plant plant-parasitic nematodes. *Plant and Soil* 98(3), 337–345. DOI: 10.1007/BF02378355.

Stapleton, J.J. and DeVay, J.E. (1986) Soil solarization: A non-chemical approach for management of plant pathogens and pests. *Crop Protection* 5(3), 190–198. DOI: 10.1016/0261-2194(86)90101-8.

Stapleton, J., Duncan, R.A. and Johnson, R. (1998) Soil disinfestation with cruciferous amendments and sub-lethal heating: Effects on *Meloidogyne incognita*, *Sclerotium rolfsii* and *Pythium ultimum*. In: *Proceeding of the 7th International Congress of Plant Pathology*, Edinburgh, UK, pp. 737–742.

Stapleton, J.J., Lear, B. and DeVay, J.E. (1987) Effect of combining soil solarization with certain nematicides on target and nontarget organisms and plant growth. *Annals of Applied Nematology* 1, 107–112.

Starr, J.L., Shim, M.Y., Lee, J.T.A. and Simpson, C.E. (1996) Additive effects of *Meloidogyne arenaria* and *Sclerotium rolfsii* on peanut. *Journal of Nematology* 28, 99–106.

Stirling, G.R. (1991) *Biological Control of Plant Parasitic Nematodes: Progress, Problems and Prospects*. CAB International, Wallingford, UK.

Stirling, G.R. (2014) *Biological Control of Plant-Parasitic Nematodes: Soil Ecosystem Management in Sustainable Agriculture*. CAB International, Wallingford, UK.

Tedford, E.C., Jaffee, B.A., Muldoon, A.E., Anderson, C.E. and Westerdahl, B.B. (1993) Parasitism of *Heterodera schachtii* and *Meloidogyne javanica* by *Hirsutella rhossiliensis* in microplots over two growing seasons. *Journal of Nematology* 25(3), 427–433.

Thies, J.A., Ariss, J.J., Hassell, R.L., Olson, S., Kousik, C.S. *et al*. (2010) Grafting for management of southern root-knot nematode, *Meloidogyne incognita*, in watermelon. *Plant Disease* 94(10), 1195–1199. DOI: 10.1094/PDIS-09-09-0640.

Thies, J.A., Merrill, S.B. and Corley, W.L. (2019) Hot water treatment for control of root-knot nematodes in cucumbers. *HortScience* 54(6), 1028–1032.

Thomason, U. (1987) Challenges facing nematology: Environmental risks with nematicides and the need for new approaches. In: Veech, J.A. and Dickson, D.W. (eds) *Vistas on Nematology*. Society of Nematologists, Hyattsville, MD, pp. 469–476.

Thorne, G. (1961) *Principles of Nematology*. McGraw-Hill, New York.

Tribe, H.T. (1980) Prospects for the biological control of plant-parasitic nematodes. *Parasitology* 81(3), 619–639. DOI: 10.1017/S0031182000062016.

Tzortzakakis, E.A., Adam, M.A.M. and Liopa-Tsakalidi, A. (2016) Description of the first-stage juveniles of *Xiphinema cretense* and *X. herakliense* – distribution of *Xiphinema* and *Longidorus* species in olive orchards and grapevines in Crete, Greece. *Hellenic Plant Protection Journal* 9, 73–77.

UC IPM Online (2024a) *Models: Insects, Mites, Diseases, Plants, and Beneficials*. Statewide Integrated Pest Management Program. Agriculture and Natural Resources, University of California. Available at: http://ucipm.ucdavis.edu/MODELS/index.html (accessed 28 March 2024).

UC IPM Online (2024b) *Phenology Model Database. Columbia Root Knot Nematode*. Statewide integrated pest management program. Agriculture and Natural Resources, University of California. Available at: http://www.ipm.ucdavis.edu/PHENOLOGY/mn-columbia_root_knot.html (accessed 28 March 2024).

UC IPM Online (2024c) *Run Models and Calculate Degree-days*. Statewide Integrated Pest Management Program. Agriculture and Natural Resources, University of California. Available at: http://www.ipm.ucdavis.edu/WEATHER/ddretrieve.html (accessed 30 March 2024).

Van Gundy, S.D., Perez, B.J.G., Stolzy, L.H. and Thomason, I.J. (1974) A pest management approach to the control of *Pratylenchus thornei* on wheat in Mexico. *Journal of Nematology* 6, 107–1165.

Vetrivelkalai, P. and Subramanian, S. (2006) Plant parasitic nematode populations as influenced by onion based cropping sequences. *Indian Journal of Nematology* 36(1), 10–14.

Wallace, H.R. (1978) The diagnosis of plant diseases of complex etiology. *Annual Review of Phytopathology* 16(1), 379–402. DOI: 10.1146/annurev.py.16.090178.002115.

Walters, T.W., Pinkerton, J.N., Riga, E., Zasada, I.A., Particka, M. *et al*. (2009) Managing plant-parasitic nematodes in established red raspberry fields. *HortTechnology* 19(4), 762–768. DOI: 10.21273/HORTTECH.19.4.762.

Weller, D.M. (1988) Biological control of soilborne plant pathogens in the rhizosphere with bacteria. *Annual Review of Phytopathology* 26(1), 379–407. DOI: 10.1146/annurev.py.26.090188.002115.

Westerdahl, B.B. (2011) Cultural methods for managing nematodes on vegetables and ornamentals. *Acta Horticulturae* 911(911), 185–198. DOI: 10.17660/ActaHortic.2011.911.18.

Westerdahl, B.B. (2020) Evaluation of trap cropping for management of root-knot nematode on annual crops. *Acta Horticulturae* 1270(1270), 141–146. DOI: 10.17660/ActaHortic.2020.1270.15.

Westerdahl, B.B. (2021) Using trap crops to manage plant parasitic nematodes on vegetable crops. *Acta Horticulturae* 1326(1326), 265–270. DOI: 10.17660/ActaHortic.2021.1326.35.

Westerdahl, B.B. (2024) Adoption of biological nematicides. In: *XX International Plant Protection Congress*, Athens, Greece, July 3.

Westerdahl, B.B. and Frate, C.A. (2007) Parasitic nematodes in alfalfa. *Irrigated Alfalfa Management for Mediterranean and Desert Zones*. University of California Division of Agriculture and Natural Resources Publication 8297, Oakland, CA. Available at: https://alfalfa.ucdavis.edu/sites/g/files/dgvnsk12586/files/media/documents/UCAlfalfa8297Nematodes-reg.pdf (accessed 30 March 2024).

Westerdahl, B.B., Carlson, H.L., Grant, J., Radewald, J.D., Welch, N. *et al*. (1992) Management of plant-parasitic nematodes with a chitin-urea soil amendment and other materials. *Journal of Nematology* 24, 669–680.

Westerdahl, B.B., Giraud, D., Etter, S., Riddle, L.J. and Anderson, C.A. (1998) Problems associated with crop rotation for management of *Pratylenchus penetrans* on Easter lily. *Journal of Nematology* 30(4S), 581–589.

Westerdahl, B.B., Khan, M.R., Nyström, Å.O. and Ahmad, I. (2023) Nematode problems in sugarcane and sugar beet and their sustainable management. In: Khan, M.R. and Quintanilla, M. (eds) *Nematode Diseases of Crops and Their Sustainable Management*. Academic Press, Cambridge, MA, pp. 565–594. DOI: 10.1016/B978-0-323-91226-6.00019-5.

Wu, D., Wang, W., Yao, Y., Li, H., Wang, Q. *et al*. (2023) Microbial interactions within beneficial consortia promote soil health. *Science of The Total Environment* 900, 165801. DOI: 10.1016/j.scitotenv.2023.165801.

Yun, H.M., Jang, J.Y. and Lee, K.C. (2021) Efficacy of hot water treatment against root-knot nematodes in tomato seedlings. *Nematology* 23(5), 673–682.

Zare, R., Gams, W. and Evans, H.C. (2001) A revision of *Verticillium* section *Prostrata*. V. The genus *Pochonia, with notes on Rotiferophthora*. *Nova Hedwigia* 73(1/2), 51–86. DOI: 10.1127/nova.hedwigia/73/2001/51.

Špak, J., Kolářová, L., Lewis, J. and Fenwick, G.R. (1993) The effect of glucosinolates (mustard oil glycosides) and products of their enzymic degradation on the infectivity of turnip mosaic virus. *Biologia Plantarum* 35(1), 73–78. DOI: 10.1007/BF02921123.

14 Nematodes as Bioindicators of Heavy Metal Contamination and Soil Salinity

Ebrahim Shokoohi[1]*, Mujeebur Rahman Khan[2] and Zafar Handoo[3]
[1]Department of Biochemistry, Microbiology, and Biotechnology, University of Limpopo, Sovenga, South Africa; [2]Department of Plant Protection, Faculty of Agricultural Sciences, Aligarh Muslim University, Aligarh, India; [3]Mycology and Nematology Genetic Diversity and Biology Laboratory, USDA, ARS, Beltsville, MD, USA

Abstract
Nematodes are the most diverse multicellular animals worldwide. Terrestrial nematodes live in all types of soil, and their response to contaminants and environmental disruptions varies greatly. The community structure, diversity, abundance and metabolic footprint of soil nematode populations are significant indicators of soil health and are closely associated with the constituents of soil and its environment. Several species belonging to Enoplida, Chromadorida and Monhysterida in marine environments have been shown to be influenced by heavy metal pollution. For instance, species of *Araelaimus* and *Hypodontolaimus* have been found to be sensitive to mercury and zinc. Plant-parasitic nematodes may also be affected by metal contamination in the soil. However, bacterivores may serve to be the most reliable indicators of heavy metals. Studies have shown that populations of predators and omnivore nematodes form strong correlations with heavy metal contamination. The abundance of *Achromadora*, *Heterocephalobus*, *Ironus*, *Prismatolaimus* and *Monhystera* has been found to be strongly correlated with mercury levels. Salinity is another important abiotic stress in agricultural fields but it has a low effect on nematodes and is, therefore, not useful as a bioindicator. However, *Scottnema lindsayae* and *Plectus antarcticus* were found to be abundantly present in soils rich in nitrate. Meta-analysis has shown that the maturity index of a nematode community may serve as the best indicator of heavy metals. This chapter presents a critical analysis of the scope and use of nematodes as bioindicators of heavy metal contamination and salinity in the soil, and offers up-to-date relevant information.

Introduction

Bioindicators, by definition, are living organisms or biological responses that serve as a reliable means of detecting the presence of environmental contaminants. Generally, bioindicators are sensitive to a contaminant and develop typical symptom(s) or measurable response(s), such as changes in growth or behavior. By monitoring bioindicators, researchers and environmentalists can obtain valuable insights into the health of ecosystems and the impact of human activity on the natural world (Mothersill and Seymour, 2016). Soil ecosystems harbor a diverse range of biological organisms, including microbes and microfauna and mesofauna such as fungi, bacteria, algae, protozoa, arthropods and nematodes. These organisms perform important

*Corresponding author: Ebrahim.shokoohi@ul.ac.za

DOI: 10.1079/9781800625228.0014

functions in nutrient cycling, organic matter decomposition, soil structure formation, etc., which ultimately contribute to soil health. Since these organisms are very small in size they are highly sensitive to abnormal fluctuations in the environmental matrices, hence they can be used as bioindicators. However, the selection of a bioindicator requires a more comprehensive approach that considers multiple factors. These include the organism's sensitivity to environmental changes, ease of detection and quantification, ability to link its presence or absence to specific environmental stressors, and ecological relevance. A thorough assessment of these factors is necessary to ensure that a good bioindicator provides accurate and relevant information about the health of the soil or an ecosystem (Neher, 2001). Soil health refers to the ability of soil to function as a living system, sustaining the biological activities, maintaining the quality of air and water, and promoting the health of plants, including humans and animals (Doran *et al.*, 1996). Soil is home to a wide variety of organisms, and it can be challenging to identify which ones are the best bioindicators. A recent study has indicated that protozoa are more sensitive than nematodes to heavy metals, and that nematodes are more affected directly by soil physicochemical properties than heavy metals (Wang *et al.*, 2023). However, among the soil organisms, nematodes stand out as bioindicators due to their unique morphological characteristics and adaptability. These tiny roundworms are easily visible under a microscope, making them ideal candidates for studying soil samples. Further, they can be easily extracted from soil, which makes them highly practical for fieldwork. Additionally, nematodes are highly sensitive to changes in soil status, such as pollution, ecological imbalances and climate change. Therefore, they are a valuable tool for measuring the impact of various factors on the soil environment (Geetanjali *et al.*, 2002). In addition, lower concentrations of heavy metals may promote nematode pathogenesis, leading to greater damage to the plant (Bisessar *et al.*, 1983; Khan *et al.*, 1988, 1996a). The marine and soil nematodes are ecological distinct groups, hence their responses to heavy metals and salinity are discussed separately below.

Reaction of Marine Nematodes to Heavy Metals

Water bodies including oceans, rivers and other aquatic environments have been increasingly facing the threat of heavy metal contamination worldwide (Khan *et al.*, 1996b, 2006). Metal contamination causes serious adverse effects on the ecosystem and poses a significant risk to the health of entire flora and fauna (Okereafor *et al.*, 2020). The issue of metal contamination is wide in prevalence and its detection methods are cumbersome and laboratory based. Hence, it becomes crucial to find reliable bioindicators that could make us aware about soil or water quality with regard to heavy metal contamination. By doing so, we can take proactive steps to save the environment and conserve it for future generations.

Certain types of nematodes, specifically *Daptonema*, *Sabatieria*, *Terschellingia*, *Theristus* and *Paradontophora*, have been found to indicate heavy metals in water bodies by their presence or population fluctuations (Gyedu-Ababio *et al.*, 1999; Moreno *et al.*, 2008). Higher populations of *Daptonema* sp. were found at sites with contaminations of diesel fuel. This species of nematode is categorized as 'opportunistic' and plays a crucial role of biomonitoring the marine environments. Therefore, the presence of *Daptonema* sp. in an area suggests that there may be diesel contamination but also indicates minimal levels of metal pollution (Somerfield *et al.*, 1994). *Pomponema* sp. had a negative correlation with diesel contamination and its population may be very low under diesel contamination. Consequently, this nematode is classified as a diesel-sensitive species, indicating its vulnerability to petroleum contamination (Ng *et al.*, 2022). A research study conducted in Mozambique has shown a strong positive correlation of *Terschellingia* sp. with a range of heavy metals, including cadmium, cobalt, chromium, copper, iron, manganese, nickel, vanadium and aluminium (Soko and Gyedu-Ababio, 2019). These findings indicate the potential of using *Terschellingia* sp. as an indicator of metal-contaminated environments. Further, this nematode also accumulates substantial concentrations of these metals in its body.

The results of nematode diversity analysis conducted on three rivers in Germany have revealed some interesting insights (Heininger *et al.*, 2007). The study found that the sites heavily contaminated with high levels of aluminium and cadmium were mostly colonized by omnivorous nematodes such as *Tobrilus* and predatory nematodes such as *Mononchus*, with relatively high abundance. On the other hand, the sites with low to medium levels of contaminates experienced dominance of bacterivorous nematodes such as *Monhystera* and *Daptonema* or omnivorous nematodes such as *Dorylaimus* (Heininger *et al.*, 2007). Free-living nematodes in aquatic systems can play a vital role in indicating water contamination due to their sensitivity to pollution, especially heavy metals (Balsamo *et al.*, 2012; Semprucci *et al.*, 2015). Ridall and Ingels (2021) noted several marine nematodes which have been affected by heavy metals (Table 14.1). Members of the order Chromadorida, followed by Enoplida, Desmodorida, Monhysterida and Araeolaimida, were the most dominantly detected nematodes in the marine system. The results of a study carried out in Taiwan revealed that the presence of four specific types of nematodes in the aquatic environment, namely *Daptonema*, *Innocuonema*, *Axonolaimus* and *Pomponema*, can be indicative of potential contamination by metals, diesel and other organic substances (Ng *et al.*, 2022). These contaminants are associated with various human activities in the area, such as the discharge of domestic waste, the use of diesel-powered machinery or vehicles, and the application of antifouling coatings on boats and other marine structures. The study has suggested that these nematodes could serve as a helpful tool for monitoring environmental contamination and identifying potential sources of pollution in aquatic ecosystems (Ng *et al.*, 2022).

Kang *et al.* (2023) demonstrated that higher levels of mercury, chromium, copper, zinc, arsenic, lead, vanadium, manganese and nickel can influence nematode diversity in the Xihe river in China. A positive correlation was recorded between Enoplida and mercury, Triplonchida and manganese, and Plectida and chromium, whereas Mononchida showed a negative correlation to mercury (Kang *et al.*, 2023). It was further discovered that the presence of mercury had an adverse impact on the development of *Caenorhabditis elegans*. In contrast, copper and zinc proved to have a beneficial effect on the reproductive capacity of *C. elegans*. Nevertheless, the presence of vanadium and zinc seemed to have a detrimental impact on the behavior of *C. elegans* (Kang *et al.*, 2023). This suggests that contamination levels in the rivers may directly affect the types of nematodes that inhabit there. A *Propanagrolaimus* species was recorded in Molepo dam, situated in Limpopo Province, South Africa; the species has been found to be intriguingly uncorrelated with any of the water parameters (Shokoohi and Masoko, 2024).

Reaction of Soil Nematodes to Heavy Metals

Heavy metal contamination may otiosely affect soil health and its function. Park *et al.* (2011) detected significant differences in pH, electrical conductivity (EC), available phosphorus, and cadmium, lead, zinc and nickel in soil contaminated with mine drainage from an abandoned mine over non-contaminated soil. The density of long-living nematodes was low in the contaminated vs non-contaminated soil, and their abundance, maturity, richness, diversity and structural indices were also low in the contaminated soil. However, the enrichment index was not significantly affected. This suggests that the heavy metals disturbed the soil ecosystem and suppressed the biological activity. To identify nematode genera, trophic groups and functional guilds as useful heavy metal indicators in soil, the nematode community was investigated at five alluvial meadows receiving water from the Litavka river in the Czech Republic contaminated with lead, zinc and arsenic from a smelter (Renčo *et al.*, 2022). The abundance of bacterivore, predatory, omnivore, plant-parasitic and fungivore nematodes was significantly decreased with the metal levels at four sites, being lowest at the highest polluted site.

Zeng *et al.* (2020) investigated the response of the soil nematode community to multiple metal mining activities containing 13.60, 7.69, 8.56, 2.35, 2.96 and 3.11 times average levels of copper and zinc, and the total metals. They

Table 14.1. Some marine nematodes and their reaction to heavy metals.

Order	Family	Genus	Cu	Zn	Co	Pb	Hg	Cr	Ni	Cr, Cu, Ni	Cu, Pb	Co, Zn
Araeolaimida	Axonolaimidae	*Axonolaimus*	R	–	–	–	–	–	–	–	R	–
		Odontophora	–	–	–	–	–	–	R	–	–	–
	Diplopeltidae	*Araeolaimus*	–	–	–	–	S	–	–	–	–	–
Chromadorida	Chromadoridae	*Hypodontolaimus*	–	S	–	–	–	–	–	–	–	–
		Ptycholaimellus	R	–	–	–	–	–	–	–	–	–
		Prochromadorella	–	–	–	–	R	–	–	–	–	–
		Neochromadora	–	–	S	–	–	–	R	–	–	–
	Cyatholaimidae	*Paralongicyatholaimus*	S	–	–	–	–	–	–	–	–	–
		Marylynnia	–	–	R	–	R	–	–	–	–	R
	Selachinematidae	*Gammanema*	–	–	–	–	–	–	–	S	–	–
Desmodorida	Desmodoridae	*Leptonemella*	–	–	–	–	–	S	S	S	–	–
		Spirinia	–	–	R	–	–	–	–	–	–	–
	Microlaimidae	*Microlaimus*	S	–	–	–	–	–	–	–	–	–
	Monoposthiidae	*Monoposthia*	S	–	–	–	–	–	–	–	–	–
Enoplida	Enchelidiidae	*Ditlevsenella*	–	–	–	–	–	–	–	S	–	–
	Oncholaimidae	*Oncholaimellus*	–	–	S	–	–	–	–	R	–	–
		Pontonema	–	–	–	–	–	–	–	S	–	–
		Viscosia	–	R	R	–	–	–	–	S	–	R
		Oncholaimus	–	R	S	R			R	–	–	–
	Tripyloididae	*Bathylaimus*	–	–	–	–	–	R	R	–	–	–
		Tripyloides	R	–	–	–	–	–	–	–	–	–

Continued

Table 14.1. Continued

Order	Family	Genus	Cu	Zn	Co	Pb	Hg	Cr	Ni	Cr, Cu, Ni	Cu, Pb	Co, Zn
Monhysterida	Linhomoeidae	*Eleutherolaimus*	R	–	–	–	–	–	–	–	–	–
		Terschellingia	R	–	–	–	–	–	–	–	–	–
	Xyalidae	*Xyala*	–	R	–	–	–	–	–	S	–	–
		Promonhystera	–	–	R	–	–	–	–	–	–	–
		Theristus	R	–	–	–	–	–	–	S	–	–
		Daptonema	R	–	–	–	–	R	R	–	–	–

R: resistant; S: susceptible; –: not studied. [Amended from Ridall and Ingels (2021) and used with permission under Open Access license CC BY 4.0.]

noticed that heavy metals stimulated microbial biomass and nematode biomass consumption but the nematode group community was negatively affected. In another study, examining the impact of distance from a metal source, the soil population of free-living nematodes was affected most at the site nearest to the source, having a relatively low population density and domination of insensitive taxa, and this effect gradually decreased with the increase in distance from the source (Šalamún *et al.*, 2012). Zhang *et al.* (2007) also examined the effect of copper and zinc emanating from a smelter on soil nematode communities with regard to distance from the source. Nematode abundance significantly increased with an increase in distance from the source. *Paratylenchus* was the dominant genus among 20 nematode genera recorded at the sampling sites. They observed negative correlations for nematode channel ratio with total copper and zinc and structure index with total copper.

Soil application of sewage sludge containing high concentrations of copper, zinc +copper and zinc resulted in a decline in free-living nematodes with regard to genus/family richness and maturity indices (MI and MI2–5) (Georgieva *et al.*, 2002). Suppression or elimination was recorded for persisters (*K*-strategists) in favor of colonizers (*r*-strategists), omnivorous nematodes (Dorylaimida: Dorylaimina) and predatory nematodes (Dorylaimida, Mononchida). The abundance of plant nematodes was increased in the zinc-treated plots, apparently due to the suppressive effects of the metals on some nematode antagonists and predators. The abundance of some bacterivore nematodes was also promoted. In another study, structure footprint, community footprint, abundance of plant feeders, bacteria feeders and omnivores/predators, and taxonomic richness were decreased with the elevated level of metal contamination (Chauvin *et al.*, 2020). Overall, higher abundance of the fungal-feeding nematodes was recorded at the very high contamination level, showing their relatively lower sensitivity to heavy metals.

Several investigations have been performed using the nematode community or specific species dominance as a bioindicator of various environments (Wasilewska, 1997; Neher, 2001; Yeates, 2003; Renčo *et al.*, 2022; Shokoohi, 2023; Young and Unc, 2023). Pollution caused by heavy metals such as cadmium, lead, arsenic, mercury, zinc, chromium, etc., can harm crop cultivation in agricultural lands (Bisessar *et al.*, 1983; Khan *et al.*, 1987). These toxic metals can accumulate in the soil, water and air, thereby contaminating crops and making them unfit for human consumption, besides inflicting a significant decline in quantity and quality of the produce (Khan *et al.*, 1994; Khan and Khan, 1995, 2010).

Heavy metal contamination may affect the maturity index of soil nematodes (Neher, 2001). The most sensitive nematodes to chemicals or soil nutrients belong to the order Dorylaimida, especially *Aporcelaimellus*, *Aporcelaimus*, *Discolaimus* and *Mesodorylaimus* (Fiscus and Neher, 2002). However, the reaction of soil nematodes to heavy metals could be different. A study in New Zealand showed that lands contaminated with sewage had more Cephalobidae than Rhabditidae or Plectidae (Yeates, 2003). In fact, in those lands, copper, nickel and zinc were detected, among which nickel had suppressed the nematode populations most. The same result was obtained in another study where soil contaminated with a mixture of copper, chromium and arsenic revealed higher numbers of Cephalobidae than other free-living bacterivore nematodes. In addition, lands contaminated with copper in Wageningen in the Netherlands also showed a higher density of Cephalobidae than other free-living bacterivore nematodes. Cephalobidae members were found to be most resistant to heavy metals (Yeates, 2003). A study in industrial regions contaminated with heavy metals in Bulgaria showed that all feeding groups of nematodes had a negative relationship with cadmium, zinc, lead and copper (Mohamedova and Lecheva, 2016). However, the most sensitive groups were predatory and omnivorous nematodes (Table 14.2). The result showed that herbivores were affected mainly by cadmium, fungivores by zinc and omnivores/predators by cadmium, whereas bacterivores were slightly affected by cadmium, showing their resistance to the metal (Mohamedova and Lecheva, 2016). As a concluding remark of the study, populations of individual nematodes increased in the absence of heavy metals or at a low concentration. The maturity index, which is used to show the stability of an ecosystem, had a negative correlation with heavy metals, and

Table 14.2. Correlation between soil nematodes and heavy metals.

Index	Cu	Zn	Pb	Cd
Herbivores	−0.587*	−0.613**	−0.201	−0.626**
Fungivores	−0.424*	−0.583**	−0.191	−0.414**
Bacterivores	−0.198	−0.164	0.022	−0.297
Omnivores/predator	−0.603**	−0.526*	−0.493*	−0.637**
Richness (S)	−0.218	−0.191	−0.138	−0.513*
Maturity index (MI_{2-5})	−0.462*	−0.501*	−0.423*	−0.704**
Structure index (SI)	−0.402*	−0.322	−0.208	−0.501*

*$P \leq 0.05$; **$P \leq 0.01$. [Amended from Mohamedova and Lecheva (2016) and used with permission under Open Access license CC BY 4.0.]

therefore the number of beneficial nematodes decreased. Consequently, nitrogen fixation and decomposition of the materials in the soil decreased, leading to a lower uptake of nutrients by the plants. Additionally, heavy metals such as cadmium, zinc, lead and copper caused a decrease in the value of the structure index, affecting predatory and omnivorous nematodes in particular. Finally, a lower structure index value caused the shortening of the food web of the soil (Mohamedova and Lecheva, 2016).

Effect of Heavy Metals on Plant-parasitic Nematodes

Mining and refining of metals are major sources of heavy metal pollution worldwide (Bisessar *et al.*, 1983; Khan *et al.*, 1987, 1988). Mercury is one of the heavy metals leaking through the mine activities to the soil. Studies have shown that mercury is highly toxic to free-living and plant nematodes, and can decrease their abundance (Khan *et al.*, 1986a; Martinez *et al.*, 2019). Mercury contamination decreased the density and abundance of *Rotylenchulus*, *Rotylenchus*, *Helicotylenchus* and *Tylenchorhynchus* (Martinez *et al.*, 2019). Mercury and cobalt above 50 ppm decreased the hatching, root penetration, galling, reproduction and soil population of *Meloidogyne incognita* in tomato (Khan *et al.*, 1986a, 1994). A similar response of root-knot nematode was also observed with nickel (Khan *et al.*, 1996a, b, 2006). However, free-living nematodes, including *Achromadora*, *Heterocephalobus*, *Ironus*, *Prismatolaimus* and *Monhystera*, were affected more than the plant nematodes by mercury contamination (Martinez *et al.*, 2019), indicating their greater sensitivity (except *Pratylenchus*).

Martinez *et al.* (2019) indicated that mercury did not affect the nematode assemblage on the field. Soil physicochemical properties such as organic matter, pH, clay contents, etc., may influence the effect of mercury on nematodes (Martinez *et al.*, 2019). Recently, another study conducted in China also revealed that mercury has no effect on nematode diversity, except for *Bursaphelenchus* (Huo *et al.*, 2024). This means that most nematodes have a resistance strategy against mercury contamination. Besides, *Filenchus* and *Acrobeloides* were the only nematodes positively affected by chromium. *Aporcelaimellus*, *Mononchoides*, *Basiria*, *Ditylenchus* and *Alaimus* were affected by cadmium (Huo *et al.*, 2024). At the same time, arsenic affected the density of *Paraphelenchus*, *Nygolaimus*, *Aglenchus* and Rhabditidae members. Interestingly, none of the heavy metals affected the numbers of *Psilenchus* and *Mesodorylaimus*. The result showed that bacterivores, followed by fungivores, have a potential to be used as heavy metal bioindicators (Huo *et al.*, 2024).

A meta-analysis showed that nematodes react to increased levels of contamination, with a significant effect on their different parameters (Chauvin *et al.*, 2020). However, various feeding groups react differently to heavy metal pollution. A greater population of fungivore nematodes was found in soil experiencing higher levels of contamination (Chauvin *et al.*, 2020), but the population of bacterivore nematodes decreased with increasing levels of heavy metals (Fig. 14.1). The effect of copper, zinc and nickel

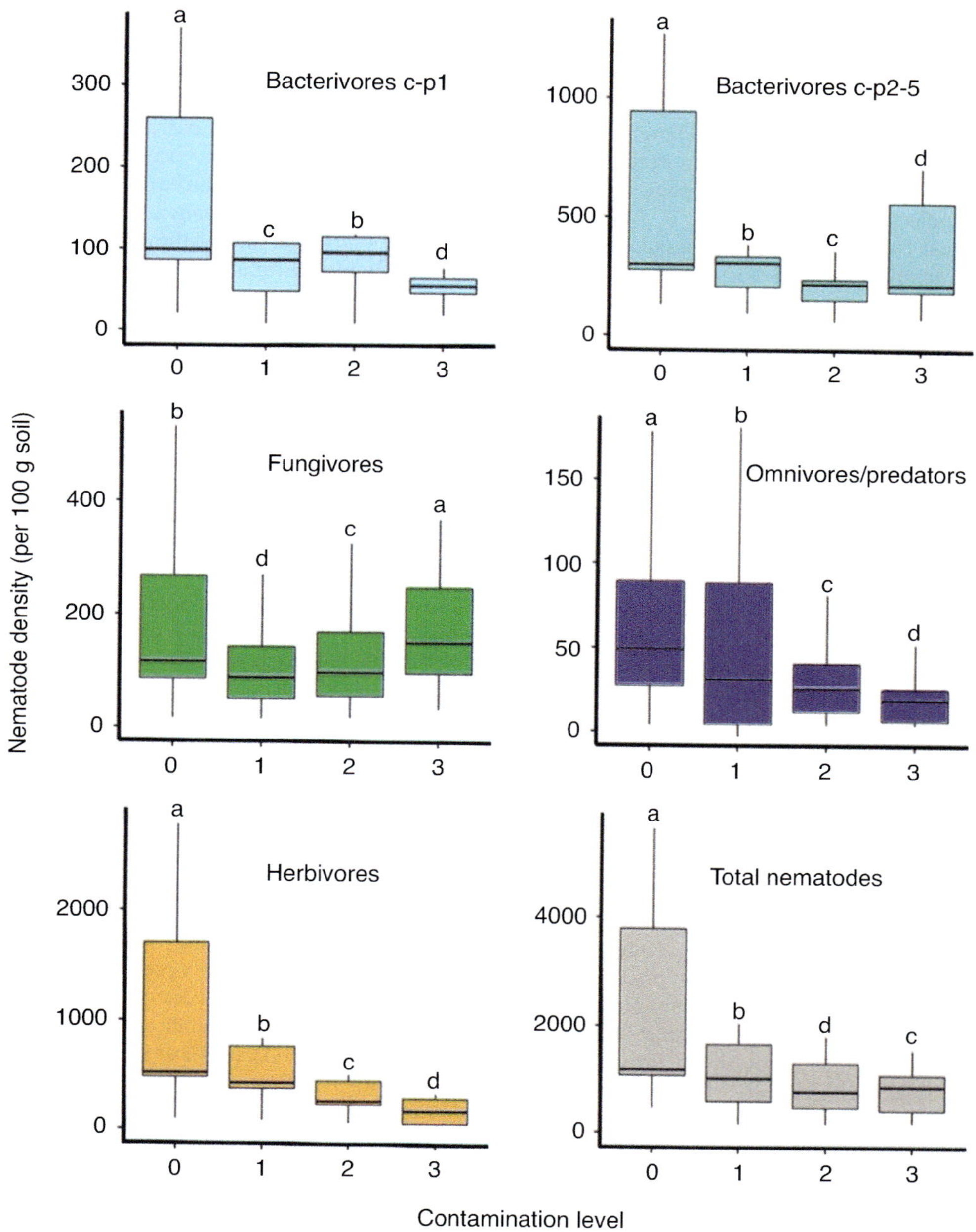

Fig. 14.1. Nematode density (per 100g soil) of various trophic groups exposed to different levels of heavy metal contamination (0 = no contamination; 1 = low contamination; 2 = medium contamination; 3 = high contamination). Different letters show significant differences at the 0.05 level (amended from Chauvin *et al*., 2020). [Figure used with permission under Open Access license CC BY 4.0.]

on plant-parasitic nematodes (*M. incognita*) and entomopathogenic nematodes (*Steinernema carpocapsae* and *Heterorhabditis bacteriophora*) was examined by exposing the nematodes to different concentrations for 24 hour (Sun *et al.*, 2016). The treatment caused mortality to infective juveniles of entomopathogenic nematodes and affected the virulence of *H. bacteriophora* against *Galleria mellonella*. In a pot experiment on cucumber, copper, zinc or chromium treatments enhanced *M. incognita* juvenile penetration into the roots.

Interestingly, the density of predators and omnivores was higher in low contamination than in higher levels of heavy metals. Overall, total nematode density decreases due to the excess amount of heavy metals. Based on meta-analysis results, bacterivore nematodes can be used as bioindicators of polluted areas with higher concentrations of heavy metals. In addition, the lack of herbivores, omnivores and predators may also be considered as an indicator of high levels of metal contamination. The meta-analysis also revealed that the enrichment index does not correspond to the increasing level of heavy metals, in which the enrichment index was without significant change due to low, medium and high contamination levels. The same scenario was applied to the maturity index and channel ratio. However, the Shannon diversity index and species richness were affected by the level of contamination, in which the highest contamination level decreased the aforementioned indices compared with zero contamination (Chauvin *et al.*, 2020). In addition, heavy metal contamination strongly affected the nematode community footprint, in which herbivores were significantly disturbed compared with other trophic groups. All nematode community indices, except the channel ratio, reacted to the heavy metal contamination. Therefore, the channel ratio is not a suitable index for use as a bioindicator. In contrast, other indices, despite their considerable sensitivity, also did not show reliability to be used as bioindicator. However, the Shannon diversity index is reliable as it showed a significant trend towards increasing levels of heavy metal contamination. However, Georgieva *et al.* (2002) and Sánchez-Moreno and Navas (2007) have shown no effect of heavy metals, including copper, lead or zinc, on bacterivore nematodes of the *c-p2* group. Hence, their evaluation should be done with caution. Among the feeding trophic groups, omnivores and predators were reported to be the most sensitive to soil heavy metal contamination (Zhang *et al.*, 2007). The community and structure footprint of Mononchida (predators) and Dorylaimida (omnivores), based on nematode biomass, were decreased due to increased levels of heavy metals, and therefore can be used as indicators of soil contamination (Chauvin *et al.*, 2020).

Renčo *et al.* (2022) studied the relationship of arsenic, cadmium, chromium, copper, nickel, lead and zinc with nematodes in the Litavka river in the Czech Republic. The results showed that only lead and zinc significantly decreased the abundance of nematodes. Furthermore, cadmium did not produce any discernible impact on any trophic groups under observation (Renčo *et al.*, 2022). The same observation was achieved with chromium with no effect, except for some herbivore nematodes. Overall, herbivore nematodes did not react to the heavy metals, indicating that they cannot be used as heavy metal bioindicators. The study on the impact of heavy metals on nematodes showed that zinc and lead had a significant impact on the maturity index. Other diversity indices were found to be unaffected. Furthermore, it was observed that the structure index was affected only by lead, indicating that this particular index is more suitable for use as a bioindicator. Overall, these findings suggest that only a few diversity indices are essential for accurately assessing environmental quality and the impact of heavy metals on soil ecosystems. A study conducted by Shokoohi (2023) in South Africa showed that there is a significant correlation between *Psilenchus* (r = 0.705) and *Rotylenchus* (r = 0.637) and copper. In a further study, the analysis of nematodes in tomato fields indicated that there is a significant correlation between copper and *Aphelenchoides* (r = –0.817), *Rotylenchus* (r = 0.963), *Pratylenchus* (r = 0.762) and *Rotylenchulus* (r = –0.926) (Shokoohi *et al.*, 2023).

Waste water is another crucial source of heavy metal contamination in agricultural lands when used for irrigation. A recent study showed that the abundance of *Panagrolaimus*, *Prismatolaimus*, *Filenchus*, *Mesorhabditis*, *Rhabditis*, *Zeldia* and *Monhystera* was increased with increasing levels of cadmium, zinc, copper, manganese, lead and potassium. However, populations of *Xiphinema*, *Thornenema*, *Nygolaimus*, *Discolaimus* and *Pratylenchus* were decreased with the above metal levels (Ghanem *et al.*, 2024). The findings of this research have revealed that heavy metal contamination in soil may negatively impact the abundance of herbivore and predatory nematodes. Metal pollution may also lead to a decrease in the sigma maturity index. However, the abundance and diversity of other types of nematodes, such as bacterivore and fungivore nematodes, responded positively to metal contamination (Ghanem *et al.*, 2024).

Reaction of Entomopathogenic Nematodes to Heavy Metals

Sun *et al.* (2016) reported the effects of toxic components from excessive fertilizer and manure applications on entomopathogenic and plant-parasitic nematode populations. They found that exposure to high levels of copper (1.5 mg/l), zinc (50 mg/l) or chromium (7.5 mg/l) for 24 hour led to increased mortality in *S. carpocapsae* and *H. bacteriophora*, while decreasing *H. bacteriophora* against *G. mellonella*. Notably, *H. bacteriophora* was more affected by these metals than *M. incognita*. In potted plant experiments, metal pretreatment decreased the biocontrol efficacy of *H. bacteriophora* by increasing *M. incognita* root penetration, although low concentrations of zinc and chromium sometimes improved the biocontrol efficacy in *S. carpocapsae*. Overall, heavy metals negatively impact nematodes directly and indirectly by reducing their biocontrol effectiveness, potentially altering nematode community structures in greenhouse systems.

Cadmium, copper, lead, zinc, manganese and magnesium impact the entomopathogenic nematode *Steinernema feltiae* under laboratory conditions. Cadmium, copper, lead and zinc were found to increase nematode larval mortality, reduce pathogenicity to *G. mellonella* and decrease reproduction. In contrast, manganese at a concentration of 400 mg/dm^3 improved nematode survival, pathogenicity and reproduction compared to distilled water (Jaworska and Gorczyca, 2002). Jarmuł and Kamionek (2001) reported the effect of lead nitrate and lead acetate on mortality and invasive properties of the infective juveniles of *S. feltiae* larvae, used in the biopreparation ‘Ovinema.’ Nematodes were exposed to lead ion concentrations of 0.40, 100, 500, 1000 and 5000 ppm over 120 hour (Jaworska *et al.*, 1997). Results showed reduced insect mortality and lower invasiveness in lead-exposed nematodes compared to controls, with minimal impact from anionic groups. The effects of lead at concentrations of 40, 100 and 500 ppm, as well as cadmium nitrate at 1, 2 and 3 ppm, were studied on the entomopathogenic nematode *S. feltiae* and *G. mellonella* larvae from a laboratory culture. The study examined the sexual structure of the nematodes developing inside the body cavity of the host as well as the survival rate of infective juveniles before leaving the insect and after migration to the external environment (Kamionek *et al.*, 2011).

Mechanism of Heavy Metal Sensitivity

Cuticle permeability is the primary factor affecting the sensitivity or resistance of nematodes to chemical uptake, including heavy metals. The cuticle is an essential layer for nematodes as it acts as a barrier between them and their environment (Hurley, 1986). However, not all nematodes have a cuticle in their mature stage, such as the *Fergusobia* nematode. Moreover, the cuticle can vary among different genera and species, and even between males and females, different stages of life and distinct areas of the body (Decraemer *et al.*, 2003). The cuticle of nematodes comprises four layers: epicuticle, cortical zone, median layer and basal zone. In addition, the epicuticle, which serves as a protective barrier against the external environment, comprises three ultra-thin layers primarily made up of a combination of lipids and proteins. These layers work together to provide mechanical strength, flexibility and permeability to the epicuticle, allowing it to regulate the exchange of nutrients and waste between the nematode and its surroundings (Decraemer *et al.*, 2003).

The layers in Mononchida (predators), Dorylaimida (omnivores), Rhabditina (bacterivores) and Tylenchida (herbivores) are different, allowing them to be tolerant or vulnerable to heavy metals in different soil types (Khan, 2008). *Caenorhabditis elegans*, a tiny nematode, possesses a remarkable natural defense mechanism that shields it from harmful chemicals, toxins and pathogens.

This mechanism is attributed to a specific type of collagen that forms a protective barrier around the nematode cuticle. The collagen acts as a shield, preventing the entry of harmful substances through the cuticle and safeguarding the organism's internal systems. This remarkable adaptation is a testament to the amazing resilience of nature and the complex mechanisms that govern the behavior of even the smallest living organisms (Sandhu *et al.*, 2021; Njume *et al.*, 2022). That is why bacterivores might be suitable bioindicators of heavy metal contamination in the soil.

In *C. elegans* (a bacterivore species), cuticle permeability is a crucial characteristic that enables the absorption of chemicals. However, this feature also puts nematodes at risk of absorbing harmful molecules that can negatively affect their physiology and behavior (Moens *et al.*, 2013). In addition, heavy metals bind proteins of the epicuticle of various groups of nematodes, causing a change in cuticle functionality (Chauvin *et al.*, 2020). It should be mentioned that concentration and exposure time also affect the permeability of the nematode cuticle. For instance, cadmium up to 160 mg/kg soil did not affect the nematode community. Experimental studies have shown that exposure to nickel, copper and zinc for a period of 1–2 weeks can significantly reduce the soil populations of omnivores and predators. This decrease occurs at a concentration of 100 mg/kg soil (Korthals *et al.*, 1996). The cuticles of two soil-dwelling nematode groups, Mononchida and Dorylaimida, have not been extensively investigated. As a result, we currently lack knowledge about the composition and structure of the collagens, as well as any natural barrier functions that the cuticle may possess. Further research in this area could provide valuable insights into the unique adaptations of these organisms to their environments.

Effect of Salinity on Nematodes

Reaction of marine nematodes to salinity

Salinity in aquatic systems refers to the amount of salt dissolved in water. Several nematodes have been studied in relationship with salinity of the water, including *Oxystomina* (Enoplida) in the oligohaline zone (0.5–5 ppt salts), *Terschellingia* (Monhysterida) and *Sabatieria* (Araeolaimida) in the mesohaline zone (5–18 ppt salts), and *Paracomesoma* (Araeolaimida), *Synonchiela* and *Odontophora* (Araeolaimida) in the polyhaline zone (18–30 ppt salts) (Adão *et al.*, 2009; Montagna *et al.*, 2013). Another study revealed that *Litoditis marina* (Rhabditidae) and *Diplolaimelloides meyli* (Monhysteridae) isolated from aquatic environments were not affected by salinity. However, juvenile development was strongly affected (Moens and Vincx, 2000). A survey in South Africa revealed that *P. siweyae* (Panagrolaimidae), isolated inside the Molepo dam in Limpopo Province, had no correlation with water salinity (Shokoohi and Masoko, 2024). Overall, the findings indicate that salinity cannot significantly affect nematode development.

Reaction of soil nematodes to salinity

Soil salinity refers to the amount of minerals and salts dissolved in the soil, and is an essential measure of soil quality for growing crops and plants. This measure is crucial for crop cultivation because excessive salt levels in soil can lead to reduced crop yields, soil degradation and even desertification. In most cases, soil water extract contains a range of mineral ions, with some being more important than others. These minerals, such as, Na^+, Cl^-, Ca^{2+}, SO_4^{-2}, HCO_3^-, K^+, Mg^{2+} and NO_3^-, are typically listed in order of importance, as they can have varying effects on plant growth and soil fertility. The salinity of the soil is measured as EC (in dS/m): salinity <4 dS/m is considered as non-saline (normal soil), whereas 4–8, 8–16 and >16 dS/m are considered as light, moderate and high salinity, respectively (Brusseau *et al.*, 2019).

The ecology and sometimes the biology of plant-parasitic nematodes are significantly influenced by salinity, depending on the salinity-affected habitats (Khan *et al.*, 1986b; Zaki *et al.*, 2012). A negative relationship was observed between salinity and soil populations of *Helicotylenchus indicus*, *Hoplolaimus indicus*, *Longidorus* sp., *Tylenchorhynchus brevilineatus*,

Tylenchus sp. and *Trichodorus* sp. (Malhotra and Chaubey, 1993). Verma *et al.* (2007) examined the effect of soil salinity due to sodium sulfate, calcium chloride and magnesium chloride (4:1:3) at 2.8, 5.6, 8.4 and 11.2 dS/m on *M. incognita* infection in cotton. The growth parameters of cotton plants from 2.2 dS/m onwards as well as the nematode galls, egg masses, fecundity, etc., from 5.6 dS/m onwards decreased with increasing EC levels over control (non-saline, 1.15 dS/m). In another experiment, Al-Sayed *et al.* (2014) examined the effect of NaCl at 500, 1000, 2000 and 3000 ppm in the irrigation water on *Meloidogyne* and *Rotylenchulus* spp. infecting eggplant, cowpea and squash. Increasing salt concentrations and the resulting high soil EC led to significant promotion in galling, egg mass production and fecundity of *M. incognita* on eggplant roots with the highest impact at 3000 ppm NaCl. However, negative correlation was recorded on cowpea roots. The population of *Rotylenchulus reniformis* increased on cowpea, but decreased on eggplant and squash with increasing salinity. In another experiment, higher galling caused by *Meloidogyne* sp. was recorded due to irrigating eggplants with 0.1% NaCl solution (Sumera *et al.*, 2015).

In a greenhouse experiment, Hamdy *et al.* (2019) observed significant decline in the soil population of *M. incognita* juveniles, galls and egg masses on okra roots with increasing salinity levels from 0.1% to 0.3% NaCl over control. The salinity levels also decreased peroxidase and catalase activities in the exposed plants. Abdelkader *et al.* (2022) investigated the effect of varying levels of soil EC, nitrogen and moisture on reproduction and infestation of cyst nematode *Heterodera* sp. on durum wheat and *Globodera* sp. on potato in three areas in Algeria. The data showed significant correlations of the above three soil characteristics with both cereal and potato cyst nematodes. The soil EC negatively affected the frequency and infestation level of *Globodera* (77%) and *Heterodera* (65%), but soil moisture and total nitrogen showed positive effects on both nematode species.

Soil salinity is crucial in agriculture because this factor affects not only nematode development but also crop production. Salinity can affect plant-parasitic nematodes by impacting various stages of their development and pathogenesis. Salinity has been verified to have an effect on *M. javanica* in tomato production (Maggenti and Hardan, 1973). In a study conducted by Riddle and Bird (1985), it was observed that *R. reniformis*, *Anguina agrostis* and *M. javanica* displayed an affinity towards some chemical elements. Out of the various salts tested, ammonium sulfate and magnesium sulfate were found to be the least attractive, while magnesium chloride was found to be the most attractive salt for the above nematodes. Khan *et al.* (1997) examined the impact of NaCl and $NaHCO_3$ on root penetration and development of *M. incognita* race two in okra and cucumber and on plant growth. Salts at 5.0 mmhos/cm suppressed root penetration of the nematode and delayed development to adults. The number of galls, egg masses and fecundity had negative correlations with the salt concentrations. However, salinity also caused a suppressive effect on plant growth (Khan *et al.*, 1997).

Salinity is also reported to induce changes in nematode behavior. Studies have shown that *M. enterolobii* and *Pratylenchus coffeae* responded negatively to the salinity levels of soil (Silva *et al.*, 2022). Increased levels of NaCl, $CaCl_2$ and $MgCl_2$ decreased the number of *P. coffeae*. The movement of *P. coffeae* was considerably slow under soil salinity. The experiments conducted by Hixson *et al.* (2005a, b) revealed that seawater or highly saline water greatly suppressed the population of *Pratylenchus* and *Belonolaimus* species. The above information is particularly useful for turfgrass managers encountering nematode issues while maintaining hilograss, *Paspalum conjugatum*. The researchers found that saline irrigation may significantly suppress the above nematodes infesting the grasses and pastures. When analyzing the root length of hilograss, it was found that under low salinity levels (0–10 dS/m), *B. longicaudatus* damaged the roots and caused stunted growth, while at 15–25 dS/m the roots were not affected. This revealed that the ability of nematode larvae in parasitizing the roots was negatively affected by increasing salinity levels (>15 dS/m). The above studies highlight the scope of exploitation of saline irrigation for the management of nematodes in turfgrass, which may also be seen as an effective and environmentally friendly solution to the nematode problem.

Reaction of entomopathogenic nematodes and free-living nematodes to salinity

Soil salinity may also affect entomopathogenic and free-living nematodes. The host-finding ability of *Steinernema glaseri* and *H. bacteriophora* when exposed to soil salts was altered with salinity levels (Thurston *et al.*, 1994). Nielsen *et al.* (2011) examined the impact of salinity ranging from 0 to 50 dS/m on the survival and infectivity of *S. riobrave*. They found that the salinity levels used did not have a significant effect on either the survival or infectivity of the symbiotic bacterium present in the nematode, indicating that *S. riobrave* may be adaptable to a wide range of salinity conditions. An investigation on the impact of NaCl, $MgSO_4$, KNO_3 and NaCl + $MgSO_4$ on two free-living bacterivores nematodes, i.e. *Scottnema lindsayae* (Cephalobidae) and *Plectus antarcticus* (Plectidae), showed that these nematodes are highly sensitive to salinity and did not survive exposure to different concentrations of NO_3 (Nkem *et al.*, 2006).

Besides salinity, soil moisture is also critical for the survival of free-living nematodes. *Scottnema lindsayae* can survive at low soil moisture, and that is a reason for tolerance to soil salinity, as in low moisture, dissolved salts are lower than at high moisture conditions. Salts present in soil water affect osmoregulation of the body through cuticle permeability (Khan, 2008). Poage *et al.* (2008) showed that *S. lindsayae* survival was negatively correlated with soil salinity, such as nitrate. A strong positive correlation was recorded with soil EC and mortality of *S. lindsayae*. Pen-Mouratov *et al.* (2010) also observed a positive correlation of bacterivore nematodes, including *Wilsonema*, *Acrobeles*, *Chiloplacus*, *Cervidellus* and *Cephalobus*, with total soluble nitrogen of the soil. It has been found in China that soil EC generally induces a positive effect on the distribution and dynamics of bacterivore and fungivore nematodes in reclaimed farmland (Yang *et al.*, 2021). However, a significant negative correlation between *Trypilina* (r = –0.862), *Prismatolaimus* (r = –0.728), *Nanidorus* (r = –0.864), *Xiphinema* (r = –0.862), *Pratylenchus* (r = –0.888) and *Rotylenchus* (r = –0.765) and a positive correlation with *Rotylenchulus* (r = –0.826) in tomato fields in South Africa was observed (Shokoohi, 2023). A negative correlation (r = –0.688) between *Acrobeloides* and nitrate in the soil was also observed in the study. *Tylencholaimus mirabilis*, which was only detected in a forest soil, showed a positive significant relationship (r=0.646) with soil phosphate (Shokoohi *et al.*, 2023).

Several genes have been found to interfere during the interaction of nematodes with salts (Xie and Zhang, 2022). *L. marina*, which is euryhaline nematode (living in a wide range of salinity), possesses a special mechanism which has demonstrated the survival of free-living bacterivore nematodes under salinity stress. When nematodes are exposed to a saline environment, it creates a stressful condition that can paralyze them. However, nematodes have developed a mechanism of osmotic regulation that enables them to resist the stress of the saline environment. To do so, many genes, such as osmolyte biosynthesis genes, transthyretin-like family genes, V-type H^+-transporting ATPase and potassium channel genes, are expressed in hypoosmotic and hyperosmotic conditions, which help nematodes adapt to the salty environment. All these genes work together to combat the salinity stress response as well as long-term acclimation processes in *L. marina* (Xie and Zhang, 2022). Although these genes have not been thoroughly examined in other nematodes, it should be noted that bacterivorous nematodes, regardless of their specific trophic group, display an exceptional ability to function as bioindicators for salinity levels. This makes them an invaluable tool for monitoring and assessing changes in the salinity of aquatic environments.

Conclusion and Future Prospects

Nematode taxa, especially those that feed on bacteria, fungi and other organisms, can serve as excellent bioindicators due to their high tolerance to environmental pollution. These tiny creatures, which range in size from 0.1 to 2 mm, are found in almost every ecosystem on Earth, including soil, freshwater and marine environments. Further, the nematodes that use an *r*-strategy, characterized by a short lifespan and high reproductive rates, have been observed to be more vulnerable to harsh environmental conditions that can disrupt their behavior. By utilizing character-based indices, the structure of the nematode community can be studied and the impact of environmental disturbances can be evaluated. This classification system is a valuable tool for understanding the ecology and dynamics of soil nematodes, making it possible to use them as bioindicators. The research suggests that certain families of nematodes, such as Chromadorida, Enoplida and Monhysterida in aquatic environments, and Cephalobidae and Rhabditidae in soil environments, can be used as reliable bioindicators to evaluate the extent of pollution in these environments. Furthermore, the maturity index is a highly effective tool for measuring heavy metal pollution in various soil types. However, nematode populations in soil are also affected by salinity. The bacterivore nematodes, including *Scottnema* and *Plectus*, have been found to be highly sensitive to salinity (nitrate levels) in soil. As a result, they are regarded as excellent bioindicators for NO_3. On the other hand, herbivore nematodes are known to be impacted by both heavy metals and salinity, but they are not suitable for use as bioindicators due to their life cycle fluctuations. However, it is essential to ascertain their relationship with soil and plants carefully as a considerable number of nematodes species, especially from Tylenchidae and Anguinidae, may feed on roots instead of fungi. However, the use of nematodes as bioindicators requires extensive research and careful consideration of the environment in which they are studied. For example, aquatic environments pose unique challenges due to the constantly changing substrates and the swift water flow, which make it challenging to trace heavy metals. The nematodes can move quickly with the water flow, making it even more difficult to study their behavior. On the other hand, soil environments provide a more stable substrate for nematodes to thrive, and tracking their relationship with heavy metals is comparatively more straightforward. Finally, the use of nematodes as bioindicators is a promising approach in environmental studies, but it requires careful consideration of the environment, and long-term studies are required before making a conclusion.

Acknowledgements

Mention of trade names or commercial products in this publication is solely for the purpose of providing specific information and does not imply recommendation or endorsement by the US Department of Agriculture. USDA is an equal opportunity provider and employer.

References

Abdelkader, M., Mohammed, B., Mohammed, R. and Antoniolli, Z.I. (2022) Effects of soil salinity on cereal cyst nematodes (*Heterodera*) and potato golden nematodes (*Globodera*). *Plant Archives* 22(2), 235–244. DOI: 10.51470/PLANTARCHIVES.2022.v22.no2.040.

Adão, H., Alves, A.S., Patrício, J., Neto, J.M., Costa, M.J. *et al.* (2009) Spatial distribution of subtidal nematoda communities along the salinity gradient in southern European estuaries. *Acta Oecologica* 3(5), 287–300. DOI: 10.1016/j.actao.2008.11.007.

Al-Sayed, A.S., Farahat, A. and Diab, S. (2014) The effect of irrigation water salinity on the infectivity and reproduction of the root-knot and the reniform nematodes on some vegetable crops. *Egyptian Journal of Agronematology* 13(1), 1–11. DOI: 10.21608/ejaj.2014.63624.

Balsamo, M., Semprucci, F., Frontalini, F. and Coccioni, R. (2012) Meiofauna as a tool for marine ecosystem biomonitoring. In: Cruzado, A. (ed.) *Marine Ecosystems*. InTechOpen, London. DOI: 10.5772/34423.

Bisessar, S., Rinne, R.J. and Potter, J.W. (1983) Effects of heavy metals and *Meloidogyne hapla* on celery grown on organic soil near a nickel refinery. *Plant Disease* 67(1), 11–14. DOI: 10.1094/PD-67-11.

Brusseau, M.L., Pepper, I.L. and Gerba, C.P. (2019) *Environmental and Pollution Science*, 3rd edn. Academic Press, Amsterdam.

Chauvin, C., Trambolho, M., Hedde, M., Makowski, D., Cérémonie, H. *et al.* (2020) Soil nematodes as indicators of heavy metal pollution: A meta-analysis. *Open Journal of Soil Science* 10(12), 579–601. DOI: 10.4236/ojss.2020.1012028.

Decraemer, W., Karanastasi, E., Brown, D. and Backeljau, T. (2003) Review of the ultrastructure of the nematode body cuticle and its phylogenetic interpretation. *Biological Reviews* 78(3), 465–510. DOI: 10.1017/s1464793102006115.

Doran, J.W., Sarrantonio, M. and Liebig, M.A. (1996) Soil health and sustainability. *Advances in Agronomy* 56, 1–54. DOI: 10.1016/S0065-2113(08)60178-9.

Fiscus, D.A. and Neher, D.A. (2002) Distinguishing sensitivity of free-living soil nematode genera to physical and chemical disturbances. *Ecological Applications* 12(2), 565–575. DOI: 10.1890/1051-0761(2002)012[0565:DSOFLS]2.0.CO;2.

Geetanjali, M., Sandeep, K., Malhotra, A., Ansari, Z. and Chatterjee, A. (2002) The role of nematodes as bioindicators in marine and freshwater habitats. *Current Science* 82, 505–507.

Georgieva, S.S., McGrath, S.P., Hooper, D.J. and Chambers, B.S. (2002) Nematode communities under stress: The long-term effects of heavy metals in soil treated with sewage sludge. *Applied Soil Ecology* 20(1), 27–42. DOI: 10.1016/S0929-1393(02)00005-7.

Ghanem, M.F.S.A., Afzal, S., Nesar, H., Imran, Z. and Ahmad, W. (2024) Impact of metal polluted sewage water on soil nematode assemblages in agricultural settings of Aligarh, India. *Soil Ecology Letters* 6(1), 230193. DOI: 10.1007/s42832-023-0193-3.

Gyedu-Ababio, T.K., Furstenberg, J.P., Baird, D. and Vanreusel, A. (1999) Nematodes as indicators of pollution: A case study from the Swartkops River System, South Africa. *Hydrobiologia* 397, 155–169. DOI: 10.1023/A:1003617825985.

Hamdy, F.A., Gad, S.B., Abo-El-Kheer, E.-S.A.M. and Aldosari, A. (2019) Influence of three rates of salinity on okra plants (*Abelmoschus esculentus* L.) infected with root-knot nematode *in vivo*. *Archives of Phytopathology and Plant Protection* 52(3–4), 277–287. DOI: 10.1080/03235408.2019.1606488.

Heininger, P., Höss, S., Claus, E., Pelzer, J. and Traunspurger, W. (2007) Nematode communities in contaminated river sediments. *Environmental Pollution* 146(1), 64–76. DOI: 10.1016/j.envpol.2006.06.023.

Hixson, A.C., Lowe, T. and Crow, W.T. (2005a) Salt influences nematodes in seashore paspalum. *US Golf Association Green Section Record* Jan/Feb 1–13.

Hixson, A.C., Crow, W.T., McSorley, R. and Trenholm, L.E. (2005b) Saline irrigation affects *Belonolaimus longicaudatus* and *Hoplolaimus galeatus* on seashore paspalum. *Journal of Nematology* 37(1), 37–44.

Huo, Z., Li, Z., Guan, P., Shi, F., Jiang, H. *et al.* (2024) Effect of heavy metal contamination on soil nematode communities in urban brownfields. *Global Ecology and Conservation* 49, e02787. DOI: 10.1016/j.gecco.2023.e02787.

Hurley, L.C. (1986) The cuticle of the parasitic nematode, *Nematospiroides dubius*. PhD thesis, Adelaide University, Adelaide, Australia.

Jarmuł, J. and Kamionek, M. (2001) Effect of lead compounds on the intensity and extend of infection of *Galleria mellonella* L. caterpillars by entomopathogenic nematodes *Steinernema feltiae* Filipjev. *Chemia i Inżynieria Ekologiczna* 8(6), 601–607.

Jaworska, M. and Gorczyca, A. (2002) The effect of metal ions on mortality, pathogenicity and reproduction of entomopathogenic nematodes *Steinernema feltiae* filipjev (*Rhabditida, steinernematidae*). *Polish Journal of Environmental Studies* 11(5), 517–520.

Jaworska, M., Gorczyca, A., Sepiol, J. and Tomasik, P. (1997) Effect of metal ions on the entomopathogenic nematode *Heterorhabditis bacteriophora poinar* (Nematoda: Heterohabditidae) under laboratory conditions. *Water, Air, & Soil Pollution* 93(1–4), 157–166. DOI: 10.1007/BF02404753.

Kamionek, M., Jarmuł-Pietraszczyk, J., Badowska-Kozakiewicz, A. and Pezowicz, E. (2011) Influence of lead and cadmium ions on the entomopathogenic nematodes *Steinernema feltiae* Filipjev. *Ecological Chemistry and Engineering* 18(3), 365–369.

Kang, Y., Zheng, S., Wan, T., Wang, L., Yang, Q. *et al.* (2023) Nematode as a biomonitoring model for evaluating ecological risks of heavy metals in sediments from an urban river. *Ecological Indicators* 147, 110013. DOI: 10.1016/j.ecolind.2023.110013.

Khan, M.R. (2008) *Plant Nematodes – Methodology, Morphology, Systematics, Biology and Ecology*. Science Publishers, Enfield, NH. DOI: 10.1201/b10766.

Khan, M.R. and Khan, M.W. (1995) Response of lentil to copper as a soil pollutant. *Acta Botanica Indica* 22, 82–86.

Khan, M.R. and Khan, M.M. (2010) Effect of varying concentration of nickel and cobalt on the plant growth and yield of chickpea. *Australian Journal of Basic and Applied Sciences* 4(6), 1036–1046.

Khan, A.A., Khan, M.W. and Khan, M.R. (1986b) Effect of soil salinity on hatching, mortality, penetration and development of root-knot nematode, *Meloidogyne javanica*. In: *Proceedings of the National Conference on Plant Parasitic Nematodes,* New Delhi, p. 13.

Khan, M.R., Khan, M.W. and Khan, A.A. (1986a) Effect of mercury, a heavy metal, on hatching, mortality and penetration of *Meloidogyne incognita* on pepper. In: *Proceedings of the National Conference on Plant Parasitic Nematodes,* New Delhi, pp. 12–13.

Khan, M.R., Singh, S.K. and Khan, M.W. (1987) Effect of nickel on growth of lentil plants. *Indian Association for Water Pollution Control, Technical Annual* 14, 57–60.

Khan, M.R., Singh, S.K. and Khan, M.W. (1988) Response of lentil to cobalt as a soil pollutant. *Annals of Applied Biology* 112, 103–104.

Khan, M.R., Khan, M.W. and Nabi, S.T. (1994) Effect of cobalt, a soil pollutant on hatching, mortality, and penetration of root-knot nematode, *Meloidogyne incognita*. *Chemical and Environmental Research* 3, 265–269.

Khan, M.R., Khan, M.W. and Singh, K. (1996a) Effects of nickel and root-knot nematode on growth and protein content of chickpea. *Nematologia Mediterranea* 24, 87–90.

Khan, M.R., Khan, M.W. and Singh, K. (1996b) Growth performance of chickpea under the influence of nickel and cobalt as soil pollutants. *Journal of Indian Botanical Society* 75, 193–196.

Khan, M.R., Khan, M.W. and Singh, K. (1997) Management of root-knot disease of tomato by the application of fly ash. *Plant Pathology* 46, 33–43. DOI: 10.1046/j.1365-3059.1997.d01-199.x.

Khan, M.R., Khan, S.M., Mohiddin, F.A. and Askary, T.H. (2006) Effect of high nickel contamination of soil on root-knot nematode disease of tomato. *Nematropica* 36, 79–87.

Korthals, G.W., Bongers, T., Kammenga, J.E., Alexiev, A.D. and Lexmond, T.M. (1996) Long-term effects of copper and pH on the nematode community in an agroecosystem. *Environmental Toxicology and Chemistry* 15(6), 979–985. DOI: 10.1002/etc.5620150621.

Maggenti, A.R. and Hardan, A. (1973) The effects of soil salinity and *Meloidogyne javanica* on tomato. *Journal of Nematology* 5(4), 231–234.

Malhotra, S.K. and Chaubey, A.K. (1993) High salinity tolerance of eucalyptus and interactions with soil and plant nematodes of pathogenic significance. In: Lieth, H. and Masoom, A.A. (eds) *Towards the Rational Use of High Salinity Tolerant Plants. Volume 2: Agriculture and Forestry Under Marginal Soil Water Conditions*. Springer, Dordrecht, The Netherlands, pp. 239–245. DOI: 10.1007/978-94-011-1860-6_30.

Martinez, J.G., Quiobe, S.P. and Moens, T. (2019) Effects of mercury (Hg) on soil nematodes: A microcosm approach. *Archives of Environmental Contamination and Toxicology* 77(3), 421–431. DOI: 10.1007/s00244-019-00652-7.

Moens, T. and Vincx, M. (2000) Temperature and salinity constraints on the life cycle of two brackish-water nematode species. *Journal of Experimental Marine Biology and Ecology* 243(1), 115–135. DOI: 10.1016/S0022-0981(99)00113-6.

Moens, T., Braeckman, U., Derycke, S., Fonseca, G., Gallucci, F. *et al.* (2013) Ecology of free-living marine nematodes. In: Schimidt-Rhaesa, A. (ed.) *Handbook of Zoology. Volume 2: Nematoda. Gastrotricha, Cycloneuralia and Gnathifera*. De Gruyter, Berlin, pp. 109–152. DOI: 10.1515/9783110274257.109.

Mohamedova, M.S. and Lecheva, I. (2016) Soil nematode response to heavy metal pollution of industrial origin in Bulgaria. *Turkish Journal of Agriculture and Forestry* 3(2), 143–150.

Montagna, P.A., Palmer, T.A. and Beseres Pollack, J. (2013) Hydrological changes and estuarine dynamics. *SpringerBriefs in Environmental Science* 8, 1–94. DOI: 10.1007/978-1-4614-5833-3.

Moreno, M., Vezzulli, L., Marin, V., Laconi, P., Albertelli, G. *et al.* (2008) The use of meiofauna diversity as an indicator of pollution in harbours. *ICES Journal of Marine Science* 65(8), 1428–1435. DOI: 10.1093/icesjms/fsn116.

Mothersill, C. and Seymour, C. (2016) Genomic instability and the spectrum of response to low radiation doses. In: Kovalchuk, I. and Kovalchuk, O. (eds) *Genome Stability*. Academic Press, New York, pp. 601–604.

Neher, D.A. (2001) Role of nematodes in soil health and their use as indicators. *Journal of Nematology* 33(4), 161–168.

Ng, W.L., Chen, C.A., Mustafa, S., Soo, C.L., Liao, Y.C. *et al.* (2022) Free-living marine nematodes community structure in the conservation area (Chaojing Park) and its adjacent area of Keelung, Taiwan. *PLoS ONE* 17(5), e0268691. DOI: 10.1371/journal.pone.0268691.

Nielsen, A.L., Spence, K.O., Nakatani, J. and Lewis, E.E. (2011) Effect of soil salinity on entomopathogenic nematode survival and behaviour. *Nematology* 13(7), 859–867. DOI: 10.1163/138855411X562254.

Njume, F.N., Razzauti, A., Soler, M., Perschin, V., Fazeli, G. *et al.* (2022) A lipid transfer protein ensures nematode cuticular impermeability. *iScience* 25(11), 105357. DOI: 10.1016/j.isci.2022.105357.

Nkem, J.N., Virginia, R.A., Barrett, J.E., Wall, D.H. and Li, G. (2006) Salt tolerance and survival thresholds for two species of Antarctic soil nematodes. *Polar Biology* 29(8), 643–651. DOI: 10.1007/s00300-005-0101-6.

Okereafor, U., Makhatha, M., Mekuto, L., Uche-Okereafor, N., Sebola, T. *et al.* (2020) Toxic metal implications on agricultural soils, plants, animals, aquatic life and human health. *International Journal of Environmental Research and Public Health* 17(7), 2204. DOI: 10.3390/ijerph17072204.

Park, B.Y., Lee, J.K., Ro, H.M. and Kim, Y.H. (2011) Effects of heavy metal contamination from an abandoned mine on nematode community structure as an indicator of soil ecosystem health. *Applied Soil Ecology* 51, 17–24. DOI: 10.1016/j.apsoil.2011.08.006.

Pen-Mouratov, S., Genzer, N., Shukurov, N., Plakht, J. and Steinberger, Y. (2010) Seasonal effect of geomorphological chronosequence features on soil biota dynamics. *Pedosphere* 20(6), 761–770. DOI: 10.1016/S1002-0160(10)60066-5.

Poage, M.A., Barrett, J.E., Virginia, R.A. and Wall, D.H. (2008) The influence of soil geochemistry on nematode distribution, McMurdo Dry Valleys, Antarctica. *Arctic, Antarctic, and Alpine Research* 40(1), 119–128. DOI: 10.1657/1523-0430(06-051)[POAGE]2.0.CO;2.

Renčo, M., Čerevková, A. and Hlava, J. (2022) Life in a contaminated environment: How soil nematodes can indicate long-term heavy-metal pollution. *Journal of Nematology* 54(1), 20220053. DOI: 10.2478/jofnem-2022-0053.

Ridall, A. and Ingels, J. (2021) Suitability of free-living marine nematodes as bioindicators: Status and future considerations. *Frontiers in Marine Science* 8, 685327. DOI: 10.3389/fmars.2021.685327.

Riddle, D.L. and Bird, A.F. (1985) Responses of the plant parasitic nematodes *Rotylenchulus reniformis*, *Anguina agrostis* and *Meloidogyne javanica* to chemical attractants. *Parasitology* 91(1), 185–195. DOI: 10.1017/s0031182000056626.

Sánchez-Moreno, S. and Navas, A. (2007) Nematode diversity and food web condition in heavy metal polluted soils in a river basin in southern Spain. *European Journal of Soil Biology* 43(3), 166–179. DOI: 10.1016/j.ejsobi.2007.01.002.

Sandhu, A., Badal, D., Sheokand, R., Tyagi, S. and Singh, V. (2021) Specific collagens maintain the cuticle permeability barrier in *Caenorhabditis elegans*. *Genetics* 217, iyaa047. DOI: 10.1093/genetics/iyaa047.

Semprucci, F., Losi, V. and Moreno, M. (2015) A review of italian research on free-living marine nematodes and the future perspectives on their use as ecological indicators (EcoInds). *Mediterranean Marine Science* 16(2), 352–365. DOI: 10.12681/mms.1072.

Shokoohi, E. (2023) Impact of agricultural land use on nematode diversity and soil quality in Dalmada, South Africa. *Horticulturae* 9(7), 749. DOI: 10.3390/horticulturae9070749.

Shokoohi, E. and Masoko, P. (2024) Morphological and molecular analysis of *Propanagrolaimus siweyae* sp. n. (Nematoda: Panagrolaimidae) from Molepo dam, Limpopo Province, South Africa, and its relationship with water parameters. *Biologia* 79, 3573–3587. DOI: 10.1007/s11756-024-01806-2.

Shokoohi, E., Moyo, N. and Gouveia, F. (2023) Relationship of nematodes in natural and disturbed land with physicochemical properties in Magoebaskloof, Limpopo Province, South Africa. *Biologia* 78(11), 3223–3233. DOI: 10.1007/s11756-023-01468-6.

Silva, I.A.C.E., Pedrosa, E.M.R., Silva, E.F.D.F.E., Rolim, M.M. and Vicente, T.F.D.S. (2022) Motility and migration of nematodes in saline environments. *Revista Caatinga* 35(4), 865–876. DOI: 10.1590/1983-21252022v35n414rc.

Soko, M.I. and Gyedu-Ababio, T.K. (2019) The use of free-living nematodes as pollution indicator in Incomati river estuary, Mozambique. *Open Journal of Ecology* 9, 117–133. DOI: 10.4236/oje.2019.95010.

Somerfield, P.J., Gee, J.M. and Warwick, R.M. (1994) Soft sediment meiofaunal community structure in relation to a long-term heavy metal gradient in the Fal estuary system. *Marine Ecology Progress Series* 105, 79–88. DOI: 10.3354/meps105079.

Sumera, M.A., Rafat, S. and Toqeer, A. (2015) Effect of salinity and root-knot nematode on growth of eggplant (*Solanum melongena*). *FUUAST Journal of Biology* 5(1), 93–97.

Sun, Y., Bai, G., Wang, Y.-X., Zhang, Y., Pan, J. *et al.* (2016) The impact of Cu, Zn and Cr salts on the relationship between insect and plant parasitic nematodes: A reduction in biocontrol efficacy. *Applied Soil Ecology* 107, 108–115. DOI: 10.1016/j.apsoil.2016.05.015.

Thurston, G.S., Ni, Y. and Kaya, H.K. (1994) Influence of salinity on survival and infectivity of entomopathogenic nematodes. *Journal of Nematology* 26(3), 345–351.
Verma, K.K., Jain, R.K. and Sharma, S.K. (2007) Influence of soil salinity on reproduction and multiplication of root-knot nematode (meloidogyne incognita) in cotton. *Indian Journal of Nematology* 37(1), 23–26.
Wang, X., Dai, Z., Lin, J., Zhao, H., Yu, H. *et al.* (2023) Heavy metal contamination collapses trophic interactions in the soil microbial food web via bottom-up regulation. *Soil Biology and Biochemistry* 184, 109058. DOI: 10.1016/j.soilbio.2023.109058.
Wasilewska, L. (1997) Soil invertebrates as bioindicators, with special reference to soil-inhabiting nematodes. *Russian Journal of Nematology* 5, 113–126.
Xie, Y. and Zhang, L. (2022) Transcriptomic and proteomic analysis of marine nematode *Litoditis marina* acclimated to different salinities. *Genes* 13(4), 651. DOI: 10.3390/genes13040651.
Yang, L., Zhang, F. and Luo, Y. (2021) A soil nematode community response to reclamation of salinized abandoned farmland. *Zoological Studies* 60, e72. DOI: 10.6620/ZS.2021.60-72.
Yeates, G.W. (2003) Nematodes as soil indicators: Functional and biodiversity aspects. *Biology and Fertility of Soils* 37(4), 199–210. DOI: 10.1007/s00374-003-0586-5.
Young, E.H. and Unc, A. (2023) A review of nematodes as biological indicators of sustainable functioning for northern soils undergoing land-use conversion. *Applied Soil Ecology* 183, 104762. DOI: 10.1016/j.apsoil.2022.104762.
Zaki, M.J., Khan, D. and Abid, M. (2012) Nematodes in the saline environment: A mini overview. *International Journal of Biology and Biotechnology* 9(1–2), 99–113.
Zeng, Q.P., Zhu, L.A., Wang, J.Z., Cheng, J., Liu, Y. *et al.* (2020) Effect of heavy metals on soil microbial biomass, and nematode trophic groups of a paddy soil affected by long-running polymetallic mining activities in Guangdong, southern China. *Applied Ecology and Environmental Research* 18(4), 4915–4927. DOI: 10.15666/aeer/1804_49154927.
Zhang, W.D., Wang, X.F., Li, Q., Jiang, Y. and Liang, W.J. (2007) Soil nematode responses to heavy metal stress. *Helminthologia* 44(2), 87–91. DOI: 10.2478/s11687-007-0009-5.
Šalamún, P., Renčo, M., Kucanová, E., Brázová, T., Papajová, I. *et al.* (2012) Nematodes as bioindicators of soil degradation due to heavy metals. *Ecotoxicology* 21, 2319–2330.

15 Impact of Air Pollution on Nematode Infestation in Plants

Mujeebur Rahman Khan*, Tanveer Fatima Rizvi and Irfan Ahmad
Department of Plant Protection, Faculty of Agricultural Sciences, Aligarh Muslim University, Aligarh, India

Abstract
Plants are highly reliant of the surrounding environment because of their permanent attachment to the soil through roots for intake of nutrients and water and exposure to the atmosphere through foliage for gas exchange and light absorption. Apart from host vulnerability, air pollution may also affect the development of plant diseases. The emissions of sulfur dioxide, ozone, carbon dioxide and other pollutants due to various anthropogenic activities may influence host–parasite relationships. Research has demonstrated that plants under the stress of elevated levels of air pollutants may respond differently to foliar and soil-borne pathogens. The parasitism of nematodes invading underground or above-ground parts of the plant is likely to be affected directly or through host-mediated effects of air pollution. This chapter presents critical and up-to-date information on the unique interaction of air pollutants and nematodes, and explains the diverse responses of plants to nematodes under atmospheric stress.

Introduction

Natural processes such as volcanic eruptions, forest fires, decomposition of organic matter, etc., are major causes of air pollution but anthropogenic sources, especially the burning of fossil fuels in different processes, hold the major share in causing air pollution. The combustion of coal (Paraschiv and Paraschiv, 2020; Inumaru *et al.*, 2021), wood (Yli-Tuomi *et al.*, 2015; Vicente *et al.*, 2020), plant residues (Benemann, 1993; Koruba *et al.*, 2017) and petroleum products (Reed, 2015) as well as animal husbandry (Reijnders and Huijbregts, 2005; Maurya *et al.*, 2015) and deep-water paddy cultivation (Bhatia *et al.*, 2013; Gupta *et al.*, 2021; Bhagawati *et al.*, 2023) are all important sources of air pollution. During these processes, toxic gases such as oxides of sulfur, nitrogen and carbon, ozone (O_3), acid rain, methane, etc., are formed directly or indirectly, and are ultimately released into the environment (Falkowska, 2016; Kern, 2021). When the concentration of these gases exceeds the self-regulating capacity of the atmosphere, and remains continuously above the standard (safe) levels of a gas, the term air pollution is used, and the gas is referred to as an air pollutant (Heggestad *et al.*, 1986; Heck *et al.*, 1986; Khan and Khan, 1994a). All kinds of terrestrial organisms, especially those that inhabit the land or extend some of their body parts above the ground, are directly exposed to air pollution. Plants and animals, including humans, are in continuous and direct exposure to the air, hence any change

*Corresponding author: mrkhan777in@yahoo.co.in

 Nematode Disease Complexes in Agricultural Crops (ed. Mujeebur Rahman Khan)
DOI: 10.1079/9781800625228.0015

in the composition of pure air or addition of higher concentrations of gases such as SO_2, O_3, etc., may adversely affect the growth of plants and animals (Bender and Weigel, 2011; Saini *et al.*, 2019; Adak and Kour, 2021).

SO_2, O_3, acid rain, etc., are well recognized to incite specific and discernible injury symptoms and diseases in plants (Benzing *et al.*, 1992; Khan and Khan, 1993a; 1994b), and may be considered as inanimate or abiotic plant pathogens (Heck *et al.*, 1986). Hence, plants in nature, in addition to biotic or animate pathogens (fungi, bacteria, viruses, nematodes, etc.), may also suffer from the inanimate pathogens such as air pollutants (Khan and Khan, 1993b). These two groups of pathogens, the animate and inanimate, may coexist in an ecosystem, and quite possibly may interact and develop some relationship on a host plant (Khan and Khan, 2000). Inanimate pathogens (pollutants) do not have any nutritional requirement, but exert damage to plants. The plants thus damaged by pollutants may cause stress on the nutritional availability to the animate pathogens parasitizing the aerial or underground parts. The air-pollution-injured and nutritionally deficient host plant may become more vulnerable to the infection and damage caused by the animate pathogen. The interaction of animate and inanimate pathogens is realistic, and may operate in ambient conditions in agricultural fields or polyhouses close to pollution sources. Theoretically, the inanimate pathogens (air pollutants) and animate pathogens (fungi, bacteria, nematodes, etc.) coexisting in a crop pathosystem may interact in the following three ways:

- *Antagonistic relationship:* Inhibitory effects of air pollutants on animate plant pathogens. The disease may be reduced, but the host plant may or may not be benefited. The adverse effects may be caused directly on the pathogen or through the host plant.
- *Synergistic relationship:* Stimulatory effect, where the pollutant may promote the infection of the pathogen or sensitivity of the host to pollutant, leading to an increase in disease severity and plant injury. The stimulatory effect may operate directly on the pathogen or through the host by rendering the plant more susceptible to the pathogen or more sensitive to the pollutant.
- *Neutral relationship:* No effect of a pollutant on the animate pathogen, and vice versa; both act independently in the ecosystem. Their combined impact will be close to the additive effects.

The above three postulates are quite logical and may operate under ambient air pollution conditions. The present chapter critically examines these aspects and presents relevant information on the impact of air pollution on plants and pathogens, and also attempts to develop some concreate conclusions on the interaction of air pollutants and nematodes in agricultural crops. Further, we also intend to convince researchers to initiate adequate research efforts to explore the dimensions of the above relationships since the area of air pollution–pathogen interaction has not received due attention by environmental scientists and plant pathologists.

Air Pollution and Its Effects on Plants and Nematodes

Environmental pollution is caused by the rapid growth of industries and cities needing energy for various activities (Emberson *et al.*, 2001; Yang *et al.*, 2002). Despite continued emphasis and implementation of policies on the promotion of solar and other renewable energy sources to curb carbon emissions, coal and petroleum continue to be major energy sources as well as the major global polluters. The burning of coal and petroleum causes direct or indirect generation and release of air pollutants, such as SO_2, O_3, acid rain, CO_2 and particulate air pollutants (Khan and Khan, 1993a). The ambient concentrations of these pollutants often exceed the permissible limits set by different air pollution agencies. When a higher concentration of a pollutant prevails for a longer duration, it injures the plants with specific injury symptoms and causes disorders (Barret and Benedict, 1970; Hesketh, 1979; Varshney *et al.*, 1979; Khan and Khan, 1993b), and may also influence the severity of plant diseases caused by fungi, nematodes, etc. (Khan *et al.*, 1995; Khan and Khan, 2000). The impact of major air pollutants on plants and phytonematodes are summarized below.

Coal Smoke and Its Effects on Plants and Nematodes

Coal smoke is the most common type of pollution in developing and underdeveloped countries where coal is largely used to run industries (Khan and Khan, 1994c). Coal smoke is a mixture of gaseous and particulate air pollutants, containing mainly the oxides of sulfur, nitrogen and carbon, suspended particulate matter, and some other contaminants (Hesketh, 1979).

Effects of coal smoke on plants

The constituent pollutants of coal smoke individually or in combination cause negative impacts on plant health, fruit setting and their development (Khan and Khan, 1990; 1994c; Srivastava *et al.*, 2023). The particulate matter present in coal smoke can settle on plant foliage and reduce the plant's ability to photosynthesize by obstructing sunlight and clogging the stomata (Iqbal *et al.*, 2010). The relative concentrations of the constituent pollutants depend on the amount and type of coal, furnace temperature, height of the stack, efficiency of the pollution control device, and the speed and direction of the wind (Khan and Khan, 1990; 1991; 1996b). Gupta and Ghouse (1987) reported around 28% loss in productivity of okra grown in the close vicinity of a thermal power plant in Kasimpur, Aligarh (India). Later, Khan and Khan (1994c) recorded mean concentration of 141–184 μg SO_2/m^3, 76–93 μg NO_2/m^3 and 200–309 μg suspended particulate matter (SPM)/m^3 in a 2 km radius around the same thermal power plant. The coal smoke emanating from coal-fired industries such as thermal power plant spreads over a large area and affects the growing vegetation. Coal smoke has been found to adversely affect plant growth, flowering and fruit setting of a number of crop plants and fruit-yielding trees (Khan and Khan, 1995b).

Effects of coal smoke on phytonematodes

Phytonematodes are obligate parasites of plants and cannot persist in the soil in the absence of their hosts for extended periods (Khan, 2008). Most phytonematodes primarily feed on roots, but there are a few genera, such as *Anguina*, *Aphelenchoides* and *Ditylenchus*, which invade stems, leaves or inflorescences (Khan, 2023a; 2023b). Based on their parasitism, plant nematodes are categorized as ectoparasites, endoparasites and semi-endoparasite. Ectoparasitic nematodes remain outside the host and feed on the surface of the roots or shoots (Decraemer and Geraert, 2013). Endoparasites enter plant tissue and feed internally, whereas semi-endoparasites form an intermediate category (Siddiqi, 2005). All groups of plant nematodes have agricultural significance but endoparasites such as *Meloidogyne* and *Pratylenchus* are highly destructive and cause significant economic losses to crops (Khan and Jairajpuri, 2010a, b; Khan and Jairajpuri, 2012; Sikora *et al.*, 2018). Nematodes in general are aquatic organisms but have adopted a terrestrial habitat (Khan, 2008), hence the majority of nematodes are soil inhabitants (Haque and Khan, 2021).

Coal smoke may also influence the plant response to pathogens including nematodes, fungi, etc. (Khan *et al.*, 1991; Khan and Khan, 1994d; 1998a). Studies conducted at sites under the stress of coal smoke have shown that root-knot severity, reproduction and soil population of *M. incognita* were significantly higher on eggplant (Khan and Khan, 1994d), okra (Khan and Khan, 1994c) and tomato (Khan and Khan, 1995a) cultivated at the sites approximately 1 and 2 km, respectively, away from a thermal power station experiencing 141 (16–269) and 184 (43–348) μg SO_2/m^3 and 76 (19–154) and 93 (29–172) XIIg NO_2/m^3 and 309 (116–665) and 205 (94–482) μg SPM/m^3. The combined effects of the air pollution and nematodes were consistently synergistic, causing greater decrease in plant growth and yield of the aforementioned vegetable crops. In these studies, greater gas injury was noted in nematode-infected plants. Root-knot disease and nematode reproduction on the vegetables were either enhanced or not affected. However, coal smoke emanating from the ceramic and pottery units in Khurja (Uttar Pradesh, India) had growth-promotory effects on cucurbit plants and suppressive effects on the disease caused by *M. incognita*, *M. arenaria* and *M. javanica* and their soil populations (Khan *et al.*, 1991). The galls and egg mass count per root system had a positive correlation with the pollution levels from the potteries.

Sulfur Dioxide and Its Effects on Plants and Phytonematodes

Sulfur dioxide (SO_2) is a major gaseous pollutant, which is mainly formed due to burning of fossil fuels and refining of crude oils. Crops grown around a SO_2 source are exposed to the gas and exhibit foliar injuries and plant growth suppression (Rahman and Husen, 2022). SO_2 diffuses inside the plant largely through stomata, but may also enter through the cuticular surfaces (Khan and Khan, 1993a). After entering into the leaf, the gas is taken up by mesophyll cells and is converted in to sulfate (SO_4^{2-}) and sulfite (SO_3^{2-}) ions (Peter Singh *et al.*, 2012; Brahmachari and Kundu, 2017). SO_3^{2-} ions are several times more phytotoxic than SO_4^{2-} ions (Agrawal, 2003). The SO_3^{2-} ions are converted into SO_4^{2-} ions if the SO_2 concentration is not very high. The extra SO_2 is taken up as SO_4^{2-} and metabolized slowly inside the vacuole (Cram, 1990; Clarkson *et al.*, 1993). Even at relatively low atmospheric SO_2 concentrations, the sulfur content of the foliage is enhanced due to the accumulation of SO_4^{2-} in the vacuole (De Kok, 1990; De Kok and Tausz, 2001). Yang *et al.* (2006) reported that SO_2 at 0.03–1.4 µl/L increased the concentration of SO_4^{2-} ions, water-soluble non-protein thiols and shoot sulfur of Chinese cabbage but the organic sulfur vs total sulfur ratio and organic nitrogen were not affected.

The ions can also participate in the sulfur reduction pathway. The SO_3^{2-} ions react with the aldehydes and ketones of carbohydrates causing a decrease in the carbohydrate content of the exposed plants (Duccer and Ting, 1970). It has been recorded by several researchers that levels of soluble sugars in plants gradually increase due to SO_2 exposures (Koziol and Jordan, 1978; Malhotra and Sarkar, 1979). Kausar *et al.* (2010) reported that the carbohydrate content in wheat (*Triticum aestivum*) was reduced by SO_2 fumigation at 0.05, 0.10 and 0.15 ppm. Non-structural total carbohydrates were reduced following SO_2 exposure in American elm seedlings (Constantinidou and Kozlowski, 1979).

Sometimes air pollution-induced stress triggers the closure of stomata. As a result, CO_2 availability in the leaves is decreased and the rate of carbon fixation is reduced (Dhir, 2016; Adak and Kour, 2021). SO_2, after being absorbed by plant leaves, adversely affects the chlorophylls and carotenoids which directly affect the absorption of light by leaves (Joshi and Swami, 2009). Soluble sugars act as osmoprotectants and cryoprotectants, and play a crucial role in maintaining the integrity of the plasma membranes, which are vital elements of the plant body and serve as the primary energy sources for all organisms. The level of soluble sugars in a plant is a reliable indicator of its physiological activity and determines the sensitivity of the plant to an air pollutant (Tripathi and Gautam, 2007). Prado *et al.* (2000) reported that the accumulation of sugars in different plant parts was increased in response to various environmental stresses. Soluble sugars have also been found to protect against stress (Gibson, 2005).

Effects of sulfur dioxide on plants

SO_2 exposures cause chronic and acute symptoms in plants (Khan and Khan, 2011) due to the relative accumulation of SO_3^{2-} and SO_4^{2-} ions in the leaf tissue. Chronic injury is caused by continuous exposure of plants to low concentrations of SO_2 (below 100 ppb). The formation and accumulation of SO_3^{2-} ions at these concentrations is slow, hence they are oxidized to SO_4^{2-} ions and subsequently utilized by the plant cells (Marier and Dibbs, 1974; Alvarez *et al.*, 2000). The oxidation process continues until sufficient SO_3^{2-} ions accumulate, causing leaf chlorosis. Excessive accumulation of SO_4^{2-} ions may also become toxic to plants (Kaiser *et al.*, 1989; Singh *et al.*, 2018). The cells remain intact, but the photosynthetic pigments undergo bleaching, as a result of which the leaves turn yellowish or chlorotic, and the lower leaf surface may exhibit a silvering or bronzing effect without any tissue death (Varshney *et al.*, 1979; Khan and Khan, 2010). Khan *et al.* (2007) reported that mustard germplasm developed foliar yellowing and mild chlorosis at 75 ppb SO_2 in an open-top chamber study. Acute injury occurs due to exposure to higher concentrations of SO_2 for a shorter duration and is characterized by necrotic lesions at the margins or between the mid ribs of the leaf tissues and they become greyish-green water-soaked areas (Peter Singh *et al.*, 2012; Khan *et al.*, 2015; Jha and Yadav, 2023; Rizvi and Khan, 2023). At an advanced stage, the dead or necrotic lesions become detached, giving the leaf a ragged appearance. If a severe injury occurs, an abscission layer is formed at the petiole base, causing the leaves to detach (Jacobson and Hill, 1970).

There are scattered reports on the impact of SO_2 under chronic and acute injuries (Khan and Khan, 2000). Long-term exposures at 34 and 58 µg SO_2/m^3 caused foliar injury in *Pinus sylvestris*, and a linear relationship between SO_2 concentration and needle necrosis was recorded (Shaw *et al.*, 1993). The SO_2 symptoms were recorded in consecutive years on the same trees, indicating genetic sensitivity of pine. The SO_2 symptoms were reproduced in a similar experiment conducted under simulated conditions. SO_2 exposures at 100 ppb SO_2 for 3 hour on alternate days caused a slight chlorosis on the leaves of tomato cv. Pusa Ruby (Khan and Khan, 1993a). However, SO_2 at 200 ppb caused discernible foliar chlorosis with browning. SO_2 at 393–524 µg/m^3 for 3 weeks incited yellow, brown and necrotic spots on the leaves of Timothy grass (*Phleum pratense*) in closed fumigation chambers (Clapperton and Reid, 1994). SO_2 at 170 µg/m^3 incited initial signs of SO_2 injury within 2 weeks. Higher concentrations of SO_2 and longer exposure periods were found to be linearly corelated with injury severity on the leaves and flowers of *Calendula officinalis* (Padhi *et al.*, 1995) and *Zinnia* (Rath *et al.*, 1995). Typical necrotic reddish-brown lesions and interveinal browning on the leaves of rice seedlings were observed due to SO_2 exposures (Rakwal *et al.*, 2003). The lesions resembled hypersensitive reactions. Intermittent exposures at 200 and 300 µg SO_2/m^3 incited chlorosis in pumpkin leaves with a trace of injury at 100 µg/m^3 (Khan *et al.*, 1996b) (Fig. 15.1). Similar symptoms of 50–200 ppb SO_2 were also recorded on cowpea (Khan and Khan, 1996a), tomato, eggplant, okra (Khan and Khan, 1994c; 1995a), black mustard (Khan *et al.*, 2007), bottle gourd, cucumber, pumpkin, sponge guard and bitter gourd (Rizvi and Khan, 2023). Besides concentration, the sensitivity of plants to SO_2 also varies with seasonal fluctuations (Cormis, 1973), physiological condition and age of the plants (Tamm and Cowling, 1976).

Fig. 15.1. Pumpkin leaf showing yellowing and chlorotic lesions due to exposure to 100 ppb SO_2 for 5 h/day on alternate days for 75 days.

Effects of sulfur dioxide on phytonematodes

Plant nematodes of either parasitic group are not likely to come under direct exposure to SO_2. Further, there is meagre possibility that a significant amount of SO_2 diffuses in the soil and adversely affects the nematode by penetrating its cuticle. Hence, nematodes are affected indirectly by SO_2 exposures through the host (Khan and Khan, 2000). The changes caused by SO_2 in the host or soil, the latter of which may have a lower contribution, may directly affect nematode pathogenesis (Heagle, 1973; Heagle *et al.*, 1979). Research on SO_2 and nematodes interactions have shown both inhibitory and promotory impacts of the gas on nematodes, which are briefly described below.

Inhibitory effects on nematodes

There is sufficient evidence to suggest that SO_2 may suppress the pathogenesis of plant nematodes and the effect may vary with gas concentration and host susceptibility (Table 15.1). Khan and Khan (1993a) observed an inhibitory

Table 15.1. Effect of coal smoke and sulfur dioxide on plant-parasitic nematodes.

Pollutant concentration	Nematode	Host	Effect	Reference
Coal smoke from a thermal power plant containing 141 (16–269) and 184 (43–348) µg SO_2/l, 76 (19–154) and 93 (29–172) XIlg NO_2/m^3 and 309 (116–665) and 205 (94–482) µg SPM/m^3	*Meloidogyne incognita* race 1	Eggplant, okra and tomato	Synergistic interaction with greater suppressions at the polluted sites	Khan and Khan, 1994c; 1994d; 1995a
Exposure at 0.2 ppm SO_2	*Anguina tritici*	Wheat	Greater reduction in plant growth	Kausar *et al*., 2005
Exposure at 0.05, 0.1 and 0.2 ppm SO_2	*Anguina tritici*	Wheat	Antagonistic interaction	Kausar and Khan, 2009
Intermittent exposure at 655 µg/m^3 SO_2	*Aphelenchoides fragariae*, *Belonolaimus longicaudatus*	Soybean	No effect	Weber *et al*., 1979
Intermittent exposure at 655 µg/m^3 SO_2	*Heterodera glycines*, *Paratrichodorus minor*	Soybean	Decrease in nematode reproduction	Weber *et al*., 1979
286 and 571 µg/m^3 SO_2	*Meloidogyne incognita*	Tomato	Synergistic interaction	Khan and Khan, 1993a
100 or 200 µg/m^3 SO_2	*Meloidogyne incognita*	Tomato	Greater reduction in plant growth, but nematode pathogenesis not affected	Khan *et al*., 1995
0.1 ppm SO_2	*Meloidogyne incognita*	Cowpea	Root galls and egg masses significantly increased	Khan and Khan, 1996a
571 µg/m^3 SO_2	*Meloidogyne incognita*	Tomato	Greater gas injury and appeared early in infected plants	Khan and Khan, 1993a
50 and 100 ppb SO_2, for 5 hour on alternate days	*Meloidogyne incognita*	Tomato	Nematode reproduction was decreased	Khan and Khan, 1997
100 ppb SO_2	*Meloidogyne javanica*	Soybean	Greater galling	Singh *et al*., 1997
300 µg/m^3 SO_2	*Meloidogyne javanica*	Tomato	Lower fecundity	Khan *et al*., 1995
0.1 and 0.2 ppm	*Meloidogyne javanica*	Pea	Lower galling	Singh *et al*., 1995
Intermittent exposure at 655 µg/m^3 SO_2	*Pratylenchus penetrans*	Soybean	Enhanced reproduction	Weber *et al*., 1979

effect of SO_2 on *M. incognita* infecting tomato plants at higher concentrations, such as 571 µg/m^3. Intermittent SO_2 exposures at 655 µg/m^3 were found to adversely affect the soil population and reproduction of *Heterodera glycines* and *Paratrichodorus minor* infecting soyabean (Weber *et al.*, 1979). Root galling in pea (*M. javanica*) was also recorded to be less on plants exposed to 0.1 and 0.2 ppm SO_2 (Singh *et al.*, 1995). A significant decrease in root galls, egg masses and eggs per egg mass of *M. javanica* or *M. incognita* was recorded due to exposures of cowpea (Khan and Khan, 1996a), tomato (Khan and Khan, 1993a), cucumber (Rumiani *et al.*, 2016), bottle gourd, pumpkin and sponge gourd (Rizvi and Khan, 2023) to 200–600 μg SO_2/m^3. In another study, 100 ppb SO_2 caused foliar leaf chlorosis in cowpea cvs V-38–1 and V-218 (Khan *et al.*, 1996b). The cowpea plants infected with *M. incognita* developed leaf chlorosis 1–2 weeks earlier than non-infected plants.

Kausar *et al.* (2005) investigated the interaction between SO_2 (0.2 ppm) and *Anguina tritici* on wheat varieties under closed-top exposure chambers. The SO_2 exposures suppressed the growth of all wheat varieties, as well as seed gall formation. The impact of 0.05, 0.1 and 0.2 ppm SO_2 concentrations was also examined on the survival of *A. tritici* larvae (Kausar and Khan, 2009). SO_2 at the concentrations used caused mortality to *A. tritici* juveniles, which increased with the concentration and exposure duration of the gas.

Stimulatory effects on nematodes

Research has shown that SO_2 exposure may also stimulate the pathogenesis of plant nematodes depending on the nematode species, host cultivar and gas concentration (Table 15.1). Bassus (1968) correlated air pollution with the increase in soil nematode populations in the vicinity of a smelter plant. He observed a higher soil population density of saprophagous and predacious nematodes in the root zone of pine trees that were injured by SO_2 in comparison to the pine trees situated at a distant location from the smelter showing only moderate SO_2 injury. Weber *et al.* (1979) found that exposure of soybean plants to intermittent doses of 655 μg SO_2/m^3 increased the reproduction of *Pratylenchus penetrans*. However, *Aphelenchoides fragariae* and *Belonolaimus longicaudatus* were not affected by this exposure. Shew *et al.* (1982) observed that SO_2 promoted the population growth and reproduction of *P. penetrans* on tomato plants. However, other research has shown that concentrations above 100 ppb (286 μg SO_2/m^3) do not generally favor the nematode pathogens.

A synergistic interaction between SO_2 and *M. incognita* on tomato cv. Pusa Ruby was recorded when plants were inoculated with 2000 J_2/kg soil and exposed to 286 μg SO_2/m^3 (Khan and Khan, 1993a). The gas exposures caused slight leaf chlorosis, but nematode-infected plants sustained severe chlorosis. The decrease in plant growth and yield variables of tomato was also significantly greater in the plants infected with *M. incognita*. However, in another study, it was observed that the number of galls, egg masses and eggs per egg mass (fecundity) of *M. incognita* were not affected in cowpea plants exposed to 200 μg SO_2/m^3 but synergistic interaction between the nematode and 200 μg SO_2 was recorded on plant growth and dry matter parameters of cowpea cvs V-38–1 and V-218 apparently due to greater uptake of SO_2 by the plants evidenced by greater foliar chlorosis in the nematode-infected plants. The exposed plants also experienced greater galling and egg mass production (Khan and Khan, 1996a). In a study examining the simultaneous effects of SO_2 on soybeans in the presence of *M. javanica* (2000 J_2/kg soil), Singh *et al.* (1997) observed significantly greater severity of nematode disease in soybean plants exposed to 100 ppb SO_2. The research described above has revealed that the interaction between SO_2 and plant nematodes is largely concentration dependent; lower levels of SO_2 such as 75 or 100 ppb promoted the infection and development of root-knot and root-lesion nematodes on susceptible hosts such as tomato and cucurbits (Rizvi and Khan, 2023).

Carbon Dioxide and Its Effects on Plants and Nematodes

Atmospheric carbon dioxide (CO_2) levels have been on a continuous increase from 315 ppm in 1959 to present day levels above 420 ppm. The future projections for ambient CO_2 concentrations reveal a continuous rise of CO_2 up to 500–1000 ppm by the year 2100 (Lemonnier and Ainsworth, 2018). Controlled and ambient condition experiments as well as potential-impact meta-analytical experiments have enumerated significant impacts of higher levels of CO_2 on plants and their physiology (Ellsworth *et al.*, 2017; Thompson *et al.*, 2017; Bao *et al.*, 2023).

Effect of carbon dioxide on plants

The elevated level of atmospheric CO_2 is an integral aspect of climate change. Several researchers have shown that a 10–20% increase in CO_2 level may significantly enhance biomass production in plants (Craine *et al.*, 2001; De Souza *et al.*, 2008; Reich *et al.*, 2014; Dahal and Vanlerberghe, 2018; Khan and Rizvi, 2020). Ellsworth *et al.* (2017) found that CO_2 assimilation increased by 19% under high CO_2 conditions. A 300 ppm rise in atmospheric CO_2 typically boosts plant growth by 30% (Mishra *et al.*, 2019).

Lamichaney *et al.* (2021) evaluated the effects of 600 ppm CO_2 on chickpea cv. JG 14 in open-top chambers during 2017–18 and 2018–19. Elevated CO_2 enhanced plant height, leaf and stem biomass, and seed output by 11–18%, pods by 6–10% and seeds per plant by 8–9% over control. However, the exposures suppressed seed protein (7%), total phenol (13%) and thiobarbituric acid reactive compounds (12%) while increasing the starch content and water uptake rate. Despite these alterations, germination, seed vigor, physical characteristics, total soluble sugars and antioxidant enzyme activity were all similar to seeds grown under ambient conditions with normal CO_2 levels. These findings indicate that CO_2 up to 600 ppm can enhance chickpea growth and productivity, although it may affect the nutritional quality. Elevated CO_2 levels are consistently reported in increasing the rate of photosynthesis (15–40%) and decreasing transpiration and stomatal conductance (14–22%) in plants, allowing plants to maintain high photosynthesis rates while minimizing water loss (Ainsworth and Rogers, 2007; Khan and Rizvi, 2020). This reduction in stomatal conductance results in a decrease in overall plant water use, typically by 5–20%. Consequently, elevated CO_2 can impact the hydrological cycle, increasing soil moisture and runoff (Leakey *et al.*, 2009). Elevated CO_2 also leads to changes in the chemical composition of plant tissues, resulting in a significant increase in the sugars and starch per unit leaf area (Ainsworth and Long, 2005; Ainsworth, 2008).

Effects of carbon dioxide on phytonematodes

Since plant nematodes are obligate parasites that require live plant tissue for feeding, reproduction and survival, CO_2-induced changes will greatly increase the availability of root mass for feeding, leading to significant increases in nematode reproduction and ultimately the population in the soil as well as the plant tissue. A study by Zhou *et al.* (2022) showed that elevated atmospheric CO_2 can lead to an average increase of 22% abundance of plant nematodes in croplands and grasslands. This was attributed to higher CO_2 levels which typically decreased the stomatal conductance, leading to reduced water absorption from the soil and more soil water. Secondly, increased CO_2 can boost above-ground shoot and root growth, resulting in more plant litter as well as moisture under the trees (Zhou *et al.*, 2022).

Plant-parasitic and free-living nematodes have been found to respond differently to CO_2 (Banerjee and Hallem, 2020). Tomato cultivars susceptible (Pusa Ruby) and resistant (SL-120) to *M. incognita* race one were exposed to 550 ppm CO_2. Giant cells were formed in both genotypes, but peroxidase and superoxide dismutase activity increased significantly, and the nematode was able to develop and complete its life cycle in both susceptible and resistant cultivars under elevated CO_2 levels (Berliner *et al.*, 2012; 2014; 2023). Khan (2015) and Rizvi *et al.* (2015) observed significant increases in disease severity, reproduction and soil populations of *M. incognita* on cucurbits under 450–500 ppm concentration of CO_2 in open-top chambers.

Fly Ash and Its Effects on Plants and Nematodes

Global electricity generation relies heavily on coal to run thermal power stations (Wang *et al.*, 2016; Li *et al.*, 2018). Around 240–300 million tonnes of coal is burnt annually to generate electricity worldwide (Nyashina *et al.*, 2018; Dan *et al.*, 2021). During coal burning, fly ash as a waste product is produced in huge amounts, around 30–40% of the coal depending on the ash content of the coal and the burning temperature in the furnace (Sharonova *et al.*, 2008; Duan *et al.*, 2012). During 2018, 130 million tonnes of fly ash was produced globally (Yao *et al.*, 2015; Giergiczny, 2019). Fly ash is collected in huge ash ponds, from where it is transported for disposal purposes, e.g. landfill, brick production, cement production, etc. (Dwivedi and Jain, 2014; Yousuf *et al.*, 2020). A considerable amount of fly ash also reaches agricultural fields by deliberate application to the soil as a fertilizer or for neutralizing soil acidity because fly ash is alkaline in nature and contains high concentrations of several plant nutrients (Khan and Khan, 1997; Sahu *et al.*, 2017; Varshney *et al.*, 2022). Fly ash is a grey to black powder comprising fine particles typically above 20 µm in diameter and has a bulk density of 0.54–0.86 g/cm^3 (Yao *et al.*, 2015; Lanzerstorfer, 2018).

Effects of fly ash on plants

Fly ash is quite rich in phosphorus, potassium, magnesium, zinc, iron, manganese, etc. (excluding nitrogen). Fly ash can improve soil texture and fertility, which may promote the growth and yield of plants (Khan *et al.*, 1996b; Yeledhalli *et al.*, 2007; Dhadse *et al.*, 2008; Parab *et al.*, 2012). Fly ash also contains important oxide components including SiO_2, Al_2O_3, CaO, MgO, Na_2O and TiO_2 (Adriano *et al.*, 1980; Ahmaruzzaman, 2010; Bartoňová, 2015). A considerable proportion of fly ash escapes from the stacks and forms a major constituent of coal smoke. The fly ash particles become heavier by absorbing moisture present in the air, and fall on the plant foliage and soil. Khan and Khan (1996b) observed that 30–40% fly ash in soil improved the soil characteristics and plant growth and yield of tomato. Totawat *et al.* (2002) found that fly ash application enhanced the physical and chemical properties of soil and increased wheat yield. Fly ash application also improved the leaf chlorophylls of potato plants due to the nutrients present in the fly ash (Lee, 2007). A significant improvement was reported in the plant growth and yield of cucumber (Pasha *et al.*, 1990), winter ornamentals (Khan, 2001, 2003; Khan and Abdusslam, 2001) and rice (Sikka and Kansal, 1995) by soil amendment with fly ash at varying levels of 10%, 20% and 30%.

The effects of field application of fly ash have also been evaluated on crop production, and it was found that soil application of 2–3 tonne fly ash/ha enhanced the yield of tomato, chilli and eggplant by 19–27% (Khan and Ghadirpour, 1999). Similar plant growth-promoting effects were also recorded on wheat crop (Tripathy and Sahu, 1997). Seed germination of mung bean and okra was found to be increased in soil amended with fly ash, with the highest enhancement in 25% fly ash, while the 50% fly ash treatment gave the greatest increase in plant growth (Vinay and Dwivedi, 1999). However, 60–80% fly ash caused a negative impact on plant growth, yield, chlorophylls, protein content, etc., of *Leucaena leucocephala* (Gupta *et al.*, 2000). Seed germination and plant growth of pea and garden vetch increased in soil amended with 25% and 50% fly ash, while above 50% fly ash treatments became phytotoxic (Singh *et al.*, 1994; Kumar *et al.*, 2000). Soil amendment with 5–20% fly ash enhanced the grain and straw yield of pearl millet (Grewal *et al.*, 2001). The beneficial effects of fly ash application in soil have also been observed on potatoes (Raghav, 2006), rice (Lee *et al.*, 2006; Nayak *et al.*, 2015), maize (Singh and Sukul, 2019), tomato (Khan and Khan, 1996c; Dominic *et al.*, 2022), cucumber (Ajaz and Tiyagi, 2003), pepper (Katiyar *et al.*, 2012), pumpkin (Ahmad *et al.*, 2021), wheat (Singh *et al.*, 2002; Kausar and Khan, 2009; Dhindsa *et al.*, 2016) and cowpea (Rai and Kehri, 2017), and a number of seasonal ornamental plants (Khan and Abdusslam, 2001; Khan, 2003).

Effects of fly ash on phytonematodes

The application of fly ash can significantly impact the physicochemical characteristics of soil, and therefore may potentially influence the parasitism of plant-pathogenic fungi and nematodes (Khan and Ghadirpour, 1999; Khan and Singh, 2001). Khan *et al.* (1997) observed that fly ash application in soil (30–40%) impeded the capability of *M. incognita* to penetrate the roots and decreased disease severity in tomato plants. A decrease in galling and egg mass production of *M. incognita* was recorded in lentils due to 60–70% fly ash treatments (Singh, 1989). Another study reported that increasing levels of fly ash inhibited *M. javanica* development in soybean roots (Singh *et al.*, 2011). Amended soil containing 20–60% fly ash showed an adverse effect on the root penetration of *M. incognita* juveniles and the number of galls and egg masses per root system of tomato (Khan *et al.*, 1997) and pumpkin (Ahmad and Khan, 2016). The development of *M. incognita* or *M. javanica* from J_2 to adult was affected by 10–20% fly ash levels in soil (Ahmad *et al.*, 2021; Bhat *et al.*, 2023; 2024). Tarannum *et al.* (2001) observed that hatching and penetration of *M. javanica* juveniles were significantly inhibited and their development was delayed in chickpea roots. Similarly, Khan and Iram (2005) reported toxic effects of fly ash extracts (30–60%) on egg hatching of *M. incognita* and *M. javanica*. Application of fly ash in *M. incognita*-infested soil decreased the severity of root knot, reproduction and soil populations of the nematode by 35–47% in eggplant, tomato and chilli (Khan and Ghadirpour, 1999).

Fig. 15.2. Leaf of *Liquidambar styraciflua* showing brown necrotic lesions due to exposure to acid rain at pH 2.5. [Figure from Rodríguez-Sánchez *et al.* (2020) used with permission under Open Access license CC BY 4.0.]

Acid Rain and Its Effects on Plants and Nematodes

Acid rain is a consequent effect of fossil fuel burning. Rainwater has always been very important for all kinds of living flora and fauna. Normally, rainwater is slightly acidic with a pH of 5.0–5.5 (Dinrifo *et al.*, 2010; Gaddamwar, 2011), but release of gaseous pollutants such as oxides of sulfur and nitrogen from different anthropogenic sources causes a substantial decrease in the pH of rainwater, and it is commonly called 'acid rain' (Khan and Khan, 1994b). These gases react with water vapor present in the atmosphere to form H_2SO_4 and/or HNO_3. The annual mean pH of rainwater in Virginia was recorded to be 4.3, being around ten times more acidic than normal rain (Avery *et al.*, 2001). In 1950s, acid rain in the form of fog caused the death of thousands of civilians in London (Clark *et al.*, 2001). Acid rain presents a highly significant environmental problem in countries which excessively consume coal and petroleums (Menz and Seip, 2004; Macaulay *et al.*, 2020).

Effects of acid rain on plants

Acid rain causes direct effects on plants (Ahmad and Khan, 2019). The exposed plants may develop interveinal chlorosis and necrosis (Fig. 15.2) with leaf folding and premature senescence (Khan and Khan, 1994e). A field-based experiment showed a significant decline in plant growth and yield of wheat varieties Malviya 213 and Sonalika due to exposure to acid rain at pH 5.0–3.0 (Singh and Agrawal, 2004). Wyrwicka and Skłodowska (2006) reported that acid rain caused a decline in growth and overall health of cucumber plants. Eguagie (2015) reported that acid rain adversely impacted the development of ovules and decreased the protein content of bean seeds (Chehregani and Kavianpour, 2007). The harmful effects of acid rain were also observed on chilli (Pal and Kumar, 2000), tomato (Dursun *et al.*, 2002), mung bean, broad bean (Murray *et al.*, 2004), radish (Varshney *et al.*, 2005), potato (Kausar *et al.*, 2005), mustard (Agrawal *et al.*, 2005) and coriander (Kazim, 2007).

Khan and Khan (1994a) reported that simulated acidic rain (SAR) at pH 3.2 caused wilting, chlorosis of leaf margins and white-to-tan colored lesions on the leaf surface. SAR also suppressed plant growth, flowering and fruiting in tomato plants by 13–24%. Fruiting in chilli, tomato and eggplant was significantly decreased by SAR at pH 3.0 (Verma *et al.*, 2010). SAR with increasing acidity levels caused chlorosis, necrosis and leaf folding and reduced various plant growth variables of *Vigna radiata* (Odiyi and Bamidele, 2014), *Capsicum annuum* (Bamidele and Eguagie, 2015) and *V. unguiculata* (Odiyi and Eniola, 2015). Similarly, SAR at pH 3.0 significantly reduced the photosynthesis, transpiration and stomatal conductance in many plant species (Sicher *et al.*, 2015), and induced necrosis, folding and death of leaves, and suppressed plant growth and biomass production, protein, carbohydrates, chlorophylls, etc., of pumpkin plants.

Effects of acid rain on phytonematodes

SAR has been reported to adversely affect the phytonematodes attacking underground parts or above-ground parts, as the acidic rainwater may bathe the plant surface as well as percolate into the soil. Hence, acid rain may affect plants and nematodes directly and may substantially influence the host–parasite relationship. A 66% inhibition in the reproduction and population of root-knot nematode *M. hapla* was observed due to treatments with SAR at pH 3.2 in comparison to pH 6.0 on *Phaseolus vulgaris* (Shriner, 1978). Khalil and Shah (1979) found that the root-knot disease caused by *M. hapla* was significantly decreased in bean plants treated with acidic water at pH 3.2. Khan and Khan (1994b) reported that root galls, egg masses, eggs per egg mass and root penetration of *M. incognita* were suppressed by SAR at pH 3.2 in tomato plants but significantly increased at pH 5.6 over pH 6.8. The combined effects of various doses of SAR at pH 3.0, 4.0 and 5.0 and *A. tritici* led to the suppression of the nematode population and significant enhancement in the severity of seed gall disease in wheat (Kausar and Khan, 2009). Ahmad and Khan (2019) examined the combined impact of varying levels of SAR and 2000 J_2 of *M. incognita* on pumpkin crops and recorded antagonistic effects on the nematode disease and reproduction. However, Bolla and Fitzsimmons (1988) noticed greater white pine specific nematodes (*Bursaphelenchus xylophilus*) on Scot pine seedlings due to exposure to SAR. Asai and Futai (2005) showed that the population of *B. xylophilus* was enhanced in Japanese black pine seedlings (*Pinus thumbergii*) exposed to SAR at pH 3.0 and 2.0.

Ozone and Its Effects on Plants and Nematodes

There are two types of ozone: 'Good-O_3' and 'Bad-O_3.' 'Good-O_3' implies the ozone present in the stratosphere which absorbs ultraviolet (UV) light and protects the Earth from its adverse effects. 'Bad-O_3' is the gas which is formed on the Earth's surface and is called tropospheric ozone. This ozone is injurious to plants and animal. Ozone is typically a secondary air pollutant and is formed by the photochemical reaction between oxides of nitrogen, carbon monoxide and methane or non-methane volatile organic compound under sunlight (UV light). These chemicals are released mainly along with exhausts from automobiles and smoke from industries. The buildup of tropospheric ozone is faster in summer during warmer periods of the day. In this chapter, 'ozone' or 'O_3' implies tropospheric ozone.

Effects of ozone on plants

Ozone is a most injurious gaseous pollutant to all kinds of vegetation (Heck *et al.*, 1986). The plant injury and damage caused by ozone was probably first reported by Richards and Taylor (1965). A year later, Heggestad (1966) demonstrated ozone as the cause of weather fleck disease in tobacco. Like the other gaseous pollutants, ozone diffuses into the plant leaf through the stomata. Thereafter, it reaches palisade and mesophyll cells, and reacts with leaf pigments causing their bleaching or discoloration (Hasan *et al.*, 2021; Nowroz *et al.*, 2024). The stomata in resistant plants may partially close during ozone exposure, whereas in susceptible cultivars they remain open during exposure (Mansfield, 1986; Khan *et al.*, 1996a). The injury on leaves may be characterized by stippling, mottling, chlorosis, etc., but the most characteristic symptoms is the formation of pinhead-like grey-to-brown spots usually on the upper leaf surface (Heck *et al.*, 1982; Khan *et al.*, 1996b; Juráň *et al.*, 2021). The spot or lesions may become light tan, red or black depending on the ozone dose and plant sensitivity. The leaves of citrus, grapes, pines, etc., may dehisce prematurely (Saborit, 2009; Zhu *et al.*, 2019; Rahimi *et al.*, 2020). The injured mesophyll cells may remain alive; but their chloroplasts and chlorophylls are disrupted (Khan and Khan, 1994d; Calatayud and Barreno, 2004; Tiwari *et al.*, 2018). Ozone may also produce light green chlorotic irregularly dispersed spots with a few large areas or completely covering the leaf followed by leaf fall (Villányi *et al.*, 2013). Environmental factors are critical in determining plant sensitivity to ozone (Cho *et al.*, 2011; Grulke and Heath, 2020).

Research shows that leaf injury may become visible at 100 ppb O_3 for 7 h/day for 3 days under ambient conditions or 75–100 ppb O_3 for 5–7 hour on alternate days for 15 days under control conditions (Khan and Khan, 1998a; 1998b; 1998c). Various researchers have reported suppression of photosynthesis in plants under ozone stress at high or low concentrations (Reich, 1983; Noormets *et al.*, 2010). Ozone exposure has been found to suppress growth under ambient or glasshouse conditions in various crops such as clover (González-Fernández *et al.*, 2008), soyabean (Li *et al.*, 2021; Pei *et al.*, 2024), potato (Vandermeiren *et al.*, 2005; Szpunar-Krok *et al.*, 2020), carrot (Paulikienė *et al.*, 2020), cotton (Zouzoulas *et al.*, 2009; Chaudhary and Rathore, 2021), pepper (Horvitz and Cantalejo, 2012; Ummat *et al.*, 2018), tomato (Mina *et al.*, 2010; Díaz-López *et al.*, 2022) and a number of other crops (Heck *et al.*, 1982; Emberson, 2020; Sahu, 2022). Research has further shown that <60 ppb O_3 may reduce plant growth and yield up to 34% and 24% under ambient and controlled conditions, respectively (Morgan *et al.*, 2003). Ozone and SO_2 often co-occur especially in areas experiencing extensive combustion of petroleum. Both gases in co-occurrence cause greater phytotoxic effects (Weber *et al.*, 1979; Reinert, 1984; Khan and Khan, 1994b).

Effect of ozone on phytonematodes

Ozone can also affect plant pathogens including fungi, bacteria and nematodes (Heck *et al.*, 1986; Khan and Khan, 1994a, 1998b, 1999). Ozone exposures have been found to influence the nematodes infesting a number of host plants (Heagle *et al.*, 1993). The exposure of soybean plants infected with *H. glycines* and *P. minor* to 490 and 655 μg O_3/m^3 suppressed nematode development and reproduction (Weber *et al.*, 1979). However, *B. longicaudatus* infesting soyabean was not affected by these concentrations. The reproduction of *P. penetrans* in soybean plants was also not affected by 490 or 655 μg O_3/m^3 (Weber *et al.*, 1979). The development of *A. fragariae* was prevented due to exposure to 490 and 655 μg O_3/m^3 (Weber *et al.*, 1979). The ozone exposures caused necrotic injuries and damage to the begonia leaves. The injury symptoms apparently were greater on the leaves which were prominently damaged by *A. fragariae*.

Bisessar and Palmer (1984) studied the influence of ozone on *M. hapla* on tobacco treated or not treated with acetylene diurea (EDU). The tobacco plants grown in ambient air containing 80 ppb O_3 exhibited suppression of plant growth irrespective of whether inoculated or not inoculated with *M. hapla*. However, nematode-inoculated plants showed significantly greater gas injury than uninoculated plants. Interestingly, when ozone-injured plants inoculated with the nematode were sprayed with EDU they developed 20% fewer galls compared to the non-sprayed plants. Similarly, Singh *et al.* (1996b) also observed a synergistic interaction between ozone and *M. incognita* on soybean. They recorded severe chlorosis and necrosis on the leaves of nematode-infected plants exposed to 0.14 ppm O_3 for 3 hour on alternate days. The disease severity in terms of galls per root system was increased on ozone-exposed plants. However, nematode reproduction, such as egg masses per root system and fecundity, was decreased. Exposure of soybean to acute doses of ozone reduced the reproduction of soybean cyst nematode and stubby-root nematode but not sting nematode (Weber *et al.*, 1979). Similar effects were observed for the foliar nematode *A. fragariae* on begonia.

Khan and Khan (1998a) studied the interaction of *M. incognita* and episodic levels of O_3, i.e. 10–20–10, 20–50–20 and 50–100–50 ppb O_3 for 2, 3 and 2 hour (total 7 h/day), every third day for 90 days on tomato. The highest ozone concentration incited chlorosis and necrosis in the form of scattered small oval spots on both nematode-infected and uninfected plants. However, the infected plants developed ozone injury 13% greater than uninfected plants at 100 ppb. The ozone-exposed plant and nematode-infested plants showed extensive opening of leaf stomata. The galls, egg masses and soil population of *M. incognita* were increased by 9–17% in the root zone of tomato plants exposed to 100 ppb O_3 over 20 ppb. However, fecundity of the nematode was decreased by 6.6% and 11% at 50 and 100 ppb O_3, respectively. Singh *et al.* (1996a) also recorded a synergistic interaction between *M. javanica* and ozone on pea that led to a significant decline in leaf chlorophylls, carotenoids, nitrogen content and protein content of chickpea.

Weber *et al.* (1979) observed that combinations of ozone and SO_2 acted only in an additive manner and did not affect nematode pathogenesis. However, Khan and Khan (1997) reported the synergistic interaction between ozone (50, 75 and 100 ppb), SO_2 (50, 75 and 100 ppb) and root-knot nematode on tomato cv. Pusa Ruby. They observed that the symptoms of the leaf injury caused by the combination of ozone and SO_2 were more severe on nematode-infected tomato plants. The inoculation with *M. incognita* resulted in a 14.4% decrease in the yield of tomato, whereas inoculated-exposed plants exhibited around 27% yield decline. Shew *et al.* (1982) evaluated the interactive effects of *P. penetrans*, ozone and SO_2 on tomato cv. Walter. The plants were exposed to air containing 0.2 and 0.4 μl O_3/l and 0.2, 0.4 or 0.8 μl SO_2/l, either singly or combined. The nematode checked the adverse effects of the gases on the growth of axillary shoots and dry weight of stems and shoots of tomato plants. On the other hand, the mixture of ozone (0.2 μl) and SO_2 (0.8 μl/l) promoted nematode reproduction. The rest of the O_3/SO_2 treatments did not significantly influence the nematode disease or its reproduction.

Conclusion and Future Prospects

In the developing and underdeveloped countries in Asia, Africa and the Americas, coal continues to fulfil the major energy requirements to run power-generating units and industries. Combustion of coal and petroleum causes the generation and release of high levels of SO_x, NO_x, CO_x, O_3, acid rain, particulate air pollutants, etc., into the environment. The crops grown around the pollution sources exhibit visible foliar injury and suppression of plant growth and yield, which are generally not recognized by the crop growers. In addition, air pollutants can potentially influence host–pathogen interactions as well the severity of the diseases caused by fungi, nematodes, etc. However, limited research has been done on air pollutant and plant nematode interactions and the resulting disease complexes. Hence, concerted research effects are required to examine and ascertain their relationships. Even fundamental aspects like the varying response of nematodes to air pollutants and the specific mechanism(s) that govern these interactions are not known. It is necessary to devise and conduct experiments that address these issues to obtain persuasive explanations on the mechanism of biotic–abiotic stress interactions. Secondly, host–parasite relationships where harmful gaseous exposures increase disease severity require more thorough investigations to ascertain the impact. The reports revealing the higher susceptibility of plants to nematodes in polluted environments are alarming and warrant adequate attention of researchers to these important issues. It is also not understood how do nematode-infected plants become more sensitive to SO_2 or ozone. If such synergistic relationships are ascertained, the levels of SO_2, ozone, etc., considered safer for crop production will need to be revised.

References

Adak, P. and Kour, N. (2021) A review on the effects of environmental factors on plants tolerance to air pollution. *Journal of Environmental Treatment Techniques* 9(4), 839–848.

Adriano, D.C., Page, A.L., Elseewi, A.A., Chang, A.C. and Straughan, I. (1980) Utilization and disposal of fly ash and other coal residues in terrestrial ecosystems: A review. *Journal of Environmental Quality* 9(3), 333–344. DOI: 10.2134/jeq1980.00472425000900030001x.

Agrawal, M. (2003) Plant responses to atmospheric sulphur. In: Abrol, Y.P. and Ahmad, A. (eds) *Sulphur in Plants*. Springer, Dordrecht, The Netherlands, pp. 279–293. DOI: 10.1007/978-94-017-0289-8_15.

Agrawal, S., Raghav, D. and Khan, A.A. (2005) An evaluation of the impact of simulated acid rain on the growth of mustard in pots. *Test of Agrochemicals and Cultivars* 21, 25–26.

Ahmad, G. and Khan, A.A. (2016) Fly ash effect on hatching, mortality and penetration of root-knot nematode (*Meloidogyne incognita*) in pumpkin roots. *International Journal of Environment* 5(3), 66–73. DOI: 10.3126/ije.v5i3.15705.

Ahmad, G. and Khan, A.A. (2019) Eco-friendly use of fly ash for the management of root-knot nematode and acid rain in pumpkin crop. *The International Journal of Biotechnology* 8(1), 93–103. DOI: 10.18488/journal.57.2019.81.93.103.

Ahmad, G., Khan, A.A. and Mohamed, H.I. (2021) Impact of the low and high concentrations of fly ash amended soil on growth, physiological response, and yield of pumpkin (*Cucurbita moschata* Duch Ex Poiret L.). *Environmental Science and Pollution Research* 28(14), 17068–17083.

Ahmaruzzaman, M. (2010) A review on the utilization of fly ash. *Progress in Energy and Combustion Science* 36(3), 327–363.

Ainsworth, E.A. (2008) Rice production in a changing climate: A meta-analysis of responses to elevated carbon dioxide and elevated ozone concentration. *Global Change Biology* 14(7), 1642–1650. DOI: 10.1111/j.1365-2486.2008.01594.x.

Ainsworth, E.A. and Long, S.P. (2005) What have we learned from 15 years of free-air CO_2 enrichment (FACE)? A meta-analytic review of the responses of photosynthesis, canopy properties and plant production to rising CO_2. *New Phytologist* 165(2), 351–372.

Ainsworth, E.A. and Rogers, A. (2007) The response of photosynthesis and stomatal conductance to rising [CO2]: Mechanisms and environmental interactions. *Plant, Cell and Environment* 30(3), 258–270. DOI: 10.1111/j.1365-3040.2007.01641.x.

Ajaz, S. and Tiyagi, S. (2003) Effect of different concentrations of fly ash on the growth of cucumber plant, cucumis sativus. *Archives of Agronomy and Soil Science* 49(4), 457–461. DOI: 10.1080/03650340310001596428.

Alvarez, E., Blanco, J., Knapp, C., Olivares, J. and Salvador, L. (2000) Pilot plant performance of a SO_2 to SO_3 oxidation catalyst for flue-gas conditioning. *Catalysis Today* 59(3–4), 417–422. DOI: 10.1016/S0920-5861(00)00307-2.

Asai, E. and Futai, K. (2005) Effects of inoculum density of pinewood nematode on the development of pine wilt disease in Japanese black pine seedlings pretreated with simulated acid rain. *Forest Pathology* 35(2), 135–144. DOI: 10.1111/j.1439-0329.2004.00396.x.

Avery, G.B.J., Tang, Y., Kieber, R.J. and Willey, J.D. (2001) Impact of recent urbanization on formic and acetic acid concentrations in coastal North Carolina rainwater. *Atmospheric Environment* 35(19), 3353–3359. DOI: 10.1016/S1352-2310(00)00328-9.

Bamidele, J.F. and Eguagie, M.O. (2015) Ecophysiological response of *Capsicum annuum* L. exposed to simulated acid rain. *Nigerian Journal of Biotechnology* 30(1), 48. DOI: 10.4314/njb.v30i1.6.

Banerjee, N. and Hallem, E.A. (2020) The role of carbon dioxide in nematode behaviour and physiology. *Parasitology* 147(8), 841–854. DOI: 10.1017/S0031182019001422.

Bao, X., Zhou, W., Xu, L. and Zheng, Z. (2023) A meta-analysis on plant volatile organic compound emissions of different plant species and responses to environmental stress. *Environmental Pollution* 318, 120886. DOI: 10.1016/j.envpol.2022.120886.

Barret, R.W. and Benedict, H.M. (1970) Sulfur dioxide. In: Jacobson, J.A. and Hill, A.C. (eds) *Recognition of Air Pollution Injury – A Pictorial Atlas*. Air Pollution Control Association, Pittsburgh, PA.

Bartoňová, L. (2015) Unburned carbon from coal combustion ash: An overview. *Fuel Processing Technology* 134, 136–158. DOI: 10.1016/j.fuproc.2015.01.028.

Bassus, V.W. (1968) Über wirkungen von industrieexhalaten auf den nematodenbesatz im boden von kiefernwäldern. *Pedobiologia* 8(1), 289–295. DOI: 10.1016/S0031-4056(23)00334-7.

Bender, J. and Weigel, H.J. (2011) Changes in atmospheric chemistry and crop health: A review. *Agronomy for Sustainable Development* 31(1), 81–89. DOI: 10.1051/agro/2010013.

Benemann, J.R. (1993) Utilization of carbon dioxide from fossil fuel-burning power plants with biological systems. *Energy Conversion and Management* 34(9–11), 999–1004. DOI: 10.1016/0196-8904(93)90047-E.

Benzing, D.H., Arditti, J., Nyman, L.P., Temple, P.J. and Bennett, J.P. (1992) Effects of ozone and sulfur dioxide on four epiphytic bromeliads. *Environmental and Experimental Botany* 32(1), 25–32. DOI: 10.1016/0098-8472(92)90026-X.

Berliner, J., Ganguly, A.K., Kamra, A., Sirohi, A. and Pathak, H. (2012) Effect of elevated carbon dioxide on life cycle of root-knot nematode, *Meloidogyne incognita* in tomato. *Indian Journal of Nematology* 42(2), 111–117.

Berliner, J., Ganguly, A.K., Sirohi, A. and Kamra, A. (2014) Effect of elevated CO_2 on host–parasite relationship of tomato and root-knot nematode, *Meloidogyne incognita* race 1. *Annals of Plant Protection Sciences* 22(1), 178–182.

Berliner, J., Ganguly, A.K., Kamra, A., Sirohi, A. and Dinesh, V.P. (2023) Effect of elevated carbon dioxide on population growth of root-knot nematode, *Meloidogyne incognita* in tomato. *Indian Phytopathology* 76(1), 309–315. DOI: 10.1007/s42360-022-00584-8.

Bhagawati, B., Khan, M.R., Neog, P.P. and Kurulkar, U. (2023) Major nematode problems in direct seeded rice and their management. In: Khan, M.R. (ed.) *Novel Biological and Biotechnological Applications in Plant Nematode Management*. Springer Nature, Singapore, pp. 515–537.

Bhat, A.A., Shakeel, A., Kesba, H.H., Handoo, Z.A. and Khan, A.A. (2023) Elucidating the role of fly ash in root-knot nematode (*Meloidogyne incognita*) suppression on okra (*Abelmoschus esculentus* L.): Insights into cellular viability and host defence system. *Physiological and Molecular Plant Pathology* 126, 102042. DOI: 10.1016/j.pmpp.2023.102042.

Bhat, A.A., Shakeel, A., Bhat, A.H., Alajmi, M.F., Khan, A.A. *et al.* (2024) Exploiting fly ash as an ecofriendly pesticide/nematicide on *Abesmoschus esculuntus*: Insights into soil amendment-induced antioxidant fight against nematode mediated ROS. *Chemosphere* 358, 142143. DOI: 10.1016/j.chemosphere.2024.142143.

Bhatia, A., Jain, N. and Pathak, H. (2013) Methane and nitrous oxide emissions from Indian rice paddies, agricultural soils and crop residue burning. *Greenhouse Gases: Science and Technology* 3(3), 196–211. DOI: 10.1002/ghg.1339.

Bisessar, S. and Palmer, K.T. (1984) Ozone, antioxidant spray and *Moidogyne hapla* effects on tobacco . *Atmospheric Environment* 18(5), 1025–1027. DOI: 10.1016/0004-6981(84)90080-5.

Bolla, R.I. and Fitzsimmons, K. (1988) Effect of simulated acid rain on *Bursaphelenchus xylophilus* infection of pine seedlings. *Journal of Nematology* 20(4), 590–598.

Brahmachari, S. and Kundu, S. (2017) SO_2 stress: Its effect on plants, plant defense responses and strategies for developing enduring resistance. *International Journal of Advanced Research in Science, Engineering and Technology* 4(7), 303–309.

Calatayud, A. and Barreno, E. (2004) Response to ozone in two lettuce varieties on chlorophyll a fluorescence, photosynthetic pigments and lipid peroxidation. *Plant Physiology and Biochemistry* 42(6), 549–555. DOI: 10.1016/j.plaphy.2004.05.002.

Chaudhary, I.J. and Rathore, D. (2021) Assessment of ozone toxicity on cotton (*Gossypium hirsutum* L.) cultivars: Its defensive system and intraspecific sensitivity. *Plant Physiology and Biochemistry* 166, 912–927. DOI: 10.1016/j.plaphy.2021.06.054.

Chehregani, A. and Kavianpour, F. (2007) Effects of acid rain on the developmental stages of ovules and seed proteins in bean plants (*Phaseolus vulgaris* L.). *American Journal of Plant Physiology* 2(6), 367–372. DOI: 10.3923/ajpp.2007.367.372.

Cho, K., Tiwari, S., Agrawal, S.B., Torres, N.L., Agrawal, M. *et al.* (2011) Tropospheric ozone and plants: absorption, responses, and consequences. *Reviews of Environmental Contamination and Toxicology* 212, 61–111. DOI: 10.1007/978-1-4419-8453-1_3.

Clapperton, M.J. and Reid, D.M. (1994) Effects of sulphur dioxide (SO_2) on growth and flowering of SO_2-tolerant and non-tolerant genotypes of *Phleum pratense*. *Environmental Pollution* 86(3), 251–258. DOI: 10.1016/0269-7491(94)90165-1.

Clark, W.C., Jäger, J., Cavender-Bares, J. and Dickson, N.M. (2001) Acid rain, ozone depletion, and climate change: An historical overview. In: Social Learning Group, d (ed.) *Learning to Manage Global Environmental Risks: A Comparative History of Social Responses to Climate Change, Ozone Depletion, and Acid Rain*. MIT Press, Cambridge, MA.

Clarkson, D.T., Hawkesford, M.J. and Davidian, J.C. (1993) Membrane and long distance transport of sulphate. In: Kok, L.J., Stulen, I., Rennenberg, H., Brunold, C. and Rauser, W.E. (eds) *Sulfur Nutrition and Assimilation in Higher Plants; Regulatory Agricultural and Environmental Aspects*. SPB Academic Publishing, The Hague, The Netherlands, pp. 3–19.

Constantinidou, H.A. and Kozlowski, T.T. (1979) Effects of sulfur dioxide and ozone on *Ulmus americana* seedlings. I. Visible injury and growth. Canadian Journal of Botany 57(2), 170–175. DOI: 10.1139/b79-025.

Cormis, L. (1973) Effect of sulfurous air pollution on the vegetation. *Annales Des Mines*Nov/Dec, 23–32.

Craine, J.M., Wedin, D.A. and Reich, P.B. (2001) The response of soil CO_2 flux to changes in atmospheric CO_2, nitrogen supply and plant diversity. Global Change Biology 7(8), 947–953. DOI: 10.1046/j.1354-1013.2001.00455.x.

Cram, W.J. (1990) Uptake and transport of sulphate. In: Rennenberg, H., Brunold, C., De Kok, L.J. and Stulen, I. (eds) *Sulfur Nutrition and Assimilation in Higher Plants; Regulatory Agricultural and Environmental Aspects*. SPB Academic Publishing, The Hague, The Netherlands, pp. 3–11.

Dahal, K. and Vanlerberghe, G.C. (2018) Growth at elevated CO_2 requires acclimation of the respiratory chain to support photosynthesis. *Plant Physiology* 178(1), 82–100. DOI: 10.1104/pp.18.00712.

Dan, A.K., Bhattacharjee, D., Ghosh, S., Behera, S.S., Bindhani, B.K. *et al.* (2021) Prospective utilization of coal fly ash for making advanced materials. In: Jyothi, R.K. and Parthi, P.K. (eds) *Clean Coal Technologies: Beneficiation, Utilization, Transport Phenomena and Prospective*. Springer, Cham, Switzerland, pp. 511–531. DOI: 10.1007/978-3-030-68502-7_20.

De Kok, L.J. (1990) Sulfur metabolism in plants exposed to atmospheric sulfur. In: Rennenberg, H., Brunhold, Ch., Kok, L.J. and Stulen, I. (eds) *Sulfur Nutrition and Sulfur Assimilation in Higher Plants; Fundamental, Environmental and Agricultural Aspects*. SPB Academic Publishing, The Hague, The Netherlands, pp. 111–130.

De Kok, L.J. and Tausz, M. (2001) The role of glutathione in plant reaction and adaptation to air pollutants. In: Grill, D., Tsausz, M. and Kok, L.J. (eds) *Significance of Glutathione to Plant Adaptation to the Environment*. Kluwer, Dordrecht, The Netherlands, pp. 185–201. DOI: 10.1007/0-306-47644-4_8.

Decraemer, W. and Geraert, E. (2013) Ectoparasitic nematodes. In: Perry, R.N. and Moens, M. (eds) *Plant Nematology*. CAB International, Wallingford, UK, pp. 179–216.

De Souza, A.P., Gaspar, M., Silva, E.A., Ulian, E.C., Waclawovsky, A.J. *et al.* (2008) Elevated CO_2 increases photosynthesis, biomass and productivity, and modifies gene expression in sugarcane. *Plant, Cell & Environment* 31(8), 1116–1127. DOI: 10.1111/j.1365-3040.2008.01822.x.

Dhadse, S., Kumari, P. and Bhagia, L.J. (2008) Fly ash characterization, utilization and government initiatives in India – a review. *Journal of Scientific and Industrial Research* 67(1), 11–18.

Dhindsa, H.S., Sharma, R.D. and Kumar, R. (2016) Role of fly ash in improving soil physical properties and yield of wheat (*Triticum aestivum*). *Agricultural Science Digest – A Research Journal* 36(2), 97–101. DOI: 10.18805/asd.v36i2.10626.

Dhir, B. (2016) Air pollutants and photosynthetic efficiency of plants. *Plant Responses to Air Pollution* 71–84.

Dinrifo, R.R., Babatunde, S.O.E., Bankole, Y.O. and Demu, Q.A. (2010) Physico-chemical properties of rain water collected from some industrial areas of Lagos state Nigeria. *European Journal of Scientific Research* 41(3), 383–390.

Dominic, S., Hussain, A.I., Saleem, M.H., Alshaya, H., Jan, B.L. *et al.* (2022) Variation in the primary and secondary metabolites, antioxidant and antibacterial potentials of tomatoes, grown in soil blended with different concentration of fly ash. *Plants* 11(4), 551. DOI: 10.3390/plants11040551.

Duan, L., Liu, D., Chen, X. and Zhao, C. (2012) Fly ash recirculation by bottom feeding on a circulating fluidized bed boiler co-burning coal sludge and coal. *Applied Energy* 95, 295–299. DOI: 10.1016/j.apenergy.2012.02.063.

Duccer, W.M. and Ting, I.P. (1970) Air pollution oxidants – their effects on metabolic processes in plants. *Annual Review of Plant Physiology* 21(1), 215–234. DOI: 10.1146/annurev.pp.21.060170.001243.

Dursun, A., Kumlay, A.M., Yildirim, E. and Guvvenc, I. (2002) Effect of simulated acid rain on plant growth and yield of tomato. *Acta Horticulture* 579, 245–248. DOI: 10.17660/ActaHortic.2002.579.40.

Dwivedi, A. and Jain, M.K. (2014) Fly ash – waste management and overview: A review. *Recent Research in Science and Technology* 6(1), 30–35.

Díaz-López, M., Siles, J.A., Ros, C., Bastida, F. and Nicolás, E. (2022) The effects of ozone treatments on the agro-physiological parameters of tomato plants and the soil microbial community. *Science of the Total Environment* 812, 151429. DOI: 10.1016/j.scitotenv.2021.151429.

Eguagie, M.O. (2015) Effects of simulated acid rain on the growth, yield and mineral nutrient relations of *Solanum lycopersicum* L. *European Journal of Biotechnology and Bioscience* 3(11), 15–18.

Ellsworth, D.S., Anderson, I.C., Crous, K.Y., Cooke, J., Drake, J.E. *et al.* (2017) Elevated CO_2 does not increase eucalypt forest productivity on a low-phosphorus soil. *Nature Climate Change* 7(4), 279–282. DOI: 10.1038/nclimate3235.

Emberson, L. (2020) Effects of ozone on agriculture, forests and grasslands. *Philosophical Transactions of the Royal Society A* 378(2183), 20190327. DOI: 10.1098/rsta.2019.0327.

Emberson, L.D., Ashmore, M.R., Murray, F., Kuylenstierna, J.C.I., Percy, K.E. *et al.* (2001) Impacts of air pollutants on vegetation in developing countries. *Water, Air, and Soil Pollution* 130(1/4), 107–118. DOI: 10.1023/A:1012251503358.

Falkowska, L. (2016) Environmental characteristics of gaseous pollutants and related adverse health effects. In: Pastuszka, J.S. (ed.) *Synergic Influence of Gaseous, Particulate, and Biological Pollutants on Human Health*. CRC Press, Boca Raton, FL, pp. 3–38. DOI: 10.1201/b19592-3.

Gaddamwar, A.G. (2011) Analytical study of rain water for the determination of polluted or unpolluted zone. *International Journal of Environmental Sciences* 1(6), 1317–1322.

Gibson, S.I. (2005) Control of plant development and gene expression by sugar signaling. *Current Opinion in Plant Biology* 8(1), 93–102.

Giergiczny, Z. (2019) Fly ash and slag. *Cement and Concrete Research* 124, 105826. DOI: 10.1016/j.cemconres.2019.105826.

González-Fernández, I., Bass, D., Muntifering, R., Mills, G. and Barnes, J. (2008) Impacts of ozone pollution on productivity and forage quality of grass/clover swards. *Atmospheric Environment* 42(38), 8755–8769. DOI: 10.1016/j.atmosenv.2008.08.030.

Grewal, K.S., Yadav, P.S., Mehta, S.C. and Oswal, M.C. (2001) Direct and residual effect of fly ash application to soil on crop field and soil properties. *Crop Research* 21, 60–65.

Grulke, N.E. and Heath, R.L. (2020) Ozone effects on plants in natural ecosystems. *Plant Biology* 22, 12–37. DOI: 10.1111/plb.12971.

Gupta, M.C. and Ghouse, A.K.M. (1987) The effects of coal-smoke pollutants on the growth, yield and leaf epidermal features of *Abelmoschus esculentus* Moench. *Environmental Pollution* 43(4), 263–270. DOI: 10.1016/0269-7491(87)90179-5.

Gupta, K., Kumar, R., Baruah, K.K., Hazarika, S., Karmakar, S. *et al.* (2021) Greenhouse gas emission from rice fields: A review from Indian context. *Environmental Science and Pollution Research International* 28(24), 30551–30572. DOI: 10.1007/s11356-021-13935-1.

Gupta, M., Kumar, A. and Yunus, M. (2000) Effect of fly-ash on metal composition and physiological responses in *Leucaena leucocephala* (Lamk.) De Wit. *Environmental Monitoring and Assessment* 61(3), 399–406. DOI: 10.1023/A:1006169716006.

Haque, Z. and Khan, M.R. (2021) *Handbook of Invasive Plant-Parasitic Nematodes*. CAB International, Wallingford, UK.

Hasan, M.M., Rahman, M.A., Skalicky, M., Alabdallah, N.M., Waseem, M. *et al.* (2021) Ozone induced stomatal regulations, MAPK and phytohormone signaling in plants. *International Journal of Molecular Sciences* 22(12), 6304.

Heagle, A.S. (1973) Interactions between air pollutants and plant parasites. *Annual Review of Phytopathology* 11(1), 365–388. DOI: 10.1146/annurev.py.11.090173.002053.

Heagle, A.S., Spencer, S. and Letchworth, M.B. (1979) Yield response of winter wheat to chronic doses of ozone. *Canadian Journal of Botany* 57(19), 1999–2005. DOI: 10.1139/b79-250.

Heagle, A.S., Miller, J.E., Sherrill, D.E. and Rawlings, J.O. (1993) Effects of ozone and carbon dioxide mixtures on two clones of white clover. *New Phytologist* 123(4), 751–762. DOI: 10.1111/j.1469-8137.1993.tb03786.x.

Heck, W.W., Taylor, O.C., Adams, R., Bingham, G., Miller, J. *et al.* (1982) Assessment of crop loss from ozone. *Journal of the Air Pollution Control Association* 32(4), 353–361. DOI: 10.1080/00022470.1982.10465408.

Heck, W.W., Heagle, A.S. and Shriner, D.S. (1986) Effect on vegetation: Native, crops, forests. In: Stern, A.S. (ed.) *Air Pollution*. Academic Press, New York, pp. 247–350. DOI: 10.1016/B978-0-08-092605-6.50011-9.

Heggestad, H.E. (1966) Ozone as a tobacco toxicant. *Journal of the Air Pollution Control Association* 16(12), 691–694. DOI: 10.1080/00022470.1966.10468534.

Heggestad, H.E., Bennett, J.H., Lee, E.H. and Douglass, L.W. (1986) Effects of increasing doses of sulfur dioxide and ambient ozone on tomatoes: Plant growth, leaf injury, elemental composition, fruit yields, and quality. *Phytopathology* 76(12), 1338. DOI: 10.1094/Phyto-76-1338.

Hesketh, H.E. (1979) Introduction to air pollution. In: Wang, L.K. and Pereira, N.C. (eds) *Air and Noise Pollution Control*. Humana Press, Totowa, NJ, pp. 3–39. DOI: 10.1007/978-1-4612-6236-7_1.

Horvitz, S. and Cantalejo, M.J. (2012) Effects of ozone and chlorine postharvest treatments on quality of fresh-cut red bell peppers. *International Journal of Food Science & Technology* 47(9), 1935–1943. DOI: 10.1111/j.1365-2621.2012.03053.x.

Inumaru, J., Hasegawa, T., Shirai, H., Nishida, H., Noda, N. *et al*. (2021) Fossil fuels combustion and environmental issues. In: Ozawa, M. and Asano, H. (eds) *Advances in Power Boilers (JSME Series in Thermal and Nuclear Power Generation)*. Elsevier, Amsterdam, pp. 1–56.

Iqbal, M., Mahmooduzzafar, M., Nighat, F. and Khan, P.R. (2010) Photosynthetic, metabolic and growth responses of *Triumfetta rhomboidea* to coal-smoke pollution at different stages of plant ontogeny. *Journal of Plant Interactions* 5(1), 11–19. DOI: 10.1080/17429140902996165.

Jacobson, S. and Hill, A. (1970) *Recognition of Air Pollution Injury to Vegetation – A Pictorial Atlas, Air Pollution Control Association*. Air Pollution Control Association, Pittsburgh, PA.

Jha, S. and Yadav, A. (2023) Plants response to SO_2 or acid deposition. In: Husen, A. (ed.) *Plants and Their Interaction to Environmental Pollution*. Elsevier, New York, pp. 99–108.

Joshi, P.C. and Swami, A. (2009) Air pollution induced changes in the photosynthetic pigments of selected plant species. *Journal of Environmental Biology* 30(2), 295–298.

Juráň, S., Grace, J. and Urban, O. (2021) Temporal changes in ozone concentrations and their impact on vegetation. *Atmosphere* 12(1), 82. DOI: 10.3390/atmos12010082.

Kaiser, G., Martinoia, E., Schröppel-Meier, G. and Heber, U. (1989) Active transport of sulfate into the vacuole of plant cells provides halotolerance and can detoxify SO_2. *Journal of Plant Physiology* 133(6), 756–763. DOI: 10.1016/S0176-1617(89)80085-9.

Katiyar, D., Singh, A., Malaviya, P., Pant, D., Singh, P. *et al*. (2012) Impact of fly-ash-amended soil on growth and yield of crop plants. *International Journal of Environment and Waste Management* 10(2/3), 150. DOI: 10.1504/IJEWM.2012.048362.

Kausar, S. and Khan, A.A. (2009) Effect of SO_2 on mortality of seed gall nematode, *Anguina tritici* and interactive effect of both on wheat. *Trends in Biosciences* 2, 44–46.

Kausar, S., Khan, A.A. and Raghav, D. (2005) Interaction of sulfur dioxide and seed gall nematode, *Anguina tritici*, on wheat. *Journal of Food, Agriculture and Environment* 3, 27–49.

Kausar, S., Hussain, M.A. and Khan, A. (2010) Response of simulated acid rain on morphological, biochemical and leaf epidermal characteristics of wheat. *Trends in Biosciences* 3(1), 34–36.

Kazim, I. (2007) Response of coriander to sulphur dioxide and acid rain. MSc dissertation, Aligarh Muslim University, Aligarh, India.

Kern, D.J. (2021) *Atmospheric Chemistry: An Overview – Ozone, Acid Rain, and Greenhouse Gases. Building STEM Skills Through Environmental Education*. IGI Global, Hershey, PA, pp. 172–218. DOI: 10.4018/978-1-7998-2711-5.ch007.

Khalil, S. and Shah, H.M. (1979) Effect of soil moisture and pH on root-knot disease of tobacco. *Tobacco Research* 5, 129–134.

Khan, M.R. (2001) Use of fly ash for the cultivation of ornamental plants for domestic purpose. In: *Proceedings RAWM-2001*, Indian Association for Environmental Management, pp. 85–89.

Khan, M.R. (2003) Effect of soil application of fly ash on the growth and flowering of seasonal ornamental plants. *Environmental Biology and Conservation* 7, 81–84.

Khan, M.R. (2008) *Plant Nematodes – Methodology, Morphology, Systematics, Biology and Ecology*. CRC Press, Boca Raton, FL. DOI: 10.1201/b10766.

Khan, M.R. (2015) Influence of environmental contamination and climate change on the development of nematode diseases of plants. In: *National Symposium on Nematode Management: A Challenge to Indian Agriculture in the Changing Climate*, Yashwantrao Chavan Academy of Development, Pune, India, January 8–10, p. 8.

Khan, M.R. (2023a) Nematode pests of agricultural crops, a global overview. In: Khan, M.R. (ed.) *Novel Biological and Biotechnological Applications in Plant Nematode Management*. Springer, Singapore, pp. 3–45. DOI: 10.1007/978-981-99-2893-4_1.

Khan, M.R. (2023b) Plant nematodes, an underestimated constraint in the global food production. In: Khan, M.R. and Quintanilla, M. (eds) *Nematode Diseases of Crops and Their Sustainable Management*. Academic Press, Cambridge, MA, pp. 3–26. DOI: 10.1016/B978-0-323-91226-6.00009-2.

Khan, M.R. and Abdusslam, A. (2001) Effects on plant height and flowering of winter ornamental plants when grown in mediums of soil and fly ash. *Annals of Applied Biology (Supplement Test of Agrochemicals and Cultivars)* 22, 44–45.

Khan, M.R. and Ghadirpour, M.H. (1999) Effects of field application of fly ash on the growth and yield of chilli, eggplant and tomato in the presence and absence of root-knot nematode. *Journal of Plant Diseases and Protection* 106(6), 624–632.

Khan, A.A. and Iram, A. (2005) Evaluation of fly ash-extract on hatching and mortality of root-knot nematodes. *Indian Journal of Nematology* 44(1), 115–117.

Khan, M.R. and Jairajpuri, S. (2010a) *Nematode Infestations Part I: Food Crops*. National Academy of Sciences, Allahabad, India.

Khan, M.R. and Jairajpuri, S. (2010b) *Nematode Infestations Part II: Industrial Crops*. National Academy of Sciences, Allahabad, India.

Khan, M.R. and Jairajpuri, S. (2012) *Nematode Infestations Part III: Horticultural Crops*. National Academy of Sciences, Allahabad, India.

Khan, M.R. and Khan, M.W. (1990) Impact of ambient air pollution on vegetable crops. In: *Proceedings 83rd Annual Meeting Air and Waste Management Association.* Paper no. 90-187.9.

Khan, M.R. and Khan, M.W. (1991) Impact of air pollutants emanating from a thermal power plant on tomato. *Journal of the Indian Botanical Society* 70, 373–378. DOI: 10.1016/0269-7491(91)90083-9.

Khan, M.R. and Khan, M.W. (1993a) The interaction of SO_2 and root-knot nematode on tomato. *Environmental Pollution* 81(2), 91–102. DOI: 10.1016/0269-7491(93)90072-v.

Khan, M.W. and Khan, M.R. (1993b) Relationship of plant pathogenic microbes with air pollution. In: Mukerji, G.K. and Singh, V.P. (eds) *Frontiers in Applied Microbiology*. Rastogi and Co, Meerut, India, pp. 153–165.

Khan, M.W. and Khan, M.R. (1994a) Effect of ozone on the conidia of *Sphaerotheca fuliginea*. Indian Phytopatholoy 47, 172–176.

Khan, M.R. and Khan, M.W. (1994b) Effects of simulated acid rain and root-knot nematode on tomato. *Plant Pathology* 43(1), 41–49. DOI: 10.1111/j.1365-3059.1994.tb00551.x.

Khan, M.R. and Khan, M.W. (1994c) Single and interactive effects of root-knot nematode and coal-smoke on okra. *New Phytologist* 126(2), 337–342. DOI: 10.1111/j.1469-8137.1994.tb03953.x.

Khan, M.R. and Khan, W.M. (1994d) Single and interactive effects of O_3 and SO_2 on tomato. *Environmental and Experimental Botany* 34(4), 461–469. DOI: 10.1016/0098-8472(94)90029-9.

Khan, M.R. and Khan, M.W. (1994e) Smoke pollution and the effects of root-knot nematodes on the growth of eggplants. *Plant Pathology* 43(4), 683–693. DOI: 10.1111/j.1365-3059.1994.tb01607.x.

Khan, M.R. and Khan, M.W. (1995a) Effects of NH_3 and root-knot nematode on tomato. *Agriculture, Ecosystems & Environment* 53(1), 71–81. DOI: 10.1016/0167-8809(94)00553-Q.

Khan, M.R. and Khan, M.W. (1995b) Response of okra to ambient air pollution. *Acta Botanica Indica* 2, 7–14.

Khan, M.R. and Khan, M.W. (1996a) Development of root-knots and root nodules on cowpea as influenced by sulphur dioxide. *Nematologia Mediterranea* 24, 33–35.

Khan, M.R. and Khan, M.W. (1996b) Impact of coal-smoke pollution on eggplant in ambient condition. *Journal of the Indian Botanical Society* 75, 187–192.

Khan, M.R. and Khan, M.W. (1996c) The effect of fly ash on plant growth and yield of tomato. *Environmental Pollution* 92(2), 105–111. DOI: 10.1016/0269-7491(95)00098-4.

Khan, M.R. and Khan, M.W. (1997) Effect of root-knot nematode on the sensitivity of tomato to sulphur dioxide and ozone. *Environmental and Experimental Botany* 37, 11–26.

Khan, M.R. and Khan, M.W. (1998a) Effects of sulfur dioxide on the development of powdery mildew of cucumber. *Environmental and Experimental Botany* 40(3), 265–273. DOI: 10.1016/S0098-8472(98)00043-4.

Khan, M.R. and Khan, M.W. (1998b) Interactive effects of ozone and powdery mildew (*Sphaerotheca fuliginea*) on bottle gourd (*Lagenaria siceraria*). *Agriculture, Ecosystems & Environment* 70(2–3), 109–118. DOI: 10.1016/S0167-8809(98)00102-9.

Khan, M.R. and Khan, M.W. (1998c) Interactive effects of ozone and root-knot nematode on tomato. *Agriculture, Ecosystems & Environment* 70(2–3), 97–107. DOI: 10.1016/S0167-8809(98)00103-0.

Khan, M.R. and Khan, M.W. (1999) Effects of intermittent ozone exposures on powdery mildew of cucumber. *Environmental and Experimental Botany* 42(3), 163–171. DOI: 10.1016/S0098-8472(99)00029-5.
Khan, M.R. and Khan, M.W. (2000) Sulphur dioxide effects on plants and pathogens. In: Iqbal, M., Shrivastava, P.S. and Siddiqi, T.O. (eds) *Environmental Hazards. Plants and People*. C.B.S. Publishers, New Delhi, pp. 118–136.
Khan, M.R. and Khan, M.M. (2010) Effect of intermittent exposures of SO_2 on the leaf blight caused by alternaria brassicicola on Indian mustard. *Agriculture, Ecosystems & Environment* 139(4), 728–735. DOI: 10.1016/j.agee.2010.11.002.
Khan, M.R. and Khan, M.M. (2011) Plants response to diseases in sulphur dioxide stressed environment. *Plant Pathology Journal* 10(1), 1–12. DOI: 10.3923/ppj.2011.1.12.
Khan, M.R. and Rizvi, T.F. (2020) Effect of elevated levels of CO_2 on powdery mildew development in five cucurbit species. *Scientific Reports* 10(1), 4986. DOI: 10.1038/s41598-020-61790-w.
Khan, M.R. and Singh, W.N. (2001) Effects of soil application of fly ash on the fusarial wilt on tomato cultivars. *International Journal of Pest Management* 47(4), 293–297. DOI: 10.1080/09670870110052059.
Khan, M.R., Khan, A.A. and Khan, M.W. (1995) Response of some vegetable crops to sulphur dioxide. In: *Annals of Applied Biology (Supplement Test of Agrochemicals and Cultivars)*, Vol. 128. pp. 98–99.
Khan, M.R., Khan, M.W. and Khan, A.A. (1996a) Evaluation of the sensitivity of some vegetable crops to ozone. *Annals of Applied Biology* 128(s1), 94–95.
Khan, M.R., Khan, M.W. and Singh, K. (1996b) Single and concomitant effects of sulphur dioxide and *Meloidogyne javanica* on pumpkin. *Nematologia Mediterranea* 23, 9–13.
Khan, M.R., Khan, M.W. and Singh, K. (1997) Management of root-knot disease of tomato by the application of fly ash. Plant Pathology 46, 33–43. DOI: 10.1046/j.1365-3059.1997.d01-199.x.
Khan, M.R., Mohiddin, F.A. and Khan, M.M. (2007) Effect of low levels of SO_2 on the growth and yield of indigenous germplasm of black mustard. Environmental Biology and Conservation (India) 12, 53–57.
Khan, M.R., Khan, M.M. and Mohiddin, F.A. (2015) Development of Alternaria leaf spot of Indian mustard caused by *Alternaria brassicae* under the stress of low levels of sulfur dioxide. Agriculture, Ecosystems & Environment 199, 154–163. DOI: 10.1016/j.agee.2014.09.001.
Khan, M.W., Khan, M.R. and Khan, A.A. (1991) Effect of air pollution caused by ceramic and pottery industries on powdery mildew and root knot nematode on cucurbits. *Journal of the Indian Botanical Society* 70, 373–378.
Koruba, D., Piotrowski, J.Z. and Latosińska, J. (2017) Biomass – alternative renewable energy source to the fossil fuels. *E3S Web of Conferences* 14, 02015.
Koziol, M.J. and Jordan, C.F. (1978) Changes in carbohydrate levels in red kidney bean (*Phaseolus vulgaris* L.) exposed to sulphur dioxide. *Journal of Experimental Botany* 29(5), 1037–1043.
Kumar, V., Zacharia, K.A. and Goswami, G. (2000) Fly ash use in agriculture: A perspective. In: *Proceedings of the 2nd International Conference on Flyash Disposal and Utilization. Volume 1*, Central Board of Irrigation and Power, New Delhi, pp. 1–13.
Lamichaney, A., Tewari, K., Basu, P.S., Katiyar, P.K. and Singh, N.P. (2021) Effect of elevated carbon-dioxide on plant growth, physiology, yield and seed quality of chickpea (*Cicer arietinum* L.) in Indo-gangetic plains. *Physiology and Molecular Biology of Plants* 27, 251–263.
Lanzerstorfer, C. (2018) Heavy metals in the finest size fractions of road-deposited sediments. *Environmental Pollution* 239, 522–531. DOI: 10.1016/j.envpol.2018.04.063.
Leakey, A.D.B., Ainsworth, E.A., Bernacchi, C.J., Rogers, A., Long, S.P. *et al.* (2009) Elevated CO_2 effects on plant carbon, nitrogen, and water relations: Six important lessons from FACE. *Journal of Experimental Botany* 60(10), 2859–2876. DOI: 10.1093/jxb/erp096.
Lee, C.H. (2007) Effect of fly ash on productivity of tomato and improvement of soil. *Korean Journal of Plant Resources* 20(1), 93–98.
Lee, H., Ha, H.S., Lee, C.H., Lee, Y.B. and Kim, P.J. (2006) Fly ash effect on improving soil properties and rice productivity in Korean paddy soils. *Bioresource Technology* 97(13), 1490–1497. DOI: 10.1016/j.biortech.2005.06.020.
Lemonnier, P. and Ainsworth, E.A. (2018) Crop responses to rising atmospheric [CO2] and global climate change. In: S.S., Y., Redden, R.J., Hatfield, J.L., Ebert, A.W. and Hunter, D. (eds) *Food Security and Climate Change*. Wiley, New York, pp. 51–69. DOI: 10.1002/9781119180661.
Li, C., Gu, X., Wu, Z., Qin, T., Guo, L. *et al.* (2021) Assessing the effects of elevated ozone on physiology, growth, yield and quality of soybean in the past 40 years: A meta-analysis. *Ecotoxicology and Environmental Safety* 208, 111644. DOI: 10.1016/j.ecoenv.2020.111644.

Li, D., Wu, D., Xu, F., Lai, J. and Shao, L. (2018) Literature overview of chinese research in the field of better coal utilization. *Journal of Cleaner Production* 185, 959–980. DOI: 10.1016/j.jclepro.2018.02.216.

Macaulay, B.M., Owoeye, J.A., Abiya, S.F., Raji, J.I. and Hung, Y.T. (2020) Acid rain: A growing global concern. *Handbook of Environment and Waste Management* 59–93. DOI: 10.1142/9789811207136_0003.

Malhotra, S.S. and Sarkar, S.K. (1979) Effects of sulphur dioxide on sugar and free amino acid content of pine seedlings. *Physiologia Plantarum* 47(4), 223–228. DOI: 10.1111/j.1399-3054.1979.tb06517.x.

Mansfield, T.A. (1986) The physiology of stomata: New insight into old problems. In: Sutcliffe, J.F., Steward, E.F. and Dale, J.E. (eds) *Plant Physiology. A Treatise*. Academic Press, New York, pp. 155–224.

Marier, P. and Dibbs, H.P. (1974) The catalytic conversion of SO_2 to SO_3 by fly ash and the capture of SO_2 and SO_3 by CaO and MgO. *Thermochimica Acta* 8(1–2), 155–165. DOI: 10.1016/0040-6031(74)85082-3.

Maurya, D.P., Singla, A. and Negi, S. (2015) An overview of key pretreatment processes for biological conversion of lignocellulosic biomass to bioethanol. *3 Biotech* 5, 597–609. DOI: 10.1007/s13205-015-0279-4.

Menz, F.C. and Seip, H.M. (2004) Acid rain in Europe and the United States: An update. *Environmental Science & Policy* 7(4), 253–265. DOI: 10.1016/j.envsci.2004.05.005.

Mina, U., Kumar, P. and Varshney, C. (2010) Effect of ozone exposure on growth, yield and isoprene emission from tomato (L.) plants. *Journal of Fruit and Ornamental Plant Research* 72(1), 35–48.

Mishra, A.K., Agrawal, S.B. and Agrawal, M. (2019) Rising atmospheric carbon dioxide and plant responses: Current and future consequences. In: Choudhary, K.K. and Kumar, A. (eds) *Climate Change and Agricultural Ecosystems*. Elsevier/Woodhead Publishing, Oxford, UK, pp. 265–306.

Morgan, P.B., Ainsworth, E.A. and Long, S.P. (2003) How does elevated ozone impact soybean? A meta-analysis of photosynthesis, growth and yield. *Plant, Cell & Environment* 26(8), 1317–1328. DOI: 10.1046/j.0016-8025.2003.01056.x.

Murray, A.M., Percival, D.C. and Stratton, G.W. (2004) Impact of simulated acid rain on photochemistry, morphology, and yield of the wild blueberry (*Vaccinium angustifolium* Ait.). *Canadian Journal of Plant Science* 84(3), 877–880. DOI: 10.4141/P03-033.

Nayak, A.K., Raja, R., Rao, K.S., Shukla, A.K., Mohanty, S. *et al.* (2015) Effect of fly ash application on soil microbial response and heavy metal accumulation in soil and rice plant. *Ecotoxicology and Environmental Safety* 114, 257–262. DOI: 10.1016/j.ecoenv.2014.03.033.

Noormets, A., Kull, O., Sôber, A., Kubiske, M.E. and Karnosky, D.F. (2010) Elevated CO_2 response of photosynthesis depends on ozone concentration in aspen. *Environmental Pollution* 158(4), 992–999. DOI: 10.1016/j.envpol.2009.10.009.

Nowroz, F., Hasanuzzaman, M., Siddika, A., Parvin, K., Caparros, P.G. *et al.* (2024) Elevated tropospheric ozone and crop production: Potential negative effects and plant defense mechanisms. *Frontiers in Plant Science* 14, 1244515. DOI: 10.3389/fpls.2023.1244515.

Nyashina, G.S., Kurgankina, M.A. and Strizhak, P.A. (2018) Environmental, economic and energetic benefits of using coal and oil processing waste instead of coal to produce the same amount of energy. Energy Conversion and Management 174, 175–187. DOI: 10.1016/j.enconman.2018.08.048.

Odiyi, B.O. and Bamidele, J.J.F. (2014) Effects of simulated acid rain on growth and yield of Cassava *Manihot esculenta* (Crantz). Journal of Agricultural Science 6(1), 96–101. DOI: 10.5539/jas.v6n1p96.

Odiyi, B.O. and Eniola, A.O. (2015) The effect of simulated acid rain on plant growth component of cowpea (*Vigna unguiculata*) L. Walps. Jordan Journal of Biological Sciences 8(1), 51–54. DOI: 10.12816/0026948.

Padhi, S.K., Rath, S. and Kar, M. (1995) Response of calendula to sulfur dioxide fumigation. Orissa Journal of Horticulture 23, 119–122.

Pal, S. and Kumar, N. (2000) Effect of simulated acidic rain on yield and carbohydrate contents of green pepper (*Capsicum annuum* L.). *Advances in Plant Sciences* 13(1), 85–88.

Parab, N., Mishra, S. and Bhonde, S.R. (2012) Prospects of bulk utilization of fly ash in agriculture for integrated nutrient management. *Bulletin of the National Institute of Ecology* 23, 31–46.

Paraschiv, S. and Paraschiv, L.S. (2020) Trends of carbon dioxide (CO_2) emissions from fossil fuels combustion (coal, gas and oil) in the EU member states from 1960 to 2018. *Energy Reports* 6, 237–242. DOI: 10.1016/j.egyr.2020.11.116.

Pasha, M.J., Khan, M.W. and Siddiqui, Z.A. (1990) Effect of soil amendment with fly ash of thermal power plant origin on root-knot nematode on cucumber. *Nematologica* 36, 381.

Paulikienė, S., Venslauskas, K., Raila, A., Žvirdauskienė, R. and Naujokienė, V. (2020) The influence of ozone technology on reduction of carrot loss and environmental impact. *Journal of Cleaner Production* 244, 118734. DOI: 10.1016/j.jclepro.2019.118734.

Pei, J., Liu, P., Feng, Z., Chang, M., Wang, J. *et al*. (2024) Long-term trajectory of ozone impact on maize and soybean yields in the United States: A 40-year spatial-temporal analysis. *Environmental Pollution* 344, 123407. DOI: 10.1016/j.envpol.2024.123407.

Peter Singh, L., Gill, S.S., Gill, R. and Tuteja, N. (2012) Mechanism of sulfur dioxide toxicity and tolerance in crop plants. In: Tuteja, N., Gill, S.S., Tiburcio, A.F. and Tuteja, R. (eds) *Improving Crop Resistance to Abiotic Stress*. Wiley-VCH, Weinheim, Germany, pp. 133–163. DOI: 10.1002/9783527632930.

Prado, F.E., Boero, C., Gallardo, M.R.A. and González, J.A. (2000) Effect of nacl on growth germination and soluble sugars content in *Chenopodium quinoa* willd. seeds. *Botanical Bulletin of Academia Sinica* 41, 27–34.

Raghav, D. (2006) Studies on the effects of particulate air pollutants application on potato (*Solanum tuberosum* L.). PhD thesis, Aligarh Muslim University, Aligarh, India.

Rahimi, J., Jhalegar, M.J., Meti, S., Haveri, N., Nanjappanavar, A. *et al*. (2020) Effect of ozone and sulphur dioxide as postharvest treatment to control Rhizopus rot and quality maintenance of table grape (*Vitis vinifera* L.). *International Journal of Chemical Studies* 8(5), 1816–1824. DOI: 10.22271/chemi.2020.v8.i5y.10565.

Rahman, S. and Husen, A. (2022) Impact of sulphur dioxide deposition on medicinal plants' growth and production of active constituents. In: Husen, A. (ed.) *Environmental Pollution and Medicinal Plants*. CRC Press, Boca Raton, FL, pp. 65–88. DOI: 10.1201/9781003178866-4.

Rai, P. and Kehri, H.K. (2017) Effect of fly ash application on the growth, nodulation and mycorrhization in cowpea (*Vigna sinensis* (L.) var. *kanchan*). *Journal of the Indian Botanical Society* 96(1 and 2), 1–17.

Rakwal, R., Agrawal, G.K., Kubo, A., Yonekura, M., Tamogami, S. *et al*. (2003) Defense/stress responses elicited in rice seedlings exposed to the gaseous air pollutant sulfur dioxide. *Environmental and Experimental Botany* 49(3), 223–235. DOI: 10.1016/S0098-8472(02)00072-2.

Rath, S., Padhi, S.K., Kar, M. and Ghosh, P.K. (1995) Response of *Zinnea* to sulfur dioxide exposure. *Journal of Ornamental Horticulture* 2, 42–45.

Reed, R.J. (2015) Liquid fossil fuels from petroleum. In: Kutz, M. (ed.) *Mechanical Engineers' Handbook. Volume 4: Energy and Power*. Wiley, New York. DOI: 10.1002/9781118985960.

Reich, P.B. (1983) Effects of low concentrations of O_3 on net photosynthesis, dark respiration, and chlorophyll contents in aging hybrid poplar leaves. *Plant Physiology* 73(2), 291–296. DOI: 10.1104/pp.73.2.291.

Reich, P.B., Hobbie, S.E. and Lee, T.D. (2014) Plant growth enhancement by elevated CO_2 eliminated by joint water and nitrogen limitation. *Nature Geoscience* 7(12), 920–924. DOI: 10.1038/ngeo2284.

Reijnders, L. and Huijbregts, M.A.J. (2005) Life cycle emissions of greenhouse gases associated with burning animal wastes in countries of the European Union. *Journal of Cleaner Production* 13(1), 51–56. DOI: 10.1016/j.jclepro.2003.08.003.

Reinert, R.A. (1984) Plant response to air pollutant mixtures. *Annual Review of Phytopathology* 22(1), 421–442. DOI: 10.1146/annurev.py.22.090184.002225.

Richards, B.L. and Taylor, O.C. (1965) Significance of atmospheric ozone as a phytotoxicant. *Journal of the Air Pollution Control Association* 15(5), 191–193. DOI: 10.1080/00022470.1965.10468360.

Rizvi, T.F. and Khan, M.R. (2023) Biological and physiological responses of root-knot disease development on five cucurbits exposed to different concentrations of sulfur dioxide. *Toxics* 11(4), 334.

Rizvi, T.F., Khan, M.R. and Ansari, R.A. (2015) The plant growth, physiology, and root-knot disease of sponge guard under the influence of elevated levels of carbon dioxide. In: *11th National Symposium on Dynamics of Crop Protection: Challenges in Agri-horticultural Ecosystems Facing Climate Change*, M.P. University of Agriculture and Technology, Udaipur, India, p. 135.

Rodríguez-Sánchez, V.M., Rosas, U., Calva-Vásquez, G. and Sandoval-Zapotitla, E. (2020) Does acid rain alter the leaf anatomy and photosynthetic pigments in urban trees. *Plants* 9(7), 862.

Rumiani, M., Karegar, A., Hamzehzarghani, H. and Banihashemi, Z. (2016) Effect of elemental sulfur on the root-knot nematode, *Meloidogyne incognita*, activities in cucumber plants. *Iranian Journal of Plant Pathology* 55, 149–160.

Saborit, J. (2009) Effects of air pollution on citrus tree. *Tree and Forestry Science Biotechnology* 3, 92–104.

Sahu, M.P. (2022) Thiourea can mitigate the adverse effect of ozone on crop productivity. In: Aftab, T. (ed.) Sustainable Management of Environmental Contaminants: Eco-Friendly Remediation Approaches. Springer, Cham, Switzerland, pp. 509–532. DOI: 10.1007/978-3-031-08446-1_20.

Sahu, G., Bag, A.G., Chatterjee, N. and Mukherjee, A.K. (2017) Potential use of flyash in agriculture: A way to improve soil health. *Journal of Pharmacognosy and Phytochemistry* 6(6), 873–880.

Saini, D.K., Garg, S.K. and Kumar, M. (2019) Major air pollutants and their effects on plant and human health: A review. Plant Archives 19(2), 3273–3278.

Sharonova, O.M., Anshits, N.N., Yumashev, V.V. and Anshits, A.G. (2008) Composition and morphology of char particles of fly ashes from industrial burning of high-ash coals with different reactivity. *Fuel* 87(10–11), 1989–1997. DOI: 10.1016/j.fuel.2007.10.007.

Shaw, P.J.A., Holland, M.R., Darrall, N.M. and McLeod, A.R. (1993) The occurrence of SO_2-related foliar symptoms on Scots pine (*Pinus sylvestris* L.) in an open-air forest fumigation experiment. New Phytologists 123, 143–152. DOI: 10.1111/j.1469-8137.1993.tb04540.x.

Shew, B.B., Reinert, R.A. and Barker, K.R. (1982) Response of tomato, *Lycopersicon esculentum* cv. walker to ozone and sulfur dioxide and infection by *Pratylenchus penetrans*. Phytopathology 72, 822–826. DOI: 10.1094/Phyto-72-822.

Shriner, D.S. (1978) Effects of simulated acidic rain on host–parasite interactions in plant diseases. *Phytopathology* 68(2), 213. DOI: 10.1094/Phyto-68-213.

Sicher, R., Bunce, J., Barnaby, J. and Bailey, B. (2015) Water-deficiency effects on single leaf gas exchange and on C4 pathway enzymes of maize genotypes with differing abiotic stress tolerance. *Photosynthetica* 53(1), 3–10. DOI: 10.1007/s11099-015-0074-9.

Siddiqi, M.R. (2005) *Tylenchida: Parasites of Plants and Insects*, 3rd edn. CAB International, Wallingford, UK.

Sikka, R. and Kansal, B.D. (1995) Effect of fly-ash application on yield and nutrient composition of rice, wheat and on pH and available nutrient status of soils. *Bioresource Technology* 51(2–3), 199–203. DOI: 10.1016/0960-8524(94)00119-L.

Sikora, R.A., Claudius-Cole, B. and Sikora, E.J. (2018) Nematode parasites of food legumes. In: Sikora, R.A., Coyne, D., Hallmann, J. and Timper, P. (eds) *Plant Parasitic Nematodes in Subtropical and Tropical Agriculture*, 3rd edn. CAB International, Wallingford, UK, pp. 290–345. DOI: 10.1079/9781786391247.0290.

Singh, J. (1989) Neonatal development altered by maternal sulfur dioxide exposure. *Neurotoxicology* 10(3), 523–527.

Singh, B. and Agrawal, M. (2004) Impact of simulated acid rain on growth and yield of two cultivars of wheat. *Water, Air & Soil Pollution* 152(1–4), 71–80. DOI: 10.1023/B:WATE.0000015331.02874.df.

Singh, L. and Sukul, P. (2019) Effects of organic manure, inorganic fertilizers and fly ash on physical and electrochemical properties of soil under maize cultivation. *Plant Archives* 19, 2797–2800.

Singh, C.B., Oswal, M.C. and Grewal, K.S. (2002) Impact of fly ash application on consumptive and water use efficiency in wheat (*Triticum aestivum*) under different soils. *The Indian Journal of Agricultural Sciences* 72(7), 396–399.

Singh, K., Khan, M.W. and Khan, M.R. (1995) Effect of SO_2 on growth, nodulation and root-knot disease of pea. *Journal of the Indian Botanical Society* 75, 49–54.

Singh, K., Khan, M.W. and Khan, M.R. (1996a) Interactive effects of meloidogyne javanica and ozone on leaf pigments, nitrogen and protein contents of pea. *Nematologia Mediterranea* 24, 185–187.

Singh, K., Khan, M.W. and Khan, M.R. (1996b) Plant growth and root-knot disease of soybean under the stress of SO_2 and O_3 gases mixture. *Nematologia Mediterranea* 24, 261–264.

Singh, K., Khan, M.W. and Khan, M.R. (1997) Concomitant effects of SO_2 on soybean in presence or absence of root-knot nematode and/or root-nodule bacteria. *Journal of the Indian Botanical Society* 76, 191–196.

Singh, K., Khan, A.A. and Sufiuddin, A. (2011) Effect of fly ash on growth, yield and root-knot disease of soybean. *Nematologia Mediterranea* 39, 127–131.

Singh, N., Singh, S.N., Yunus, M. and Ahmad, K.J. (1994) Growth response and element accumulation in *Beta vulgaris* L. raised in fly-ash-amended soils. *Ecotoxicology* 3(4), 287–298. DOI: 10.1007/BF00117993.

Singh, R., Parihar, P. and Prasad, S.M. (2018) Simultaneous exposure of sulphur and calcium hinder as toxicity: Upregulation of growth, mineral nutrients uptake and antioxidants system. *Ecotoxicology and Environmental Safety* 161, 318–331. DOI: 10.1016/j.ecoenv.2018.05.060.

Srivastava, S., Singh, R. and Arya, P. (2023) Effect of coal-smoke pollution on plants growth, metabolism and production. In: Husen, A. (ed.) *Plants and Their Interaction to Environmental Pollution*. Elsevier, New York, pp. 125–137. DOI: 10.1016/B978-0-323-99978-6.00011-X.

Szpunar-Krok, E., Jańczak-Pieniążek, M., Migut, D., Skrobacz, K., Piechowiak, T. *et al*. (2020) Physiological and biochemical properties of potato (*Solanum tuberosum* L.) in response to ozone-induced oxidative stress. *Agronomy* 10(11), 1745. DOI: 10.3390/agronomy10111745.

Tamm, C.O. and Cowling, E.B. (1976) Acid precipitation and forest vegetation. In: Dochinger, L.S. and Seliga, T.A. (eds) *Proceedings of the First International Symposium on Acid Precipitation and the Forest Ecosystem*., US Department of Agriculture, Northeastern Forest Experiment Station, pp. 845–855.

Tarannum, A., Khan, A.A., Diva, I. and Khan, B. (2001) Impact of fly ash on hatching, penetration and development of root-knot nematode, *Meloidogyne javanica* in chick pea (*Cicer arietinum* L.). *Nematology Mediterranea* 29, 215–218.

Thompson, M., Gamage, D., Hirotsu, N., Martin, A. and Seneweera, S. (2017) Effects of elevated carbon dioxide on photosynthesis and carbon partitioning: A perspective on root sugar sensing and hormonal crosstalk. *Frontiers in Physiology* 8, 578. DOI: 10.3389/fphys.2017.00578.

Tiwari, S., Agrawal, M., Tiwari, S. and Agrawal, M. (2018) Effect of ozone on physiological and biochemical processes of plants. In: Tiwari, S. and Agrwal, M. (eds) *Tropospheric Ozone and Its Impacts on Crop Plants: A Threat to Future Global Food Security*. Springer, Dordrecht, The Netherlands, pp. 65–113.

Totawat, K.L., Nagar, G.L., Jat, S.L. and Jangir, R.K. (2002) Effect of flyash on the performance of wheat on ustochrepts of sub-humid plains of India. In: *17 World Congress of Soil Science*, Bangkok, Thailand, pp. 14–21 (Vol. 17).

Tripathi, A.K. and Gautam, M. (2007) Biochemical parameters of plants as indicators of air pollution. *Journal of Environmental Biology* 28(1), 127.

Tripathy, A. and Sahu, R.K. (1997) Effect of coal fly ash on growth and yield of wheat. *Journal of Environmental Biology* 18, 131–135.

Ummat, V., Singh, A.K. and Sidhu, G.K. (2018) Effect of aqueous ozone on quality and shelf life of shredded green bell pepper (*Capsicum annuum*). *Journal of Food Processing and Preservation* 42(10), 10. DOI: 10.1111/jfpp.13718.

Vandermeiren, K., Black, C., Pleijel, H. and Temmerman, L. (2005) Impact of rising tropospheric ozone on potato: Effects on photosynthesis, growth, productivity and yield quality. *Plant, Cell & Environment* 28(8), 982–996. DOI: 10.1111/j.1365-3040.2005.01316.x.

Varshney, A., Dahiya, P., Sharma, A., Pandey, R. and Mohan, S. (2022) Fly ash application in soil for sustainable agriculture: An Indian overview. *Energy, Ecology and Environment* 7(4), 340–357. DOI: 10.1007/s40974-022-00241-w.

Varshney, C.K., Garg, J.K., Lauenroth, W.K. and Heitschmidt, R.K. (1979) Plant responses to sulfur dioxide pollution. *CRC Critical Reviews in Environmental Control* 9(1), 27–49. DOI: 10.1080/10643387909381667.

Varshney, P., Kausar, S. and Khan, A.A. (2005) An evaluation of the susceptibility of radish in pots to simulated acid rain. *Test of Agrochemical Cultivar* 27, 29–30.

Verma, A., Tewari, A. and Azami, A. (2010) An impact of simulated acid rain of different pH levels on some major vegetable plant in India. *Report and Opinion* 2(4), 38.

Vicente, E.D., Vicente, A.M., Evtyugina, M., Oduber, F.I., Amato, F. *et al*. (2020) Impact of wood combustion on indoor air quality. *Science of the Total Environment* 705, 135769. DOI: 10.1016/j.scitotenv.2019.135769.

Villányi, V., Déri, H., Péli, E. and Csintalan, Z. (2013) Differences in histological and physiological traits of ozone sensitive and resistant bean strains. *Open Life Sciences* 8(4), 386–397. DOI: 10.2478/s11535-013-0140-2.

Vinay, S. and Dwivedi, S.K. (1999) Effect of fly ash on seed germination, plant growth and chlorophyll content of two crops of economic importance. *Acta Botanica Indica* 27(2), 145–150.

Wang, K., Wang, S., Liu, L., Yue, H., Zhang, R. *et al*. (2016) Environmental co-benefits of energy efficiency improvement in coal-fired power sector: A case study of Henan province, China. *Applied Energy* 184, 810–819. DOI: 10.1016/j.apenergy.2016.06.059.

Weber, D.E., Reinert, R.A. and Barker, K.R. (1979) Ozone and sulfur dioxide effects on reproduction and host–parasite relationships of selected plant-parasitic nematodes. *Phytopathology* 69(6), 624. DOI: 10.1094/Phyto-69-624.

Wyrwicka, A. and Skłodowska, M. (2006) Influence of repeated acid rain treatment on antioxidative enzyme activities and on lipid peroxidation in cucumber leaves. *Environmental and Experimental Botany* 56(2), 198–204. DOI: 10.1016/j.envexpbot.2005.02.003.

Yang, L., Stulen, I., Kok, L.J. and Zheng, Y. (2002) SO_2, NO_x and acid deposition problems in China – impact on agriculture. *Phyton* 42(3), 255–264.

Yang, L., Stulen, I. and Kok, L.J. (2006) Sulfur dioxide: Relevance of toxic and nutritional effects for Chinese cabbage. *Environmental and Experimental Botany* 57, 236–245.

Yao, Z.T., Ji, X.S., Sarker, P.K., Tang, J.H., Ge, L.Q. *et al.* (2015) A comprehensive review on the applications of coal fly ash. *Earth-Science Reviews* 141, 105–121.

Yeledhalli, N.A., Prakash, S.S., Gurumurthy, S.B. and Ravi, M.V. (2007) Coal fly ash as modifier of physico-chemical and biological properties of soil. *Karnataka Journal of Agricultural Sciences* 20(3), 531–534.

Yli-Tuomi, T., Siponen, T., Taimisto, R.P., Aurela, M., Teinila, K. *et al.* (2015) Impact of wood combustion for secondary heating and recreational purposes on particulate air pollution in a suburb in Finland. *Environmental Science and Technology* 49(7), 4089–4096.

Yousuf, A., Manzoor, S.O., Youssouf, M., Malik, Z.A. and Khawaja, K.S. (2020) Fly ash: Production and utilization in India – an overview. *Journal of Materials and Environmental Science* 11(6), 911–921.

Zhou, J., Wu, J., Huang, J., Sheng, X., Dou, X. *et al.* (2022) A synthesis of soil nematode responses to global change factors. *Soil Biology and Biochemistry* 165, 108538. DOI: 10.1016/j.soilbio.2021.108538.

Zhu, X., Jiang, J., Yin, C., Li, G., Jiang, Y. *et al.* (2019) Effect of ozone treatment on flavonoid accumulation of *Satsuma mandarin* (*Citrus unshiu* Marc.) during ambient storage. *Biomolecules* 9(12), 821. DOI: 10.3390/biom9120821.

Zouzoulas, D., Koutroubas, S.D., Vassiliou, G. and Vardavakis, E. (2009) Effects of ozone fumigation on cotton (*Gossypium hirsutum* L.) morphology, anatomy, physiology, yield and qualitative characteristics of fibers. *Environmental and Experimental Botany* 67(1), 293–303. DOI: 10.1016/j.envexpbot.2009.05.016.

16 Impact of Climate Change on Plant Nematodes and Diseases Complexes

Milad Rashidifard[1]*, Sunil Singh[2], Gerhard Du Preez[3] and Samad Ashrafi[1,4]*
[1]Department of Zoology and Entomology, University of the Free State, South Africa; [2]School of Pacific Arts, The University of the South Pacific, Fiji; [3]Unit for Environmental Sciences and Management, North-West University, Potchefstroom, South Africa; [4]Institute for Crop and Soil Science, Julius Kühn Institute (JKI) – Federal Research Centre for Cultivated Plants, Braunschweig, Germany

Abstract

Food security and nutrition availability are globally threatened by climate change and require immediate actions to protect crops from emerging abiotic and biotic challenges. This is in association with other activities to reach the Sustainable Development Goal of a world without hunger by 2030. As climate variability develops, plant–nematode interactions must be observed more closely to study the potentially new mechanisms involved in nematode parasitism, tolerance and resistance. Further, climate change may also influence intricate interactions between nematodes and other pathogens, and the development of disease complexes. Critical actions in protecting the plants via developing more adaptable resistance cultivars, effective multifaceted nematode biocontrol agents, innovative agricultural practices and sustainable farming strategies will be required under climate change. A holistic management approach, where a comprehensive assessment is employed to identify the new patterns of nematode distribution, pathogenicity and ecology in emerging climatic conditions, can help to ensure the best practices to overcome the challenges. This chapter highlights the future challenges for nematode control and the importance of nematode management under global climatic variabilities.

Introduction

The United Nations proposed the 2030 Agenda for Sustainable Development (www.un.org/sustainabledevelopment/development-agenda) for a global sustainable future based on 17 Sustainable Development Goals (SDGs) (www.un.org/sustainabledevelopment). Of these, Climate Action and Zero Hunger are most relevant to sustainability in food security.

According to the definition provided by the UN Food and Agriculture Organization, 'To be sustainable, agriculture must meet the needs of present and future generations, while ensuring profitability, environmental health, and social and economic equity' (www.fao.org/sustainability). However, agricultural sustainability is jeopardized by climate variability (www.fao.org/documents/card/en/c/cb1928en), typically presented as rising temperature, rainfall changes, extreme weather, increasing and prolonged drought, elevated CO_2, and changes to sea levels (Wheeler and von Braun, 2013). Climate change is also characterized by fluctuations in

*Corresponding authors: RashidifardM@ufs.ac.za; samad.ashrafi@julius-kuehn.de

 Nematode Disease Complexes in Agricultural Crops
(ed. Mujeebur Rahman Khan)
DOI: 10.1079/9781800625228.0016

temperature, changes in precipitation patterns and the unpredictability of extreme weather conditions.

Sünnemann *et al.* (2023) showed that the factors associated with global warming will negatively affect soil functionality in both croplands and grasslands. This is in addition to the intensification of land use due to global food demands, which threaten services provided by soil including carbon sequestration, microbial diversity and water storage (Kopittke *et al.*, 2019). Changes in temperature and rainfall patterns can impact plant growth and crop production (Lesk *et al.*, 2016). This may alter the interaction between pathogens and their host plants, resulting in variability in disease severity and the development of new disease situations (Velásquez *et al.*, 2018). This may lead to shifts in the selection of crop cultivars or species, leading to an increase in crop diseases and pests, including plant-parasitic nematodes (PPN) and disease complexes. Climate changes can negatively affect soil microbes and subsequently, microbial biodiversity and their contribution to the delivery of ecosystem services, such as nutrient cycling and water regulation (Bardgett and van der Putten, 2014; Kopittke *et al.*, 2019; Trivedi *et al.*, 2022). Changes in soil ecosystems may also lead to alteration in disease suppressiveness (Banerjee and van der Heijden, 2023). Any potential changes in microbial communities of disease-suppressive soils could exponentially increase the pressures of pests and pathogens on plant health and growth, especially in an ecosystem already exposed to global warming-associated factors such as higher temperature or prolonged dryness (Sikora *et al.*, 2021).

Global warming (rising temperature) and changes in rainfall, drought and humidity patterns may directly influence population dynamics, reproduction rates, virulence and invasion, abundance, and epidemiology of PPN (Colagiero and Ciancio, 2011; Bristol *et al.*, 2023; Khanal and Land, 2023). Higher temperatures under changing climatic conditions can increase the severity of plant disease (Cohen and Leach, 2020). For example, higher temperatures can influence the development and reproduction rate of *Globodera* spp. (Jones *et al.*, 2017). Another study on two nematode species, *Rotylenchulus reniformis* and *Meloidogyne floridensis*, showed that higher temperatures in soil (<32°C) significantly impacted the survival, reproduction, virulence and disease severity of *M. floridensis*, while *R. reniformis* was not affected (Khanal and Land, 2023).

Prolonged drought alters microbial abundance and distribution that could result in a shift of microbial communities, including nematodes (Franco *et al.*, 2019; Jansson and Hofmockel, 2020). In general, nematode communities are sensitive to environmental changes, such as temperature and soil moisture (Wilschut and Geisen, 2021). Long-term drought or increased frequency of dry years can lead to an increase in populations of PPN, e.g. due to a decrease in the populations of nematode predators. This shift in predator–prey balance can subsequently reduce plant performance via higher levels of plant infection with PPN (Stevnbak *et al.*, 2012; Franco *et al.*, 2019).

Climate change would also alter the insect communities and dispersion (Harvey *et al.*, 2023). This may potentially reshape the incidence of insect-transmitted PPN, including one of the most devastating species, the pinewood nematode *Bursaphelenchus xylophilus* (Giblin-Davis *et al.*, 2003; Jones *et al.*, 2013). Potentially new nematode incidence could endanger forests worldwide when this group of PPN are vectored to new habitats.

Our understanding of how soil ecosystems will respond to newly introduced nematodes under changed climatic factors is still growing. We are still not fully aware of how soil ecosystems in temperate regions will respond/react to newly introduced PPN and their interactions with other soil microbes, and how host plants will be challenged with nematodes in this altered ecosystem. Perhaps more experiments under new conditions will be needed to predict nematode behavior.

In this chapter, we highlight the future challenges for nematode control and the importance of nematode management under global climatic variabilities. An overview of the ongoing recommendations describing the impact of climate change on PPN is described in the flow diagram in Fig. 16.1.

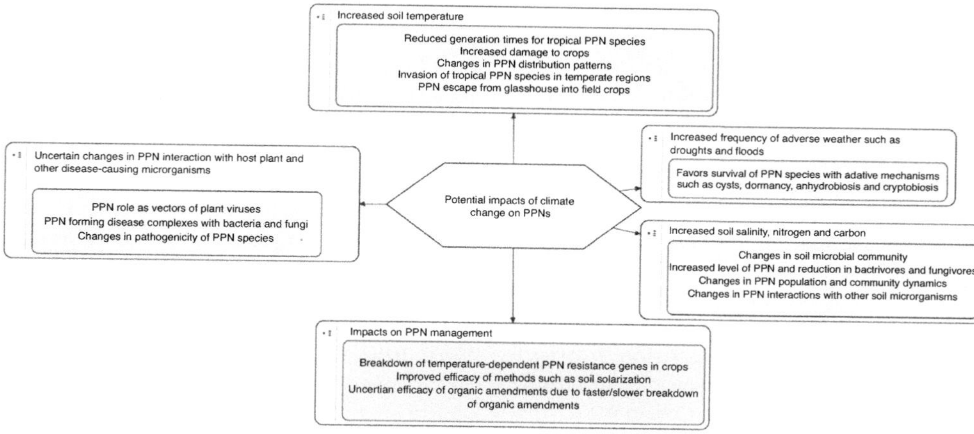

Fig. 16.1. An outline of the potential impacts of climate change on plant-parasitic nematodes (PPNs).

Food Production Under Climate Change in the Presence of Nematode Pests

Multiple studies have reported on the effect of climate-related factors on PPN, including increased temperature (Ghini *et al.*, 2008; Jones *et al.*, 2017), CO_2 enrichment (Sticht *et al.*, 2009; Berliner *et al.*, 2023) and altered rainfall (Fleming *et al.*, 2016). Climate-related responses in PPN communities are not fixed but vary depending on the three-way interactions between hosts, parasites and the environment. More specifically, factors related to the host (e.g. population density and adaptability) and the PPN (e.g. specificity and complexity of life cycle) are influenced by changes in the environment (Dutta and Phani, 2023). For example, a simulation study by Gendron St-Marseille *et al.* (2019) predicted that the increase in temperatures and the linked production of soybeans in Quebec (Canada) would result in up to five generations of soybean cyst nematodes per growing season by 2050. Another example is the change in the distribution of the emerging PPN *Meloidogyne enterolobii* in subtropical and tropical regions. According to Pan *et al.* (2023), climate change will result in this nematode's range of ideal habitats expanding towards higher latitudes in future climate scenarios.

It is projected that altered environmental conditions will intensify PPN damage by promoting their abundance, spread, reproduction and generation. However, opposing effects such as sex reversal, entering cryptobiosis and decreased survival may counterbalance this impact (Dutta and Phani, 2023). A meta-analysis on soil nematode responses to global change factors concluded that global warming had only minor effects on soil nematode communities (Zhou *et al.*, 2022). The authors suggested that a possible explanation for this is that the positive effects of increased temperatures on plant growth were counteracted by the negative effects of increased temperatures on soil moisture. Nonetheless, under simulated and experimental conditions, increased pathogenic potential has been recorded for *Meloidogyne incognita* in coffee plantations (Ghini *et al.*, 2008), increased abundance of *Globodera rostochiensis* in potatoes (Jones *et al.*, 2017), and increased juvenile and egg production of *Heterodera carotae* in carrots (Colagiero and Ciancio, 2011). Interestingly, the Zhou *et al.* (2022) analysis showed that elevated atmospheric CO_2 can significantly increase (22% on average) PPN abundance especially in croplands and grasslands. This was attributed to higher CO_2 levels which typically decrease plant stomatal conductance, leading to reduced water absorption from the soil and more soil water. Second, increased CO_2 can boost above-ground plant growth resulting in more plant litter (Zhou *et al.*, 2022).

There are also more indirect effects of climate change on PPN pressure and prevalence. Beneficial nematode communities play a major role in the ecological functioning of soils by influencing decomposition, nutrient cycling, and pest and disease regulation, among other functions (Creamer *et al.*, 2022). Siebert *et al.* (2020) studied the effects of simulated climate change on nematode communities and functional groups, revealing clear ecological implications. For example, the interaction between future climate scenarios and intensive land use will likely result in increased nematode populations of opportunistic and PPN species at the cost of reduced diversity and structure within the nematode community (Siebert *et al.*, 2020). This clearly poses a threat to the ability of soil ecosystems to contribute towards pest and disease regulation.

Emerging Patterns of Nematode Pathogenicity and Biology with Changing Climates

Climate change is an imminent threat to global food security due to its impacts on agriculture, forestry, fisheries and aquaculture (Bezner Kerr *et al.*, 2022). It is anticipated that climate change will differently affect crops and associated pests and diseases in tropical, subtropical and temperate areas (Skendžić *et al.*, 2021; Bezner Kerr *et al.*, 2022). Except for central Asia, much of the remaining areas in the world are already or will experience negative impacts on the production of major food crops due to climate change (Bezner Kerr *et al.*, 2022). A changing climate can also have a multitude of impacts on pests and diseases of crops (Sutherst *et al.*, 2011; Nethi and Prasad, 2012; Lamichhane *et al.*, 2015; Bradford *et al.*, 2019; Skendžić *et al.*, 2021).

Plant-parasitic nematodes are a major constraint to crop production worldwide and cause damages of over US$157 billion per annum (Abad *et al.*, 2008). They attack all major food crops including cereals, pulse, vegetables oilseed, fibers beverage crops, fruit crops, ornamentals, forests and inhabit a wide range of environments. Out of the over 4100 described PPN species (Decraemer and Hunt, 2006), the majority have limited distributions and are known from mainly agricultural areas (Sikora *et al.*, 2018; Zasada *et al.*, 2019). Of these, a few hundred species are widely distributed, affecting economically important crops worldwide in tropical, subtropical and temperate regions (Bebber *et al.*, 2014; Singh *et al.*, 2014).

Soil temperature, moisture, pH and soil carbon are major factors that affect the life cycle of PPN and can be used to determine their likely distribution and damage potential (Singh *et al.*, 2022; Pan *et al.*, 2023). In this section, we consider the known and likely impacts of climate change on PPN distribution, abundance and damage to crops.

Under controlled conditions, it has been demonstrated that several PPN species (e.g. *M. floridensis*, *R. reniformis*, *Ditylenchus weischeri* and *D. dipsaci*) have reduced generation times and as a result their population levels may increase quickly in the presence of a suitable host and warmer climatic conditions (Leach *et al.*, 2009; Hajihassani *et al.*, 2017; Khanal and Land, 2023). Using CLIMEX models (www.hearne.software/Software/CLIMEX-DYMEX/Editions), which use existing species distribution records and species phenology such as generation times, cold stress, heat stress, wet stress and dry stress, it is estimated that there will be range expansion and distribution shifts among several species of PPN (Yeates *et al.*, 1998; Singh *et al.*, 2022).

Certain nematode species, such as *Meloidogyne* spp., *Pratylenchus* spp., *Helicotylenchus* spp. and *D. dipsaci*, have experienced a shift towards northern latitudes since the 1960s (Bebber *et al.*, 2013). *M. enterolobii* is predicted to shift towards higher latitudes and large areas in Africa, South America, Asia and North America between 30°S to 30°N that could be suitable for its survival (Pan *et al.*, 2023). Under a warmer climate, tropical PPN species pose a greater risk of establishment and damage in subtropical and temperate areas. Plant-parasitic nematode species such as *Meloidogyne* which are currently major pests under glasshouse conditions (Wesemael *et al.*, 2011; Gamon and Lenne, 2012) could establish and spread in areas under natural cultivation in temperate regions. In addition, changes in cultivation practices and the growing of novel crops in new geographical areas as a result of changing climatic conditions are likely to provide opportunities for the invasion of PPN species (Sheppard *et al.*, 2011; Pan *et al.*, 2023).

Experimental studies on *M. enterolobii*, *M. floridensis* and *M. incognita* under increased soil temperatures showed that they were able to complete their life cycle in shorter time frames at 30°C (Velloso *et al.*, 2022). In addition to the effect on life cycle duration, temperature also influences the effectiveness of resistance genes in plants and under increased temperatures expression of resistance genes is affected (Trudgill and Blok, 2001) thus rendering the plants susceptible to attack and damage. The above observations reveal that under changing climate conditions different nematode species may dominate and continue to cause damage to crops.

Plant-parasitic species are known for their ability to quickly adapt to changing soil conditions and have various survival mechanisms (anhydrobiosis, cryptobiosis, cyst formation and dormancy stages) which enable them to survive adverse conditions such as prolonged droughts, temperature extremes, flooded conditions and increased soil salinity (Wharton 1995, 2004; Perry and Moens, 2011; Renčo *et al.*, 2022)

There are reports of increased damage from previously less destructive species of PPN in new areas. For instance, the foliar nematode *Litylenchus crenatae mccannii* has been reported to cause damage to beech trees (beech leaf disease) in the state of Ohio in the USA and has more recently spread to western and northern Pennsylvania, New York and Ontario in Canada (Ewing *et al.*, 2019; Carta *et al.*, 2020; Marra and LaMondia, 2020). In Japan this species is a pest of beech and causes galls; however, the disease is not as severe (Kanzaki *et al.*, 2019). Whether the increase in pathogenicity of *L. crenatae mccannii* is due to genetic differences or climatic factors is not yet known. Similarly, *B. xylophilus*, which causes pine wilt disease, is not as damaging in its native range in North America but causes major damage to pine trees in Asia (Mota and Vieira, 2008). *B. xylophilus* continues to spread in Europe (Vicente *et al.*, 2012) and greater parts of Asia, and is predicted to further increase from its current geographical range under climate change conditions (Li *et al.*, 2022a, 2022b; Ouyang *et al.*, 2022). Therefore, it is plausible that nematode species distributions as well as their pathogenicity can alter with changing climatic conditions.

Based on different stages in their life cycle, PPN exhibit different levels of adaptation to changing climatic conditions and availability of resources such as host plants, and competition with other species. For example, under resource constraints, it is known that *Meloidogyne* spp. are able to alter the sex ratio of their progeny, i.e. more males vs more females (Triantaphyllou, 1973; Castagnone-Sereno *et al.*, 2013). While there is still uncertainty on how exactly PPN populations may behave under changing climatic conditions, existing studies (Wharton, 2004; Perry and Moens, 2011) indicate that PPN have a variety of adaptation mechanisms which may enable them to cope or even survive better under future climate conditions.

Ecological Solutions to Nematode Pest Regulation

Novel and innovative technologies will likely play a major role in responding to climate-driven changes in PPN prevalence and pressure. However, promoting ecological intensification through enhancing natural processes is also recognized as fundamental to building resilience against the future effects of climate change in agricultural systems (Tittonell, 2014; Raj *et al.*,

2021). Disease and pest regulation is one of four major soil ecosystem functions, the others being carbon and climate regulation, water regulation and purification, and nutrient cycling (Zwetsloot *et al.*, 2021; Creamer *et al.*, 2022). According to Creamer *et al.* (2022) two main subfunctions influence disease and pest regulation. The first is disease suppression and pest control, which involves more direct processes including antibiosis, competition, predation, parasitism and microbial grazing. The second subfunction is plant health promotion where the ability of a crop to resist and build defenses against pests and diseases is promoted. Most scientists agree that greater ecological functioning coincides with promoting soil health (Bünemann *et al.*, 2018; Creamer *et al.*, 2022), which is also captured in the most widely used and accepted definition of soil health, namely 'the continued capacity of a soil to function as a vital living ecosystem that sustains plants, animals and humans' (NRCS, 2023).

However, promoting soil health and ecological functioning is not straightforward with different proposed approaches and systems including agroecology (Wezel *et al.*, 2020), conservation (Thierfelder *et al.*, 2018) and regenerative agriculture (Lal, 2020), and organic farming (Tully and McAskill, 2020). There is much debate about the real-world benefits of these systems. This is because often claims are made with little or no concrete scientific evidence. Another issue is that the success of any implemented agricultural system will be highly contextual and based on the local environment (Giller *et al.*, 2021). Thus, rather than advocating for a particular system, our discussion will center on highlighting selected practices that are scientifically validated to enhance soil health and ecosystem functioning, with a specific focus on nematode pest regulation. We recognize, however, that the effectiveness of these practices is contingent upon the specific environmental conditions of each locality.

Considering the previously mentioned processes linked to disease suppression and pest control, some play an important role in nematode pest regulation. These include parasitism of PPN (e.g. nematophagous fungi), predation of PPN and ecological competition for resources. Healthy and structured soil ecosystems are characterized by higher-level trophic organisms, e.g. carnivorous nematodes, mites and tardigrades, capable of general suppression of PPN (Timper, 2014). An experimental study by Timper *et al.* (2021) evidenced a 55% reduction in *M. incognita* J_2 by the native soil community, as well as a positive relationship between the relative abundance of omnivore and carnivore nematodes and J_2 suppression. Unfortunately, carnivorous nematodes are often sensitive to perturbations including tillage and agrochemical use (Timper, 2014; Bongiorno *et al.*, 2019). Therefore, reduced tillage and the judicious use of agrochemicals are highly recommended for promoting soil health and ecosystem functioning (Thierfelder *et al.*, 2018). Crop diversification also shows the potential to promote food web structure and reduce PPN prevalence and pressure. A meta-analysis by Puissant *et al.* (2021) revealed that cover cropping increased the abundance of omnivore and predator nematodes by 80%, while crop rotation reduced the abundance of PPN by 47%.

Nematode Management with Novel Approaches Under Climate Change

Nematodes have evolved from free-living marine species to sedentary endoparasites (Poinar, 2014), making them among the most versatile living organisms on the Earth. Plant-parasitic nematodes, having the capacity to co-evolve with the environment and to colonize almost all ecosystems, seem to keep adapting themselves to the changing environment so they can colonize their hosts and reproduce efficiently (Holterman *et al.*, 2019; Sapir, 2021). Based on the current arguments, some of which were presented earlier in this chapter, nematodes most likely will have the ability to colonize host plants in the warmer soil and given the higher susceptibility of crops due to heat stress, crop damage caused by nematode infection will likely be substantially higher than the current estimated amount.

Research aiming at nematode management in newly emerging climatic conditions should consider the effect of soil moisture and temperature as nematode biology and physiology are very dependent on these two factors. Therefore, fluctuation in any of the two factors may have a strong effect on nematode biological activity, more specifically their reproductive potential, which certainly leads to different outcomes of management strategies. Rashidifard *et al.* (2019) reported that increasing the average temperature in a greenhouse by 2°C resulted in a substantially higher reproduction rate recorded for *M. enterolobii*, *M. incognita* and *M. javanica* populations. It is worth mentioning that nematodes require certain heat units to complete their life cycle, commonly referred to as degree-days; increasing temperature means that nematodes reach the required degree-days quicker and have a shorter life cycle duration (Collett *et al.*, 2024). Previous research has indicated resistance breakdown in tobacco and tomato cultivars carrying resistant genes against root-knot nematode at temperatures above 32°C (Abdul-Baki *et al.*, 1996; Pollok *et al.*, 2023). Hence, given the high adaptability of nematodes to environmental changes (Khanal and Land, 2023), increasing global temperature could lead to higher abundance of certain nematode pests due to quicker population buildup in the soil. On the other hand, loss of resistance in host plants could lead to greater nematode damage to agriculture production. Some adverse conditions may also negatively impact nematode reproduction, survival abilities and pathogenicity. Therefore, while certain alterations could work in favor of nematodes, others may have conflicting outcomes, making it difficult to make a definitive conclusion and prediction about future trends. Under these fluctuating environmental factors, sustainable control techniques must be relied upon as the focus has shifted towards smart management tactics that promote soil health and have minimum negative impact on beneficial organisms besides reducing crop damage.

Integrated Nematode Management

Integrated nematode management (INM) is an approach that employs a combination of various management practices such as botanical, biological, chemical, cultural and physical methods to contain the population density of PPN under the economic threshold level (Dutta and Phani, 2023; Sikora *et al.*, 2023). Within an INM framework, nematologists have already highlighted the needs that could help moving forward towards nematode control strategies, that are more sustainable and have less negative impact and damage to the ecosystem (Desaeger *et al.*, 2021; Sikora *et al.*, 2021). It is crucial to select a combination of tactics based on available resources and environmental factors to meet the needs of growers and their agricultural production systems. The use of cover crops, especially those with nematode-antagonistic potential (e.g. grain sorghum, millets, oil radish, Mulato grass, sorghum-sudangrass, ryegrass and daikon radish), could be a multi-purpose tactic to reduce nematode infestation (Asmus *et al.*, 2008; Acharya *et al.*, 2021; Khanal and Harshman, 2022), enhance soil fertility, prevent soil erosion, improve nutrient and water availability, and more importantly to boost soil microbial diversity and activities (Sharma *et al.*, 2018).

Alteration of environmental factors such as temperature might cause changes in the interaction between PPN and their antagonistic microbes (bacteria and fungi) in the soil. An impairment in the attachment of *Pasteuria penetrans* endospores to the cuticle of J_2 of *M. arenaria* has been reported when the nematode is exposed to 35–40°C or when the bacteria incubated at 50°C (Freitas *et al.*, 1997). In contrast, a higher abundance of *Pochonia chlamydosporia* propagules and consequently increased egg parasitism against *M. incognita* were reported as a result of elevated soil temperature (Luambano *et al.*, 2015). Additionally, altered climate factors such as elevated temperature and reduced humidity will influence the abundance of nematodes and antagonistic microbes in the soil, and their exposure to their natural enemies, which may result in a mismatch between them (Dutta and Phani, 2023). Despite the evidence from relevant research, it is challenging to have a solid conclusion concerning the future trend for any modification in soil suppressiveness or biocontrol potential of nematode-antagonistic agents in response to climate change as these microorganisms are part of naturally coevolving interactions with nematodes and their host plants (Rafaluk-Mohr *et al.*, 2018).

Chemical Nematicides

The efficacy and persistence of chemical nematicides in soil can be affected by fluctuating temperatures and irregular precipitation as a result of climate change which eventually impacts PPN management (Delcour *et al.*, 2015). Higher demand for using chemical nematicides by growers is expected due to the high incidence and breakout of PPN under climate change. Only a few chemical nematicides such as fluensulfone, fluopyram, fluazaindolizine, tioxazafen, ethane dinitrile and spirotetramat with a restricted range of crops are still available on the market (Desaeger *et al.*, 2020). Hence, generating knowledge with respect to the effects of climate change on these products seems to be crucial. Furthermore, despite their effectiveness in reducing PPN pressure in a short time, their application may not be sustainable given the current global call to reduce the application of chemical pesticides. A comprehensive search for more sustainable synthetic chemicals (Gaberthüel *et al.*, 2021), especially with a focus on natural nematicidal compounds (Li and Zhang, 2023), can potentially result in the discovery of environmentally friendly new bioactive compounds that could sustainably safeguard plant health.

Conclusion and Future Prospects

Food safety and global food security require sustainable agriculture to meet the needs of present and future generations. At present, it remains difficult to predict the alteration of population dynamics of PPN and its impact on global food security under new climate variability, but a picture is emerging that a shift in nematode distribution patterns happens, where it might affect different cropping systems, especially in temperate regions. Altered climatic conditions will impact the abundance and diversity of nematode pests and jeopardize the efficacy of current control measures. This situation requires extra effort to establish a sophisticated pest monitoring system using novel tools or improved existing methods which ultimately results in having adapted INM in place for the changed condition (Sikora *et al.*, 2021): development of prediction models, adaptation of cropping system according to the ongoing climatic changes, improvement of tools and techniques used in precision agriculture, supporting suppressive soils by paying more attention to the role of microbial communities in self-regulation of agroecosystems, and use of new molecular approaches and genome editing techniques to accelerate breeding processes for development of new resistant varieties and nematode control. Discovery of new nematode-antagonistic beneficial microbes, development and commercialization of biocontrol agents and improvement of application techniques to ensure sufficient effectiveness, and development of nature-based nematicidal compounds or safer and more selective chemical nematicides will adopt more comprehensive INM to new climatic changes. Biosecurity measures preventing human-mediated introduction of invasive PPN species into new geographical areas and regions could also be beneficial in avoiding crop losses.

Climate change is a socio-scientific issue. Therefore, public awareness and farmer training on the impact of new climatic conditions on the emergence of new nematode diseases in a post-chemical era would empower us to face and manage the incoming challenges using more innovative approaches and to build a more sustainable world.

References

Abad, P., Gouzy, J., Aury, J.-M., Castagnone-Sereno, P., Danchin, E.G.J. *et al.* (2008) Genome sequence of the metazoan plant-parasitic nematode *Meloidogyne incognita*. *Nature Biotechnology* 26(8), 909–915. DOI: 10.1038/nbt.1482.

Abdul-Baki, A.A., Haroon, S.A. and Chitwood, D.J. (1996) Temperature effects on resistance to *Meloidogyne* spp. in excised tomato roots. *HortScience* 31(1), 147–149. DOI: 10.21273/HORTSCI.31.1.147.

Acharya, K., Yan, G. and Plaisance, A. (2021) Effects of cover crops on population reduction of soybean cyst nematode (*Heterodera glycines*). *Plant Disease* 105(4), 764–769. DOI: 10.1094/PDIS-08-20-1778-RE.

Asmus, G.L., Inomoto, M.M. and Cargnin, R.A. (2008) Cover crops for reniform nematode suppression in cotton: Greenhouse and field evaluations. *Tropical Plant Pathology* 33(2), 85–89. DOI: 10.1590/S1982-56762008000200001.

Banerjee, S. and van der Heijden, M.G.A. (2023) Soil microbiomes and one health. *Nature Reviews Microbiology* 21(1), 6–20. DOI: 10.1038/s41579-022-00779-w.

Bardgett, R.D. and van der Putten, W.H. (2014) Belowground biodiversity and ecosystem functioning. *Nature* 515(7528), 505–511. DOI: 10.1038/nature13855.

Bebber, D., Ramotowski, M. and Gurr, S. (2013) Crop pests and pathogens move polewards in a warming world. *Nature Climate Change* 3(11), 985–988. DOI: 10.1038/nclimate1990.

Bebber, D.P., Holmes, T. and Gurr, S.J. (2014) The global spread of crop pests and pathogens. *Global Ecology and Biogeography* 23(12), 1398–1407. DOI: 10.1111/geb.12214.

Berliner, J., Ganguly, A.K., Kamra, A., Sirohi, A. and Dinesh, V.P. (2023) Effect of elevated carbon dioxide on population growth of root-knot nematode, *Meloidogyne incognita* in tomato. *Indian Phytopathology* 76(1), 309–315. DOI: 10.1007/s42360-022-00584-8.

Bezner Kerr, R., Naess, L.O., Allen-O'Neil, B., Totin, E., Nyantakyi-Frimpong, H. *et al.* (2022) Interplays between changing biophysical and social dynamics under climate change: Implications for limits to sustainable adaptation in food systems. *Global Change Biology* 28(11), 3580–3604. DOI: 10.1111/gcb.16124.

Bongiorno, G., Bodenhausen, N., Bünemann, E.K., Brussaard, L., Geisen, S. *et al.* (2019) Reduced tillage, but not organic matter input, increased nematode diversity and food web stability in European long-term field experiments. *Molecular Ecology* 28(22), 4987–5005. DOI: 10.1111/mec.15270.

Bradford, J.B., Schlaepfer, D.R., Lauenroth, W.K., Palmquist, K.A., Chambers, J.C. *et al.* (2019) Climate-driven shifts in soil temperature and moisture regimes suggest opportunities to enhance assessments of dryland resilience and resistance. *Frontiers in Ecology and Evolution* 7, 358. DOI: 10.3389/fevo.2019.00358.

Bristol, D., Hassan, K., Blankinship, J.C. and Nielsen, U.N. (2023) Responses of nematode abundances to increased and reduced rainfall under field conditions: A meta-analysis. *Ecosphere* 14(1), e4364. DOI: 10.1002/ecs2.4364.

Bünemann, E.K., Bongiorno, G., Bai, Z., Creamer, R.E., Deyn, G. *et al.* (2018) Soil quality – a critical review. *Soil Biology and Biochemistry* 120, 105–125. DOI: 10.1016/j.soilbio.2018.01.030.

Carta, L.K., Handoo, Z.A., Li, S., Kantor, M., Bauchan, G. *et al.* (2020) Beech leaf disease symptoms caused by newly recognized nematode subspecies *Litylenchus crenatae mccannii* (Anguinata) described from *Fagus grandifolia* in North America. *Forest Pathology* 50(2), e12580. DOI: 10.1111/efp.12580.

Castagnone-Sereno, P., Danchin, E.G.J., Perfus-Barbeoch, L. and Abad, P. (2013) Diversity and evolution of root-knot nematodes, genus *Meloidogyne*: New insights from the genomic era. *Annual Review of Phytopathology* 51(1), 203–220. DOI: 10.1146/annurev-phyto-082712-102300.

Cohen, S.P. and Leach, J.E. (2020) High temperature-induced plant disease susceptibility: More than the sum of its parts. *Current Opinion in Plant Biology* 56, 235–241. DOI: 10.1016/j.pbi.2020.02.008.

Colagiero, M. and Ciancio, A. (2011) Climate changes and nematodes: Expected effects and perspectives for plant protection. *Redia* 94, 113–118.

Collett, R.L., Rashidifard, M., Marais, M., Daneel, M. and Fourie, H. (2024) Insights into the life-cycle development of *Meloidogyne enterolobii, M. incognita* and *M. Javanica* on tomato, soybean and maize. *European Journal of Plant Pathology* 168(1), 137–146. DOI: 10.1007/s10658-023-02741-9.

Creamer, R.E., Barel, J.M., Bongiorno, G. and Zwetsloot, M.J. (2022) The life of soils: Integrating the who and how of multifunctionality. *Soil Biology and Biochemistry* 166, 108561. DOI: 10.1016/j.soilbio.2022.108561.

Decraemer, W. and Hunt, D.J. (2006) Structure and classification. In: Perry, R.N. and Moens, M. (eds) *Plant Nematology*. CAB International, Wallingford, UK, pp. 3–32. DOI: 10.1079/9781845930561.0003.

Delcour, I., Spanoghe, P. and Uyttendaele, M. (2015) Literature review: Impact of climate change on pesticide use. *Food Research International* 68, 7–15. DOI: 10.1016/j.foodres.2014.09.030.

Desaeger, J., Wram, C. and Zasada, I. (2020) New reduced-risk agricultural nematicides – rationale and review. *Journal of Nematology* 52(1), 1–16. DOI: 10.21307/jofnem-2020-091.

Desaeger, J., Sikora, R.A. and Molendijk, L.P.G. (2021) Outlook: A vision of the future of integrated nematode management. In: Sikora, R.A., Desaeger, J. and Molendijk, L.P.G. (eds) *Integrated Nematode Management: State-of-the-Art and Visions for the Future*. CAB International, Wallingford, UK, pp. 475–483. DOI: 10.1079/9781789247541.0065.

Dutta, T.K. and Phani, V. (2023) The pervasive impact of global climate change on plant–nematode interaction continuum. *Frontiers in Plant Science* 14, 1143889. DOI: 10.3389/fpls.2023.1143889.

Ewing, C.J., Hausman, C.E., Pogacnik, J., Slot, J. and Bonello, P. (2019) Beech leaf disease: An emerging forest epidemic. *Forest Pathology* 49(2), e12488. DOI: 10.1111/efp.12488.

Fleming, T.R., McGowan, N.E., Maule, A.G. and Fleming, C.C. (2016) Prevalence and diversity of plant parasitic nematodes in Northern Ireland grassland and cereals, and the influence of soils and rainfall. *Plant Pathology* 65(9), 1539–1550. DOI: 10.1111/ppa.12525.

Franco, A.L.C., Gherardi, L.A., Tomasel, C.M., Andriuzzi, W.S., Ankrom, K.E. *et al.* (2019) Drought suppresses soil predators and promotes root herbivores in mesic, but not in xeric grasslands. *Proceedings of the National Academy of Sciences of the United States of America* 116(26), 12883–12888. DOI: 10.1073/pnas.1900572116.

Freitas, L.G., Mitchell, D.J. and Dickson, D.W. (1997) Temperature effects on the attachment of *Pasteuria penetrans* endospores to *Meloidogyne arenaria* race 1. *Journal of Nematology* 29(4), 547–555.

Gaberthüel, M., Slaats, B. and Goll, M. (2021) What does it take to develop a nematicide today and for the future? In: Sikora, R.A., Desaeger, J. and Molendijk, L.P.G. (eds) *Integrated Nematode Management: State-of-the-Art and Visions for the Future*. CAB International, Wallingford, UK, pp. 439–445. DOI: 10.1079/9781789247541.0061.

Gamon, A. and Lenne, N. (2012) *Meloidogyne chitwoodi* and *Meloidogyne fallax* in France: Initial management experiences. *EPPO Bulletin* 42(1), 122–126. DOI: 10.1111/j.1365-2338.2012.02529.x.

Gendron St-Marseille, A.-F., Bourgeois, G., Brodeur, J. and Mimee, B. (2019) Simulating the impacts of climate change on soybean cyst nematode and the distribution of soybean. *Agricultural and Forest Meteorology* 264, 178–187. DOI: 10.1016/j.agrformet.2018.10.008.

Ghini, R., Hamada, E., Pedro Júnior, M.J., Marengo, J.A. and Gonçalves, R. (2008) Risk analysis of climate change on coffee nematodes and leaf miner in Brazil. *Pesquisa Agropecuária Brasileira* 43(2), 187–194. DOI: 10.1590/S0100-204X2008000200005.

Giblin-Davis, R.M., Davies, K.A., Morris, K. and Thomas, W.K. (2003) Evolution of parasitism in insect-transmitted plant nematodes. *Journal of Nematology* 35(2), 133–141.

Giller, K.E., Hijbeek, R., Andersson, J.A. and Sumberg, J. (2021) Regenerative agriculture: An agronomic perspective. *Outlook on Agriculture* 50(1), 13–25. DOI: 10.1177/0030727021998063.

Hajihassani, A., Tenuta, M. and Gulden, R.H. (2017) Influence of temperature on development and reproduction of *Ditylenchus weischeri* and *D. dipsaci* on yellow pea. *Plant Disease* 101(2), 297–305. DOI: 10.1094/PDIS-04-16-0479-RE.

Harvey, J.A., Tougeron, K., Gols, R., Heinen, R., Abarca, M. *et al.* (2023) Scientists' warning on climate change and insects. *Ecological Monographs* 93(1), e1553. DOI: 10.1002/ecm.1553.

Holterman, M., Schratzberger, M. and Helder, J. (2019) Nematodes as evolutionary commuters between marine, freshwater and terrestrial habitats. *Biological Journal of the Linnean Society* 128(3), 756–767. DOI: 10.1093/biolinnean/blz107.

Jansson, J.K. and Hofmockel, K.S. (2020) Soil microbiomes and climate change . *Nature Reviews. Microbiology* 18(1), 35–46. DOI: 10.1038/s41579-019-0265-7.

Jones, J.T., Haegeman, A., Danchin, E.G.J., Gaur, H.S., Helder, J. *et al.* (2013) Top 10 plant-parasitic nematodes in molecular plant pathology. *Molecular Plant Pathology* 14(9), 946–961. DOI: 10.1111/mpp.12057.

Jones, L.M., Koehler, A.K., Trnka, M., Balek, J., Challinor, A.J. *et al.* (2017) Climate change is predicted to alter the current pest status of *Globodera pallida* and *G. rostochiensis* in the United Kingdom. *Global Change Biology* 23(11), 4497–4507. DOI: 10.1111/gcb.13676.

Kanzaki, N., Ichihara, Y., Aikawa, T., Ekino, T. and Masuya, H. (2019) *Litylenchus crenatae* n. sp. (Tylenchomorpha: Anguinidae), a leaf gall nematode parasitising *Fagus crenata* Blume. *Nematology* 21(1), 5–22.

Khanal, C. and Harshman, D. (2022) Evaluation of summer cover crops for host suitability of *Meloidogyne enterolobii*. *Crop Protection* 151, 105821. DOI: 10.1016/j.cropro.2021.105821.

Khanal, C. and Land, J. (2023) Study on two nematode species suggests climate change will inflict greater crop damage. *Scientific Reports* 13(1), 14185. DOI: 10.1038/s41598-023-41466-x.

Kopittke, P.M., Menzies, N.W., Wang, P., McKenna, B.A. and Lombi, E. (2019) Soil and the intensification of agriculture for global food security. *Environment International* 132, 105078. DOI: 10.1016/j.envint.2019.105078.

Lal, R. (2020) Regenerative agriculture for food and climate. *Journal of Soil and Water Conservation* 75(5), 123A–124A. DOI: 10.2489/jswc.2020.0620A.

Lamichhane, J.R., Barzman, M., Booij, K., Boonekamp, P., Desneux, N. *et al.* (2015) Robust cropping systems to tackle pests under climate change. A review. *Agronomy for Sustainable Development* 35(2), 443–459. DOI: 10.1007/s13593-014-0275-9.

Leach, M., Agudelo, P. and Gerard, P. (2009) Effect of temperature on the embryogenesis of geographic populations of *Rotylenchulus reniformis*. *Journal of Nematology* 41(1), 23–27.

Lesk, C., Rowhani, P. and Ramankutty, N. (2016) Influence of extreme weather disasters on global crop production. *Nature* 529(7584), 84–87. DOI: 10.1038/nature16467.

Li, G.H. and Zhang, K.-Q. (2023) Natural nematicidal metabolites and advances in their biocontrol capacity on plant parasitic nematodes. *Natural Product Reports* 40(3), 646–675. DOI: 10.1039/d2np00074a.

Li, H., Xing, L., Liu, X., Pu, Y., Yang, Y. *et al.* (2022a) Potential impact of climate change on the distribution of the pinewood nematode *Bursaphelenchus xylophilus* in Chongqing, China. *Pakistan Journal of Zoology* 54(2), 809–816. DOI: 10.17582/journal.pjz/20190912070900.

Li, Z., Tao, J. and Zong, S. (2022b) Cold tolerance in pinewood nematode *Bursaphelenchus xylophilus* promoted multiple invasion events in mid-temperate zone of China. *Forests* 13(7), 1100. DOI: 10.3390/f13071100.

Luambano, N.D., Manzanilla-López, R.H., Kimenju, J.W., Powers, S.J., Narla, R.D. *et al.* (2015) Effect of temperature, pH, carbon and nitrogen ratios on the parasitic activity of *Pochonia chlamydosporia* on *Meloidogyne incognita*. *Biological Control* 80, 23–29. DOI: 10.1016/j.biocontrol.2014.09.003.

Marra, R.E. and LaMondia, J.A. (2020) First report of beech leaf disease, caused by the foliar nematode, *Litylenchus crenatae mccannii*, on American Beech (*Fagus grandifolia*) in connecticut. *Plant Disease* 104(9), 2527. DOI: 10.1094/PDIS-02-20-0442-PDN.

Mota, M.M. and Vieira, P.R. (2008) *Pine Wilt Disease: A Worldwide Threat to Forest Ecosystems*. Springer, Dordrecht, The Netherlands. DOI: 10.1007/978-1-4020-8455-3.

Nethi, S. and Prasad, J.S. (2012) Plant–nematode interactions: Consequences of climate change. In: Shanker, A.K., Shanker, C. and Maheshwari, M. (eds) *Crop Stress and Its Management: Perspectives and Strategies*. Springer, Dordrecht, The Netherlands, pp. 547–564. DOI: 10.1007/978-94-007-2220-0_17.

NRCS (2023) *Soil Health*. US Department of Agriculture. Available at: https://www.nrcs.usda.gov/conservation-basics/natural-resource-concerns/soils/soil-health (accessed 30 March 2024).

Ouyang, X., Chen, A., Li, Y., Han, X. and Lin, H. (2022) Predicting the potential distribution of pine wilt disease in China under climate change. *Insects* 13(12), 1147. DOI: 10.3390/insects13121147.

Pan, S., Peng, D. -l., Li, Y. -m., Chen, Z. -j., Zhai, Y. -y *et al.* (2023) Potential global distribution of the guava root-knot nematode *Meloidogyne enterolobii* under different climate change scenarios using MAXEnt ecological niche modeling. *Journal of Integrative Agriculture* 22(7), 2138–2150. DOI: 10.1016/j.jia.2023.06.022.

Perry, R.N. and Moens, M. (2011) Survival of parasitic nematodes outside the host. In: Perry, R.N. and Wharton, D.A. (eds) *Molecular and Physiological Basis of Nematode Survival*. CAB International, Wallingford, UK, pp. 1–27. DOI: 10.1079/9781845936877.0001.

Poinar, G. (2014) Evolutionary history of terrestrial pathogens and endoparasites as revealed in fossils and subfossils. *Advances in Biology* 2014, 181353. DOI: 10.1155/2014/181353.

Pollok, J.R., Johnson, C.S., Eisenback, J.D., David Reed, T. and Adamo, N. (2023) Effect of soil temperature on reproduction of root-knot nematodes in flue-cured tobacco with homozygous and/or resistance genes. *Journal of Nematology* 55(1), 20230032. DOI: 10.2478/jofnem-2023-0032.

Puissant, J., Villenave, C., Chauvin, C., Plassard, C., Blanchart, E. *et al.* (2021) Quantification of the global impact of agricultural practices on soil nematodes: A meta-analysis. *Soil Biology and Biochemistry* 161, 108383. DOI: 10.1016/j.soilbio.2021.108383.

Rafaluk-Mohr, C., Ashby, B., Dahan, D.A. and King, K.C. (2018) Mutual fitness benefits arise during coevolution in a nematode-defensive microbe model. *Evolution Letters* 2(3), 246–256. DOI: 10.1002/evl3.58.

Raj, A., Jhariya, M.K., Khan, N., Banerjee, A. and Meena, R.S. (2021) Ecological intensification for sustainable development. In: Jhariya, M.K., Meena, R.S. and Banerjee, A. (eds) *Ecological Intensification of Natural Resources for Sustainable Agriculture*. Springer, Singapore, pp. 137–170. DOI: 10.1007/978-981-33-4203-3_5.

Rashidifard, M., Marais, M., Daneel, M.S. and Fourie, H. (2019) Reproductive potential of south African thermophilic *Meloidogyne* populations, with special reference to *Meloidogyne enterolobii*. *Nematology* 21(9), 913–921. DOI: 10.1163/15685411-00003263.

Renčo, M., Adámek, M., Jílková, V. and Devetter, M. (2022) Post-fire recovery of soil nematode communities depends on fire severity. *Diversity* 14(12), 1116. DOI: 10.3390/d14121116.

Sapir, A. (2021) Why are nematodes so successful extremophiles? *Communicative & Integrative Biology* 14(1), 24–26. DOI: 10.1080/19420889.2021.1884343.

Sharma, P., Singh, A., Kahlon, C.S., Brar, A.S., Grover, K.K. *et al*. (2018) The role of cover crops towards sustainable soil health and agriculture – a review paper. *American Journal of Plant Sciences* 09(9), 1935–1951. DOI: 10.4236/ajps.2018.99140.

Sheppard, A.W., Gillespie, I., Hirsch, M. and Begley, C. (2011) Biosecurity and sustainability within the growing global bioeconomy. *Current Opinion in Environmental Sustainability* 3(1–2), 4–10. DOI: 10.1016/j.cosust.2010.12.011.

Siebert, J., Ciobanu, M., Schädler, M. and Eisenhauer, N. (2020) Climate change and land use induce functional shifts in soil nematode communities. *Oecologia* 192(1), 281–294. DOI: 10.1007/s00442-019-04560-4.

Sikora, R.A., Coyne, D., Hallmann, J. and Timper, P. (2018) *Plant Parasitic Nematodes in Subtropical and Tropical Agriculture*, 3rd edn. CAB International, Wallingford, UK. DOI: 10.1079/9781786391247.0000.

Sikora, R.A., Padgham, J. and Desaeger, J. (2021) The unpredictability of adapting integrated nematode management to climate variability. In: Sikora, R.A., Desaeger, J. and Molendijk, L.P.G. (eds) *Integrated Nematode Management: State-of-the-Art and Visions for the Future*. CAB International, Wallingford, UK, pp. 463–471. DOI: 10.1079/9781789247541.0064.

Sikora, R.A., Helder, J., Molendijk, L.P.G., Desaeger, J., Akker, S. *et al*. (2023) Integrated nematode management in a world in transition: Constraints, policy, processes, and technologies for the future. *Annual Review of Phytopathology* 61(1), 209–230. DOI: 10.1146/annurev-phyto-021622-113058.

Singh, S.K., Paini, D.R., Ash, G.J. and Hodda, M. (2014) Prioritising plant-parasitic nematode species biosecurity risks using self organising maps. *Biological Invasions* 16(7), 1515–1530. DOI: 10.1007/s10530-013-0588-7.

Singh, S.K., Kriticos, D.J., Ota, N. and Hodda, M. (2022) Potential distribution and biosecurity risks from three economically important plant-parasitic nematodes. *Annals of Applied Biology* 180(3), 371–382. DOI: 10.1111/aab.12739.

Skendžić, S., Zovko, M., Živković, I.P., Lešić, V. and Lemić, D. (2021) The impact of climate change on agricultural insect pests. *Insects* 12(5), 440. DOI: 10.3390/insects12050440.

Stevnbak, K., Maraldo, K., Georgieva, S., Bjørnlund, L., Beier, C. *et al*. (2012) Suppression of soil decomposers and promotion of long-lived, root herbivorous nematodes by climate change. *European Journal of Soil Biology* 52, 1–7. DOI: 10.1016/j.ejsobi.2012.04.001.

Sticht, C., Schrader, S., Giesemann, A. and Weigel, H.-J. (2009) Sensitivity of nematode feeding types in arable soil to free air CO_2 enrichment (FACE) is crop specific. *Pedobiologia* 52(5), 337–349. DOI: 10.1016/j.pedobi.2008.12.001.

Sünnemann, M., Beugnon, R., Breitkreuz, C., Buscot, F., Cesarz, S. *et al*. (2023) Climate change and cropland management compromise soil integrity and multifunctionality. *Communications Earth & Environment* 4(1), 394. DOI: 10.1038/s43247-023-01047-2.

Sutherst, R.W., Constable, F., Finlay, K.J., Harrington, R., Luck, J. *et al*. (2011) Adapting to crop pest and pathogen risks under a changing climate. *WIREs Climate Change* 2(2), 220–237. DOI: 10.1002/wcc.102.

Thierfelder, C., Baudron, F., Setimela, P., Nyagumbo, I., Mupangwa, W. *et al*. (2018) Complementary practices supporting conservation agriculture in Southern Africa. A review. *Agronomy for Sustainable Development* 38(2), 16. DOI: 10.1007/s13593-018-0492-8.

Timper, P. (2014) Conserving and enhancing biological control of nematodes. *Journal of Nematology* 46(2), 75–89.

Timper, P., Strickland, T.C. and Jagdale, G.B. (2021) Biological suppression of the root-knot nematode *Meloidogyne incognita* following winter cover crops in conservation tillage cotton. *Biological Control* 155, 104525. DOI: 10.1016/j.biocontrol.2020.104525.

Tittonell, P. (2014) Ecological intensification of agriculture – sustainable by nature. *Current Opinion in Environmental Sustainability* 8, 53–61. DOI: 10.1016/j.cosust.2014.08.006.

Triantaphyllou, A.C. (1973) Environmental sex differentiation of nematodes in relation to pest management. *Annual Review of Phytopathology* 11(1), 441–462. DOI: 10.1146/annurev.py.11.090173.002301.
Trivedi, P., Batista, B.D., Bazany, K.E. and Singh, B.K. (2022) Plant–microbiome interactions under a changing world: Responses, consequences and perspectives. *The New Phytologist* 234(6), 1951–1959. DOI: 10.1111/nph.18016.
Trudgill, D.L. and Blok, V.C. (2001) Apomictic, polyphagous root-knot nematodes: Exceptionally successful and damaging biotrophic root pathogens. *Annual Review of Phytopathology* 39(1), 53–77. DOI: 10.1146/annurev.phyto.39.1.53.
Tully, K.L. and McAskill, C. (2020) Promoting soil health in organically managed systems: A review. *Organic Agriculture* 10(3), 339–358. DOI: 10.1007/s13165-019-00275-1.
Velásquez, A.C., Castroverde, C.D.M. and He, S.Y. (2018) Plant–pathogen warfare under changing climate conditions. *Current Biology* 28(10), R619–R634. DOI: 10.1016/j.cub.2018.03.054.
Velloso, J.A., Maquilan, M.A.D., Campos, V.P., Brito, J.A. and Dickson, D.W. (2022) Temperature effects on development of *Meloidogyne enterolobii* and *M. floridensis*. *Journal of Nematology* 54(1), 20220013. DOI: 10.2478/jofnem-2022-0013.
Vicente, C., Espada, M., Vieira, P. and Mota, M. (2012) Pine wilt disease: A threat to European forestry. *European Journal of Plant Pathology* 133(1), 89–99. DOI: 10.1007/s10658-011-9924-x.
Wesemael, W.M.L., Viaene, N. and Moens, M. (2011) Root-knot nematodes (*Meloidogyne* spp.) in Europe. *Nematology* 13(1), 3–16. DOI: 10.1163/138855410X526831.
Wezel, A., Herren, B.G., Kerr, R.B., Barrios, E., Gonçalves, A.L.R. *et al*. (2020) Agroecological principles and elements and their implications for transitioning to sustainable food systems. A review. *Agronomy for Sustainable Development* 40(6), 40. DOI: 10.1007/s13593-020-00646-z.
Wharton, D.A. (1995) Cold tolerance strategies in nematodes. *Biological Reviews of the Cambridge Philosophical Society* 70(1), 161–185. DOI: 10.1111/j.1469-185X.1995.tb01442.x.
Wharton, D.A. (2004) Survival strategies. In: Gaugler, R. and Bilgrami, A.L. (eds) *Nematode Behaviour*. CAB International, Wallingford, UK, pp. 371–400.
Wheeler, T. and von Braun, J. (2013) Climate change impacts on global food security. *Science* 341(6145), 508–513. DOI: 10.1126/science.1239402.
Wilschut, R.A. and Geisen, S. (2021) Nematodes as drivers of plant performance in natural systems. *Trends in Plant Science* 26(3), 237–247. DOI: 10.1016/j.tplants.2020.10.006.
Yeates, G.W., Boag, B., Evans, K.A. and Neilson, R. (1998) Impact of climatic changes on the distribution of *Paratrichodorus minor* (Nematoda: Trichodoridae) as estimated using "CLIMEX". *Nematologica* 44(3), 293–301. DOI: 10.1163/005425998X00053.
Zasada, I.A., Kitner, M., Wram, C., Wade, N., Ingham, R.E. *et al*. (2019) Trends in occurrence, distribution, and population densities of plant-parasitic nematodes in the Pacific Northwest of the United States from 2012 to 2016. *Plant Health Progress* 20(1), 20–28. DOI: 10.1094/PHP-11-18-0077-RS.
Zhou, J., Wu, J., Huang, J., Sheng, X., Dou, X. *et al*. (2022) A synthesis of soil nematode responses to global change factors. *Soil Biology and Biochemistry* 165, 108538. DOI: 10.1016/j.soilbio.2021.108538.
Zwetsloot, M.J., Leeuwen, J., Hemerik, L., Martens, H., Simó Josa, I. *et al*. (2021) Soil multifunctionality: Synergies and trade-offs across European climatic zones and land uses. *European Journal of Soil Science* 72(4), 1640–1654. DOI: 10.1111/ejss.13051.

Index

Note: Page numbers in **bold** type refer to **figures;** Page numbers in *italics* type refer to *tables*

Printed and bound by CPI Group (UK) Ltd, Croydon, CR0 4YY

13/09/2026

14961113-0005